Applied Microbiology and Biotechnology

Applied Microbiology and Biotechnology

Edited by **Dean Watson**

SYRAWOOD
PUBLISHING HOUSE
New York

Published by Syrawood Publishing House,
750 Third Avenue, 9th Floor,
New York, NY 10017, USA
www.syrawoodpublishinghouse.com

Applied Microbiology and Biotechnology
Edited by Dean Watson

International Standard Book Number: 978-1-68286-288-9 (Hardback)

Printed in China.

Contents

	Preface	IX
Chapter 1	**Response surface modeling for hot, humid air decontamination of materials contaminated with *Bacillus anthracis* ΔSterne and *Bacillus thuringiensis* Al Hakam spores** Edward J Prokop, John R Crigler, Claire M Wells, Alice A Young and Tony L Buhr	1
Chapter 2	**Genome features of *Pseudomonas putida* LS46, a novel polyhydroxyalkanoate producer and its comparison with other *P. putida* strains** Parveen K Sharma, Jilagamazhi Fu, Xiangli Zhang, Brian Fristensky, Richard Sparling and David B Levin	16
Chapter 3	**Common gas phase molecules from fungi affect seed germination and plant health in *Arabidopsis thaliana*** Richard Hung, Samantha Lee, Cesar Rodriguez-Saona and Joan W Bennett	34
Chapter 4	**Printed paper-based arrays as substrates for biofilm formation** Anni Määttänen, Adyary Fallarero, Janni Kujala, Petri Ihalainen, Pia Vuorela and Jouko Peltonen	41
Chapter 5	**Deletion of glucose oxidase changes the pattern of organic acid production in *Aspergillus carbonarius*** Lei Yang, Mette Lübeck and Peter S Lübeck	53
Chapter 6	**Contamination of water resources by pathogenic bacteria** Pramod K Pandey, Philip H Kass, Michelle L Soupir, Sagor Biswas and Vijay P Singh	62
Chapter 7	**Production of aroma compounds from whey using *Wickerhamomyces pijperi*** Naoki Izawa, Miyuki Kudo, Yukako Nakamura, Harumi Mizukoshi, Takahiro Kitada and Toshiro Sone	78
Chapter 8	**Phenotypic and metabolic traits of commercial *Saccharomyces cerevisiae* yeasts** Catarina Barbosa, Patrícia Lage, Alice Vilela, Arlete Mendes-Faia and Ana Mendes-Ferreira	87
Chapter 9	**Paracentrin 1, a synthetic antimicrobial peptide from the sea-urchin *Paracentrotus lividus*, interferes with staphylococcal and *Pseudomonas aeruginosa* biofilm formation** Domenico Schillaci, Maria Grazia Cusimano, Angelo Spinello, Giampaolo Barone, Debora Russo, Maria Vitale, Daniela Parrinello and Vincenzo Arizza	101

Chapter 10 **Diverse allosteric and catalytic functions of tetrameric D-lactate dehydrogenases from three Gram-negative bacteria** **110**
Nayuta Furukawa, Akimasa Miyanaga, Misato Togawa, Masahiro Nakajima and Hayao Taguchi

Chapter 11 **Screening of liquid media and fermentation of an endophytic *Beauveria bassiana* strain in a bioreactor** **122**
Rieke Lohse, Desiree Jakobs-Schönwandt and Anant V Patel

Chapter 12 **Surface functionalization by covalent immobilization of an innovative carvacrol derivative to avoid fungal biofilm formation** **133**
Aïcha Gharbi, Thibaut Legigan, Vincent Humblot, Sébastien Papot and Jean-Marc Berjeaud

Chapter 13 **Cultivation and detection of endophytic aerobic methanotrophs isolated from *Sphagnum* species as a perspective for environmental biotechnology** **145**
Zofia Stępniewska and Agnieszka Kuźniar

Chapter 14 **Unraveling the kinetic diversity of microbial 3-dehydroquinate dehydratases of shikimate pathway** **154**
Chang Liu, Yi-Ming Liu, Qing-Lan Sun, Cheng-Ying Jiang and Shuang-Jiang Liu

Chapter 15 **Physiological and growth response of rice plants (*Oryza sativa* L.) to *Trichoderma spp.* inoculants** **163**
Febri Doni, Anizan Isahak, Che Radziah Che Mohd Zain and Wan Mohtar Wan Yusoff

Chapter 16 **Identification and characterization of endophytic bacteria from corn (*Zea mays* L.) roots with biotechnological potential in agriculture** **170**
Vivian Jaskiw Szilagyi-Zecchin, Angela Cristina Ikeda, Mariangela Hungria, Douglas Adamoski, Vanessa Kava-Cordeiro, Chirlei Glienke and Lygia Vitória Galli-Terasawa

Chapter 17 **Diversity and biochemical features of culturable fungi from the coastal waters of Southern China** **179**
Li Li, Purnima Singh, Ying Liu, Shenquan Pan and Guangyi Wang

Chapter 18 **Genetic transformation of *Knufia petricola* A95 - a model organism for biofilm-material interactions** **190**
Steffi Noack-Schönmann, Tanja Bus, Ronald Banasiak, Nicole Knabe, William J Broughton, H Den Dulk-Ras, Paul JJ Hooykaas and Anna A Gorbushina

Chapter 19 **Purification and characterization of a novel lipopeptide from *Streptomyces amritsarensis* sp. nov. active against methicillin-resistant *Staphylococcus aureus*** **196**
Deepika Sharma, Santi M Mandal and Rajesh Kumari Manhas

Chapter 20 **Systematic analysis of the ability of Nitric Oxide donors to dislodge biofilms formed by *Salmonella enterica* and *Escherichia coli* O157:H7** **205**
Massimiliano Marvasi, Charles Chen, Manuel Carrazana, Ian A Durie and Max Teplitski

Chapter 21 ***Sulfobacillus thermosulfidooxidans*** **strain Cutipay enhances chalcopyrite bioleaching under moderate thermophilic conditions in the presence of chloride ion** **216**
Roberto A Bobadilla-Fazzini, Maria Paz Cortés, Alejandro Maass and Pilar Parada

Chapter 22 **Inhibitory effects of H-Ras/Raf-1-binding affibody molecules on synovial cell function** **221**
Seiji Shibasaki, Miki Karasaki, Torbjörn Gräslund, Per-Åke Nygren, Hajime Sano and Tsuyoshi Iwasaki

Chapter 23 **Identification of a gene, *FMP21*, whose expression levels are involved in thermotolerance in *Saccharomyces cerevisiae*** **230**
Toshihide Nakamura, Mami Yamamoto, Katsuichi Saito, Akira Ando and Jun Shima

Chapter 24 **Degradative actions of microbial xylanolytic activities on hemicelluloses from rhizome of *Arundo donax*** **238**
Licia Lama, Annabella Tramice, Ilaria Finore, Gianluca Anzelmo, Valeria Calandrelli, Eduardo Pagnotta, Giuseppina Tommonaro, Annarita Poli, Paola Di Donato, Barbara Nicolaus, Massimo Fagnano, Mauro Mori, Adriana Impagliazzo and Antonio Trincone

Chapter 25 **Sterilization of hydrogen peroxide resistant bacterial spores with stabilized chlorine dioxide** **247**
Anthony Friedline, Malcolm Zachariah, Amy Middaugh, Matt Heiser, Neeraj Khanna, Parag Vaishampayan and Charles V Rice

Chapter 26 **Metabolism of alkenes and ketones by *Candida maltosa* and related yeasts** **253**
Andy Beier, Veronika Hahn, Uwe T Bornscheuer and Frieder Schauer

Permissions

List of Contributors

Preface

This book has been an outcome of determined endeavour from a group of educationists in the field. The primary objective was to involve a broad spectrum of professionals from diverse cultural background involved in the field for developing new researches. The book not only targets students but also scholars pursuing higher research for further enhancement of the theoretical and practical applications of the subject.

Microorganisms are widely used across different industries for catalysis, biosynthesis and transformation of various compounds and substances into different commodities like pharmaceuticals, biomaterials, etc. This book is a compilation of chapters that discuss the most vital concepts in the interdisciplinary fields of applied microbiology and biotechnology such as production of enzymes, waste management, pharmaceutical applications of microorganisms, and bioprocess engineering, etc. The various studies that are constantly contributing towards advancing technologies and evolution of this field are examined in detail. It is an essential guide for both academicians and those who wish to pursue this discipline further.

It was an honour to edit such a profound book and also a challenging task to compile and examine all the relevant data for accuracy and originality. I wish to acknowledge the efforts of the contributors for submitting such brilliant and diverse chapters in the field and for endlessly working for the completion of the book. Last, but not the least; I thank my family for being a constant source of support in all my research endeavours.

Editor

1

Response surface modeling for hot, humid air decontamination of materials contaminated with *Bacillus anthracis* ΔSterne and *Bacillus thuringiensis* Al Hakam spores

Edward J Prokop, John R Crigler, Claire M Wells, Alice A Young and Tony L Buhr*

Abstract

Response surface methodology using a face-centered cube design was used to describe and predict spore inactivation of *Bacillus anthracis* ΔSterne and *Bacillus thuringiensis* Al Hakam spores after exposure of six spore-contaminated materials to hot, humid air. For each strain/material pair, an attempt was made to fit a first or second order model. All three independent predictor variables (temperature, relative humidity, and time) were significant in the models except that time was not significant for *B. thuringiensis* Al Hakam on nylon. Modeling was unsuccessful for wiring insulation and wet spores because there was complete spore inactivation in the majority of the experimental space. In cases where a predictive equation could be fit, response surface plots with time set to four days were generated. The survival of highly purified *Bacillus* spores can be predicted for most materials tested when given the settings for temperature, relative humidity, and time. These predictions were cross-checked with spore inactivation measurements.

Keywords: *Bacillus*; Spore; Decontamination; Hot humid air; RSM

Introduction

Conventional chemical decontamination (e.g. bleach and/or harsh oxidizers) can create health risks and damage many materials, which increases decontamination training costs, personal protection requirements, and may result in replacement, reconstruction and/or remodeling of sensitive equipment and materials. These issues increase decontamination costs and the cleanup time required after decontamination (Craig and Anderson 1995; Garverick 1994; Rutala et al. 2008; Koch et al. 2001; Herzberg 2012). For example, decontamination and site remediation following the 2001 anthrax (*B. anthracis*) attacks cost in excess of 300 million US dollars and took 34 months to complete (Canter 2005; Buhr et al. 2012a). Aircraft are of particular concern since they are a vehicle for global epidemic spread and are comprised of extremely sensitive equipment and materials (Moser et al. 1979; Olsen et al. 2003). Because there are limited choices for aircraft interior decontamination, corrosive chemicals may be improperly selected and used in the event of potential International epidemics such as influenza or SARS. The use of corrosive chemicals may then undermine the integrity of the aircraft. Thus, there is a need to develop effective decontaminants with low corrosiveness and improved materials compatibility for various materials including, but not exclusive to those found on aircraft.

Thermal inactivation of biological microbes is a well-known physical procedure that applies to both food sterilization and decontamination of inanimate objects. The Roman poet Virgil described attempts at thermal decontamination of *B. anthracis*-infected wool and hides several millennia ago (Dirckx 1981; Sternbach 2003). Cooking, grilling and pasteurization are used to inactivate food microbes (Ullmann 2007). Autoclaving is another well-known thermal inactivation procedure (Hugo 1991). Numerous manuscripts have been published describing thermal inactivation of *Bacillus* spores (Murrell and Scott 1966; Alderton and Snell 1970; Gerhardt and Marquis 1989; Melly et al. 2002; Coleman et al. 2007; Sunde et al. 2009; Zhang et al. 2010; Buhr et al. 2012b).

* Correspondence: tony.buhr@navy.mil
Naval Surface Warfare Center, Dahlgren Division, CBR Concepts and Experimentation Branch, 4045 Higley Road Suite 345, Dahlgren VA 22448-5162, USA

Three variables to control thermal decontamination are time, temperature and water content. Adjustment of these variables may be less damaging to materials compared to application of corrosive chemicals (Decker et al. 1954; Ehrlich et al. 1970; Holwitt et al. 2000). Response surface methodology (RSM) (Beauregard et al. 1992; Myers et al. 2009) was coupled with the development of test methods to control time, temperature and relative humidity while safely containing spores (Buhr et al. 2012b). Analogous to slow cooking, spore inactivation of spore-contaminated materials was tested using low temperatures and long times in order to explore the limits of sporicidal efficacy using hot, humid air. Six materials were chosen that were representative of materials found on aircraft: wiring insulation, polypropylene, aircraft performance coating on aluminum, nylon, InsulFab insulation and anti-skid tape on aluminum (Buhr et al. 2012b). Decontamination of these materials using hot, humid air showed comparable spore kill kinetics among virulent *B. anthracis* Ames, attenuated *B. anthracis* ΔSterne and *B. thuringiensis* Al Hakam spores. In order to mitigate decontamination test costs and schedule associated with BSL3 testing of *B. anthracis* Ames in a large-scale design of experiments, *B. anthracis* ΔSterne and *B. thuringiensis* Al Hakam spores were selected for decontamination testing that measured decontamination on greater than 2,500 tests and controls with ≥7 logs of spores per test (Buhr et al. 2012b). This manuscript focuses on describing the application of response surface methodology to experimental design and data analysis of the hot, humid air data. The objective is to mathematically describe the effectiveness of hot, humid air decontamination using first or second order polynomial surfaces and to determine the region of inactivating 6 logs of spores using the lower bound of a 90% confidence interval for the surface. This threshold was selected because it is a requirement for the Food and Drug Administration, Environmental Protection Agency and Department of Defense (Tomasino et al. 2008; Ryan et al. 2010; Buhr et al. 2012a). The result of this research is a mathematical description that can help guide an end-user to find acceptable operating conditions depending on the material(s) to be decontaminated.

Materials and methods

Bacillus anthracis ΔSterne and *Bacillus thuringiensis* Al Hakam spores

Bacillus anthracis ΔSterne was obtained from the Unified Culture Collection at USAMRIID (Frederick, MD, USA): Unified Culture Collection Identifier BAC1056, lot number CA062700A. *Bacillus thuringiensis* Al Hakam was provided by Johnathan Kiel at Brooks Air Force Base, San Antonio, TX (Buhr et al. 2008, 2011, 2012a, b; McCartt et al. 2011).

Coupon materials and inoculation

All substrates and substrate inoculations have been previously described (Buhr et al. 2012b) and are summarized here. A substrate was defined as an individual coupon, polypropylene or wet sample. The aircraft-related materials chosen for this study were aluminum 2024-T3 coated with Aircraft Performance Coating (APC), aluminum 2024-T3 coated with anti-skid material, InsulFab insulation, wiring insulation, nylon webbing, and polypropylene. Each substrate was cut into 2×2 centimeter squares except polypropylene. All substrates were inoculated with ≥7 logs of spores per substrate. Spores were dried inside 50 ml conical tubes for polypropylene tests. An additional seventh substrate was wet spores; i.e., spores in 0.1% Tween 80, but never dried onto a surface.

Environmental test chamber setup and RSM

The environmental chamber validation and RSM have been previously described (Buhr et al. 2012b) and are summarized here. Envirotronics (Grand Rapids, MI) LH0010 environmental test chambers were used to control temperature and relative humidity during testing. Three test conditions were selected for each of the three test parameters of temperature, relative humidity and time. The highest (strongest) selected combination of conditions was expected to result in complete spore inactivation, the lowest (mildest) selected combination of conditions was expected to result in complete spore survival, and the midpoint selected was expected to result in partial spore inactivation. The test conditions were equally spaced in an attempt to generate a spore inactivation response curve for each strain/substrate combination. The temperature limits considered took into account the goal of completely inactivating spores at the lowest temperature possible in order to meet materials temperature qualification limits. Based on preliminary decontamination test screening of virulent *B. anthracis* Ames, attenuated *B. anthracis* ΔSterne and *B. thuringiensis* Al Hakam spores, the upper limit for temperature and relative humidity was set at 77°C, 90% relative humidity because the data indicated that complete spore inactivation would be achieved within 7 days at this combination. For the lower limits, the temperature was set at 60°C, 60% relative humidity, conditions where spores were known to survive for multiple days. Initially, temperatures of 140, 155 and 170 °F were selected, then converted to Celsius and rounded to 60, 68 and 77°C. The final test conditions for relative humidity were 60, 75, and 90%. The final conditions for time were 1, 4, and 7 days. Each temperature/relative humidity combination was above the dew point to prevent condensation.

Test methods and control-driven experimental design

Test methods and experimental design have been previously described (Buhr et al. 2012b) and are summarized here. Test methods were developed to contain spores or spore-inoculated substrates within filter-capped 50 ml conical tubes. The 0.2 μM filters permitted hot, humid air to pass through the filter cap and the filter containing spores. This method was critical for proof-of-principle testing with virulent *B. anthracis* Ames spores since a fan circulated hot, humid air during testing. This method also permitted spore extraction in the test tube without extra handling steps of the spore-inoculated substrate.

For each test run, spores (≥7 logs of spores) from each of ten independent spore preparations (five spore preparations for *B. anthracis* ΔSterne spores and another five spore preparations for *B. thuringiensis* Al Hakam) were inoculated on the test substrates and the control substrates. Spore survival was quantified for 70 tests and 70 room temperature controls for each test run with a total of 19 independent test runs. For each test run, there were ten substrates for each of the six materials plus the wet control (7 total) for test samples, and the six materials plus the wet control (7 total) for the room temperature control samples. There were a total of 1,330 independent tests and 1,330 independent room temperature controls for the entire study.

Quantification of spore survival and calculations

Spore quantification and spore survival calculations have been described (Buhr et al. 2012b), and are summarized here. Spore-inoculated substrates were exposed to specified temperature, relative humidity and time, or left at 22 ± 3°C and ambient relative humidity. After decontamination, spores were extracted in ten ml of extraction medium followed by dilution plating and scoring for growth. Wet spore controls, incubated at 22 ± 3°C and ambient relative humidity, served as the 100% spore recovery reference values in order to correct for spore extraction efficiency from different substrates. The number of surviving spores (CFU ml−1) from each hot, humid air-treated test substrate was divided by the extraction percentage for that substrate in order to determine the number of surviving spores (CFU ml−1). This spore concentration was multiplied by 10 ml to determine the total number of spores surviving (total CFU) for each test sample. A $\log_{10}$ transformation ($\log_{10}$ (total CFU + 1)) of the total surviving spores was then calculated.

Response surface methodology and test design

Response surface methodology is a statistical procedure used to understand, improve, and optimize a process. The objective is to model a response variable in terms of one or more independent predictor variables. Second order polynomial models (Equation 1)

$$y = \beta_0 + \sum_{j=1}^{k}\beta_j x_j + \sum_{j=1}^{k}\beta_{jj} x_j^2 + \sum_{i<j}\sum_{j=2}^{k}\beta_{ij} x_i x_j$$

are most commonly used in RSM because they are flexible in describing data in which there is curvature. They also make for straightforward locating of optima. RSM is typically used sequentially. An experiment is first done to screen potential predictor variables. If it is determined that the current settings of the predictors are near an optimum, supplementary runs are done to fit a second order model. If the experimental region is not near an optimum, the method of steepest ascent can be used to find the optimum or arrive at a region of operability should an optimum be unattainable. In other situations, the region of operability is the entire experimental region. Such was the case for this study. The experiment used in this situation was the face-centered cube design (FCD) (Myers and Montgomery 2002). The final design in terms of coded and original predictor variables is given in Table 1.

As Table 1 shows, there were three predictors, temperature (T), relative humidity (RH), and days (D). These were transformed to the coded variables, x_1, x_2, and x_3 in order to better determine the relative size of their effects. The codes were $x_1 = \frac{T-68.\bar{3}}{8.3}$ $x_2 = \frac{RH-75}{15}$ and $x_3 = \frac{D-4}{3}$.

The measured response, p(x1, x2, x3), was the proportion of spores inactivated divided by the average number of spores tested. Since the response was a proportion, the proper link function is the logit (or log odds) (Agresti 1996, 2002; Hosmer and Lemeshow 2000); i.e., $y = log\left(\frac{p}{1-p}\right)$

Table 1 Face-centered cube design matrix with coded and original units

Run	X_1	X_2	X_3	Temperature (°C)	Rel. humidity (%)	Days	Block
3	−1	−1	1	60.0	60	1	0
9	−1	−1	1	60.0	60	7	0
12	−1	1	−1	60.0	90	1	0
6	−1	1	1	60.0	90	7	0
1	0	0	0	68.3	75	4	0
4	0	0	0	68.3	75	4	0
8	0	0	0	68.3	75	4	0
11	0	0	0	68.3	75	4	0
13	0	0	0	68.3	75	4	0
2	1	−1	−1	76.7	60	1	0
10	1	−1	1	76.7	60	7	0
7	1	1	−1	76.7	90	1	0
5	1	1	1	76.7	90	7	0
14	−1	0	0	60.0	75	4	1
16	0	−1	0	68.3	60	4	1
18	0	0	−1	68.3	75	1	1
19	0	0	1	68.3	75	7	1
17	0	1	0	68.3	90	4	1
15	1	0	0	76.7	75	4	1

where p is shorthand for p(x1, x2, x3) and log denotes the natural logarithm function. If a particular material had complete spore inactivation on one or two observations, p was conservatively imputed as the minimum detection level, $\left(\frac{0.1}{avg.\ no.\ spores\ tested}\right)$, to bound the logit away from infinity. In some cases, however, too many coupons had complete inactivation and could not be modeled using Equation 1.

An attempt to fit a second order model was made for each strain on each material. ANOVA tables for the sequential sum of squares were used to determine which group of terms contributed significantly to the models. The groups were first-order terms (x_1, x_2, x_3), pure quadratic terms (x^2_1, x^2_2, x^2_3), and interaction terms (x_1x_2, x_1x_3, x_2x_3). If a group of terms was deemed significant, all terms from that group were kept in the model (Lenth 2009). A blocking variable, b, was included in each model since the observations were gathered in two phases, one phase for the factorial and center points, the second phase for the face-centered points. Information concerning individual parameter estimates, standard error and statistical significance is provided in separate ANOVA tables.

Each model was subjected to diagnostic measures consisting of the F-test for lack of fit, examination of raw residuals and jackknifed residuals vs. fitted values, and examination of normal plots based on both raw residuals and jackknifed residuals (Venables and Ripley 1999). These diagnostics were crucial for outlier detection and in determining for which cases a model could not be fit. Analysis was performed using R along with the rsm package (Lenth 2009).

Results

In the sections that follow equations were fitted for each strain/substrate combination and graphically compare the two strains when possible. There is no physical reason to use four days when contrasting the graphs other than that it is the median value for days. Likewise, there is no particular reason to fix days and let temperature and relative humidity vary other than to keep the manuscript concise. The interested reader can use their favorite software and the fitted equations to construct whatever graphs he or she wishes.

Aircraft Performance Coating on aluminum (APC)

B. anthracis ΔSterne on APC

All test variables were at the lowest settings during test run three, and there was no or minimal spore inactivation. This observation could not be modeled using Equation 1. Hence, data from test run three was omitted and the model was refitted. A second order model with pure quadratic terms and two-way interactions was needed. The F-test for lack of fit had a P-value of 0.110 which was large enough to not reject the model. R^2 and adjusted R^2 for the model were 0.9548 and 0.8901, respectively. Table 2 shows the parameter estimates, standard errors, t-values, and P-values. All second order terms were significant. The fitted equation for inactivation of *B. anthracis* ΔSterne on APC is $\hat{Y} = 17.5984 - 0.7847b + 6.7224x_1 + 6.2954x_2 + 6.7968x_3 - 2.6850x_1x_2 - 3.4490x_1x_3 - 2.8576x_2x_3 - 3.6926x^2_1 - 3.3049x^2_2 - 2.7348x^2_3$, where $\hat{Y}$ is the log odds of spore inactivation. Figure 1 shows a slice of the response surface at four days and block (b) fixed at its mean value of 0.3333.

B. thuringiensis Al Hakam on APC

All test variables were at the lowest settings during test run three, and there was no or minimal spore inactivation. This observation could not be modeled using Equation 1. Hence, data from test run three was omitted and the model was refitted. A second order model with pure

Table 2 Parameter estimates for *B. anthracis* ΔSterne and *B. thuringiensis* Al Hakam on APC

Model term	Parameter estimate	Standard error	t-value	P-value
B. anthracis ΔSterne				
Intercept	17.5984	0.9562	18.405	3.4646×10^{-7}
Block (b)	−0.7847	1.1276	−0.696	0.5089
x_1	6.7224	0.8281	8.118	8.3006×10^{-5}
x_2	6.2954	0.8281	7.602	0.0001
x_3	6.7968	0.8281	8.208	7.7356×10^{-5}
x_1x_2	−2.685	0.9636	−2.786	0.0271
x_1x_3	−3.449	0.9636	−3.579	0.0090
x_2x_3	−2.8576	0.9636	−2.965	0.0209
x^2_1	−3.6926	1.3148	−2.809	0.0262
x^2_2	−3.3049	1.3148	−2.514	0.0402
x^2_3	−2.7348	1.3148	−2.08	0.0761
B. thuringiensis Al Hakam				
Intercept	16.8775	1.5466	10.912	1.1998×10^{-5}
Block (b)	−0.7869	1.8239	−0.431	0.6791
x_1	7.5240	1.3394	5.617	0.0008
x_2	7.5517	1.3394	5.638	0.0008
x_3	7.2767	1.3394	5.433	0.0010
x_1x_2	−3.6367	1.5587	−2.333	0.0524
x_1x_3	−3.5525	1.5587	−2.279	0.0567
x_2x_3	−2.7313	1.5587	−1.752	0.1232
x^2_1	−1.4004	2.1266	−0.658	0.5313
x^2_2	−4.6715	2.1266	−2.197	0.0640
x^2_3	−3.3409	2.1266	−1.571	0.1602

Df = 1 for estimates.

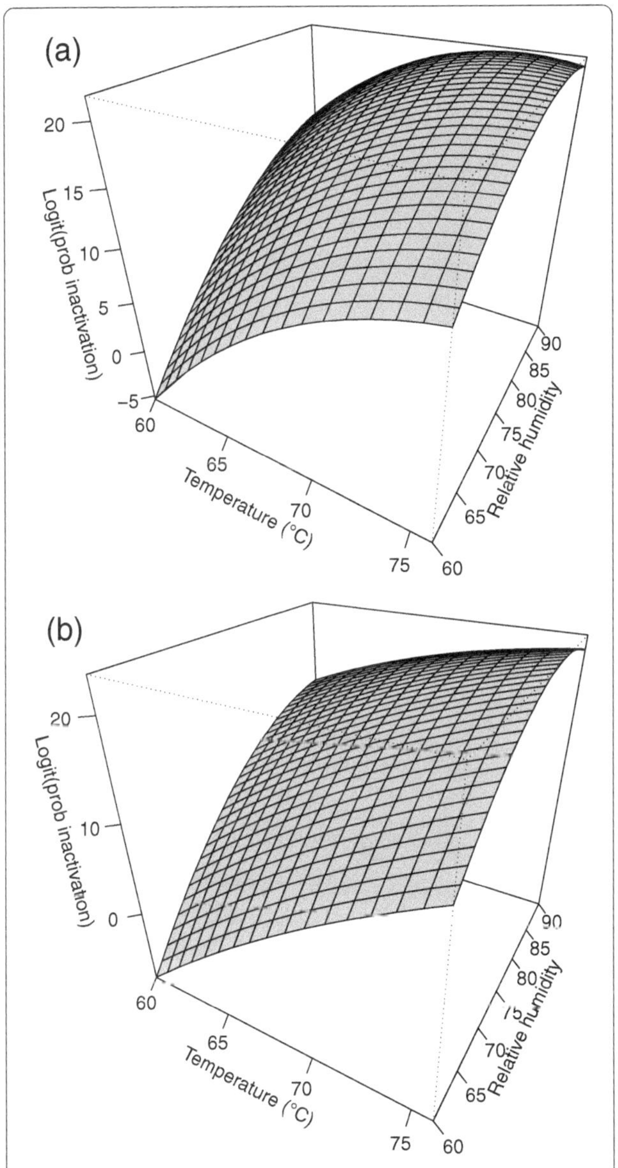

Figure 1 Response surface plot for (a) *B. anthracis* ΔSterne and (b) *B. thuringiensis* Al Hakam on APC at 4 days.

quadratic terms and two-way interactions was needed. The F-test for lack of fit had a P-value of 0.289 which does not indicate lack of fit. R^2 and adjusted R^2 for the model were 0.9079 and 0.7763, respectively. Table 2 shows the parameter estimates, standard errors, t-values, and P-values. The fitted equation for inactivation of *B. thuringiensis* Al Hakam on APC is $\hat{Y} = 16.8775 - 0.7869b + 7.5240x_1 + 7.5517x_2 + 7.2767x_3 - 3.6367x_1x_2 - 3.5525x_1x_3 - 2.7313x_2x_3 - 1.4004x^2_1 - 4.6715x^2_2 - 3.3409x^2_3$, where $\hat{Y}$ is the log odds of spore inactivation. Figure 1 shows a slice of the response surface at four days and block (b) fixed at its mean value of 0.3333.

Antiskid

B. anthracis ΔSterne on antiskid

A second order model with pure quadratic terms was needed. Two-way interaction terms did not contribute significantly to the model ($p = 0.846$) and were not included. The F-test for lack of fit had a P-value of 0.270 which did not indicate lack of fit. R^2 and adjusted R^2 for the model were 0.8833 and 0.8091, respectively. Table 3 shows the parameter estimates, standard errors, t-values, and P-values. The pure quadratic term for days is the significant second order term. The fitted equation for inactivation of *B. anthracis* ΔSterne on antiskid is $\hat{Y} = 16.4509 - 0.9795b + 4.4938x_1 + 3.9625x_2 + 3.8248x_3 - 1.7352x^2_1 - 1.0571x^2_2 - 2.8849x^2_3$, where $\hat{Y}$ is the log odds of spore inactivation. Figure 2 shows a slice of the response surface at four days and block (b) fixed at its mean value of 0.3158.

B. thuringiensis Al Hakam on antiskid

Run number one, a center point, was difficult to model and an outlier in the residual and normal plots. This run was omitted and the model refitted. Only a first order model was needed. Neither two-way interactions nor pure quadratic effects were significant ($p = 0.560$ and $p = 0.513$, respectively). The F-test for lack of fit had a P-value of 0.652 which did not indicate lack of fit. R^2 and adjusted R^2 for the model were 0.8984 and 0.8671, respectively. Table 3 shows the parameter estimates, standard errors, t-values, and

Table 3 Parameter estimates for *B. anthracis* ΔSterne and *B. thuringiensis* Al Hakam on antiskid

Model term	Parameter estimate	Standard error	t-value	P-value
B. anthracis ΔSterne				
Intercept	16.4509	1.2098	13.598	3.1872×10^{-8}
Block (b)	−0.9795	1.4042	−0.698	0.5000
x_1	4.4938	0.8555	5.253	0.0003
x_2	3.9625	0.8555	4.632	0.0007
x_3	3.8248	0.8555	4.471	0.0009
x^2_1	−1.7352	1.6443	−1.055	0.3139
x^2_2	−1.0571	1.6443	−0.643	0.5335
x^2_3	−2.8849	1.6443	−1.755	0.1071
B. thuringiensis Al Hakam				
Intercept	8.9575	0.6919	12.946	8.3652×10^{-9}
Block (b)	3.2641	1.1984	2.724	0.0174
x_1	5.2788	0.7579	6.965	9.8472×10^{-6}
x_2	4.4240	0.7579	5.837	5.8096×10^{-5}
x_3	3.7842	0.7579	4.993	0.0002

Df = 1 for estimates.

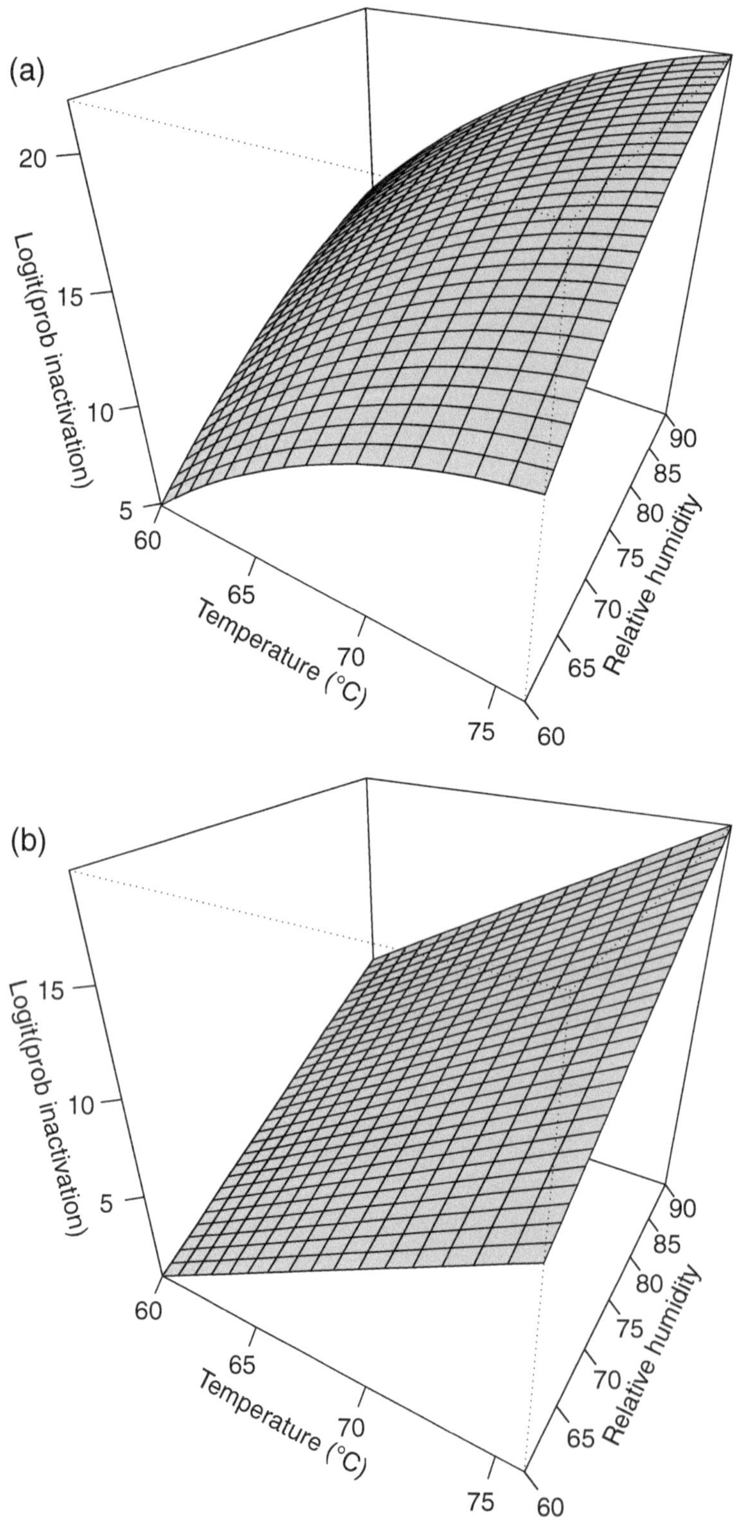

Figure 2 Response surface plot for (a) *B. anthracis* ΔSterne and (b) *B. thuringiensis* Al Hakam on antiskid at 4 days.

P-values. The fitted equation for inactivation of *B. thuringiensis* Al Hakam on antiskid is $\hat{Y} = 8.9575 + 3.2641b + 5.2788x_1 + 4.4240x_2 + 3.7842x_3$, where $\hat{Y}$ is the log odds of spore inactivation. Figure 2 shows a slice of the response surface at four days and block (b) fixed at its mean value of 0.3333.

InsulFab

B. anthracis ΔSterne on InsulFab

Run number one, a center run, was an outlier and deleted from the analysis. A second order model with pure quadratic terms was used in this case. Two-way interaction terms did not contribute significantly to the model ($p = 0.890$) and were not included. The F-test for lack of fit had a P-value of 0.261 which did not indicate lack of fit. R^2 and adjusted R^2 for the model were 0.8275 and 0.7068, respectively. Table 4 shows the parameter estimates, standard errors, t-values, and P-values. The pure quadratic term for relative humidity is the significant second order term. Although 0.176 is a somewhat high P-value for significance, it is preferable to allow for the possibility of some curvature of the response surface. The fitted equation for inactivation of *B. anthracis* ΔSterne on InsulFab is $\hat{Y} = 17.0315 - 2.0862b + 3.0930x_1 + 3.5849x_2 + 3.7208x_3 - 1.6774x^2_1 - 2.9997x^2_2 - 1.6517x^2_3$, where $\hat{Y}$ is the log odds of spore inactivation. Figure 3 shows a slice of the response surface at four days and block (b) fixed at its mean value of 0.3333.

B. thuringiensis Al Hakam on InsulFab

Only a first order model was needed. Neither two-way interactions nor pure quadratic effects were significant ($p = 0.9997$ and $p = 0.9386$, respectively). The F-test for lack of fit had a P-value of 0.340 which did not indicate lack of fit. R^2 and adjusted R^2 for the model were 0.7594 and 0.6906, respectively. Table 4 shows the parameter estimates, standard errors, t-values, and P-values. The fitted equation for inactivation of *B. thuringiensis* Al Hakam on Insulfab is $\hat{Y} = 12.0283 + 2.1611b + 5.7042x_1 + 4.3161x_2 + 3.7813x_3$, where $\hat{Y}$ is the log odds of spore inactivation. Figure 3 shows a slice of the response surface at four days and block (b) fixed at its mean value of 0.3158.

Table 4 Parameter estimates for *B. anthracis* ΔSterne and *B. thuringiensis* Al Hakam on InsulFab

Model term	Parameter estimate	Standard error	t-value	P-value
B. anthracis ΔSterne				
Intercept	17.0315	1.6808	10.1330	1.4082×10^{-6}
Block (b)	2.0862	1.8155	1.1490	0.2772
x_1	3.0930	1.0630	2.9010	0.0156
x_2	3.5849	1.0630	3.3720	0.0071
x_3	3.7208	1.0630	3.5000	0.0057
x^2_1	−1.6774	2.0586	−0.815	0.4341
x^2_2	−2.9997	2.0586	−1.457	0.1757
x^2_3	−1.6517	2.0586	−0.802	0.4410
B. thuringiensis Al Hakam				
Intercept	12.0283	1.0831	11.1050	2.5117×10^{-8}
Block (b)	2.1611	1.9274	1.1210	0.2810
x_1	5.7042	1.2349	4.6190	0.0004
x_2	4.3161	1.2349	3.4950	0.0036
x_3	3.7813	1.2349	3.0620	0.0084

Df = 1 for estimates.

Nylon

B. anthracis ΔSterne on nylon

B. anthracis ΔSterne spores on nylon were the most difficult to inactivate and to model. Using residual plots, three problematic test runs, one, three and fifteen, were identified. These observations were omitted and the model refitted with the caveat that the results for nylon are not as dependable as the results for other coupons. A second order model with pure quadratic terms and two-way interactions was needed. The F-test for lack of fit had a P-value of 0.141 which did not indicate lack of fit. R^2 and adjusted R^2 for the model were 0.9972 and 0.9917, respectively. Table 5 shows the parameter estimates, standard errors, t-values, and P-values. The fitted equation for inactivation of *B. anthracis* ΔSterne on nylon is $\hat{Y} = 0.07582 + 2.1487b + 6.0157x_1 + 4.6424x_2 + 1.3926x_3 + 4.2282x_1x_2 - 0.8748x_1x_3 - 0.3612x_2x_3 + 5.3117x^2_1 + 0.6467x^2_2 - 2.3881x^2_3$, where $\hat{Y}$ is the log odds of spore inactivation. Figure 4 shows a slice of the response surface at four days and block (b) fixed at its mean value of 0.3125.

B. thuringiensis Al Hakam on nylon

Test runs one and fourteen were difficult to fit. These runs were omitted and the model was refit. Interestingly, the number of days of treatment was not a significant predictor for this coupon ($p = 0.664$), so it was excluded from the model. A second order model with pure quadratic terms and two-way interactions was used. The F-test for lack of fit had a P-value of 0.698 which did not indicate lack of fit. R^2 and adjusted R^2 for the model were 0.9778 and 0.9644, respectively. Table 5 shows the parameter estimates, standard errors, t-values, and P-values. The fitted equation for inactivation of *B. thuringiensis* Al Hakam on nylon is $\hat{Y} = -0.2778 + 0.1487b + 4.5440x_1 + 4.0575x_2 + 3.9837x_1x_2 + 1.5441x^2_1 + 2.6875x^2_2$, where $\hat{Y}$ is the log odds of spore inactivation. Figure 4 shows the response surface with block (b) fixed at its mean value of 0.2941.

Wiring insulation

B. anthracis ΔSterne on wiring insulation

Most observations for wiring insulation had complete inactivation of spores and could not be modeled using RSM. All runs from Table 1 had complete inactivation except for runs two, three, nine, and twelve. The notable thing about these four points is that they all had two or more predictors at the low setting; i.e., at least two of x_1, x_2, or x_3 were equal to −1 for these four observations.

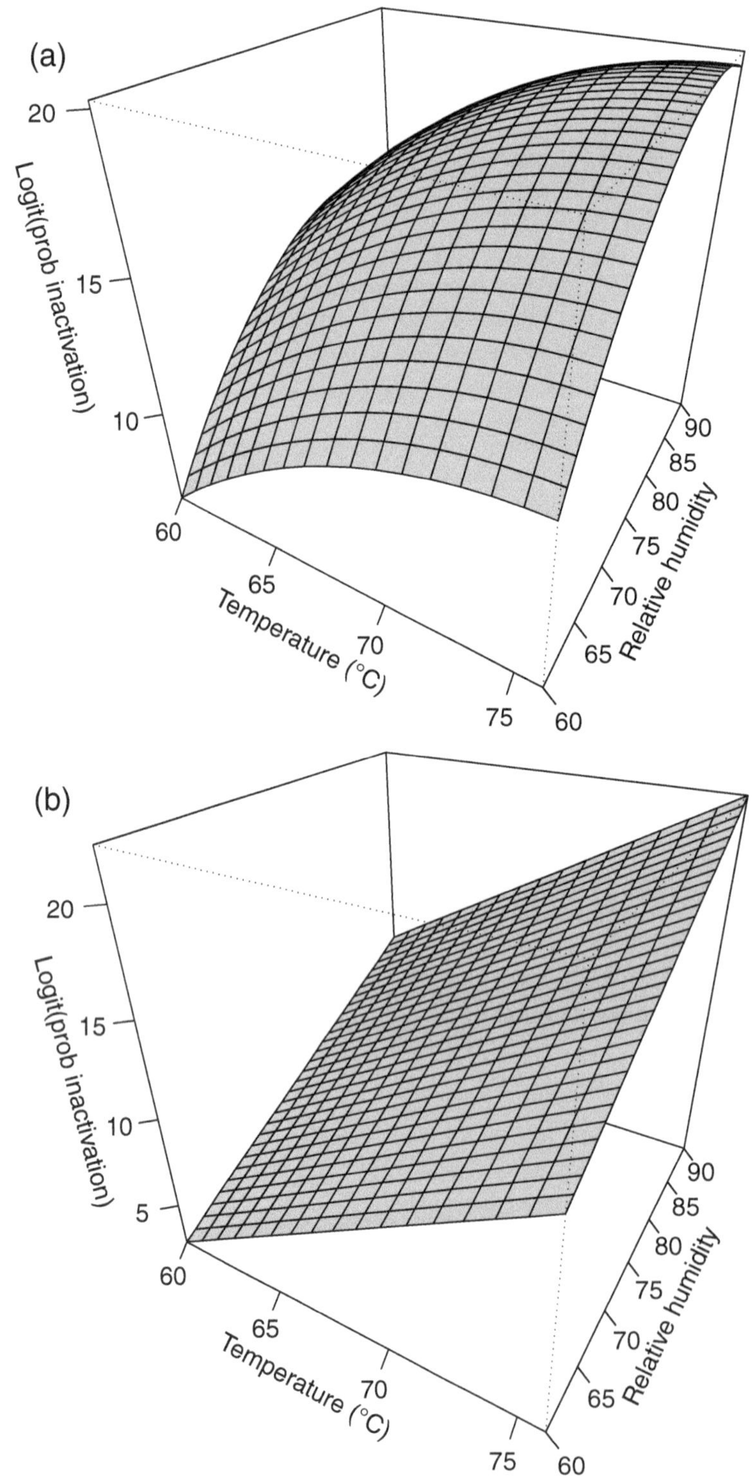

Figure 3 Response surface plot for (a) *B. anthracis* ΔSterne and (b) *B. thuringiensis* Al Hakam on InsulFab at 4 days.

This might be considered as the "back corner" of the predictor space. Excepting these four points, the model would be $\hat{Y} = \infty$ or, equivalently, prob(inactivation) = 1.

B. thuringiensis Al Hakam on wiring insulation

The results for *B. anthracis* ΔSterne apply to *B. thuringiensis* Al Hakam.

Table 5 Parameter estimates for *B. anthracis* ΔSterne and *B. thuringiensis* Al Hakam on nylon

Model term	Parameter estimate	Standard error	t-value	P-value
B. anthracis ΔSterne				
Intercept	0.0758	0.3034	0.2500	0.8126
Block (b)	2.1487	0.3841	5.5930	0.0025
x_1	6.0157	0.2809	21.4170	4.1152×10^{-6}
x_2	4.6424	0.2405	19.3010	6.8867×10^{-6}
x_3	1.3926	0.2405	5.7900	0.0022
x_1x_2	4.2282	0.2809	15.0540	2.3431×10^{-5}
x_1x_3	−0.8748	0.2809	−3.1140	0.0264
x_2x_3	−0.3611	0.2809	−1.2860	0.2549
$x^2{}_1$	5.3117	0.4940	10.7540	0.0001
$x^2{}_2$	0.6467	0.4222	1.5320	0.1862
$x^2{}_3$	−2.3881	0.4222	−5.6560	0.0024
B. thuringiensis Al Hakam				
Intercept	−0.2778	0.5317	−0.5220	0.6127
Block (b)	0.1487	0.6584	0.2260	0.8259
x_1	4.544	0.3822	11.8900	3.1850×10^{-7}
x_2	4.0575	0.3473	11.6830	3.7552×10^{-7}
x_1x_2	3.9837	0.3883	10.2600	1.2560×10^{-6}
$x^2{}_1$	1.5441	0.7735	1.9960	0.0739
$x^2{}_2$	2.6875	0.7264	3.7000	0.0041

Df = 1 for estimates.

Polypropylene

B. anthracis ΔSterne on polypropylene

Run number one, a center run, was an outlier and deleted from the analysis. A second order model with pure quadratic terms was used in this case. Table 6 shows that the significant pure quadratic term is relative humidity. Adding two-way interaction terms did not contribute significantly to the model (p = 0.641) and were not included. The F-test for lack of fit had a P-value of 0.187 which did not indicate lack of fit. R^2 and adjusted R^2 for the model were 0.8713 and 0.7811, respectively. Table 6 shows the parameter estimates, standard errors, t-values, and P-values. The fitted equation for inactivation of *B. anthracis* ΔSterne on polypropylene is $\hat{Y} = 6.4025 + 6.4256b + 6.9452x_1 + 2.2431x_2 + 4.3604x_3 - 1.0017x^2{}_1 + 4.8915x^2{}_2 - 2.3429x^2{}_3$, where $\hat{Y}$ is the log odds of spore inactivation. Figure 5 shows a slice of the response surface at four days and block (b) fixed at its mean value of 0.3333.

B. thuringiensis Al Hakam on polypropylene

Complete inactivation of *B. thuringiensis* Al Hakam spores was measured for test runs five, seven, fifteen, seventeen and nineteen. These data prevented the fitting of a second order model. Going beyond the usual RSM framework, however, a piecewise model was fitted using a second order model with pure quadratic terms for the points of incomplete activation and mapping the observations in the design space enclosed by runs five, seven, fifteen, and seventeen to ∞ or, equivalently, prob (inactivation) = 1. The region of incomplete inactivation was fitted as $\hat{Y} = 1.4271 + 0.7180b + 1.7227x_1 - 0.1465x_2 + 0.3641x_3 - 0.0792x^2{}_1 + 0.8467x^2{}_2 - 0.5264x^2{}_3$, where $\hat{Y}$ is the log odds of spore inactivation. Pure quadratic terms were significant for that part of the model. Table 6 shows that relative humidity is the significant second order term. Two-way interactions were not significant (p = 0.449). The F-test for lack of fit had a P-value of 0.452 which did not indicate lack of fit. Table 6 shows the parameter estimates, standard errors, t-values, and P-values. Figure 5 shows a slice of the response surface at four days and block (b) fixed at its mean value of 0.2308. We used $\hat{Y}$ =20 instead of $\hat{Y} = \infty$ to keep the scale in the figure reasonable but still show the piecewise behavior of the model

$$\hat{Y} = \begin{cases} \infty, if(x_1 \geq 0, x_2 \geq 1, x_3 \leq x_1, x_3 \geq -x_1) or (x_1 \geq 1, x_2 \geq 0, x_3 \leq x_2, x_3 \geq -x_2) \\ 1.4271 + 0.7180b + 1.7227x_1 - 0.1465x_2 + 0.3641x_3 - 0.0792x_1^2 + 0.8467x_2^2 - 0.5264x_3^2 \end{cases}$$

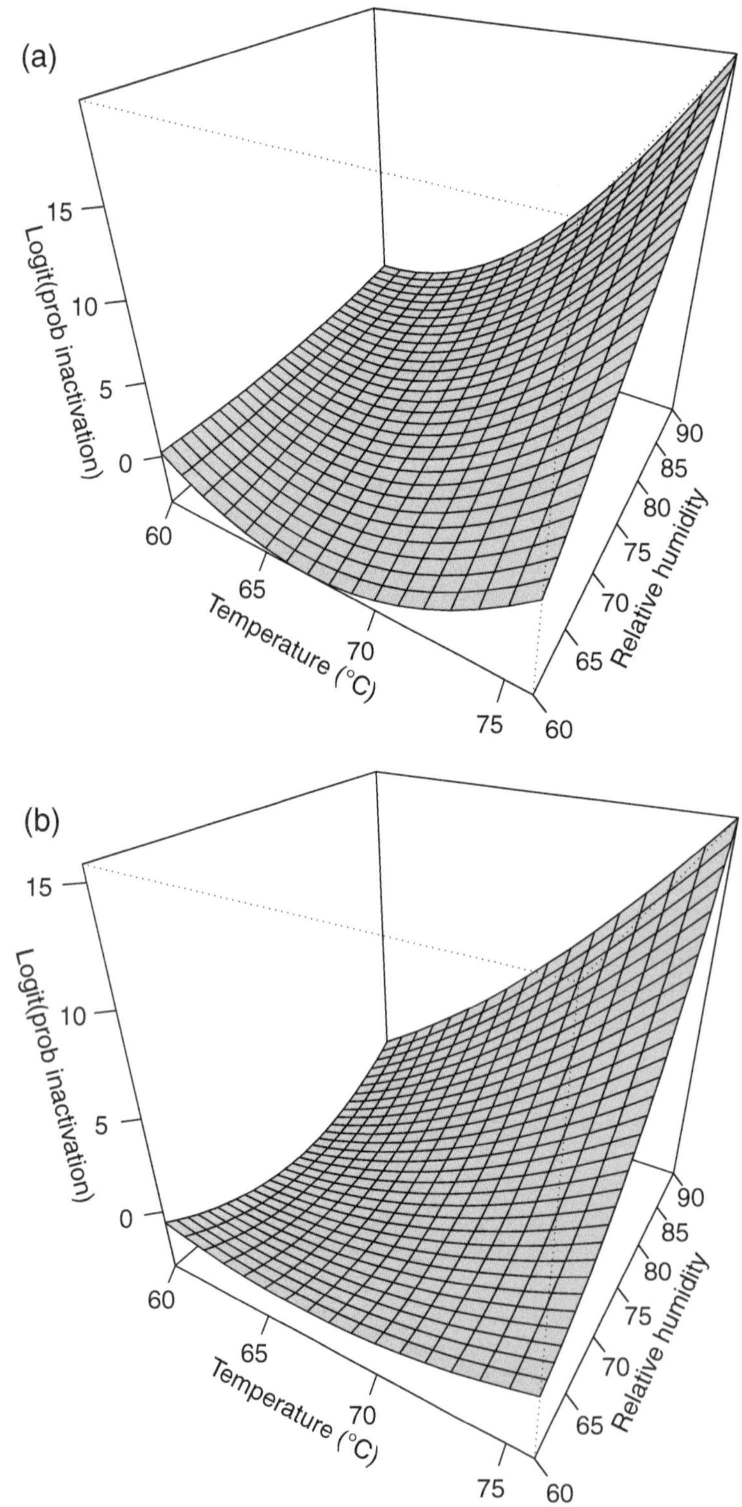

Figure 4 Response surface plot for (a) *B. anthracis* ΔSterne and (b) *B. thuringiensis* Al Hakam on nylon at 4 days.

Table 6 Parameter estimates for *B. anthracis* ΔSterne and *B. thuringiensis* Al Hakam on polypropylene

Model term	Parameter estimate	Standard error	t-value	P-value
B. anthracis ΔSterne				
Intercept	6.4025	1.8516	3.4578	0.0061
Block (b)	6.4256	2.0000	3.2128	0.0093
x_1	6.9452	1.1711	5.9306	0.0001
x_2	2.2431	1.1711	1.9154	0.0844
x_3	4.3604	1.1711	3.7234	0.0040
x^2_1	−1.0017	2.2678	−0.4417	0.6681
x^2_2	4.8915	2.2678	2.1570	0.0564
x^2_3	−2.3429	2.2678	−1.0331	0.3259
B. thuringiensis Al Hakam				
Intercept	1.4271	0.1529	9.3320	0.0002
Block (b)	0.7180	0.2202	3.2600	0.0224
x_1	1.7227	0.1529	11.2640	9.6332×10^{-5}
x_2	−0.1465	0.1529	−0.9580	0.3822
x_3	0.3641	0.1249	2.9160	0.0332
x^2_1	−0.0792	0.2823	−0.2810	0.7903
x^2_2	0.8467	0.2823	2.9990	0.0301
x^2_3	−0.5264	0.2777	−1.8960	0.1165

Df = 1 for estimates.

A response with such a large jump is not the most desirable model and demonstrates the need for further experimentation using less extreme values of the predictors on polypropylene. Nevertheless, it contrasts the difference in behavior of polypropylene versus the other materials tested.

Wet solution

B. anthracis ΔSterne in wet solution

Most observations for wet solution had complete inactivation of spores, and could not be modeled using RSM. All *B. anthracis* ΔSterne runs from Table 1 had complete inactivation except for runs three, six, and nine. The notable thing about these three points is that they all had temperature at the low setting.

B. thuringiensis Al Hakam in wet solution

All *B. thuringiensis* Al Hakam test runs from Table 1 had complete inactivation except for runs one, three, six, nine, and twelve. Except for run one, a center point, these observations had temperature at the low setting.

Contour plots in two dimensions

Decontamination requirements were a ≥6-log (10^6) spore inactivation out of a ≥6-log (10^6) spore challenge (Buhr et al. 2012b). This translates into a probability of spore inactivation equal to 0.999999 or a log odds of spore inactivation equal to $log\left(\frac{0.999999}{1-0.999999}\right) \approx 13.82$. The lower bound from a two-sided 90% confidence surface for the log odds of inactivation was used for the contour plots (Myers and Montgomery 2002). In cases where a model could be fit using Equation 1, a contour plot for temperature vs. relative humidity is given in original units with days fixed at 4. (The plot for *B. thuringiensis* Al Hakam on nylon was omitted since the region of ≥6-log spore inactivation was barely visible on the graph.) These are given in Figures 6, 7, 8, and 9.

Discussion

Control-driven test method improvements and the use of multiple independent spore preparations with a single protocol useful for both *B. anthracis* ΔSterne and *B. thuringiensis* Al Hakam (Buhr et al. 2012b) allowed for the application of a statistically based experimental design, specifically RSM. This use of RSM permitted subsequent mathematical analysis and modeling of the response, generating a predictive capability valuable to potential users of hot, humid air decontamination technology. These advancements in data analyses and interpretation, when coupled with decreased test time and cost (Buhr et al. 2012b), serve to better support the transition of decontamination technologies from the laboratory to the field.

In all modeled strain-material pairs but one, all three predictors (temperature, relative humidity, and time) were significant and each had an effect on spore inactivation. The exception was *B. thuringiensis* Al Hakam spores dried onto nylon webbing material, where time was not significant. Interestingly, for *B. anthracis* ΔSterne on two materials (InsulFab and polypropylene), relative humidity was the most significant second order pure quadratic term.

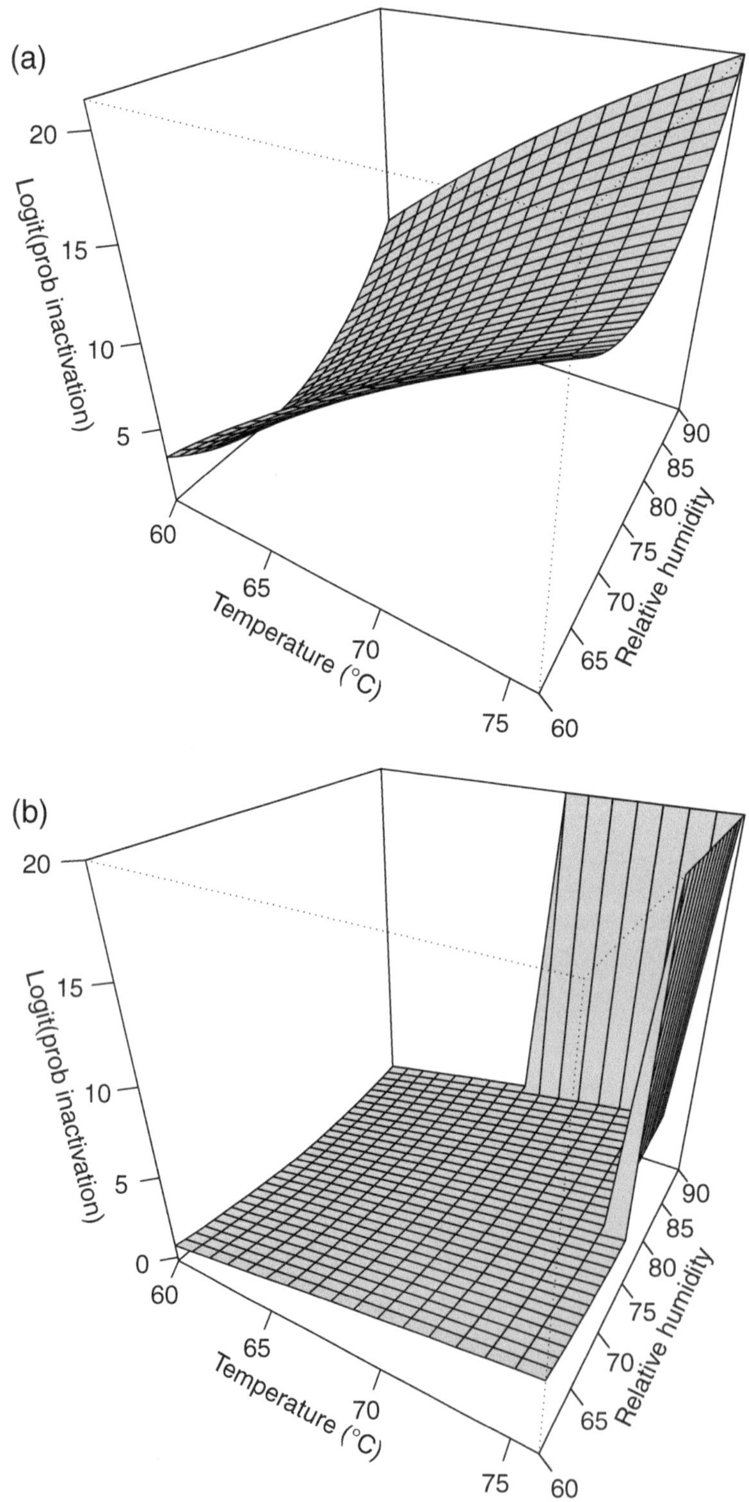

Figure 5 Response surface plot for (a) ***B. anthracis*** **ΔSterne and (b)** ***B. thuringiensis*** **Al Hakam on polypropylene at 4 days.**

The most difficult materials to model were nylon webbing and polypropylene (dried plastic tubes) where a piecewise modeling strategy was required to fit *B. thuringiensis* Al Hakam on polypropylene. Nylon was the only material for which multiple runs were omitted based on examination of residual plots. Hot, humid air

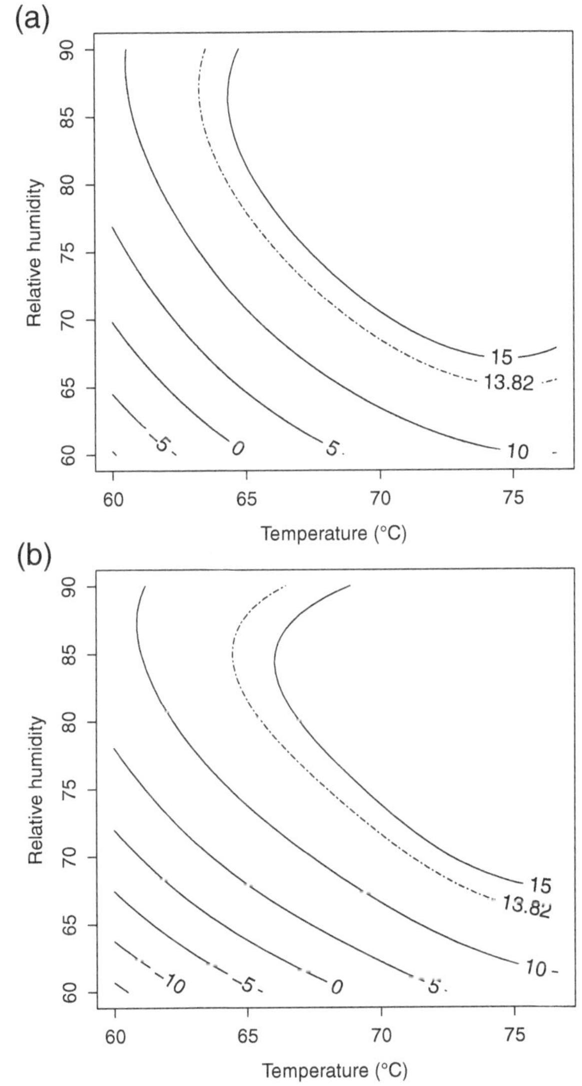

Figure 6 90% lower confidence contour for spore inactivation of (a) *B. anthracis* ΔSterne and (b) *B. thuringiensis* Al Hakam on APC.

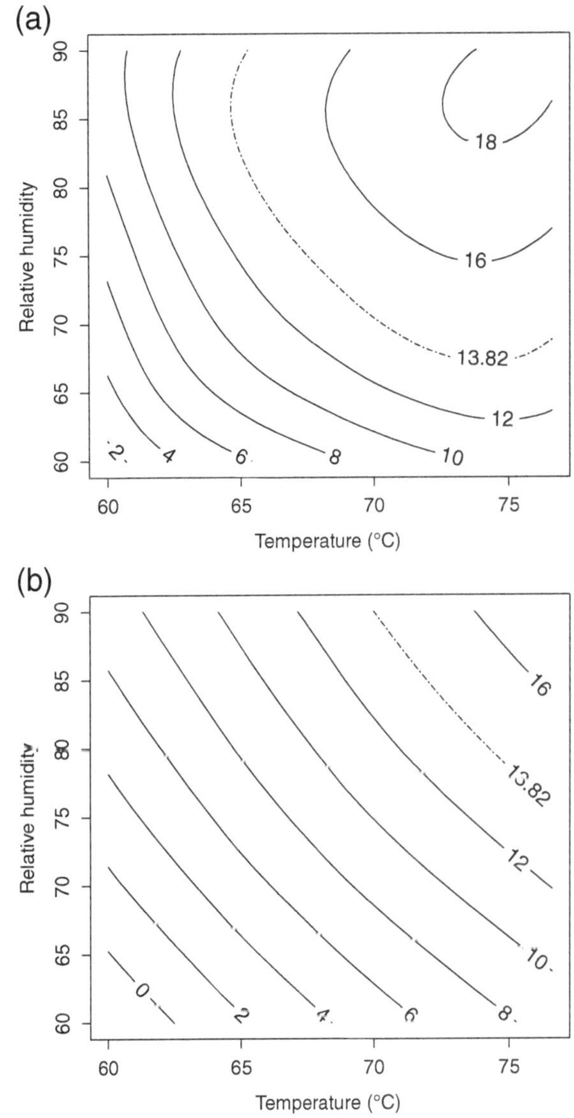

Figure 7 90% lower confidence contour for spore inactivation of (a) *B. anthracis* ΔSterne and (b) *B. thuringiensis* Al Hakam on antiskid.

was least effective at inactivating spores on nylon. This may be due to the porous and/or hydrophobic properties of nylon.

For wet spores and those dried onto wiring insulation, the majority of test runs showed complete inactivation. For wet spores of both strains, the cases without complete inactivation were at the lowest setting for temperature with only one exception. For spores of both strains on wiring insulation, the conditions without complete inactivation all had two or more predictors at the lowest settings.

The mathematical models were cross-checked with published spore survival results for high purity spores of *B. anthracis* ΔSterne and *B. thuringiensis* Al Hakam (Buhr et al. 2012b). The published data set consisted of total of 210 combinations of strain/substrate/temperature/relative humidity/time including the six substrates and the wet control (seven total). Data was compiled from 1,330 test samples and 1,330 room temperature controls over 19 independent test runs (Buhr et al. 2012b). These published data were cross-checked with the contour graphs. Spore inactivation data of less than 6-log was below the 90% lower statistical confidence line (higher practical confidence), and spore inactivation data of greater than 6-log was above the 90% lower statistical confidence line. Thus each spore inactivation/decontamination datum was located at a logical position in the contour graphs. Both the

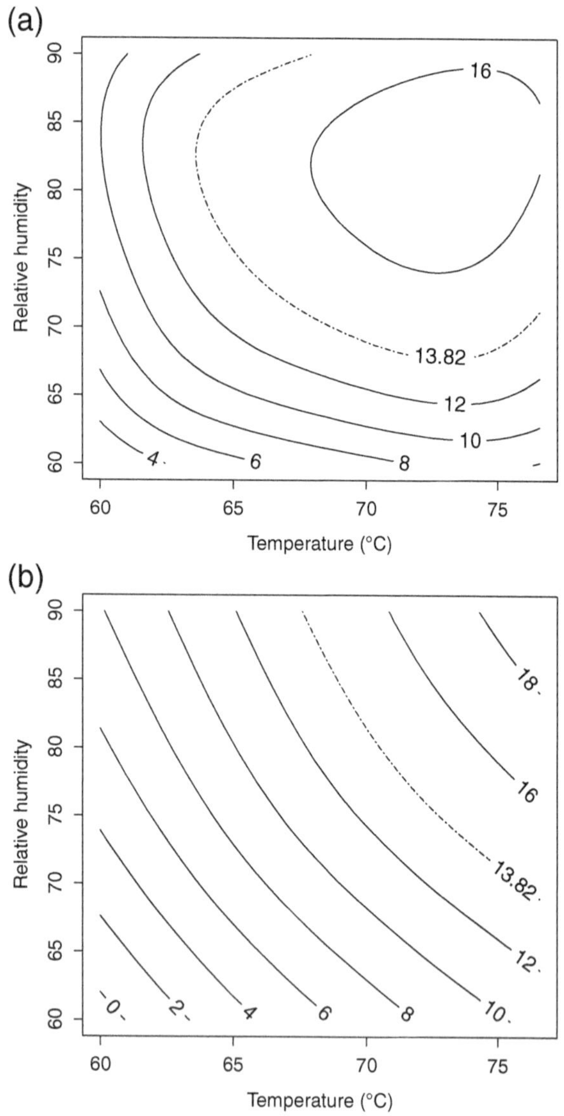

Figure 8 **90% lower confidence contour for spore inactivation of (a) *B. anthracis* ΔSterne and (b) *B. thuringiensis* Al Hakam on InsulFab.**

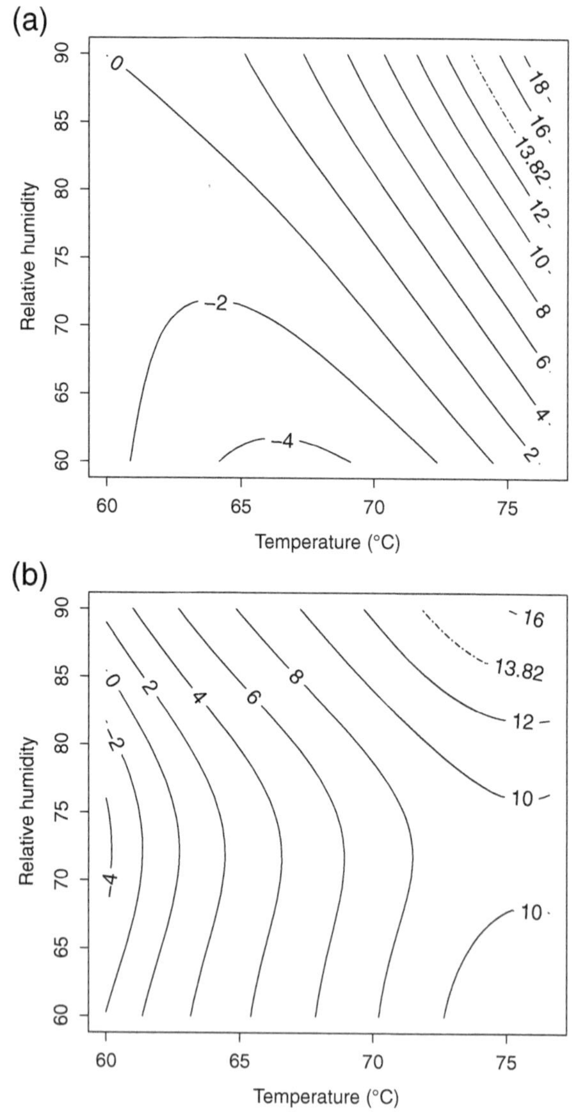

Figure 9 **90% lower confidence contour for spore inactivation of *B. anthracis* ΔSterne on (a) nylon and (b) polypropylene.**

published decontamination results and the mathematical models showed that *B. anthracis* ΔSterne and *B. thuringiensis* Al Hakam spores were similarly decontaminated, although *B. thuringiensis* Al Hakam spores survived slightly better on some substrates, such as polypropylene, under the mildest temperature/relative humidity/time conditions.

This work demonstrates the use of statistical methods to design biological decontamination experiments at the limits of decontamination technology and to analyze the resultant data. The fitted equations specific to strain and material permit a broad range of end-users with varying requirements and constraints the ability to predict the success of hot, humid air treatments on specific materials based on achievable temperature and relative humidity conditions and permissible decontamination times.

Competing interests

A patent application has been submitted on behalf of the United States Navy regarding the mathematical models in order to protect the United States government's investment in this technology.

Acknowledgements

This work was supported through funding provided by the Defense Threat Reduction Agency Joint Science and Technology Office, Protection and Hazard Mitigation Capability Area (Project Number BA08PHM113). We thank Dr. Clifton Sutton of the Statistics Department at George Mason University for statistical advice and technical guidance. We thank Aaron Kota for editorial support.

References

Agresti A (1996) Categorical data analysis. Wiley, New York, NY

Agresti A (2002) An introduction to categorical data analysis, 2nd edn. Wiley, Hoboken, NJ

Alderton G, Snell N (1970) Chemical states of bacterial spores: heat resistance and its kinetics at intermediate water activity. Appl Microbiol 19(4):565–572

Beauregard MR, Mikulak RJ, Olson BA (1992) A practical guide to statistical quality improvement: opening up the statistical toolbox. Van Nostrand Reinhold Company, New York, NY

Buhr T, McPherson DC, Gutting BW (2008) Analysis of broth-cultured *Bacillus atrophaeus* and *Bacillus cereus* spores. J Appl Microbiol 105:1604–1613

Buhr T, Young AA, Minter ZA, Wells CM, Shegogue DA (2011) Decontamination of a hard surface contaminated with *Bacillus anthracis* ΔSterne and *B. anthracis* Ames spores using electrochemically generated liquid-phase chlorine dioxide ($eCLO_2$). J Appl Microbiol 111:1057–1064

Buhr T, Canter D, Kempter J (2012a) Biological decontamination. In: Bizzigotti GO, Lee SJ (eds) Handbook of chemical and biological warfare agent decontamination. ILM Publications, Glendale, AZ

Buhr T, Young AA, Minter ZA, Wells CM, McPherson DC, Hooban C, Johnson C, Prokop EJ, Crigler JR (2012b) Hot, humid air decontamination of materials contaminated with *Bacillus anthracis* ΔSterne and *Bacillus thuringiensis* Al Hakam spores. J Appl Microbiol 113:1037–1051

Canter DA (2005) Remediating anthrax-contaminated sites: learning from the past to protect the future. Chemical Health and Safety 12(4):13–19

Coleman WH, Chen D, Yong-qing L, Cowan AE, Setlow P (2007) How moist heat kills spores of *Bacillus subtilis*. J Bacteriol 189(23):8458–8466

Craig BD, Anderson DS (1995) Handbook of corrosion data, 2nd edn. ASM international, Materials Park, OH

Decker HM, Citek FJ, Harstad JB, Gross NH, Piper FJ (1954) Time temperature studies of spore penetration through an electric air sterilizer. Appl Microbiol 2(1):33

Dirckx JH (1981) Virgil on anthrax. Am J Dermatopathol 3(2):191–196

Ehrlich R, Miller S, Walker RL (1970) Relationship between atmospheric temperature and survival of airborne bacteria. Appl Microbiol 19(2):245–249

Garverick L (1994) Corrosion in the petrochemical industry. ASM international, Materials Park, OH

Gerhardt P, Marquis RE (1989) Spore thermoresistance mechanisms. In Smith I, Slepecky RA, Setlow P (eds) Regulation of Prokaryotic Development. Am Soc Microbiol, Washington, DC

Herzberg E (2012) DoD releases a report on the effects of corrosion on the cost and availability of air force aircraft and missiles. In: CorrDefense. Inside DoD 8(1). http://corrdefense.nacc.org/corrdefense_spring_2012/DoD_1.asp. Accessed 15 Nov 2012

Holwitt E, Kiel JL, Alls JL, Morales PJ, Gifford H (2000) Thermal sensitivity of biowarfare simulants. AeroSense, SPIE, pp 31–39

Hosmer DW, Lemeshow S (2000) Applied logistic regression, 2nd edn. Wiley, New York, NY

Hugo W (1991) A brief history of heat and chemical preservation and disinfection. J Appl Bacteriol 71(1):9–18

Koch G, Brongers MPH, Thompson NG, Virmani YP, Payer JH (2001) Corrosion costs and preventive strategies in the United States. FHWA, Washington DC. URL http://www.nace.org/uploadedFiles/Publications/ccsupp.pdf. Accessed 14 Nov 2012

Lenth RV (2009) Response-surface methods in R, using rsm. Journal of Statistical Software 32(7):1–17, URL http://www.jstatsoft.org/v32/i07/. Accessed 1 Feb 2011

McCartt AD, Gates SD, Jeffries JB, Hanson RK, Joubert LM, Buhr T (2011) Response of *Bacillus thuringiensis* Al Hakam endospores to gas dynamic heating in a shock tube. Z Physical Chemistry 225:1367–1377

Melly E, Genest P, Gilmore M, Little S, Popham D, Driks A, Setlow P (2002) Analysis of the properties of spores of *Bacillus subtilis* prepared at different temperatures. J Appl Microbiol 92(6):1105–1115

Moser MR, Bender TR, Margolis HS, Noble GR, Kendal AP, Ritter DG (1979) An outbreak of influenza aboard a commercial airliner. Am J Epidemiol 110(1):1–6

Murrell WG, Scott WJ (1966) The heat resistance of bacterial spores at various water activities. J Gen Microbiol 43(3):411–425

Myers RH, Montgomery DC (2002) Response surface methodology, 2nd edn. Wiley, New York, NY

Myers RH, Montgomery DC, Anderson-Cook CM (2009) Response surface methodology: product and process optimization using designed experiments, 3rd edn. Wiley, Hoboken, NJ

Olsen SJ, Chang HL, Cheung TY, Tang AF, Fisk TL, Ooi SP, Kuo HW, Jiang DD, Chen KT, Lando J, Hsu KH, Chen TJ, Dowell SF (2003) Transmission of the severe acute respiratory syndrome on aircraft. N Engl J Med 349(25):2416–2422

Rutala WA, Weber DJ, Healthcare Infection Control Practices Advisory Committee (HICPAC) (2008) Guideline for disinfection and sterilization in healthcare facilities. Centers for Disease Control, Atlanta, GA, http://www.cdc.gov/hicpac/pdf/guidelines/Disinfection_Nov_2008.pdf. Accessed 1 Jun 2013

Ryan SP, Calfee MW, Wood JP, Attwood BC, Schaefer FW (2010) Research to support the decontamination of surfaces and buildings contaminated with biothreat agents. Rutala WA (ed) Disinfection, sterilization, and antisepsis: principles, practices, current issues, new research, and new technologies, APIC, Washington, DC

Sternbach G (2003) The history of anthrax. J Emerg Medicine 24(4):463–467

Sunde EP, Setlow P, Hederstedt L, Halle B (2009) The physical state of water in bacterial spores. Proc Natl Acad Sci USA 106(46):19,334–19,339

Tomasino SF, Pines RM, Cottrill MP, Hamilton MA (2008) Determining the efficacy of liquid sporicides against spores of *Bacillus subtilis* on a hard nonporous surface using the quantitative three step method: collaborative study. J AOAC Int 91(4):833–852

Ullmann A (2007) Pasteur-koch: distinctive ways of thinking about infectious diseases. Microbe 2(8):383

Venables WM, Ripley BD (1999) Modern applied statistics with S-PLUS, 3rd edn. Springer, New York, NY

Zhang P, Kong L, Setlow P, Li Y (2010) Characterization of wet-heat inactivation of single spores of *Bacillus* species by dual-trap raman spectroscopy and elastic light scattering. Appl Environ Microbiol 76(6):1796–1805

2

Genome features of *Pseudomonas putida* LS46, a novel polyhydroxyalkanoate producer and its comparison with other *P. putida* strains

Parveen K Sharma[1], Jilagamazhi Fu[1], Xiangli Zhang[2], Brian Fristensky[2], Richard Sparling[3] and David B Levin[1*]

Abstract

A novel strain of *Pseudomonas putida* LS46 was isolated from wastewater on the basis of its ability to synthesize medium chain-length polyhydroxyalkanoates (mcl-PHAs). *P.putida* LS46 was differentiated from other *P.putida* strains on the basis of cpn60 (UT). The complete genome of *P.putida* LS46 was sequenced and annotated. Its chromosome is 5,86,2556 bp in size with GC ratio of 61.69. It is encoding 5316 genes, including 7 rRNA genes and 76 tRNA genes. Nucleotide sequence data of the complete *P. putida* LS46 genome was compared with nine other *P. putida* strains (KT2440, F1, BIRD-1, S16, ND6, DOT-T1E, UW4, W619 and GB-1) identified either as biocontrol agents or as bioremediation agents and isolated from different geographical region and different environment. BLASTn analysis of whole genome sequences of the ten *P. putida* strains revealed nucleotide sequence identities of 86.54 to 97.52%. *P.putida* genome arrangement was LS46 highly similar to *P.putida* BIRD1 and *P.putida* ND6 but was markedly different than *P.putida* DOT-T1E, *P.putida* UW4 and *P.putida* W619. Fatty acid biosynthesis (*fab*), fatty acid degradation (*fad*) and PHA synthesis genes were highly conserved among biocontrol and bioremediation *P.putida* strains. Six genes in *pha* operon of *P. putida* LS46 showed >98% homology at gene and proteins level. It appears that polyhydroxyalkanoate (PHA) synthesis is an intrinsic property of *P. putida* and was not affected by its geographic origin. However, all strains, including *P. putida* LS46, were different from one another on the basis of house keeping genes, and presence of plasmid, prophages, insertion sequence elements and genomic islands. While *P. putida* LS46 was not selected for plant growth promotion or bioremediation capacity, its genome also encoded genes for root colonization, pyoverdine synthesis, oxidative stress (present in other soil isolates), degradation of aromatic compounds, heavy metal resistance and nicotinic acid degradation, manganese (Mn II) oxidation. Genes for toluene or naphthalene degradation found in the genomes of *P. putida* F1, DOT-T1E, and ND6 were absent in the *P. putida* LS46 genome. Heavy metal resistant genes encoded by the *P. putida* W619 genome were also not present in the *P. putida* LS46 genome. Despite the overall similarity among genome of *P.putida* strains isolated for different applications and from different geographical location a number of differences were observed in genome arrangement, occurrence of transposon, genomic islands and prophage. It appears that *P.putida* strains had a common ancestor and by acquiring some specific genes by horizontal gene transfer it differed from other related strains.

Keywords: *Pseudomonas putida*; Comparative bioinformatics analysis; Comparative genome analysis; Polyhydroxyalkanoates; Pan-Genome; Insertion sequences; Metabolic diversity

* Correspondence: david.levin@umanitoba.ca
[1]Department of Biosystems Engineering, University of Manitoba, Winnipeg, MB R3T 2N2, Canada
Full list of author information is available at the end of the article

Introduction

The genus *Pseudomonas* consists of a very heterogeneous group of microorganisms isolated from diverse environments and belonging to the gamma (γ)-proteobacteria (Palleroni 1984). The genus has approximately 100 named species grouped into different subgroups based on multi-locus analysis (Winsor et al. 2011). The genus *Pseudomonas* contains human pathogens, plant pathogens, and heterotrophic bacteria prevalent in soil, water, and on plant surfaces. Most pseudomonads are free-living saprophytic organisms in soil or water. Because of their great metabolic diversity they play an important role in decomposition, biodegradation, and the carbon and nitrogen cycles. *Pseudomonas putida* is non-pathogenic soil bacteria and is capable of degrading xenobiotic and promoting plant growth after root colonization as well as simultaneously providing protection for the plant from pests and other harmful bacteria (Compant et al. 2005; Singleton 1994).

Pseudomonas putida KT2440, *P. putida* BIRD1, *P. putida* UW4, *P. putida* S11 were isolated from rhizosphere soil and developed as plant growth promoting rhizobacteria (Glick 1995; Matilla et al. 2011; Nakazawa 2002; Ponraj et al. 2012). *Pseudomonas putida* F1 was isolated from polluted soil and developed as a bioremediation agent (Choi et al. 2003; Eaton 1997). *Pseudomonas putida* S16 was isolated in China and identified as a nicotine degrading strain (Tang et al. 2012). *Pseudomonas putida* ND6 was identified as a naphthalene degrading bacterium, in which naphthalene degradation genes were located on a plasmid (Dennis et al. 2004; Li et al. 2002). *Pseudomonas putida* DOT-T1E was isolated as a toluene tolerant bacterium (Ramos et al. 1995), while *P. putida* W619 was isolated as an endophyte of poplar (Taghavi et al. 2005). *Pseudomonas putida* GB-1 was isolated from fresh water and identified as manganese oxidizer (Rosson and Nelson 1982). None of *P.putida* strain was isolated for PHAs production; however, *P.putida* KT2440 was reported as a PHAs producer (Haywood et al. 1989; Huisman et al. 1989).

Polyhydroxyalkanoates (PHAs) are the largest group of biologically produced biodegradable biopolyesters providing different properties and applications (Huijbert and Eggink 1996; de Smet et al. 1983). A wide variety of microorganisms accumulate PHAs as a carbon and energy storage material in carbon-excess conditions when a major nutrient (typically nitrogen or phosphorus) is limiting (Elbahloul and Steinbüchel 2009; Rehm 2010; Brandl et al. 1988). Accumulation of excess carbon is a general mechanism used by *Pseudomonas* and is essential for resource balancing (de Eugenio et al. 2010a; Escapa et al. 2012). De Smet et al. (1983) detected the inclusion bodies in *Pseudomonas olevorans* for first time when grown on octane and identified is as polymer of 3-hydroxyoctanoate. Huisman et al. (1989) studied the PHAs synthesis in fluorescent pseudomonads and confirmed that *P.putida, P.fluorescens, P.aeruginosa, P.tetosteroni* and *P. oleovorans* have the capacity to produce PHAs. Among the *P.putida* strains only *P. putida* KT2440, which was isolated as a plant growth promoting bacterium, has been studied for PHAs production.

Pseudomonas putida LS46 was isolated from wastewater on the basis of its ability to synthesize novel medium chain-length polyhydroxyalkanoates (mcl-PHAs), which accumulated 22% - 56% of cell dry weight when cultured, under batch culture conditions in media containing glucose, waste vegetable fryer oils or fatty acids (Sharma et al. 2011). Analysis of *P. putida* LS46 16S rDNA displayed more than 99% nucleotide sequence identity to other *P. putida* strains. Complete genome sequence of other *P.putida* strains which were developed for plant growth promotion or degradation of xenobiotic, are available. Last genome comparison of *P. putida* strain W619 identified genes for aromatic pathways, heavy metal resistance and plant growth promotion in *P. putida* strain KT2440, F1 and GB-1 (Wu et al. 2011). Since then six new *P. putida* strains, including *P. putida* LS46 genomes have been added to the database. The objective of this work is to see how the genomes of *P.putida* identified as either as biocontrol or bioremediation agents differ from each other due to their geographic origin, environment. Keeping this in mind the genome sequence of *P. putida* LS46 was compared with the genomes of nine other *P. putida* strains for general genome features, shared gene analysis, occurrence of insertion sequences, prophages, genomic islands and for metabolic diversity including PHAs production. The present study revealed more similarity than variability among the genome of *P. putida* strains isolated from different ecological niches in different locations.

Materials and methods

Genome sequences and accession numbers

P putida LS46 was isolated from a municipal wastewater treatment plant in Winnipeg, Manitoba, Canada, and the strain was deposited with International Depository Authority of Canada (IDAC) at the National Microbiology Laboratory, Health Canada Culture Collection (NML-HCCC), WDCM number 840 (Sharma et al. 2011). The *P. putida* LS46 genome was sequenced by the Genome Canada sequencing facility at McGill University, Montreal, Quebec, using a combination of Illumina Gaii (Bennett 2002) and 454 pyrosequencing technology (Margulies et al. 2005). The draft genome sequence of *P.putida* LS46 was deposited in the NCBI GenBank database with Accession number ALPV02000000 (Sharma et al. 2013). The complete genomes of nine *P. putida* strains KT2440 (NC002947), F1 (NC009512), BIRD1 (NC017530), GB-1 (NC010322) S16 (CP003734), ND6 (CP_003589), DOT-T1E (NC_018222), UW4 (CP_003890) and W619 (NC010501) were obtained from the Joint Genome Institute database (http://img.jgi.doe.gov) and compared with the newly sequenced *P. putida* LS46. General genome features like genome size, total number of genes, number of coding sequences (CDS), number of Clusters of Orthologous Groups (COGs) were compared among the genomes of the 10 *P. putida* strains.

Phylogenetic relationship among *Pseudomonas* strains

The *P. putida* LS46 chaparonin gene (*cpn*60) universal target (UT) sequence was obtained from the *P. putida* LS46 genome sequence and aligned with the *cpn*60 UT sequences of 29 *Pseudomonas* strains available in the Integrated Microbial Genome database Markowitz et al. 2006). The sequences were aligned using Bioedit (Thompson et al. 1994) and phylogenetic analysis was conducted using MEGA5 (Tamura et al. 2011). Phylogenetic tree was constructed using the neighbor-joining method (Saitou and Nei 1987).

Occurrence of prophage, insertion sequence and genomic islands

The ten *P. putida* genomes were analyzed for presence of prophages using the PHAST (PHAge Search Tool) prophage finder database (Zhou et al. 2011). Insertion Sequence (IS) elements in the *P. putida* genomes were identified using the ISfinder tool (Siguier et al. 2006; Varani et al. 2011). Genomic islands in *P.putida* genomes were identified using Island finder web based tools (Langille and Brinkman 2009).

Whole genome alignments, BLASTn and Pan-genome analysis

Whole genome identity was calculated from pair-wise comparison of genomes using BLASTn analysis. Comparative synteny Dot plot analysis of the *P. putida* strains was carried out and nine plots of comparison between *P. putida* LS46 and nine *P. putida* strains were obtained (Huang and Zhang 2004). Maps showing *P. putida* strains were prepared using the Gview server *P. putida* LS46 was compared with the other eight *P. putida* strains using Pan genome analysis from Gview server (Petkau et al. 2010). Shared genes in different strains were calculated using the Phylogenetic profile program using a single gene profile from the Integrated Microbial Genome server. The number of homologous genes in genome 1 in comparison to genome 2 was counted. Reciprocally, the numbers of homologous genes in genome 2 in comparison to genome 1 were counted. Percent shared genes were calculated by adding the homologous genes in genome 1 and 2, dividing by the total number of genes in genome 1 + 2, and multiplied by 100 (Palmer et al. 2012).

Comparison of metabolic genes

Genes related with different metabolic functions were identified by searching for homologous genes in *P.putida* genomes by using BLAST analysis of find gene in IMG web page using default parameter (Integrated Microbial Genome). Homologous genes for aromatic compound degradation, heavy metal resistance, manganese oxidation, nicotinic acid degradation, iron scavenging were identified using BLASTn analysis of IMG web page. Likewise TonB dependent receptor and dioxygenase genes were identified in different *P.putida* strains.

Results

Phylogenetic relationship of *P. putida* LS46 to other *Pseudomonas* strains

A polyhydroxyalkanoate producing bacterium was isolated from wastewater and was identified as a strain of *Pseudomonas putida* on the basis of 16S rRNA gene sequence. The16S rRNA gene sequence analysis revealed more than 99% nucleotide sequence identity to other *P. putida* strains. Protein-encoding genes are known to provide higher levels of taxonomic resolution than non-protein-encoding genes like 16S rRNA gene. Therefore, a protein coding gene *cpn60* (Hsp60 or GroEL) was used for phylogenetic analysis of *Pseudomonas* species and strains. Neighbor joining trees based on *cpn60* genes divided the *Pseudomonas* species into two clades, one containing *P. putida, P. entomophila, P. mendocina, P. fulva P. aeruginosa* and *P. stutzeri.* The other clade included *P. fluorescens* and *P. syringae* strains. Although *P.putida* strains were clustered in one subclade yet minor differences in *cpn60* gene sequence separated these strains from each other. *Pseudomonas putida* LS46 was closely related to *P. putida* strains ND6, F1, DOT-T1E, BIRD1, KT2440 and clustered with these strains forming a sub-clade (Figure 1). *Pseudomonas putida* strains LS46 was more distantly related to *P.putida* strains GB1, S16 and W619. *Pseudomonas putida* UW4 was not related to any other *P. putida* strains and clustered with *P. fluorescens.*

Genome features of *P. putida* LS46

The complete genome *P.putida* LS46 was assembled in 32 contigs and consists of 5,874,759 bp DNA with the G + C content of 61.69%. It was predicted to encode 5346 genes, with 5248 CDSs (98.17% of predicted genes). Coding region covers 88.48% of the whole genome. Out of 5248 CDSs, 3889 were assigned functions and 1359 CDSs encoded hypothetical proteins. Among these CDSs, 4270 (79.87%) genes could be classified into COG families composed of 22 categories. A total of seven rRNA operons including seven 5S rRNAs, seven 16S rRNAs, and eight 23S rRNAs were present on the chromosome. In addition, 76 tRNA genes that represent all 20 amino acids, and a tRNA for selenocysteine, were identified (Table 1). No plasmid was detected in *P.putida* LS46. BLASTn analysis identified 181 unique genes (less than 30% minimal identity and maximum e-value of 1 e^{-5}) in *P. putida* LS46, which were not present in other nine other *P. putida* strains. Most of the unique genes were predicted to code hypothetical proteins.

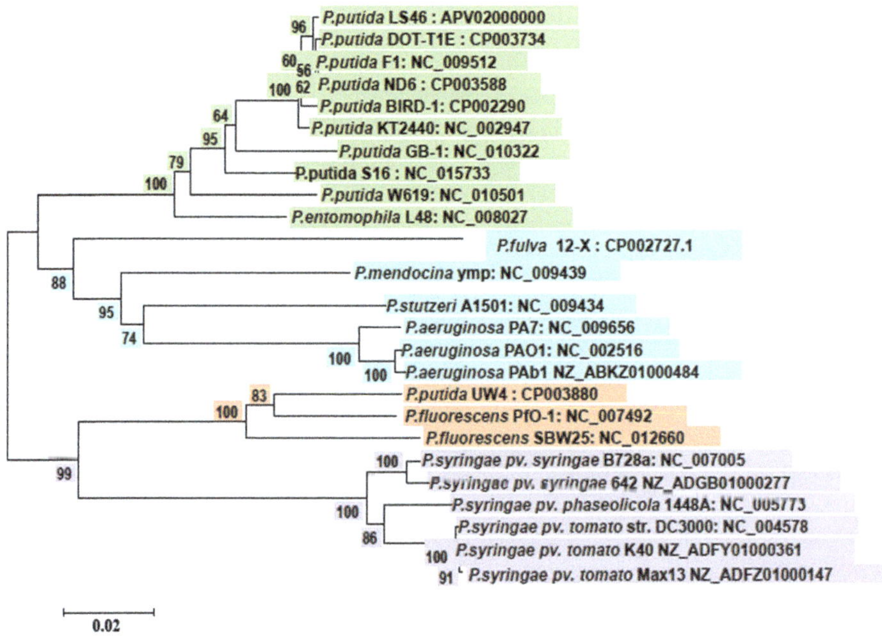

Figure 1 Phylogenetic tree depicting the relationship among *Pseudomonas* species. The tree is based on *cpn60* gene sequences, which were aligned by ClustalW and a neighbor-joining tree was generated using MEGA5 program. Bootstrap values are mentioned at the node.

Insertion Sequence in *P. putida* LS46

Blast analyses of the *P. putida* LS46 genome with IS finder database identified three complete (ISPa41, ISPs1, ISPa42) and one partial (ISPpu9) IS elements (Additional file 1: Table S1). IS element ISPa41, which belongs to IS5 family, was earlier identified in *P. aerugonosa* DK2, *P. putida*, and *P. syringae* DC3000 (IS finder database). Blasting of ISPa41 sequence of *P.putida* LS46 identified homologous sequences in *P.putida* strain GB-1, KT2440 and ND6 but not in *P.putida* BIRD1, DOT-T1E and UW4. Partial sequences of ISPa41 were also present in *P.putida* F1, S16 and W619. ISPs1 was originated from *P. syringae* pv. *syringae* plasmid

Table 1 General features of the genomes of *P. putida* strains

Strain	BIRD1	DOT-T1E	F1	GB-1	KT2440	LS46	ND6	S16	UW4	W619
Isolated in	Spain	Spain	USA	USA	Japan	Canada	China	China	Canada	USA
Genome size	5731541	6260702	5959964	6078430	6181863	5862556	6304310	5984790	6183388	5774330
% G + C	61.74	61.43	61.86	61.94	61.52	61.69	61.62	62.32	60.05%	61.44
Gene count	5046	5802	5423	5515	5481	5316	6484	5307	5517	5292
Homologous genes (%)[a]	92.62	91.39	93.75	90.17	91.32	100	85.57	89.97	82.02	89.14
% Homology[b]	95.89	97.67	97.52	90.86	95.94	100	97.51	91.41	86.59	87.06
CDS count	4960	5721	5300	5417	5350	5219	6391	5218	5423	5194
CDS %	98.30	98.60	97.73	98.22	97.61	98.17	98.57	98.32	98.30	98.15
16S rRNA	7	0	6	7	7	7	6	6	7	7
tRNA	64	58	76	74	74	76	74	70	72	75
COG count	4139	4509	4171	4267	4199	4260	4519	4313	4541	4089
% COG	82.02	77.71	76.91	77.37	76.61	78.61	69.69	81.27	82.31	77.27
Protein W FP[c]	75.23	71.35	73.45	74.75	66.30	79.50	64.24	76.37	75.37	74.79
Protein w/o FP [d]	23.07	27.25	24.29	23.37	31.31	17.29	34.33	21.95	22.93	23.26
Reference	Matilla et al. 2011	Udaondo et al. 2012	Wu et al. 2011	Wu et al. 2011	Nelson et al. 2002	Sharma et al. 2013	Li et al. 2012	Yu et al. 2011	Duan t et al. 2013	Wu et al. 2011

[a] 80% minimal nucleotide sequence identity; e-value 1 e^{-5}.
[b]BLASTn of whole genome.
[c] Proteins with function prediction (%).
[d] Proteins w/o predicted function (%).

pPSR1 and *P. syringae* pv. *savastanoi*. ISPs1 from *P. putida* LS46 had 84% nucleotide sequence identity to ISPs1 from *P. syringae* plasmid pSR1. Complete ISPsI was also present in *P.putida* GB-1 and ND6. Incomplete ISPsI was also detected in *P.putida* strain S16, F1 and W619. Complete ISPpu9 from *P. putida* KT2440 was 2043 bp but in *P.putida* LS46 partial ISPpu9 was present (1697bp). It showed 98% sequence identity to ISPpu9 from *P. putida* KT2440. The ISPpu9 integration target site is 23 bp sequence, which is a part of REP identified in *P. putida* KT2440 (Aranda-Olmedo et al. 2002). ISPpu9 was also present in *P.putida* BIRD1 but was absent in *P.putida* F1, GB1, DOT, ND6 UW4 and W619.

ISPa42, which is synonymous to ISPsy10, encodes a Tn3-like transposable element, which is thought to originate from *P. aeruginosa* DK3. ISPa42, a complete *Tn*3 element of 16941 bp in *P. aeruginosa* DK2, was detected in *P. putida* LS46, but was disrupted by an intergenic region of 19922 bp. It bifurcated it into two segments of 4765 and 12176 bp. This 17 kb transposon was a deletion derivative of Tn4661 toluene degradation transposon Tn4651. The 19.9kb inergenic region of Tn3 carried heavy metal resistance genes like czcA efflux protein (PPUTLS46_008564) *copA* (PPUTLS46_008564), *copB* (PPUTLS46_08569), copper translocating P-type ATPase (PPUTLS46_008644), blue copper domain (PPUTLS46_008654), two component metal regulator sensor signal (PPUTLS46_8669) and heavy metal sensor signal transduction histidine kinase (PPUTLS46_008674) along with sterol desaturase (PPUTLS46_008539). This transposon was unique and was not present in any other *P. putida* strains. Tn4651, a toluene transposon, carried genes for toluene degradation (*xyl*) genes (Tsuda et al. 1989). *P. putida* LS46 did not have toluene degradation ability due to deletion of toluene genes in Tn4661. A number of catabolic transposons have been identified in *Pseudomonas* species for toluene, naphthalene, chlorobenzoate, chlorobenzene and halogenated alkanoates (Wyndham et al. 1994). Absence of these catabolic transposons in *P.putida* LS46 limited it metabolic diversity.

Prediction of prophages

Prophage finder (PHAST) identified one intact prophage and one putative phage genes in *P. putida* LS46 genome. The intact prophage sequence was was present on Contig ALPV02000012 (18373 to 68052) and was 49.6 kb in size with G + C ratio of 59.64%. A total of 61 CDSs which included 40 phage proteins, 21 hypothetical proteins and one tRNA were identified in this prophage. *P.putida* LS46 prophage showed upto to 73% nucleotide sequence identity to *Pseudomonas aeruginosa* phage *Pseudomonas* vB_PaeS_PMG1 which is a virulent phage with lytic infection cycle (Krylov et al. 2012). A 79 bp attachment sites (attL and attR) were also present. This phage was also present in *P.putida* BIRD1, *P.putida* KT2440 and *P.putida* W619 genomes (Additional file 1: Table S2). However the possible prophages are decided on the basis of number of CDS shared with other prophages. *P.putida* LS46 shared 15 CDS with *Pseudomonas* vB_PaeS_PMG1 while *P.putida* BIRD1, *P.putida* KT2440 and *P.putida* W619 had 8, 14 and 15 CDS respectively. The organization of prophage *Pseudomonas* vB_PaeS_PMG1 genes was unique to *P.putida* LS46. The putative prophage was only 8.4 kb and had only 9 protein coding sequences. It was similar to SXt2_1717. Other *P. putida* genomes (from PHAST database http://phast.wishartlab.com) were predicted to contain 1–6 (intact, incomplete, and questionable) prophages (Additional file 1: Table S2). No prophage was detected in *P. putida* UW4.

Genomic islands in *P.putida* LS46

Genomic island finder using integrated method identified 22 genomic islands (GIs) in *P.putida* LS46 (Figure 2). The size of smallest genomic island was 4505 bp while largest was 78290 bp (Figure 2, Additional file 1: Table S3). Genomic island 3 was 26639 bp in size and it showed 96% homology to *P.putida* H8234 genome (Molina et al. 2013). This GI carried a transposon Tn4652 along with heavy metal resistance genes. *P. putida* H8234, a clinical isolate from France, showed low pathogenic potential compared with *P. aeruginosa* and was resistant to commonly used antibiotics. *P. putida* LS46 was also resistant to ampicillin, choloramphenicol, gentamycin and tetracycline. GI 18 contains M.XmaI and R.XmaI genes, which are component of XmaI restriction system. Two unique genes PPUTLS46_017749, which encodes an N-4 cytosine-specific methyltransferase, and PPUTLS46_017754, which encode a Type II restriction enzyme, encoded this system. This restriction system is isoschizomer of *XcyI* of *Xanthomonas campestris* pv *cynopsidis*. The *XcyI* restriction-modification system recognizes the sequence, CCCGGG, and cleaves after C1 and (Withers et al. 1992). These genes are present on the plasmid AG1 of *Xanthomonas axonopodis* pv *glycine* as well as on the plasmid pRA2 of *Pseudomonsa alcaligenes* NCIB 9867. This plasmid carried two mobile elements, Tn5563 and IS1633 along with Pac25I (XcyI) restriction-modification system (Kwong et al. 2000). This restriction system did not affect the pRA2 plasmid stability in heterologous *Pseudomonas* hosts. The other GIs carried a number hypothetical protein.

Comparative genome analysis of *P.putida* strains

The genomes of nine *P. putida* strains range in size from 5.73 to 6.3 Mb, with an average GC content of 62% (Table 1). Among *P. putida* strains only strain ND6 carried two plasmids of 117 kb and 101 kb, which encoded an additional 136 and 102 genes, respectively (Table 1). *P. putida* strains KT2440, BIRD1, GB1, and LS46 had seven copies of 16S rRNA genes, while six copies of 16S rRNA genes were present in strains F1, ND6, UW4, and S16 (Table 1).

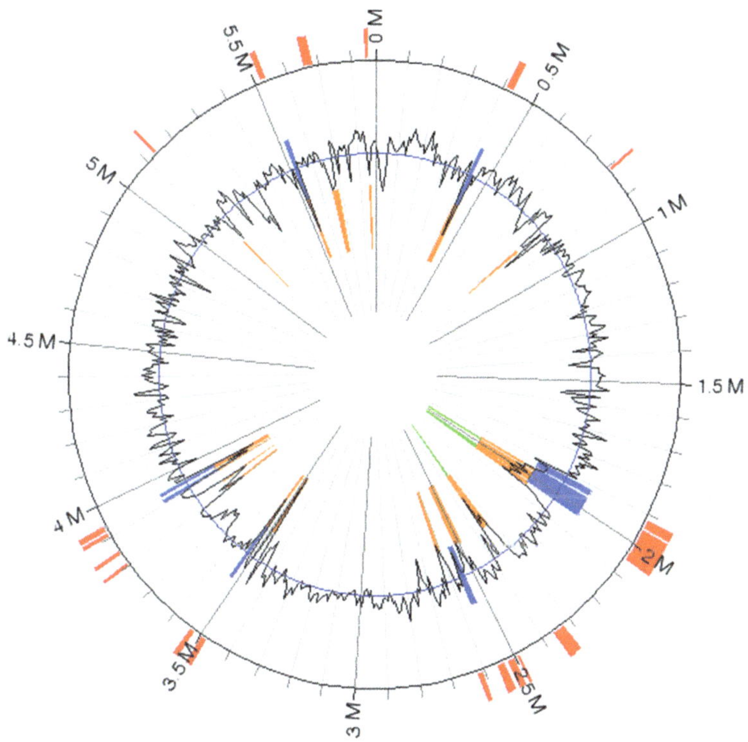

Figure 2 Occurrence of genomic islands in *P.putida* LS46 genomes. Genomic islands were identified using IslandViewer programme using Genomic island predictions were calculated for using Integrated method of IslandPick, IslandPath-DIMOB, and SIGI-HMM (Langille and Brinkman (2009)).

P. putida strains carried 4171 to 4519 COG counts, representing 69.7% to 82.0% of their genes.

Genome of nine *P.putida* strains including *P.putida* LS46 was blasted against *P.putida* KT2440 (type strain) as the reference strain. Whole genome BLAST identified several homologous regions as well as diverse region in the *P. putida* strains in comparison to *P.putida* KT2440. On the basis of whole genome BLASTn analysis of nine *P. putida* genomes had 86.6% to 97.7% nucleotide sequence identity to the *P. putida* LS46 genome (Table 2). *P. putida* DOT-T1E displayed the greatest sequence identity with *P. putida* LS46, while *P. putida* UW4 had the least sequence identity. *P.putida* DOT-T1E was isolated from Spain after enrichment with 1% toluene while *P.putida* UW4 was isolated as PGPR from Canada (Ramos et al. 1995; Glick 1995).

Genome arrangement of *P. putida* genomes

Genome arrangement of *P.putida* LS46 was compared with 9 other *P.putida* strains using Dot plot analysis (Figure 3). Dot plot analysis of *P.putida* LS46 with other *P.putida* genomes indicated the similarities in genome arrangement at the nucleotide level. The diagonal line showed the colinearity DNA strands. The blue block on the left hand indicated the translocation and inversions in the genomes.

Table 2 Percent shared genes among different *Pseudomonas putida* strains

Strain	LS46	BIRD1	DOT-T1E	F1	GB1	KT2440	ND6	S16	UW4	W619
LS46	100									
BIRD	92.54	100								
DOT-T1E	91.60	91.90	100							
F1	93.67	92.72	93.99	100						
GB1	90.13	89.36	92.16	89.83	100					
KT2440	91.28	89.38	89.96	90.32	87.90	100				
ND6	85.67	85.13	86.96	86.90	83.13	84.31	100			
S16	89.85	89.15	88.14	90.55	90.12	89.11	83.86	100		
UW4	82.05	82.54	79.64	81.77	81.82	81.91	75.32	82.20	100	
W619	89.07	88.78	88.22	89.69	87.74	87.42	83.29	89.52	81.84	100

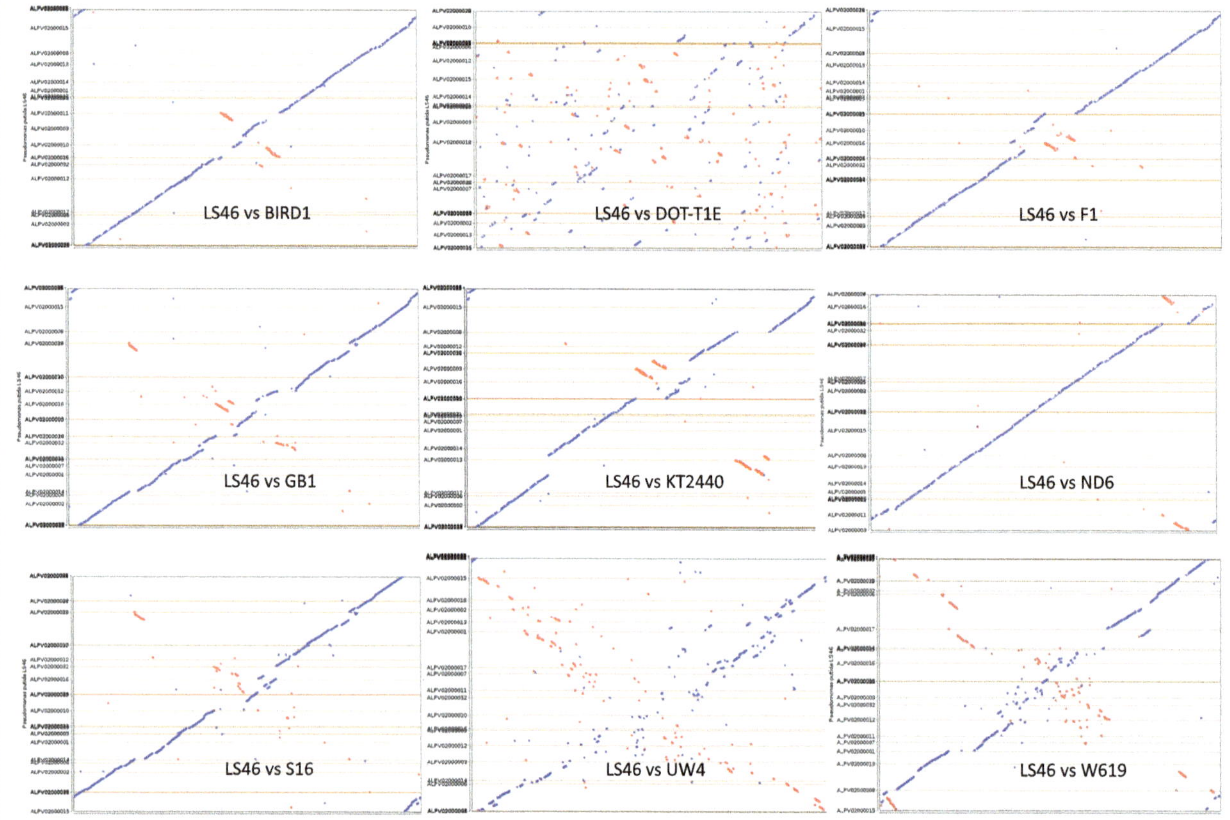

Figure 3 Comparative Synteny Dot plot of *P.putida* strains showing orthologous relationship of *P. putida* LS46 with nine *P.putida* strains. The analysis was carried out using the Dot plot from Integrated Microbial Genome (IMG) website.

The red blocks represented translocations in anti parallel strands of the genome. The high degrees of genome similarity as well as differences in the arrangement were apparent among *P.putida* genomes. The genome wide distribution of conserved region of *P. putida* strains varied from strain to strain. Organization of the *P. putida* LS46 genome was clearly different than other strains and a number of inversions and translocations were observed the genome in comparison to other *P.putida* genomes (Figure 3). On the basis of genome arrangement *P.putida* strains could be divided into two groups. In first group *P.putida* strains LS46, ND6, F1, KT2440, GB-1, S16 and BIRD1 can be placed which had significantly similarity to *P.putida* LS46 genome arrangement while in second group comprises of *P.putida* W619, DOT-T1E and UW4 which had low similarity to *P. putida* LS46. *P.putida* LS46 genome arrangement was markedly different from *P.putida* DOT-T1E, W619 and UW4 with large number translocations and inversions. *Pseudomonas putida* LS46 genome showed large conserved blocks that are also present in *P. putida* BIRD1, F1, and ND6 strains, while *P. putida* DOT-T1E had a large number of small conserved blocks. In comparison to *P. putida* LS46, the genomes of *P.putida* KT2440 and S16 genomes had more inversions.

Homologous and shared gene among *P.putida* strains

BLASTn analysis of all genes of *P. putida* LS46 (80% minimal identity e value 1^{-5}) against nine genome identified 82.02- 93.75% homologous genes encoded by the *P. putida* strains were shared by the ten genomes. *P.putida* F1 shared highest number (93.7%) of genes with *P.putida* LS46 while *P.putida* ND6 (85%) and *P.putida* UW4 (82%) least number of genes with *P.putida* LS46 (Table 1). Using single gene profiler 3271 genes were identified which were present in present in all *P.putida* strains. A total of 8786 core and unique genes were represented the pangenome of nine *P. putida* strains (excluding *P.putida* UW4). Unique region in *P. putida* genomes were identified using pangenome analysis, which identified unique genes present in only one strain (Figure 4). However, all genomes showed higher functional identity (presence of COGs) reflected by high correlation coefficients among the different genomes (r^2 = 0.94), although the distribution of different COGs categories among the different genomes was different and represented the functional diversity. The number and percentage of different COG categories varied greatly among ten *P. putida* strains. *P.putida* LS46 had highest number of COGs with unknown function. *P.putida* LS6 genome arrangement was strikingly different from *P.putida* DOT-T1E however

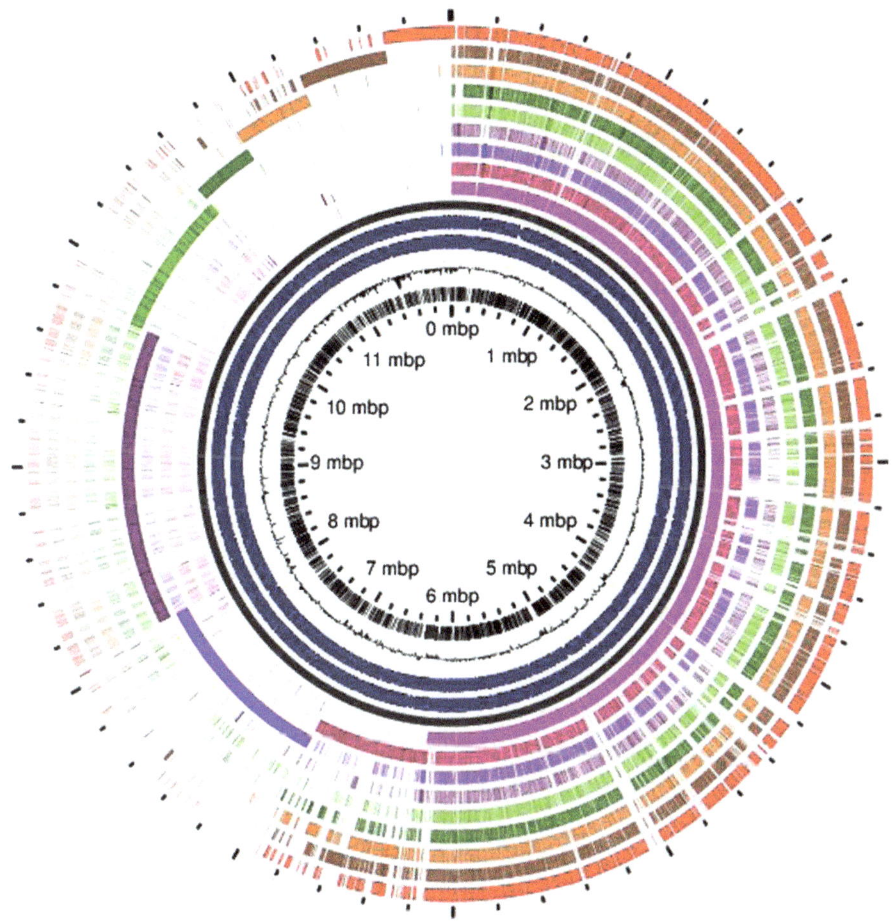

Figure 4 Pangenome analysis of nine *P. putida* strains with *P. putida* KT2440 as a reference. From inside to outside the circles. COG categories, GC content, backbone, COG in positive strand of pangenome, COG in negative strand in pangenome, *P. putida* KT2440, *P. putida* F1, *P. putida* GB-1, *P. putida* W619, *P. putida* S16, *P. putida* BIRD-1, *P. putida* ND6, *P. putida* DOT-T1E and *P. putida* LS46.

both the strains shared higher percentage of genes (91.6%). Inversely *P.putida* LS46 and *P.putida* ND6 had significantly similar genome arrangement but % of shared genes between two genomes was low (85%).

House keeping genes

The 16S rDNA is an universal marker and has been widely used for comparison between divergent bacteria. However, the resolution of 16S rRNA gene sequences at the intrageneric level is low Anzai et al. 2000; Yamamoto et al. 2000). Gene sequences for 'housekeeping' proteins provide better phylogenetic resolution and have been used to differentiate the genomes of closely related strains. Zeigler (2005) identified some house keeping genes for studying genome relatedness among different strains. Comparison of 33 house keeping genes of *P. putida* LS46 with other nine *P. putida* strains revealed a high degree of homology among these genes (Additional file 1: Table S4). However, some house keeping genes like *dnaX* (PPUTLS46_019951), *lig* (PPUTLS46_022046), *pgi* (PPUTLS46_007236), *uvrC* (PPUTLS46_012340,) *glyA* (PPUTLS46_024688), *trpS* (PPUTLS46_013403) and *trmE* (PPUTLS46_016619) were highly diverse among *P. putida* strains (Additional file 1: Table S4).

Fatty acid and Polyhydroxyalkanoate synthesis

Polyhydroxyalkanoates are produced *Pseudomonas putida* LS46 from glucose, glycerol, biodiesel glycerol, fatty acids and waste fryer oil (Sharma et al. 2011). Major components of PHAs produced by *P.putida* LS46 are either 3-hydroxyoctonoate or 3-hydroxydecanoate depending on the carbon substrate used for PHAs production (Sharma et al. 2011). There are six genes in the polyhydroxyalkanoate (PHA) synthesis operon (*pha*) in *P.putida*. These are *phaC1, phaZ, phaC2, phaD, phaF* and *phaI.* The *phaC1* and *phaC2* are PHA synthases (type II) that incorporate (R)-3-hydroxyacyl-CoA monomers into the PHA polymer (Huisman et al., 1989). The *phaZ* encodes a PHA depolymerase, which hydrolyses the PHA monomers, which can be fed into central metabolism for growth (Galán et al. 2011; de Eugenio et al., 2010a). The other three genes (*phaD, phaF* and *phaI*) are regulatory genes (Arias et al. 2013; de Eugenio et al. 2010b; Galán et al. 2011). The organization of *pha* operon was identical in all the *P. putida* strains. The PHA synthesis gene products were highly conserved among all the *P. putida* strains, with more than 90% amino acid sequence identity (Table 3). The two-polyhydroxyalkanoate synthase genes (*phaC1* and *phaC2*) had only 71% nucleotide identity and 55% aa sequence identity.

The *phaC1* and *phaC2* of different *Pseudomonas* species formed different clusters in neighbor joining tree. The *phaC1* and *phaC2* in *P.putida* strains were highly conserved but were different *phaC1* and *phaC2* from other *Pseudomonas* species. The *phaC1* and *phaC2* genes of *P. putida, P.aeruginosa, P.fluorescens, P.stutzeri. P.entomophila and P.mendocina* formed different cluster in neighbor joining tree (Figure 5). Further fatty acid biosynthesis and fatty acid degradation proteins were highly conserved among *P.putida* strains. Fatty acid biosynthesis (*fab*) and fatty acid degradation (*fad*) gene products provide the precursor for PHAs synthesis. Most of the fatty acid synthesis and degradation proteins of *P.putida* LS46 had multiple genes coding isozymes i.e. FadB had 4 isologs, FadA had 5 isologs and FadD had 7 isologs for short and long chain fatty acids. Likewise FadE (acyl-CoA dehydrogenase) had six isologs specific for small, medium and long chain fatty acids. Fab and Fad proteins of *P.putida* LS46 showed high homology to Fab and Fad proteins of other *P.putida* strains. *P.putida* LS46 can utilize fatty acids (C5-C18) for PHAs production. Two fatty acid transporters (FadL), one for short chain fatty acid (PPUTLS46_007654) and other for long fatty acid (PPUTLS46_015009) were present in *P. putida* LS46. However, it preferentially used long chain fatty acids (C6-C18) than short chain fatty acid C3-C5) for PHAs synthesis. The specificity of FadD and FadL these proteins are not known but transfer of FadD from *E.coli* and FadL from *P.putida* into *Aeromonas hydrophila* improved its ability to utilize C6 and C8 fatty acids. (Jian et al. 2010). The intermediate 3hydroxyacyl-ACP of *de novo* fatty acid synthesis is converted to 3hydroxyacyl-CoA for polymerization to PHAs with help of (R)-3-hydroxyacyl-ACP:CoA transacylase enzyme PhaG. PhaG was present in all *P.putida* strains. The absence of PhaG in PHB producers limits their ability to produce mcl-PHAs. Transfer and expression of *phaG* of *P.putida* in *Ralstonia eutropha* or *Aeromonas hydrophila* producer may confer mcl-PHAs production ability into PHB producers.Recently, a PHAs granule-associated acyl-CoA-synthetase (Acs1) has been

Table 3 Homology of proteins associated with polyhydroxyalkanoate production

Name	Locus tag	Enzyme	Product size (aa)	*P.putida* strains							
				BIRD1	DOT	F1	GB1	KT2440	ND6	S16	W619
FadD	07414	Acyl-CoA synthetase	565	99	100	99	94	99	99	94	87
FadA	04404	3-ketoacyl-CoA thiolase	392	99	100	100	99	99	100	99	100
FadAx	07424	Acetyl-CoA acetyltransferases	380	99	99	99	98	99	99	98	96
FadB	00695	3-hydroxyacyl-CoA dehydrogenase/epimerase	715	98	93	99	94	94	99	93	95
FadB2	07419	3-hydroxyacyl-CoA dehydrogenase	255	99	100	100	97	100	99	94	96
FadD2	21046	Acyl-CoA synthetase	562	99	99	99	98	99	99	96	93
FadE	00145	Acyl CoA dehydrogenase	601	98	93	99	94	94	99	93	95
FadE2	05901	Acyl CoA dehydrogenase	592	95	99	99	95	98	99	95	94
PhaC1	05621	Poly(R)-hydroxyalkanoic acid synthase	560	99	99	98	99	99	99	98	94
PhaZ	05616	Poly(3-hydroxyalkanoate) depolymerase	283	99	100	100	99	99	99	97	95
PhaC2	05611	Poly(R)-hydroxyalkanoic acid synthase	559	99	99	99	96	99	100	99	95
PhaD	05606	Transcriptional regulator	204	100	99	96	99	99	99	97	95
PhaF	05601	PHA granule associated protein	253	99	100	100	97	99	100	98	94
PhaI	05596	PHA granule associated protein	139	99	100	96	93	96	96	91	91
PhaG	13888	Hydroxyacyl-ACP:CoA transacylase	295	99	99	99	98	99	100	94	89
FabD	00250	malonyl CoA-acyl carrier protein transacylase	312	99	99	100	96	99	100	98	92
FabA	12710	3 hydroxyacyl(decanyl)-ACP dehydratase	171	100	100	100	99	100	100	98	92
FabB	12715	2 -Oxoacyl carrier protein synthase	406	100	99	99	99	99	99	99	99
FabG	23353	3-ketoacyl-(acyl-carrier-protein) reductase	450	99	99	99	96	99	99	95	90
FabG_2	00255	3-oxoacyl-(acyl-carrier-protein) reductase	450	99	100	99	99	100	99	100	99
FabG_3	15924	3-oxoacyl-(acyl-carrier-protein) reductase	450	100	100	100	97	99	99	94	91
FabZ	14589	3-hydroxyacyl-[acyl carrier protein] dehydratase	146	100	100	100	100	100	100	99	95

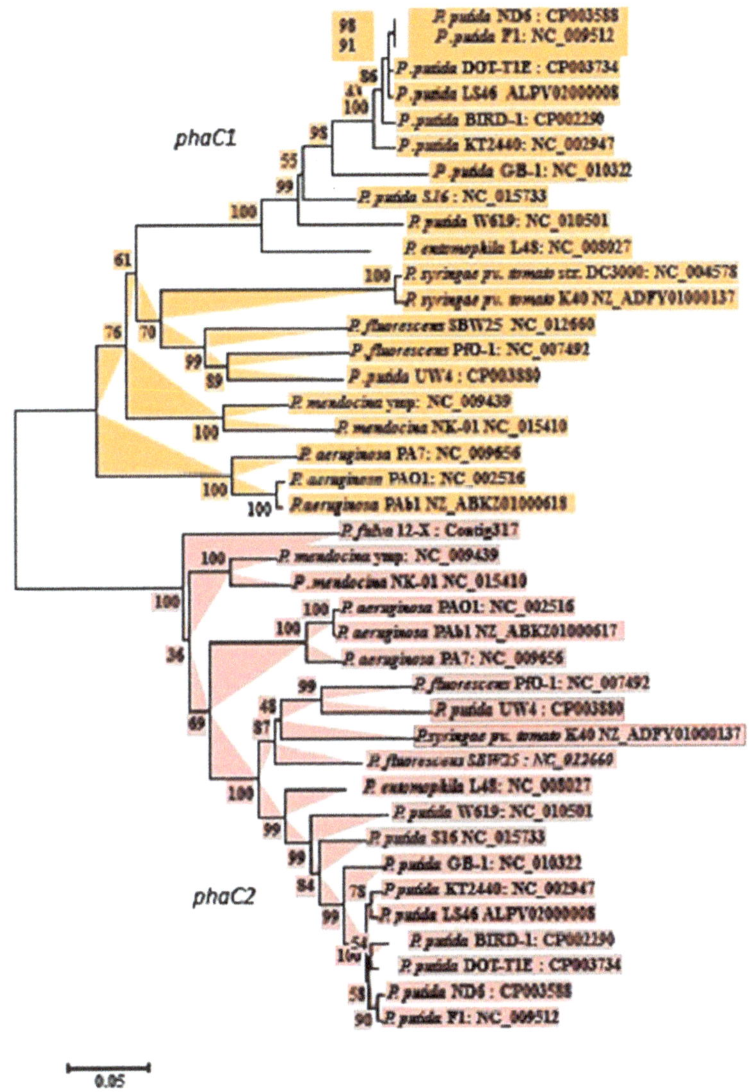

Figure 5 Phylogenetic tree depicting the relationship of *phaC1* and *phaC2* genes among *Pseudomonas* species. The *phaC* gene sequences were aligned by ClustalW and a neighbor-joining tree was generated using MEGA5 program. Bootstrap values are mentioned at the node.

identified which is highly conserved among P.putida strains. It directs the carbon flux of these central metabolites towards PHA accumulation and converts 3hydroxyalkanoic acids to 3 hydroxyacyl-CoA thioesters (Ruth et al. 2008).

Comparison of common metabolic pathways of *P. putida* strains

Metabolic diversity

The compared *P. putida* strains were isolated from diverse environment for bioremediation of xenobiotics or decomposition of other materials. Different *P. putida* strains have unique genes associated with specific functions and these genes are not present in other strains. *P. putida* KT2440 has been considered as a metabolically diverse saprophytic bacterium. Analysis of the *P. putida* KT2440 genome identified 18 dioxygenase, 51 putative hydrolases, 40 dehydrogenases, and more than 62 transferases (Jimenez et al. 2002; Molina-Henares et al. 2009; Nelson et al. 2002). By way of comparison, the *P. putida* LS46 genome had 22 dioxygenases (Additional file 1: Table S5), 75 hydrolases, 149 dehydrogenases, and 43 transferases. All the three dioxygenases, benzoate dioxygenase (PPUTLS46_007854, PPUTLS46_007859), catechol dioxygenase PPUTLS46_007879), and protochatuate dioxygenase (PPUTLS46_010694) were present in *P. putida* LS46. These dioxygenases were earlier identified in *P.putida* KT2440 and are involved in metabolism of aromatic compounds (Nelson et al. 2002). *P.putida* LS46 had 22 dioxygenases genes. One of the dioxygenase genes (PPUTLS46_007879 glyoxalase/bleomycin resistance protein/dioxygenase) was unique to *P. putida* LS46 and was not present in other *P.putida* strains (Additional file 1: Table S5).

The majority of the genes for degradation of ferulate, coniferyl, and coumuryl alcohol, aldehydes and acids, p-hydroxybenzoate, and protocatechuate present in *P. putida* KT2440 were also present in *P. putida* LS46 (Additional file 1: Table S6). These genes were reported to have role in degradation of aromatic/aliphatic sulphonate, benzoate, toulate, catechol, hippurate, maleate, phenylalanine, phenylacetic acidprotocatechuate, quinate, taurine etc. The notable exceptions in *P. putida* LS46, however, were ferulic acid transferase (PP_3354), transferuloyl CoA hydratase (PP_3358), and vanillate dehydrogenase (PP_3357). These genes were present in *P. putida* F1 (Pput_2404, Pput_2400, Pput_2401) and *P. putida* W619 (PputW619_2051, PputW619_2047, PputW619_2048), and are associated with transformations of ferulic acid to vanillin and vanillin to protocatechuate. Another gene, which was only reported in *P. putida* KT2440, and that is missing from all other *P. putida* strains, was propanediol dehydrogenase (PP_2803).

P. putida F1 is another metabolically diverse strain that has been demonstrated to oxidize toluene, 3-hydro xyphenyl propionate, and cymene (Finette and Gibson 1988; Zylstra et al. 1988). The genes that encode the enzymes for these reactions were also present in *P. putida* DOT-T1E, which encodes the genes (TIE_4277-T1E-4240) for degradation of toluene, 3- hydroxyphenylpropoinate, and cymene (Additional file 1: Table S8), but absent in *P. putida* LS46 and 7 other *P. putida* strains. As reported earlier the genes for degradation of 3 hydroxyphenylpropionate were also present in *P.putida* W619 (Wu et al. 2011).

P. putida KT2440 can use aromatic or aliphatic sulphonates as sulphur sources. The genes for degradation of aromatic or aliphatic sulphonates are encoded in the *ssuFBCDEA* operon (PP_0241-PP_0235) in *P. putida* KT2440. This operon was present in *P. putida* LS46 (PPUTLS46_024748, PPUTLS4_025163, 025168, 025173, 025178, 025183) and all other *P. putida* strains (Additional file 1: Table S6). Three chlorohyrolases i.e. atrazine chlorohydrolase (PP5036, PP2584) and hydroxydechloroatrazine ethylaminohydrolase (PP3209) detected in *P. putida* KT2440 were also identified in *P. putida* LS46 (PPUTLS46_10994, PPUTLS46_005456, and PPUTLS46_008014). However, chloride channel protein of *P. putida* KT2440 (PP3959) was absent in *P. putida* LS46. Pak et al. (2000) identified xenobiotic reductase from *P. fluorescens* for the transformation of 2, 4, 6-trinitritoluene (TNT). Its homologue was identified in *P. putida* KT2440 (PP_0920) (Nelson et al. 2002) and *P. putida* LS46 (PPUTLS46_022986). These genes was also present in other eight *P. putida* strains, but was absent from *P. putida* UW4.

Pseudomonas putida DOT-T1E is a solvent-tolerant strain able to grow in the presence of > 1% (v/v) toluene in the culture medium. Its multidrug efflux pumps have been found to play a major role in toluene-tolerance (Rojas et al. 2001). This ability is mainly conferred by an efflux pump encoded in a self-transmissible 133 kb plasmid named pGRT1 (Molina et al. 2011). No plasmid was detected in *P. putida* LS46, but the genes encoding an efflux pump (*sepA, sepB, sepC*), were present in the *P. putida* LS46 (PPUTLS46_0133778, 013773, 013768) genome, as reported earlier for *P. putida* strains KT2440, GB1 and W619 (Wu et al. 2011).

Nicotinic acid and nicotine degradation

Nicotinic acid (NA) is a carboxylic derivative of pyridine that is widely distributed in nature as part of pyridine cofactors (NAD and NADP) and alkaloids (e.g., nicotine and anabasine), and it is essential (vitamin B3) for those organisms that are not able to carry out its synthesis. In bacteria and fungi, NA is used as a carbon source. The biochemical pathways involved in the degradation of NA was elucidated (Kaiser et al. 1996). Jiménez et al. (2008) identified a *nic* gene cluster (PP_3939 - PP_3948) in *P. putida* KT2440 responsible for aerobic nicotine degradation. These genes were also present in *P. putida* strain F1, GB1, and W619 as reported earlier Jiménez et al. (2008). All the genes of *nic* operon were also present in *P. putida* LS46, BIRD1, ND6, DOT-T1E, and these genes were homologous to genes (89.6% to 100% nucleotide identity) from *P. putida* KT2440 (Additional file 1: Table S7).

Recently, the genome sequence of another nictotine degrading *P. putida*, strain S16, was released (Tang et al. 2012). *P. putida* S16 degrades nicotine through pyrrolidine pathway. A gene cluster containing six genes encoding PPS_0380 HSP hydroxylase (*hspB*), PPS_4060 maleate isomerase (*iso*), PPS_4059 NFM deformylase (*nfo*), PPS_4058 DHP dioxygenase (*hpo*), PPS_4057 maleamate amidase (*ami*), and PPS_4061 6-hydroxynicotinate 3-monooxygenase (*hna*) was identified in *P. putida* S16 and this gene cluster is designated as *nic*2 cluster. This gene cluster was present on a large genomic island. Both nicotine and nicotinic acid produced 2,5- dihydroxypyridine (2,5-DHP) as an intermediate, which is converted to N-formylmalaemic acid. The N-formylmalaemic acid is degraded to maleic acid and then to fumaric acid as in nicotinic degradation pathway (Wang et al. 2007). BLAST analysis indicated that these genes were also present in *P. putida* LS46 (PPUTLS46_011665 (*ami*), 11670 (*iso*), 11675 (*nfo*), 11685 (*hpo*) and showed 36.63, 69.35, 55.64 43.73% homology to respective proteins of *P. putida* S16. Other *P.putida* strains showed low homology to the *P. putida* S16 *nic*2 cluster (36.6-65-3%). No homologue of PPS_0380 HSP hydroxylase (*hspB*) was present in *P. putida* LS46 (Additional file 1: Table S8). The *nic* cluster genes of *P. putida* KT2440 showed high nucleotide sequence identity (>95%) to corresponding genes of *P.putida* LS46, as reported earlier in *P. putida* strain W619, F1, and GB1 (Molina et al. 2011) [59]. *P. putida* KT2440, however, is not able to degrade nicotine. The *nic* cluster of *P. putida* S16 had a GC content of only 48% compared with the 62% GC content of the rest of the *P. putida* genome sequence.

The *nic* gene cluster for the nicotinic acid degradation pathway of *P. putida* KT2440 was also detected in *P.putida* BIRD1, *P. putida* ND6, and *P. putida* DOT-T1E, but was absent in the genome of *P.putida* strain UW4.

Heavy metal resistance

The level of heavy metal tolerance is very high in *P. putida* W619 in comparison to other *P. putida* strains (Canovas et al. 2003; Taghavi et al. 2009). In *P. putida* W619, copper, cobalt, zinc, cadmium, arsenate, mercury, nickel chromate, and molybdenum resistance genes are present in two genomic regions (1 and 31). Heavy metal tolerance of *P. putida* LS46 has not been investigated, but this strain encodes genes for heavy metal resistance present in genomic region 1 of *P. putida* W619. These genes are involved in copper, cobalt, zinc, and cadmium resistance. However, the genes located on genomic region 31 of *P. putida* W619, involved in copper and chromate resistance, were absent in the majority of *P. putida* strains, including *P. putida* LS46 (Additional file 1: Table S8). All *P. putida* strains lacked the genes present in genomic region 18 of *P. putida* W619, which were associated with mercury and nickel resistance.

Manganese oxidation

P. putida GB1 is a Mn (II) oxidizing bacterium (Buzzo 2011) and attempts have been made to identify the genes related with Mn (II) oxidation by transposon mutagenesis. However, the role of different genes in Mn (II) oxidation is not clear (Caspi et al. 1998; Brouwers et al. 1999). Recently, using transposon mutagenesis, Geszvain et al. (2013) and Geszvain and Tebo (2009) identified two genes, which are homologous to multi-copper oxidases (PputGB1_2447 and PputGB1_2665) in *P. putida* GB1. The genes encoding multi-copper oxidase homologues of P.putida GB1 were also present in *P. putida* LS46 (PPUTLS46_002532 and PPUTLS46_006964), as well as in *P. putida* strains KT2440, F1, ND6, W619, S16, UW4, and DOT-T1E. Only one multi-copper oxidase gene was identified in *P. putida* BIRD1 (Additional file 1: Table S8).

Iron scavenging genes in P. putida strains

It is well known that iron deficiency in bacteria limits growth Andrew et al. (2003). *Pseudomonas* produces low molecular weight iron-chelating compounds termed ferri-siderophores to scavenge iron (Wiener 2005). Ferri-siderophores are taken up via outer membrane receptors, which function as gated porin channels (Koebnik 2005; Ratledge and Dover 2000). After binding of the ferri-siderophores, transport is mediated by a complex of inner membrane-anchored proteins TonB (Wiener 2005). Fluorescent pseudomonads respond to iron-deficiency by secreting the yellow-green fluorescent peptidic siderophores called pyoverdines (Meyer 2000; Cornelis et al. 2009). Pyoverdines have a conserved chromophore with a variable peptide chain, and each *Pseudomonas* species produces a different pyoverdine (Ravel and Cornelis 2003; Visca et al. 2007). Pyoverdine is synthesized by non-ribosomal peptide synthetase (NRPS). Like other *P. putida* strains, genes for pyoverdine synthesis were also present in *P. putida* LS46 (PPUTLS46_025659 and PPUTLS46_012775). The *pvdE* gene (PPUTLS46_012775) encoded a PVD ABC transporter, as in other *P. putida* strains. *P. putida* LS46, like the other *P.putida* strains did not carry the genes for pseudomonine production, which was reported for *P. entomophila* L48 (Vodovar et al. 2006).

The number and type of TonB siderophore receptors show their diversity to survive in the different environment (Cornelis and Bodilis 2009). TonB-dependent receptor genes are not constitutively expressed and are regulated by the iron availability (Koebnik et al. 2000; Bodilis et al. 2009) and their expression provides additional mechanisms to survive under different niches. Twenty-six (26) TonB siderophore receptor genes were present in the *P. putida* LS46 genome (Additional file 1: Table S9). Across all of the *P. putida* strains, 19–53 TonB genes are present. Comparison of TonB genes (at 80% nucleotide sequence identity) of *P. putida* LS46 identified 26, 28, 26 20, 24, 26, 19, 1 and 15 homologous TonB genes in *P. putida* LS46 strain BIRD1, DOT-T1E, F1, GB1, KT2440, ND6, S16, UW4 and W619, respectively. *P.putida* strain UW4 had 11 TonB genes, which showed low homology (50% nuclotide sequence identity) to other TonB genes of *P. putida* strains. The greatest number of TonB genes (55) was present in *P. putida* strain GB1.

Discussion

Pseudomonas putida strains have been identified as root colonizing bacterium and developed as biocontrol agent. Other *P.putida* strains due to their diverse metabolic potential have been developed as bioremediation agents. Potential of *P.putida* as polyhydroxyalkanoates (PHAs) producer was identified in 1989 (Huisman et al. 1989). However, PHAs producing *P.putida* strain KT2440 was originally isolated as a root colonizing bacterium (Nelson et al. 2002). A number of *P.putida* strains were isolated and developed as bioremediation agents but no other *P.putida* strain was isolated for PHAs production. *P.putida* LS46 was isolated and screened for PHAs production and it was proved as a good PHAs producer as *P.putida* KT2440. Complete genome of this bacterium was sequenced to know the similarities and differences among biocontrol, bioremediation and PHAs producing *P.putida* strains.

P. putida LS46 strain showed more than 99% identity to 16S rRNA gene of *P.putida* strains irrespective of their geographic origin or their application. *P.putida* classifications based on protein-encoding genes like *gyrB, rec, rpoB, recN,* and *cpn60* alongwith 16S rRNA gene have earlier been proposed (Adékambi et al. 2009; Arahal et al. 2008;

Zeigler 2005; Mulet et al. 2010, 2013). Cpn60 (Hsp60 or GroEL), a highly conserved protein found in bacteria, has been widely used for phylogeny, microbial identification, microbial ecology and evolution (Hill et al. 2002, 2004). The phylogeny of *P. putida* strains based on *cpn60* confirmed the close relationship among *P. putida* strains, strengthening the 16S rRNA gene based phylogeny. However, the *cpn60* gene phylogeny was more robust and separated strains that were clustered together in the 16S rRNA gene phylogeny. The *cpn60* (UT) analysis differentiated *P. putida* W619, *P. putida* S16, and *P. putida* UW4 from other *P. putida* strains. *P. putida* W619 was isolated as an endophyte of Poplar, *P. putida* UW4 was isolated as a plant growth promoting species, and *P. putida* S16 was isolated as a nicotine degrading species. Recently *cpn60* (UT) was used for studying the phylogenetic relationship among *Aeromonas, Themoanaerobacter,* and vaginal microbiota (Miñana-Galbis et al. 2009, Verbeke et al. 2011; Schellenberg et al. 2011).

The phylogeny of *P. putida* UW4 was described on the basis of 16S rRNA genes and a multilocus approach that used four concatenated housekeeping genes (16S rRNA, *gyrB, rpoD and rpoB*). *P. putida* UW4 was found to be closely related to *P. fluorescens* rather *P. putida* strains including *P. putida* LS46 (Duan et al. 2013). It confirmed our observations on the phylogeny of *P. putida* UW4 based on the *cpn60* gene that *P. putida* UW4 was clustered with *P. fluorescens.* Earlier, Loper et al. (2012) on the basis of ten concatenated house keeping genes, showed similar phylogenetic relationships among *Pseudomonas* species. The *cpn60* gene phylogeny placed the ten sequenced *P. putida* strains and *P. entomophila* in the same clade. The phylogeny of the newly isolated *P. putida* LS46 based on *cpn60* differentiated this strain from other closely related *P.putida* strains, and based on both the 16S rRNA and *cpn60* analyses, *P. putida* LS46 was closely related to type strain *P. putida* KT2440. No plasmids have been detected in *P. putida* LS46, but two plasmids were present in *P. putida* ND6 and one plasmid was identified in *P. putida* DOT-T1E (Li et al. 2002; Udaondo et al. 2012).

Variation in genome content is thought to be a key factor in the evolution of bacteria and variation in genome arrangement may also improve the fitness of the bacteria (Silby et al. 2011a, b). Presence of duplicated genes like rRNA operons, multiple transposons, insertion sequences, prophages and genomic islands leads to genome rearrangements in bacteria that contribute to evolution (Tillier and Collins 2000; Klockgether et al. 2010). Rearrangements in genome are not random, but predominantly occur at end-points either at the origin or at the terminus of replication (Eisen et al. 2000).

Overall, strains of *P. putida* isolated from different geographical regions and from varied ecological niches had high similarity in genome structure and functions (Molina et al. 2011; Li et al. 2012; Nelson et al. 2002; Wu et al. 2011). All the *P. putida* genomes had higher level of genome similarity, but differed from other on the basis of insertion sequence, presence of prophage and genomic islands. However, Dot plot analyses of the *P. putida* strain genomes identified a number of rearrangements, which are possibly due to presence of these genetic elements, such as the presence of ISPa42 in *P. putida* LS46, which originated from *P. aeruginosa* DK3. ISPa42 may be associated with the acquisition of genes for heavy metal resistance.

In comparison to *P. putida* LS46, genome inversions were observed in *P. putida* BIRD-1 and *P. putida* W619 at the origin and terminus of replication in *P. putida* S16 and *P. putida* ND6, respectively. These observations strongly support the contention that bacterial genomes are not static and significant variations are observed even among strains within the same species. These variations are the result of genome deletions and/or gene acquisitions by horizontal gene transfer of elements such as transposon and genomic islands (Hacker and Carniel 2001, Mackiewicz et al. 2001).

P. putida LS46 and *P. putida* DOT-T1E had a maximum sequence identity of 97.7%, but their genome arrangements were very diverse (Figure 2). It appears that the genome of *P. putida* DOT-T1E has undergone a number of rearrangements without losing any major function. *P. putida* DOT-T1E was isolated as toluene degrading species, and has acquired genes for toluene degradation that are absent in *P. putida* LS46. Using BLAST analysis 81.84-93.99% genes were identified which were shared among the 10 *P. putida* strains, while the rest of the genes were unique to the strain. Ballerstedt et al. (2007), using specific gene probe of *P. putida* KT2440 in a microarray identified 67.8-100% identical genes in six *P. putida* strains. Acquisition of specific groups of genes conferred specialized functions to specific strains, but their core genomes were identical.

Three *P. putida* strains, KT2440, BIRD1, and UW4, were isolated from soil and developed as plant growth promoting bacteria. *P. putida* strains F1, ND6, DOT-T1E, S16, GB1, and W619 were identified as bioremediation agents on the basis of their capacity to degrade pollutants. *P. putida* strain LS46 was isolated from wastewater for PHA production. This strain was able to use glucose, glycerol, fatty acids, and waste fryer oil and accumulates PHAs to 20-58% of the cell dry weight. Like other *P. putida* strains, central metabolic pathways for utilization of these substrates were also present in *P. putida* LS46. The application based classification of *P. putida* as biocontrol, bioremediation, or PHA producer is not relevant because biocontrol agent BIRD-1, or bioremediation strains F1, ND6, and DOT-T1E also encode the genes for PHA synthesis and homology among these genes is very high. The most of the genes present in the manganese oxidizing strain *P. putida* GB-1 or the endophyte of poplar stain *P. putida* W619 are also present in *P. putida* LS46.

Complete genome sequence of *P. putida* LS46 was compared with *P. putida* strains isolated from different geographical regions and different niches. Two strains, *P. putida* ND6 and *P. putida* DOT-T1E isolated from China, and *P. putida* F1 isolated from USA, had the highest sequence identities to *P. putida* LS46, which was isolated in Canada. *P. putida* strains GB-1 isolated from USA, UW4 isolated in Canada, and S16 isolated from China, had low sequence identities with *P. putida* LS46. *P. putida* W619, which was isolated as an endophyte of poplar had only 87.1% sequence identity with the *P. putida* LS46 genome, while *P. putida* UW4 showed only 82.0% identity with *P. putida* LS46. Our earlier result on phylogeny of *P.putida* UW4 has clearly demonstrated that this strain is related to *P. fluorescens* rather than *P. putida*. The low genome identity of *P. putida* strains with *P. putida* UW4 is not surprising. High genome similarities among *P. putida* strains isolated from different regions indicate their common ancestry (Biello 2008).

P. putida LS46 shares a number of metabolic features with *P. putida* KT2440, *P. putida* F1, *P. putida* BIRD1, and *P. putida* GB-1, such as metabolism of aromatic compounds, manganese oxidation, root colonization, and PHAs production. All *P. putida* genomes have the ability to synthesize PHA irrespective of their applications, either as biocontrol agents or bioremediation agents. *P. putida* LS46 differs from other *P. putida* strains in number of dioxygenase genes, TonB dependent receptors, transferases, hydrolases, dehydrogenases and transferases. *P. putida* LS46, like other *P. putida* strains metabolize glucose, glycerol, and fatty acid by glycolysis, the tricarboxylic acid cycle, the pentose pathway, and β oxidation (Figure 6). At least three different metabolic pathways provide the precursors for the synthesis of PHAs. (i) Fatty acid *de novo* biosynthesis is the main route during growth on carbon sources that are metabolized to acetyl-CoA, like glucose, gluconate, glycerol etc. (ii) β-oxidation is the main pathway for PHAs production when fatty acids are used as carbon source, (iii) chain

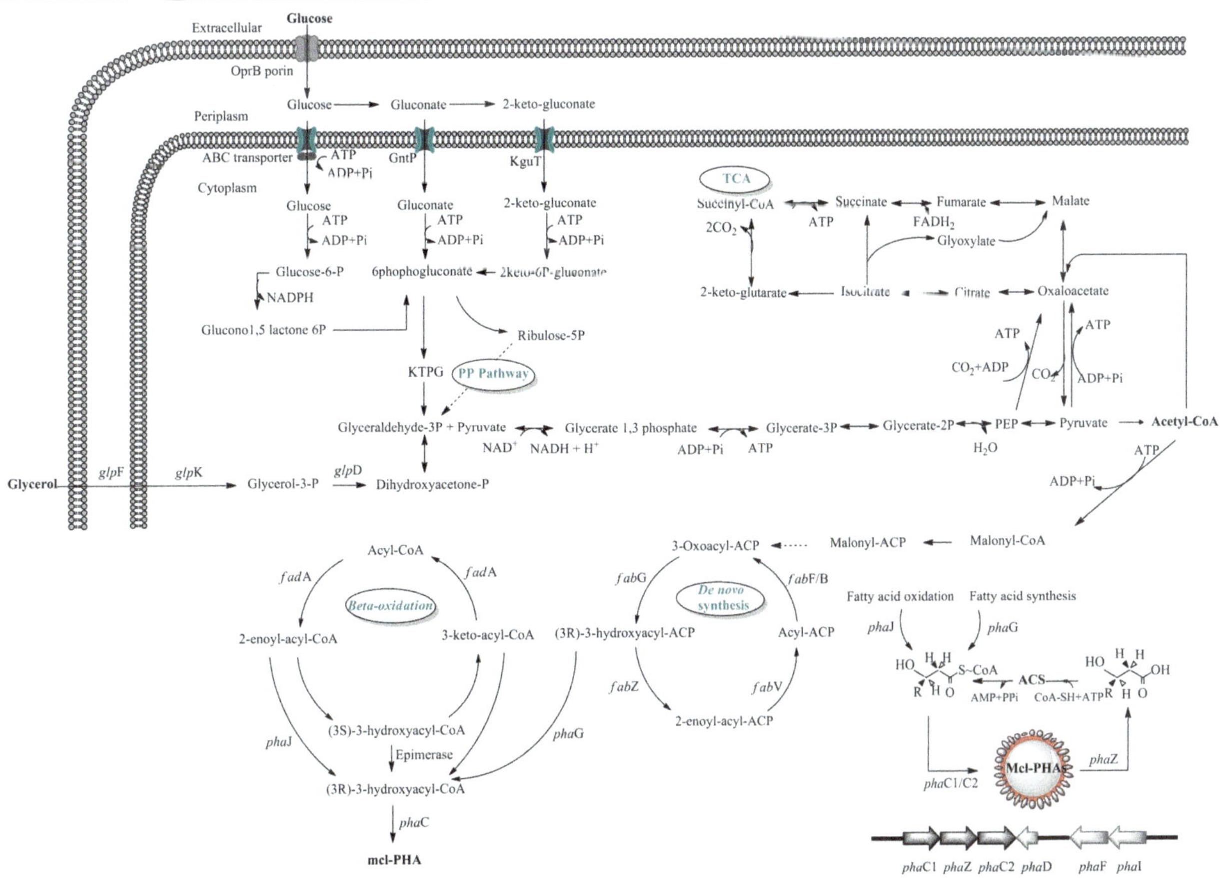

Figure 6 Proposed PHA synthesis and degradation pathways for *P. putida* LS46. Metabolic pathways involved carbon metabolism, fatty acid synthesis and degradation, and PHA synthesis are summarized. PHA can be synthesized from glucose, glycerol, or fatty acids via the PHA synthesis pathway.

elongation reactions in which acetyl-CoA moieties are condensed to 3-hydroxyacyl-CoA is involved in the PHA synthesis when small chain length fatty acids like C6 and C7 are used. Intermediates of fatty acid *de novo* synthesis (3 hydroxylacyl-ACP) as well as fatty acid degradation pathways (3-hydroxylacyl-CoA) are used as precursors for PHAs production. Rehm et al. (1998) identified a link between fatty acid degradation and fatty acid synthesis by confirming the PHAs biosynthesis in β-oxidation defection mutants (*fadB*). This enzyme converts 3hydroxyacyl-ACP to 3-hydroxyacyl-CoA, which is a substrate for polymerization to PHAs. Glucose and glycerol are transported and metabolized to acetyl-CoA, which is further used for production of various fatty acids using fatty acid biosynthesis (*fab*) genes. Fatty acids or waste fryer oil containing long chain fatty acids (C16 and C18) are utilized by β-oxidation and intermediates are used for PHAs production (Wang et al. 2012). Manipulation of fatty acid synthesis and degradation genes is known to improve PHAs production with altered monomer composition (Fiedler et al. 2002). PHAs production in *P.putida* is a part of central metabolic pathway and it was evident from high level of identity among PHAs production proteins and proteins of feeding pathways like fatty acid biosynthesis and degradation.

P. putida LS46 does not match, however, the metabolic diversity of either *P. putida* F1 or *P. putida* DOT-T1E, which encode genes for degradation of toluene, cymene, and 3-hydroxyphenylpropionate, or *P. putida* ND6, which encodes genes for naphthalene degradation. These genes may have been acquired during the evolution process. A number of transposons and insertion sequences were present in *P. putida* genomes indicating their labile nature. ISPpu10, a transposon present in REP sequence of *P. putida* KT2440 has been implicated in genome rearrangement (Ramos-Gonzalez et al. 2006). But this transposon was absent from all *P. putida* strain except *P. putida* S16. Mg (II) oxidation (*P. putida* GB1), pyoverdine production, aromatic/aliphatic sulphonate degradation (*P. putida* KT2440), and polyhydroxyalkanoate production (*P. putida* strains KT2440 and LS46) are core genes that are present in all *P. putida* strains whether these strains were isolated as biocontrol agents or as bioremediation agents (Wu et al. 2011). The genome of *P. putida* LS46 had a majority of the genes involved in plant growth promotion even it was not isolated as plant growth promoter and in this regards it was similar to *P. putida* KT2440 and *P. putida* BIRD1. Several unique genes were identified in the *P. putida* LS46 genome. The presence of prophages, insertion sequences, and genomic islands in different genomes indicate that the *P. putida* genome is not static, but changes by acquiring new genes from related and unrelated species.

Additional file

Additional file 1: Table S1. Occurrence of IS elements in different *P.putida* strains. **Table S2**: Predicted prophages in *P. putida* genomes. **Table S3**: Homology of house keeping genes encoded in the genomes of different *P. putida* strains. **Table S4**: Homology of house keeping genes encoded in the genomes of different *P. putida* strains. **Table S5**: Occurrence of different dioxygenases encoded in the *P.putida* LS46 genome and their homologues encoded in the genomes of other *P. putida* strains. **Table S6**: Presence of genes encoded in the *P. putida* KT2440 genome associated with different metabolic pathways in other *P.putida* strains. **Table S7**: Presence of genes encoded by the *P. putida* F1 genome involved in aromatic compound degradation and their homologues encoded in the genomes of other *P. putida* strains. **Table S8**: Occurrence of heavy metal tolerance genes encoded by *P. putida* W619 in different *P. putida* strains. **Table S9**: Identification of TonB receptors in *P. putida* LS46 and their homologues in other *P. putida* strains.

Competing interests
The authors declare that they have no competing interests, financial or otherwise.

Authors' contributions
The nucleotide sequence alignments, genome analyses, and preparation of the manuscript draft were carried out by PKS. JF assisted PKS with the genome analyses. The genome assembly and annotation, and uploading of files to IMG were carried out by XLZ, who was supervised by BF. RS and DBL are the co-PIs for the Genome Canada funded research program, and DBL served as PKS's direct supervisor. DBL also prepared the final draft of the manuscript for submission. All authors read and approved the final manuscript.

Acknowledgements
This work was supported by funds provided by the Natural Sciences and Engineering Research Council of Canada (NSERC), through a Strategic Programs grant (STPGP 306944–04), by Genome Canada, through the Applied Genomics Research in Bioproducts or Crops (ABC) program for the grant titled, "Microbial Genomics for Biofuels and Co-Products from Biorefining Processes", and by the Province of Manitoba, through the Manitoba Research Innovation Fund (MRIF).

Author details
[1]Department of Biosystems Engineering, University of Manitoba, Winnipeg, MB R3T 2N2, Canada. [2]Department of Plant Science, University of Manitoba, Winnipeg, MB R3T 2N2, Canada. [3]Department of Microbiology, University of Manitoba, Winnipeg, MB R3T 2N2, Canada.

References
Adékambi T, Drancourt M, Raoult D (2009) The *rpoB* gene as a tool for clinical microbiologists. Trends Microbiol 17:37–45

Andrews SC, Robinson AK, Rodriguez-Quinones F (2003) Bacterial iron homeostasis. FEMS Microbiol Rev 27:215–237

Anzai Y, Kim H, Park J, Wakabayashi H, Oyaizu H (2000) Phylogenetic affiliation of the pseudomonads based on 16S rRNA sequence. Int J Syst Evol Microbiol 50:1563–1589

Arahal DR, Sanchez E, Macian MC, Garay E (2008) Value of *recN* sequences for species identification and as a phylogenetic marker within the family "Leuconostocaceae". Int Microbiol 11:33–39

Aranda-Olmedo I, Tobes R, Manzanera M, Ramos JL, Marques S (2002) Species-specific repetitive extragenic palindromic (REP) sequences in *Pseudomonas putida*. Nucleic Acids Res 15:1826–1833

Arias S, Bassas-Galia M, Molinari G, Timmis KN (2013) Tight coupling of polymerization and depolymerization of polyhydroxyalkanoates ensures efficient management of carbon resources in *Pseudomonas putida*. Microbial Biotech. doi:10.1111/1751-7915.12040

Ballerstedt H, Volkers RJM, Mars AE, Hallsworth JE, dos Santos VA, Puchalka JJ, Eggink G, Timmis KN, de Jan AM, Bont JAM, Wery J (2007) Genomotyping of *Pseudomonas putida* strains using *P. putida* KT2440-based high-density DNA microarrays: implications for transcriptomics studies. Appl Microbiol Biotechnol 75:1133–1142

Bennett S (2002) Solexa Ltd. Pharmacogenomics 5:433–438

Biello D (2008) Do microbes make snow? Scientific America 28Feb2008

Bodilis J, Ghysels B, Osayande J, Matthijs S, Pirnay JP, Denayer S, De Vos D, Cornelis P (2009) Distribution and evolution of ferripyoverdine receptors in *Pseudomonas aeruginosa*. Environ Microbiol 11:2123–2135

Brandl H, Gross RA, Lenz RW, Fuller RC (1988) *Pseudomonas oleovorans* as a source of poly(3-hydroxyalkanoates) for potential application as biodegradable polyesters. Appl Environ Microbiol 54:1977–1982

Brouwers GJ, de Vrind JP, Corstjens PL, Cornelis P, Baysse C, de Jong EW (1999) cumA, a gene encoding a multicopper oxidase, is involved in Mn21 oxidation in *Pseudomonas putida* GB-1. Appl Environ Microbiol 65:1762–1768

Buzzo J (2011) New insights in Mn(II) oxidation in Pseudomonas putida GB-1. Oregon Health & Science University, Portland, OR, M.S. thesis

Canovas D, Cases I, de Lorenzo V (2003) Heavy metal tolerance and metal homeostasis in *Pseudomonas putida* as revealed by complete genome analysis. Environ Microbiol 5:1242–1256

Caspi R, Tebo BM, Haygood MG (1998) C-type cytochromes and manganese oxidation in *Pseudomonas putida* MnB1. Appl Environ Microbiol 64:3549–3555

Choi EN, Cho MC, Kim Y, Kim CK, Lee K (2003) Expansion of growth substrate range in *Pseudomonas putida* F1 by mutations in both *cymR* and *todS*, which recruit a ring-fission hydrolase CmtE and induce the *tod* catabolic operon, respectively. Microbiology 149:795–805

Compant S, Duffy B, Jerzy Nowak J, Clément C, Barka EA (2005) Use of plant growth-promoting bacteria for biocontrol of plant diseases: principles, mechanisms of action, and future prospects. Appl Environ Microbiol 71:4951–4959

Cornelis P, Bodilis J (2009) A survey of TonB-dependent receptors in fluorescent pseudomonads. Environ Microbiol Rep 1:256–262

Cornelis P, Matthijs S, Van Oeffelen L (2009) Iron uptake regulation in *Pseudomonas aeruginosa*. Biometals 22:15–22

de Eugenio LI, Galán B, Escapa IF, Maestro B, Sanz JM, García J, Prieto MA (2010a) The PhaD regulator controls the simultaneous expression of the *pha* genes involved in polyhydroxyalkanoate metabolism andturnover in *Pseudomonas putida* KT2442. Environ Microbiol 12:1591–1603

de Eugenio LI, Escapa IF, Morales V, Dinjaski N, Galán B, García JL, Prieto MA (2010b) The turnover of medium chain-length polyhydroxyalkanoates in *Pseudomonas putida* KT2442 and the fundamental role of PhaZ depolymerase for the metabolic balance. Environ Microbiol 12:207–221

de Smet MJ, Eggink G, Witholt B, Kingma J, Wynberg H (1983) Characterization of intracellular inclusions formed by *Pseudomonas oleovorans* during growth on octane. J Bacteriol 154:870–878

Dennis JJ, Zylstra GJ (2004) Complete sequence and genetic organization of pDTG1, the 83 kilobase naphthalene degradation plasmid from *Pseudomonas putida* strain NCIB 9816–4. J Mol Biol 341:753–768

Duan J, Jiang W, Cheng Z, Heikkila JJ, Glick BR (2013) The complete genome sequence of the plant growth-promoting bacterium *Pseudomonas* sp. PLoS ONE 8(3):e58640

Eaton RW (1997) *p* -Cymene catabolic pathway in *Pseudomonas putida* F1: cloning and characterization of DNA encoding conversion of *p* -cymene to *p* -cumate. J Bacteriol 179:3171–3180

Eisen JA, Heidelberg JF, White O, Salzberg SL (2000) Evidence for symmetric chromosomal inversions around the replication origin in bacteria. Genome Biol 1(6):RESEARCH0011

Elbahloul Y, Steinbüchel A (2009) Large-scale production of poly(3-Hydroxyoctanoic Acid) by *Pseudomonas putida* GPo1 and a simplified downstream process. Appl Environ Microbiol 75:643–651

Escapa IF, García JL, Bühler B, Blank LM, Prieto MA (2012) The polyhydroxyalkanoate metabolism controls carbon andenergy spillage in *Pseudomonas putida*. Environ Microbiol 14:1049–1063

Finette BA, Gibson DT (1988) Initial studies on the regulation of toluene degradation by Pseudomonas putida Fl. Biocatalysis 2:29–37

Fiedler S, Steinbüchel A, Rehm BH (2002) The role of the fatty acid beta-oxidation multienzyme complex from Pseudomonas oleovorans in polyhydroxyalkanoate biosynthesis: molecular characterization of the fadBA operon from *P. oleovorans* and of the enoyl-CoA hydratase genes phaJ from *P. oleovorans* and *Pseudomonas putida*. Arch Microbiol 178:149–160

Galán B, Dinjaski N, Maestro B, de Eugenio LI, Escapa IF, Sanz JM (2011) Nucleoid-associated PhaF phasin drives intracellular location and segregation of polyhydroxyalkanoate granules in *Pseudomonas putida* KT2442. Mol Microbiol 79:402–418

Geszvain K, Tebo BM (2009) Identification of a two-component regulatory pathway essential for Mn(II) oxidation in *Pseudomonas putida* GB-1. Appl Environ Microbiol 76:1224–1231

Geszvain K, McCarthy JK, Tebo BM (2013) Elimination of manganese (ii, iii) oxidation in Pseudomonas putida GB-1 by a double knockout of two putative multicopper oxidase genes. Appl Environ Microbiol 79:357–366

Glick BR (1995) The enhancement of plant growth promotion by free-living bacteria. Can J Microbiol 141:109–117

Hacker J, Carniel E (2001) Ecological fitness, genomic islands and bacterial pathogenicity. A Darwinian view of the evolution of microbes. EMBO Rep 2:376–381

Haywood GW, Anderson AJ, Dawes EA (1989) A survey of the accumulation of novel polyhydroxyalkanoates by bacteria. Biotechnol Lett 11:471–476

Hill JE, Seipp RP, Betts M, Hawkins L, Van Kessel AG, Crosby WL, Hemmingsen SM (2002) Extensive profiling of a complex microbial community by high-throughput sequencing. Appl Environ Microbiol 68:3055–3066

Hill JE, Penny SL, Crowell KG, Goh SH, Hemmingsen SM (2004) cpnDB: a chaperonin sequence database. Genome Res 14:1669–1675

Huang Y, Zhang L (2004) Rapid and sensitive dot matrix methods for genome analysis. Bioinformatics 20:460–466

Huijberts GNM, Eggink G (1996) Production of poly(3- hydroxyalkanoates) by *Pseudomonas putida* KT2442 in continuous cultures. Appl Microbiol Biotechnol 46:233–239

Huisman GW, de Leeuw O, Eggink G, Witholt B (1989) Synthesis of poly(3-hydroxyalkanoates) is a common feature of fluorescent pseudomonads. Appl Environ Microbiol 55:1949–1954

Jian J, Li ZJ, Ye HM, Yuan MQ, Chen GQ (2010) Metabolic engineering for microbial production of polyhydroxyalkanoates consisting of high 3-hydroxyhexanoate content by recombinant Aeromonas hydrophila. Bioresour Technol 101:6096–6102

Jimenez JI, Minambres B, Garcia JL, Diaz E (2002) Genomic analysis of the aromatic catabolic pathways from *Pseudomonas putida* KT2440. Environ Microbiol 4:824–841

Jiménez JI, Canales A, Jiménez-Barbero J, Ginalski K, Rychlewski L, García JL, Díaz E (2008) Deciphering the genetic determinants for aerobic nicotinic acid degradation: the *nic* cluster from *Pseudomonas putida* KT2440. Pro Nat Acad Sci 105:11329–11334

Kaiser JP, Feng Y, Bollag JM (1996) Microbial metabolism of pyridine, quinoline, acridine, and their derivatives under aerobic and anaerobic conditions. Microbiol Rev 60:483–498

Klockgether J, Munder A, Neugebauer J, Davenport CF, Stanke F, Larbig KD, Heeb S, Schöck U, Pohl TM, Wiehlmann L, Tümmler B (2010) Genome diversity of Pseudomonas aeruginosa PAO1 laboratory strains. J Bacteriol 192:1113–1121

Koebnik R (2005) TonB-dependent trans-envelope signalling: the exception or the rule? Trends Microbiol 3:343–347

Koebnik R, Locher KP, Van Gelder P (2000) Structure and function of bacterial 1225 Q40 outer membrane proteins: barrels in a nutshell. Mol Microbiol 37:239–253

Krylov SV, Kropinski AM, Pleteneva EA, Shaburova OV, Burkal'tseva MV, Miroshnikov KA, Krylov VN (2012) Properties of the new D3-like *Pseudomonas aeruginosa* bacteriophage phiPMG1: genome structure and prospects for the use in phage therapy. Genetika 48:1057–1067

Kwong SM, Yeo CC, Suwanto A, Poh CL (2000) Characterization of the endogenous plasmid from *Pseudomonas alcaligenes* NCIB 9867: DNA sequence and mechanism of transfer J. Bacteriol 182:81–90

Langille MGI, Brinkman FSL (2009) "IslandViewer: an integrated interface for computational identification and visualization of genomic islands". Bioinformatics 25:664–665

Li W, Shi J, Wang X, Han Y, Tong W, Ma L, Liu B, Cai B (2002) Complete nucleotide sequence and organization of the naphthalene catabolic plasmid pND6-1 from *Pseudomonas* sp Strain ND6. Gene 336:231–240

Li S, Zhao H, Li Y, Niu S, Cai B (2012) Complete genome sequence of the naphthalene-degrading *Pseudomonas putida* Strain ND6. J Bacteriol 194:5154–5155

Loper JE, Hassan KA, Mavrodi DV, Davis EW II (2012) Lim CK (2012) Comparative genomics of plant-associated *Pseudomonas* spp: insights into diversity and inheritance of traits involved in multitrophic interactions. PLoS Genet 8:e1002784

Mackiewicz P, Mackiewicz D, Kowalczuk M, Cebrat S (2001) Flip-flop around the origin and terminus of replication in prokaryotic genomes. Genome Biol 2(12): INTERACTIONS1004

Margulies M, Egholm M, Altman WE, Attiya S, Bader JS, Bemben LA, Berka J, Braverman MS, Chen YJ, Chen Z, Dewell SB, Du L, Fierro JM, Gomes XV, Godwin BC, He W, Helgesen S, Ho CH, Irzyk GP, Jando SC, Alenquer ML, Jarvie TP, Jirage KB, Kim JB, Knight JR, Lanza JR, Leamon JH, Lefkowitz SM, Lei M, Li J (2005) Genome sequencing in microfabricated high-density picolitre reactors. Nature 437:376–380

Markowitz VM, Korzeniewski F, Palaniappan K, Szeto E, Werner G, Padki A, Zhao X, Dubchak I, Hugenholtz P, Anderson I, Lykidis K, Ivanova N, Kyrpides NC (2006) The integrated microbial genomes (IMG) system. Nucleic Acid Res 34:D334–D348

Matilla MA, Pizarro-Tobias P, Roca A, Fernández M, Duque E, Molina L, Wu X, van der Lelie D, Gómez MJ, Segura A, Ramos JL (2011) Complete genome of the plant growth-promoting rhizobacterium *Pseudomonas putida* BIRD-1. J Bacteriol 193:1290

Meyer J (2000) Pyoverdines: pigments, siderophores and potential taxonomic markers of fluorescent *Pseudomonas* species. Arch Microbiol 174:135–142
Miñana-Galbis D, Urbizu-Serrano A, Farfán M, Fusté CM, Gaspar Lorén JG J (2009) Phylogenetic analysis and identification of *Aeromonas* species based on sequencing of the *cpn60* universal target. Int J Syst Evol Microbiol 59:1976–1983
Molina L, Duque E, Manuel J, Gómez MJ, Krell T, Lacal J, García-Puente A, García V, Matilla MA, Ramos JL, Segura A (2011) The pGRT1 plasmid of *Pseudomonas putida* DOT-T1E encodes functions relevant for survival under harsh conditions in the environment. Environ Microbiol 13:2315–2327
Molina L, Bernal P, Udaondo Z, Segura A, Ramos JL (2013) Complete genome sequence of a *Pseudomonas putida* clinical isolate, strain H8234. Genome Announc 1(4):E00496-13
Molina-Henares MA, Garcıa-Salamanca A, Molina-Henares J, de la Torre AJ, Herrera MC, Ramos JL, Duque E (2009) Functional analysis of aromatic biosynthetic pathways in *Pseudomonas putida* KT2440. Microbial Biotech 2:91–100
Mulet M, Lalucat J, Garcia-Valdes E (2010) DNA based analysis of the *Pseudomonas* species. Environ Microbiol 12:1513–1530
Mulet M, Garcıa-Valdes E, Lalucat J (2013) Phylogenetic affiliation of *Pseudomonas putida* biovar A and B strains. Res Microbiol 164:351–359
Nakazawa T (2002) Travels of a *Pseudomonas* from Japan around the world. Environ Microbiol 4:782–786
Nelson KE, Weinel C, Paulsen IT, Dodson RJ, Hilbert H, Santos VAP M d, Fouts DE, Gill SR, Pop M, Holmes M, Brinkac L, Beanan M, DeBoy RT, Daugherty S, Kolonay J, Madupu R, Nelson W, White O, Peterson J, Khouri H, Hance I, Chris Lee P, Holtzapple E, Scanlan D, Tran K, Moazzez A, Utterback T, Rizzo M, Lee K, Kosack D et al (2002) Complete genome sequence and comparative analysis of the metabolically versatile *Pseudomonas putida* KT2440. Environ Microbiol 4:799–808
Pak JW, Knoke KL, Noguera DR, Fox BG, Chambliss GH (2000) Transformation of 2,4,6-trinitrotoluene by purified xenobiotic reductase B from *Pseudomonas fluorescens* I-C. Appl Environ Microbiol 66:4742–4750
Palleroni NJ (1984) Genus I *Pseudomonas* Migula 1894. In: Krieg NR, Holt JG (eds) Bergey's manual of systematic bacteriology. Vol 2. The Williams and Wilkins Co, Baltimore, MD, USA, pp 323–379
Palmer KL, Godfrey P, Griggs A, Kos VN, Zucker J, Desjardins C, Cerqueira G, Gevers D, Walker S, Wortman J, Feldgarden M, Haas B, Birren B, Gilmore MS (2012) Comparative genomics of Enterococci: variation in *Enterococcus faecalis*, clade structure in *E. faecium*, and defining characteristics of *E. gallinarum* and *E. casseliflavus*. MBio 3:00318-11
Petkau A, Stuart-Edwards M, Stothard P, Van Domselaar G (2010) Interactive microbial genome visualization with Gview. Bioinformatics 26:3125–3126
Ponraj P, Shankar M, Ilakkiam D, Rajendhran J, Gunasekaran P (2012) Genome sequence of the plant growth-promoting rhizobacterium *Pseudomonas putida*. J Bacteriol 194:6015
Ramos JL, Duque E, Huertas MJ, Dour AH (1995) Isolation and expansion of the catabolic potential of a Pseudomonas putida strain able to grow in the presence of high concentrations of aromatic hydrocarbons. J Bacteriol 177:3911–3916
Ramos-González MI, Campos MJ, Ramos JL, Manuel Espinosa-Urgel M (2006) Characterization of the *Pseudomonas putida* mobile genetic element ISPpu*10*: an occupant of repetitive extragenic palindromic sequences J. Bacteriol 188:37–44
Ratledge C, Dover LG (2000) Iron metabolism in pathogenic bacteria. Annu Rev Microbiol 54:881–941
Ravel J, Cornelis P (2003) Genomics of pyoverdine mediated iron uptake in pseudomonads. Trends Microbiol 11:195–200
Rehm BH (2010) Bacterial polymers: biosynthesis, modifications and applications. Nat Rev Microbiol 2010(8):578–592
Rehm BH, Niels Krueger N, Steinbuechel A (1998) New metabolic link between fatty acid *de novo* synthesis and polyhydroxyalkanoic acid synthesis: the *phaG* gene from *Pseudomonas putida* KT2440 encodes a 3-hydroxyacyl-acyl carrier protein-coenzyme A transferase. J Biol Chem 273:24044–24051
Rojas A, Duque E, Mosqueda G, Golden G, Hurtado A, Ramos JL, Segura A (2001) Three efflux pumps are required to provide efficient tolerance to toluene in *pseudomonas putida* DOT-T1E. J Bacteriol 183:3967–3973
Rosson RA, Nealson KH (1982) Manganese binding and oxidation by spores of a marine *Bacillus*. J Bacteriol 151:1027–1034
Ruth K, de Roo G, Egli T, Ren Q (2008) Identification of two acyl-CoA synthetases from *Pseudomonas putida* GPo1: one is located at the surface of polyhydroxyalkanoates granules. Biomacromolecules 9:1652–1659
Saitou N, Nei M (1987) The neighbor-joining method: a new method for reconstructing phylogenetic trees. Mol Biol Evol 4:406–425
Schellenberg J, Links MG, Hill JE, Dumonceaux TJ, Kimani J, Jaoko W, Wachihi C, Mungai JN, Peters GA, Tyler S, Graham M, Severini A, Fowke KR, Ball TB, Plummer FA (2011) Molecular definition of vaginal microbiota in East African commercial sex workers. Appl Environ Microbiol 77:4066–4074
Sharma PK, Fu J, Cicek N, Sparling R, Levin DB (2011) Kinetics of medium-chain-length polyhydroxyalkanoates (mcl-PHAs) production by a novel isolate of *Pseudomonas putida* LS46. Can J Microbiol 58:982–989
Sharma PK, Fu J, Zhang XL, Fristensky BW, Davenport K, Chain PSG, Sparling R, Levin DB (2013) Draft genome sequence of medium-chain-length polyhydroxyalkanoate-producing *Pseudomonas putida* strain LS46. Genome Announc 1:e00151-13
Siguier P, Perochon J, Lestrade L, Mahillon J, Chandler M (2006) Isfinder: the reference centre for bacterial insertion sequences. Nucleic Acids Res 34:D32–D36
Silby MW, Cerdeno-Tarraga AM, Vernikos GS (2011a) Genomic and genetic analyses of diversity and plant interactions of *Pseudomonas fluorescens*. Genome Biol 10:R51
Silby MW, Winstanley C, Godfrey SAC, Levy SB, Jackson RW (2011b) *Pseudomonas* genome: diverse and adaptable. FEMS Microbiol Rev 35:652–680
Singleton I (1994) Microbial metabolism of xenobiotics: fundamental and applied research. J Chem Technol Biotechnol 59:9–23
Taghavi S, Barac T, Greenberg B, Borremans B, Vangronsveld J, van der Lelie D (2005) Horizontal gene transfer to endogenous endophytic bacteria from poplar improves phytoremediation of toluene. Appl Environ Microbiol 71:8500–8505
Taghavi S, Garafola C, Monchy S (2009) Genome survey and characterization of endophytic bacteria exhibiting a beneficial effect on growth and development of poplar trees. Appl Environ Microbiol 75:748–757
Tamura K, Peterson D, Peterson N, Stecher G, Nei M, Kumar S, MEGA5 (2011) Molecular evolutionary genetics analysis using maximum likelihood, evolutionary distance, and maximum parsimony methods. Mol Biol Evol 28:2731–2739
Tang H, Yao Y, Wang L, Yu H, Ren Y, Wu G, Xu P (2012) Genomic analysis of *Pseudomonas putida*: genes in a genome island are crucial for nicotine degradation. Sci Rep 2:377
Thompson JD, Higgins DG, Gibson TJ (1994) CLUSTAL W Improving the sensitivity of progressive multiple sequence alignment through sequence weighting, position-specific gap penalties and weight matrix choice. Nucleic Acids Res 22:4673–4680
Tillier ER, Collins RA (2000) Genome rearrangement by replication-directed translocation. Nat Genet 26:195–197
Tsuda M, Minegishi K, Iino T (1989) Toluene transposons Tn4651 and Tn4653 are class II transposons. J Bacteriol 171:1386–1393
Udaondo Z, Duque E, Fernández M, Molina L, de la Torre J, Bernal P, Niqui J, Pini C, Roca A, Matilla MA, Molina-Henares A, Hortencia Silva-Jiménez H, Gloria Navarro-Avilés G, Busch A, Lacal J, Krell T, Segura A, Juan-Luis Ramos JL (2012) Analysis of solvent tolerance in *Pseudomonas putida* DOT-T1E based on its genome sequence and a collection of mutants. FEBS Lett 586:2932–2938
Varani AM, Siguier P, Gourbeyre E, Charneau V, Chandler M (2011) ISsaga: an ensemble of web-based methods for high throughput identification and semi-automatic annotation of insertion sequences in prokaryotic genomes. Genome Biol 12:R30
Verbeke TJ, Sparling R, Hill JE, Links MG, Levin D, Dumonceaux TJ (2011) Predicting relatedness of bacterial genomes using the chaperonin-60 universal target (cpn60 UT): Application to *Thermoanaerobacter* species. Syst Appl Microbiol 34:171–179
Visca P, Imperi F, Lamont IL (2007) Pyoverdine siderophores: from biogenesis to biosignificance. Trends Microbiol 15:22–30
Vodovar N, Vallenet D, Cruveiller S (2006) Complete genome sequence of the entomopathogenic and metabolically versatile soil bacterium *Pseudomonas entomophila*. Nat Biotechnol 24:673–679
Wang SN, Liu Z, Tang HZ, Meng J, Xu P (2007) Characterization of environmentally friendly nicotine degradation by *Pseudomonas putida* biotype A strain S16. Microbiology 153:1556–1565
Wang Q, Tappel RC, Zhu C, Nomura CT (2012) Development of a new strategy for production of medium-chain-length polyhydroxyalkanoates by recombinant *Escherichia coli* via inexpensive non-fatty acid feedstocks. Appl Environ Microbiol 78:519–527
Wiener MC (2005) TonB-dependent outer membrane transport: going for Baroque? Curr Opin Struct Biol 15:394–400
Winsor GL, Lam DK, Fleming L, Lo R, Whiteside MD, Yu NY, Hancock RE, Brinkman FS (2011) *Pseudomonas* genome database: improved comparative analysis and population genomics capability for *Pseudomonas* genomes. Nucleic Acids Res 39 (Database issue):D596–D600
Withers BE, Ambroso LA, Dunbar JC (1992) Structure and evolution of the XcyI restriction-modification system. Nucl Acids Res 20:6267–6273
Wu X, Monchy S, Taghavi S, Zhu W, Ramos J, Van der Lelie D (2011) Comparative genomics and functional analysis of niche specific adaptation in *P putida*. FEMS Microbiol Rev 2011(35):299–323

Wyndham RC, Cashore AE, Nakatsu CH, Peel MC (1994) Catabolic transposons. Biodegradation 5:323–342

Yamamoto S, Kasai H, Arnold DL, Jackson RW, Vivian A, Harayama S (2000) Phylogeny of the genus *Pseudomonas*: intrageneric structure reconstructed from the nucleotide sequences of *gyrB* and *rpoD* genes. Microbiology 2000(146):2385–2394

Yu H, Tang H, Wang L, Yao Y, Wu G, Xu P (2011) Complete genome sequence of nicotine-degrading *Pseudomonas putida* strain S16. J Bacteriol 193:5541–5542

Zeigler DR (2005) Application of a *recN* sequence similarity analysis to the identification of species within the bacterial genus *Geobacillus*. Int J Syst Evol Microbiol 55:1171–1179

Zhou Y, Liang Y, Lynch KH, Dennis JJ, Wishart DS, PHAST (2011) A fast phage search tool. Nucleic Acids Res 39(Web Server issue):W347–W352

Zylstra GJ, McCombie WR, Gibson DT, Finette BA (1988) Toluene degradation by *Pseudomonas putida* F1: genetic organization of the *tod* operon. Appl Environ Microbiol 54:1498–1503

3

Common gas phase molecules from fungi affect seed germination and plant health in *Arabidopsis thaliana*

Richard Hung[1*], Samantha Lee[1], Cesar Rodriguez-Saona[2] and Joan W Bennett[1]

Abstract

Fungal volatile organic compounds (VOCs) play important ecophysiological roles in mediating inter-kingdom signaling with arthropods but less is known about their interactions with plants. In this study, *Arabidopsis thaliana* was used as a model in order to test the physiological effects of 23 common vapor-phase fungal VOCs that included alcohols, aldehydes, ketones, and other chemical classes. After exposure to a shared atmosphere with the 23 individual VOCs for 72 hrs, seeds were assayed for rate of germination and seedling formation; vegetative plants were assayed for fresh weight and chlorophyll concentration. All but five of the VOCs tested (1-decene, 2-n-heptylfuran, nonanal, geosmin and -limonene) had a significant effect in inhibiting either germination, seedling formation or both. Seedling formation was entirely inhibited by exposure to 1-octen-3-one, 2-ethylhexanal, 3-methylbutanal, and butanal. As assayed by a combination of fresh weight and chlorophyll concentration, 2-ethylhexanal had a negative impact on two-week-old vegetative plants. Three other compounds (1-octen-3-ol, 2-ethylhexanal, and 2-heptylfuran) decreased fresh weight alone. Most of the VOCs tested did not change the fresh weight or chlorophyll concentration of vegetative plants. In summary, when tested as single compounds, fungal VOCs affected *A. thaliana* in positive, negative or neutral ways.

Keywords: Volatile organic compound; *Arabidopsis thaliana*; Fungi; Seed germination; Chlorophyll concentration; Gas chromatography; Mass spectroscopy

Introduction

Volatiles organic compounds (VOCs) are low molecular mass compounds with high vapor pressure and low to medium water solubility that exist in the gaseous state at room temperature (Herrmann 2010). Approximately 250 VOCs have been identified from fungi (Chiron and Michelot 2005) as the products of both primary and secondary metabolism (Turner and Aldridge 1983; Korpi et al. 2009). These gas phase molecules are emitted in complex mixtures that vary quantitatively and qualitatively depending not only on the age and genetic profiles of the producing species but also on extrinsic variables such as substrate, temperature, moisture level and pH (Sunesson et al. 1995; Claeson et al. 2002; Matysik et al. 2008). Fungal volatiles have distinctive odorant properties and they have been studied extensively for their positive and negative sensory properties. They impart unique aromas and flavors to mold-ripened cheeses, Japanese koji and other mold-fermented food products (Steinkraus 1983; Kinderlerer 1989), and are responsible for the bouquet of gourmet mushrooms such as boletes, chanterelles and truffles (Cho et al. 2008; Fraatz and Zorn 2010). On the negative side, when foods are contaminated by molds, they produce off flavors. VOCs have been used as an indirect indicator of fungal spoilage in agricultural products (Borjesson et al. 1992; Jelen and Wasowicz 1998; Schnürer et al. 1999) and of mold contamination in water-damaged buildings (Kuske et al. 2005; Sahlberg et al. 2013). Finally, because VOCs can diffuse through the atmosphere and the soil, they are well adapted for signaling between species that share a common ecological niche. Both bacterial and fungal VOCs play competitive roles in chemical interactions between microorganisms (Beattie and Torrey 1986; Morath et al. 2012). Many plant and microbial volatile molecules function as semiochemicals, otherwise known as "infochemicals", and there is a large literature on the ability of fungal VOCs to mediate

* Correspondence: Richard.Hung@rutgers.edu
[1]Department of Plant Biology and Pathology, Rutgers, The State University of New Jersey, 59 Dudley Rd., New Brunswick, NJ 08901, USA
Full list of author information is available at the end of the article

arthropod behavior, where they have properties as synomones, allomones, and kairomones (Rohlfs et al. 2005; Mburu et al. 2011; Davis et al. 2013). The fungal VOC commonly called "mushroom alcohol" (1-octen-3-ol) is responsible for much of the musty odor associated with mold contamination and is an important insect semiochemical. It attracts many insect species, including the malaria mosquito (Takken and Knols 1999; Thakeow et al. 2008).

In contrast, the interactions between fungal VOCs and plants have not received much scientific attention (Bitas et al. 2013). Based on the observation that there is very little vegetation in areas known to have truffles (subterranean gourmet fungi), it has been hypothesized that these fungi may have the ability to suppress plant growth through their volatiles (Splivallo et al. 2007). Kishimoto et al. (2007) have shown that 1-octen-3-ol enhances resistance of mature plants of *Arabidopsis thaliana* to *Botrytis cinerea* and activates some of the same defense genes turned on by ethylene and jasmonic acid signaling, important plant hormones involved in plant defense.

We hypothesized that we could distinguish between bioactive and inactive fungal vapors by using chemical standards of individual VOCs and then exposing plants to controlled concentrations in a model habitat. We selected *A. thaliana* as our test species due to the many benefits associated with the use of a well recognized model system including but not limited to: small size, short life cycle, genetic tractability, and comprehensively researched background. Preliminary studies have also shown that tomato plants exposed to VOCs are affected in a similar fashion to *A. thaliana* exposed to the same VOCs indicating that *A. thaliana* is a good model organism for study. Similarly, the effects of fungal VOCs on plant development have been demonstrated in several plants in *Brassicaceae* family including radish, cabbage, rape, and broccoli (Ogura et al. 2000). The aim of our study was to evaluate the effect of individual fungal VOCs on seed germination, vegetative plant growth and chlorophyll concentration in a controlled environment. In this report, our specific objectives have been to create standardized model exposure habitats in order to compare the possible stimulatory and inhibitory effects of fungal VOCs from different chemical classes (e.g., alcohols, aldehydes, ketones, and so forth) and to conduct exposure studies using *A. thaliana* seeds and two-week-old vegetative plants.

Materials and methods

Plant material and seed preparation

All volatile exposure tests were done with *Arabidopsis thaliana* ecotype Columbia 7. Surface-sterilization of seeds and seedling formation studies were conducted as described previously with slight modifications (Hung et al. 2013). Surface sterilized seeds were sown on Murashige and Skoog (MS) media with vitamins, 3% sucrose, and 0.3% Gellan Gum Powder (G 434 PhytoTechnology Laboratories, Shawnee Mission, KS). In germination-seedling formation studies, seeds were sown on Petri dishes (20 seeds per plate) with 20 ml MS media and placed at 4°C in the dark for three days to stratify the seeds. Seeds used to grow plants for exposure assays of two–week old plants were sown individually in test tubes with 10 ml of MS, covered with plant tissue culture caps and then stratified as described above. After three days, the stratified seeds in their individual test tubes were placed in a growth chamber at 21°C ± 2°C with a 16 hour photoperiod for two weeks prior to exposure to VOCs.

Chemicals and exposure conditions

Authentic standards of these high purity chemicals were purchased in liquid form from Sigma-Aldrich (St. Louis, Missouri). The criteria for the selection of these compounds were: 1) the volatiles should represent different chemical classes, 2) they had been isolated from a range of fungal species including both mushrooms and molds, and 3) that they included several VOCs commonly found in soils.

The germination and vegetative exposure to VOCs were determined using the methods described previously (Hung et al. 2014; Lee et al. 2014). Seeds (in Petri plates) or two-week-old plants (in individual test tubes) were exposed in one liter culture vessels (see Additional file 1). All tests were done at a low concentration similar to the concentration of VOCs analyzed previously: one part per million (1 ppm = 1 μl/l). The desired concentration of 1 ppm in the test container was obtained by depositing a drop of the chemical standard (VOC) in liquid form onto the inside of the glass vessel. The compounds, due to their chemical properties will quickly volatilize into the gas phase in the test conditions. Before sealing the lids, a 10 × 10 cm piece of Dura Seal Cling Sealing Film (Diversified Biotech) was placed over the top of each culture vessel so as to prevent VOC leakage through the polypropylene closure. The culture vessels containing either seeds in Petri plates or two-week-old plants in test tubes were arranged randomly in the growth chamber and then placed on a one inch throw rotator at 40 rpm in order to volatilize and evenly distribute the compounds. The control plants were placed in identical conditions without any VOCs.

Scoring germination stages

The seeds were exposed to the individual VOCs for 72 hours and then examined under a binocular microscope where they were scored into three categories: no germination, germination (emergence of the radical [embryonic root]), and seedling formation (presence of the radicle, the hypocotyls and the cotyledons) (see Additional file 2). Seeds scored as "no germination" included seeds with a ruptured testa (seed coat) but without the presence of the radicle.

Plant mass and chlorophyll concentration

After exposure to the vapors of the individual VOCs, plants were removed from the test conditions, the shoot and leaves were cut away from the roots, and fresh weight of the shoots and leaves was obtained. The chlorophyll was extracted using the method of Jing et al. (2002) with some modifications. The plants were soaked overnight in 80% acetone at 4°C in darkness prior to obtaining photometric readings at 663 and 645 nm with a spectrophotometer (DU800, Beckman Coulter, Brea, CA). Total chlorophyll = 20.2 (A_{645}) + 8.02 (A_{663})(V/1,000 * w) where V = total volume of the sample, w = weight of the sample, A_{663} = absorbance at 663 nm, A_{645} = absorbance at 645 nm (Palta 1990). Each solvent extract contained one plant per treatment.

Statistical analysis

The data were analyzed and plotted using Excel software (Microsoft, Redmond, WA) and SigmaPlot (SPSS Science Inc., IL). To test the significance of the exposure studies, one-way analysis of variance (ANOVA) and Student's t-tests were performed with the aggregated data. The Student's t-tests determine if there are significant differences between two sets of data: the control and VOC exposed plants. For germination exposures, two replicate plates with 20 seeds per plate were tested for each compound, with two independent experiments, for a total of 80 seeds. For vegetative plant exposure, four plants were placed in the exposure vessel and three jars were used for each experiment. There were three independent experiments for each compound, for a total of 36 plants.

Results

Seedling formation tests

The percentage of seeds germinating and progressing to seedling formation after exposure to the 23 fungal volatiles is shown in Figure 1 and summarized in Figure 2. In our experiments, more than 75% of control seeds progressed to the seedling stage after 72 hrs. Similar rates of seedling formation were observed for seeds exposed to 1-decene, 2-n-heptylfuran, nonanal, geosmin and -limonene. In contrast, none of the seeds exposed to 2-ethylhexanal 1-octen-3-one, 3-methylbutanal, or butanal formed seedlings. Of these four inhibitory volatiles, radical protrusion (i.e. germination) was observed in only 19% of 1-octen-3-one exposed seeds, while 83% of seeds exposed to butanal exhibited formation of a radical. Seeds exposed to the other 14 volatile compounds tested had intermediate levels of germination efficiency and seedling formation (Figure 2). Of the five aldehydes we tested, three (2-ethylhexanal, 3-methylbutanal, or butanal) inhibited seedling formation by 100% and one (3-methylproponal) inhibited seedling formation by 70%. Nevertheless, no matter the extent of inhibition in the presence of the 23 VOCs tested, when removed from exposure to the VOCs after 72 hr, all seeds resumed germination and formed seedlings.

Vegetative plants tests

The fresh weight and chlorophyll concentration of control and exposed plants are given in Figure 3. The fresh weight of control plants was 22.4 mg (±6.7 SE). A significant decrease in fresh weight was observed after 72 hr exposure to 1-octen-3-ol, 2-ethylhexanal, and 2-heptylfuran, where the mean fresh weight was respectively 6.7 mg (±2.7 SE), 11.2 mg (±4.9 SE), and 6.3 mg (±5.4 SE) less than controls. The other compounds tested ((±)2-methyl-1-butanol, geosmin, 2-methylpropan-1-ol, 1-octen-3-ol, octan-1-ol, octan-3-ol, dec-1-ene, oct-1-ene, butanal, 2-ethylhexanal, 2-methylpropanal, 3-methylbutanal, nonanal, heptan-2-one, octan-2-one, oct-1-en-3-one, pentan-2-one, isothiocyanatocyclohexane, octanoic acid, +limonene, –limonene, and 2-heptylfuran) did not cause significant differences in fresh weight of exposed vegetative plants (Figure 3).

In Figure 3, we have expressed the chlorophyll concentration data as mg/gm of fresh weight of shoots and leaves. Five compounds (–2-methyl-1-butanol, 1-octen-3-ol, dec-1-ene, heptan-2-one, and isothiocyanatocyclohexane) yielded statistically significant increases in chlorophyll concentration showing, respectively, 0.02 mg/g, 0.2 mg/g, 0.18 mg/g, 0.02 mg/g, and 0.4 mg/g greater amounts than controls. Three compounds, geosmin, 2-ethylhexanal, and 1-octen-3-one caused a statistically significant decrease in chlorophyll concentration of, respectively, 0.15 mg/g, 0.35 mg/g, and 0.42 mg/g less than controls. While geosmin did not adversely affect fresh weight, it did cause a significant decrease in chlorophyll concentration as indicated above. The compound 2-ethyl-hexanal decreased both fresh weight and chlorophyll concentration. The nonracemic form of –2-methyl-1-butanol showed a small but significantly different increase in both parameters.

In conclusion, all but five of the VOCs tested had a significant effect in inhibiting either germination, seedling formation or both. Three of the compounds (2-ethylhexanal, 1-octen-3-one, and 3-methylbutanal) showed more than 50% inhibition of seed germination, while 12 of the compounds (2-methylpropan-1-ol, 2-methylpropanal, heptan-2-one, octan-2-one, isothiocyanatocyclohexane, pentan-2-one, octan-3-ol, 1-octen-3-ol, 2-ethylhexanal, 1-octen-3-one, 3-methylbutanal, and butanal) were associated with more than a 50% retardation in seedling formation. Butanal was unusual in that 83% of seeds germinated (i.e. formed a radicle) but none of these germinated seeds progressed to seedling formation. In

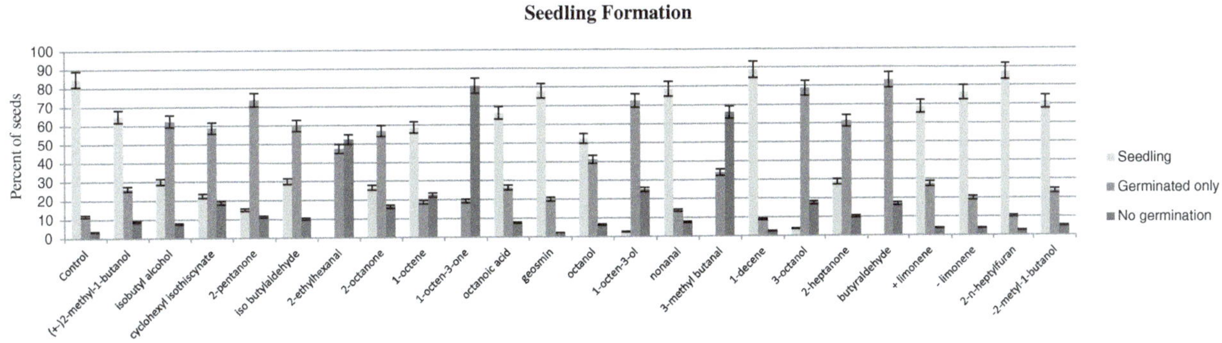

Figure 1 Percentage of seeds that have reached each stage after 72 hrs of exposure to 23 fungal VOCs. Standard error indicated in error bars.

general, the most bioactive of the VOCs we tested were aldehydes and ketones. The single, most inhibitory VOC against germination and seedling formation was oct-1-en-3-one, an eight carbon ketone that has been isolated from both molds and mushrooms (Jelen and Wasowicz 1998). Nevertheless, in all cases, VOC exposed seeds were able to resume germination and progress to seedling stage when removed from the shared atmosphere with the VOC. We conclude that the fungal VOCs we tested have a phytostatic (inhibitory), not a phytocidal (lethal) effect on seeds.

Discussion

In order to study the influence of individual fungal VOCs on plant health, we used *A. thaliana* as our test organism and developed standardized protocols for exposing seeds and young vegetative plants. We investigated a representative sample of common fungal VOCs encompassing seven alcohols, two alkenes, five aldehydes, four ketones, and a single representative isothiocyanate, carboxylic acid, and furan. In addition, we tested both isomers of the terpene 1-methyl-4-(1-methyetenyl)-cyclohexene, commonly known as limonene. We used a

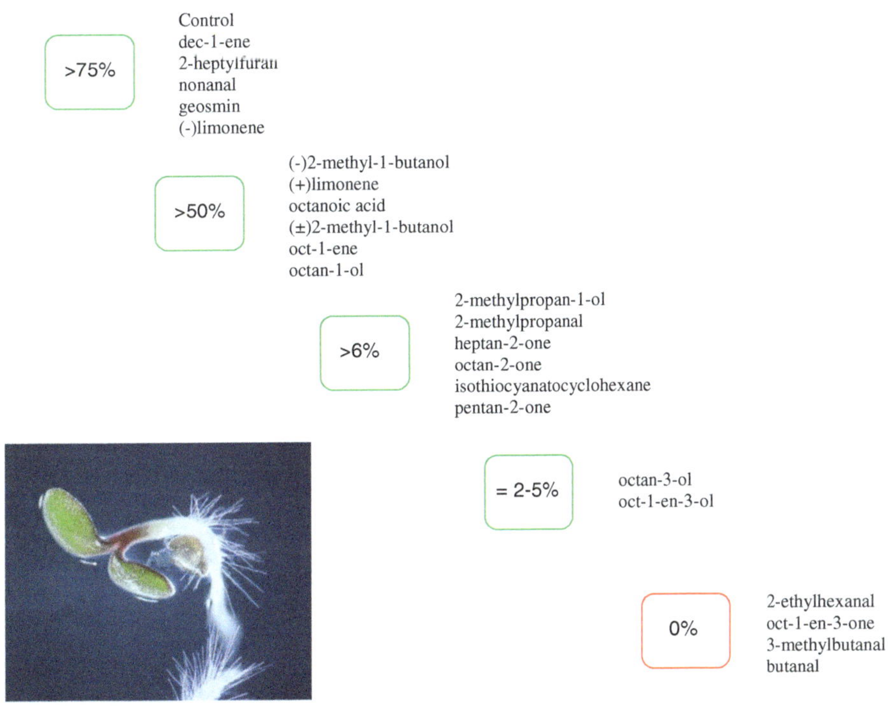

Figure 2 A summary of the categories of percent seedling formation of *A. thaliana* seeds after 72 hrs of exposure to chemical standards of 23 different volatile organic compounds.

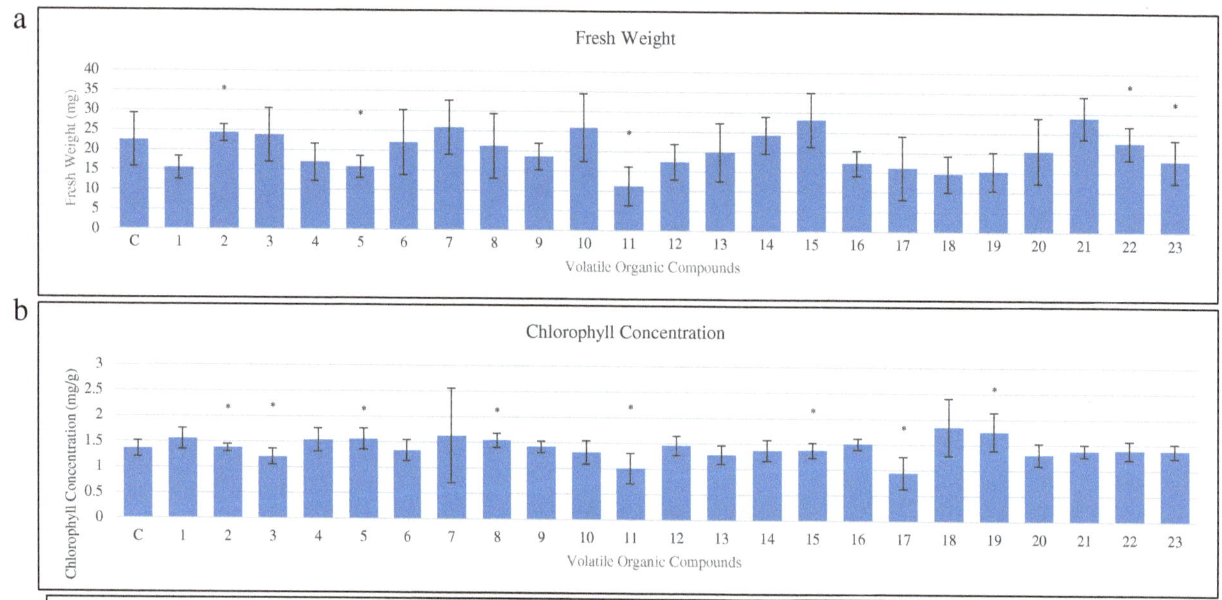

Legend

C	Control	6	octan-1-ol	12	2-methylpropanal	18	pentan-2-one
1	(±)2-methyl-1-butanol	7	octan-3-ol	13	3-methylbutanal	19	isothiocyanatocyclohexane
2	-2-methyl-1-butanol	8	dec-1-ene	14	nonanal	20	octanoic acid
3	geosmin	9	oct-1-ene	15	heptan-2-one	21	+limonene
4	2-methylpropan-1-ol	10	butanal	16	octan-2-one	22	-limonene
5	1-octen-3-ol	11	2-ethylhexanal	17	oct-1-en-3-one	23	2-heptylfuran

Figure 3 Fresh weight and chlorophyll concentration of two week old plants of *A. thaliana* exposed for 72 hrs to chemical standards of 23 different fungal volatile organic compounds. a. Fresh weight in mg. **b**. Chlorophyll concentration in mg/g. Significant values compared to control are marked with an asterisk $p < 0.04$. Standard error indicated in error bars.

72 hour exposure period, and exposed either seeds or young vegetative plants to 1 ppm of chemical standards of the 23 fungal VOCs in a contained chamber.

Seed germination assays with lettuce, cucumber and other economically important species have been widely employed as low-cost, ethically acceptable toxicity tests to screen for dangerous levels of industrial contamination in water and soils (Banks and Schultz 2005; Wang et al. 2001). Almost all of these assay studies have involved aqueous phase compounds, although there have been a few scattered reports describing inhibitory effect of plant volatiles on the germination of seeds from crop and weed species (Holm 1972; Bradow and Connick 1990). The physiological basis for this inhibitory action has not been elucidated. *A. thaliana*, though not agriculturally important, offers a number of experimental advantages for studying basic aspects of plant biology, including seed germination, because of its many genetic resources (Koornneef et al. 2002). Our development of an *A. thaliana* exposure system for studying VOC effects under controlled conditions offers the promise of being able to use this species to dissect the germination inhibition response at the molecular level.

In contrast to seed germination studies, which all report inhibition of germination by VOCs (Holm 1972; French et al. 1975; Bradow and Connick 1990). To date, published studies of the effects of VOCs on vegetative plants report VOC–associated growth stimulation by mixtures of VOCs emitted by growing bacteria or fungi (Ryu et al. 2003; Minerdi et al. 2011; Hung et al. 2013; Paul and Park 2013). Many soil dwelling microbes emit VOCs that mediate various chemical "conversations" between the rhizosphere and plants (Wenke et al. 2010, 2012). For example, plant growth promoting rhizobacteria (PGPR) produce mixtures of VOCs that enhance growth in a wide variety of species (Farag et al. 2006; Vespermann et al. 2007; Lugtenberg and Kamilova 2009). In some cases, microbial VOCs induce systemic resistance (Van Loon et al. 1998; Ryu et al. 2003) or inhibit the growth of plant pathogens (Minerdi et al. 2009, 2011). Volatiles of *Cladosporium cladosporioides* enhance growth of tobacco plants (Paul and Park 2013) and our laboratory has shown that vegetative plants of *A. thaliana* seedlings grown in a shared atmosphere with volatiles emitted by living cultures of the biocontrol fungus *Trichoderma viride*, displayed increased size and vigor (Hung et al.

2013). Fungal VOCs may contribute to the ability of certain species to outcompete neighboring plants. For example, the VOCs produced by *Muscodor yucantanensis* were toxic to the roots, and inhibited seed germination, of amaranth, tomato and barnyard grass (Macias-Rubalcava et al. 2010). In all of these cases, the growth-enhancing effects were mediated by mixtures of naturally emitted VOCs that change with both growth phase and extrinsic environmental parameters.

In our studies, we used a controlled system and exposed plants to low concentrations of individual VOCs. In this controlled habitat, the VOCs tested either had neutral or negative effects on vegetative plant growth, suggesting that the known growth enhancing effects of bacterial and fungal VOCs maybe by synergistic mixtures working in concert. A parallel example is provided by the antibiotic effects of volatiles produced by *Muscodor albus*. This species produces a mixture of VOCs that inhibit and kill a wide range of plant pathogenic fungi and bacteria (Strobel et al. 2001). Nevertheless, when *Muscodor* VOCs were tested individually, the inhibitory effects were not observed, suggesting that the antifungal activity required a suite of VOCs working in concert (Strobel et al. 2001).

It should be recognized that studies on transkingdom signalling mediated by fungal VOCs are technically difficult to conduct. Biogenic VOCs exhibit enormous heterogeneity chemically, spatially, and temporally; are found in low concentrations; and by definition have innate evaporative properties that make it difficult to investigate their impact on plant growth and development in natural settings. Our exploratory research shows that by using controlled concentrations of pure synthetic compounds in model habitats, individual VOCs can be studied one by one, thereby isolating their distinct growth-promoting, growth-inhibiting and other physiological effects. The genetic and genomic resources available for *A. thaliana* make this organism well suited for future research on the mechanistic basis of VOC mediated interactions and for analysis of the consequent biological responses. In summary, the study of gas phase fungal metabolites offers many interesting prospects for enlarging our understanding of the way in which fungi interact with plants in nature and may have potential for commercial application of VOCs in greenhouse agriculture.

Additional files

Additional file 1: Diagrams of exposure chambers for a. seeds in Petri dishes and b. plants in test tubes.

Additional file 2: Representative images of stages of seedling formation A) No germination B) Germination, radical protrusion C) Seedling formation.

Competing interests

The authors declare that they have no competing interests.

Authors' contributions

RH designed the volatile exposure studies. In addition, he participated in the volatile exposure experiments and the drafting of the paper. SL participated in the exposure experiments and the drafting of the paper. CR-S collected and analyzed the volatile organic compounds. JWB conceived the study and participated in the drafting of the paper. All authors read and approved the final manuscript.

Acknowledgements

We thank Barbra Zilinskas, Chee-kok Chin, and Prakash Masurekar for their guidance and perspective. We also thank Elvira de Lange for helpful comments on an earlier draft of the manuscript. This research was supported by funds from Rutgers, The State University of New Jersey and National Science Foundation Graduate Research Fellowship under Grant No. 0937373 to SL.

Author details

[1]Department of Plant Biology and Pathology, Rutgers, The State University of New Jersey, 59 Dudley Rd., New Brunswick, NJ 08901, USA. [2]Department of Entomology, Rutgers, The State University of New Jersey, 96 Lipman Drive, New Brunswick, NJ 08901, USA.

References

Banks MK, Schultz KE (2005) Comparison of plants for germination toxicity tests in petroleum-contaminated soils. Water Air Soil Pollut 167:211–219

Beattie SE, Torrey GS (1986) Toxicity of methanethiol produced by *Brevibacterium linens* toward *Penicillium expansum*. J Agr Food Sci 34:102–104

Bitas V, Kim H-S, Bennett J, Kang S (2013) Sniffing on microbes: diverse roles of microbial volatile organic compounds in plant health. Mol Plant Microbe Interact 26:835–843

Borjesson T, Stollman U, Schnurer J (1992) Volatile metabolites produced by six fungal species compared with other indicators of fungal growth on stored cereals. Appl Environ Microbiol 58:2599–2605

Bradow JM, Connick WJ (1990) Volatile seed germination inhibitors from plant residues. J Chem Ecol 16:645–666

Chiron N, Michelot D (2005) Odeurs de champignons: chimie et rôle dans les interactions biotiques- une revue. Cryptogam Mycol 26:299–364

Cho IH, Namgung H-J, Choi H-K, Kim YS (2008) Volatiles and key odorants in the pileus and stipe of pine-mushroom (*Tricholoma matsutake* Sing). Food Chem 106:71–76

Claeson AS, Levin JO, Blomquist G, Sunesson AL (2002) Volatile metabolites from microorganisms grown on humid building materials and synthetic media. J Environ Monit 4:667–672

Davis TS, Crippen TL, Hofstetter RW, Tomberlin JK (2013) Microbial volatile emissions as insect semiochemicals. J Chem Ecol 39:840–859

Farag MA, Ryu CM, Sumner LW, Paré PA (2006) GC-MS SPME profiling or rhizobacterial volatiles reveals prospective inducers of growth promotion and induced systemic resistance in plants. Phytochemistry 67:2262–2268

Fraatz MA, Zorn H (2010) Fungal Flavours. In: Hofrichter M (ed) The Mycota X: Industial Applications, 2nd edition. Springer-Verlag, Berlin, pp 249–264

French RC, Gale AW, Graham CL, Latterell FM, Schmitt CG, Marchetti MA, Rines HW (1975) Differences in germination response of spores of several species of rust and smut fungi to nonanal, 6-methyl-5-hepten-2-one, and related compounds. J Agric Food Chem 23:766–770

Herrmann A (2010) The chemistry and biology of volatiles. Wiley, Chichester

Holm RE (1972) Volatile metabolites controlling germination in buried weed seeds. Plant Physiol 30:293–297

Hung R, Lee S, Bennett JW (2013) The effect of low concentrations of the semiochemical 1-octen-3-ol on *Arabidopsis thaliana*. Fungal Ecol 6:19–26

Hung R, Lee S, Bennett JW (2014) The effects of low concentrations of the enantiomers of mushroom alcohol (1-octen-3-ol) on *Arabidopsis thaliana*. Mycology: Int J Fungal Biol 5:73–80

Jelen HH, Wasowicz E (1998) Volatile fungal metabolites and their relation to the spoilage of agricultural commodities. Food Rev Int 14:391–426

Jing H, Sturre MJG, Hille J, Dijkwel PP (2002) *Arabidopsis* onset of leaf death mutants identify a regulatory pathway controlling leaf senescence. Plant J 32:51–63

Kinderlerer J (1989) Volatile metabolites of filamentous fungi and their role in food flavor. J Appl Bacteriol Symp Suppl. 67:133S–144S

Kishimoto K, Matsui K, Ozawa R, Takabayashi J (2007) Volatile 1-octen-3-ol induces a defensive response in *Arabidopsis thaliana*. J Gen Plant Pathol 73:35–37
Koornneef M, Bentsink L, Hilhorst H (2002) Seed dormancy and germination. Curr Opin Plant Biol 5:33–36
Korpi A, Jarnberg J, Pasanen A-L (2009) Microbial volatile organic compounds. Crit Rev Toxicol 39:139–193
Kuske M, Romain A-C, Nicolas J (2005) Microbial volatile organic compounds as indicators of fungi. Can an electronic nose detect fungi in indoor environments? Build Environ 40:824–831
Lee S, Hung R, Schink A, Mauro J, Bennett JW (2014) Phytotoxicity of volatile organic compound. Plant Grow Reg. doi:10.1007/s10725-014-9909-9
Lugtenberg B, Kamilova F (2009) Plant-growth-promoting rhizobacteria. Annu Rev Microbiol 63:541–556
Macias-Rubalcava ML, Hernandez-Bautista BE, Oropeza F, Duarte G, Gonzalez MC, Glenn AE, Hanlin RT, Anaya AL (2010) Allelochemical effects of volatile compounds and organic extracts from *Muscodor yucatanensis*, a tropical endophytic fungus from *Bursera simaruba*. J Chem Ecol 36:1122–1131
Matysik S, Herbarth O, Mueller A (2008) Determination of volatile metabolites originating from mould growth on wall paper and synthetic media. J Microbiol Methods 75:182–187
Mburu DM, Ndung'u MW, Maniania NK, Hassanali A (2011) Comparison of volatile blends and gene sequences of two isolates of *Metarhizium anisopliae* of different virulence and repellency toward the termite *Macrotermes michaelseni*. J Exp Biol 214:956–962
Minerdi D, Bossi S, Gullino ML, Garibaldi A (2009) Volatile organic compounds: a potential direct long-distance mechanism for antagonistic action of *Fusarium oxysporum* strain MSA 35. Environ Microbiol 11:844–854
Minerdi D, Bossi S, Maffei ME, Gullino ML, Garibaldi A (2011) *Fusarium oxysporum* and its bacterial consortium promote lettuce growth and expansin A5 gene expression through microbial volatile organic compound (MVOC) emission. FEMS Microbiol Ecol 76:342–351
Morath SU, Hung R, Bennett JW (2012) Fungal volatile organic compounds: a review with emphasis on their biotechnological potential. Fungal Biol Rev 26:73–83
Ogura T, Sunairi M, Nakajima M (2000) 2-Methylisoborneol and geosmin, the main sources of soil odor, inhibit the germination of *Brassicaceae* seeds. Soil Sci 46:217–227
Palta JP (1990) Leaf chlorophyll content. Remote Sens Rev 5:207–213
Paul D, Park KS (2013) Identification of volatiles produced by *Cladosporium cladosporioides* CL-1, a fungal biocontrol agent that promotes plant growth. Sensors 13:13969–13977
Rohlfs M, Obmann BR, Petersen R (2005) Competition with filamentous fungi and its implication for a gregarious lifestyle in insects living on ephemeral resources. Ecol Entomol 30:556–563
Ryu C-M, Farag MA, Hu C-H, Reddy MS, Wei H-X, Paré PW, Kloepper JW (2003) Bacterial volatiles promote growth in *Arabidopsis*. Proc Natl Acad Sci U S A 100:4927–4932
Sahlberg B, Gunnbjörnsdottir M, Soonc A, Jogi R, Gislason T, Wieslander G, Janson C, Norback D (2013) Airborne molds and bacteria, microbial volatile organic compounds (MVOC), plasticizers and formaldehyde in dwellings in three North European cities in relation to sick building syndrome (SBS). Sci Total Environ 444:433–440
Schnürer J, Olsson J, Borjesson T (1999) Fungal volatiles as indicators of food and feeds spoilage. Fungal Genet Biol 27:209–217
Splivallo R, Novero M, Bertea CM, Bossi S, Bonfante P (2007) Truffle volatiles inhibit growth and induce an oxidative burst in *Arabidopsis thaliana*. New Phytol 175:17–24
Steinkraus KH (1983) Industrial Applications of Oriental Fungal Fermentations. In: Smith JE, Berry DR, Kristiansesn B (ed) The Filamentous Fungi Vol. 4 Fungal Technology. Edward Arnold, London, pp 171–189
Strobel GA, Dirkse E, Sears J, Markworth C (2001) Volatile antimicrobials from *Muscodor albus*, a novel endophytic fungus. Microbiology 147:2943–2950
Sunesson A-L, Vaes WHJ, Nilsson C-A, Blomquist GR, Andersson B, Carlson R (1995) Identification of volatile metabolites from five fungal species cultivated on two media. Appl Environ Microbiol 61:2911–2918
Takken W, Knols BG (1999) Odor-mediated behavior of Afrotropical malaria mosquitoes. Annu Rev Entomol 44:131–157
Thakeow P, Angeli S, Weißbecker B, Schütz S (2008) Antennal and behavioral responses of *Cis boleti* to fungal odor of *Trametes gibbosa*. Chem Senses 33:379–387
Turner WB, Aldridge DC (1983) Fungal Metabolites II. Academic Press, London
Van Loon LC, Bakker PAHM, Pieterse CMJ (1998) Systemic resistance induced by rhizosphere bacteria. Annu Rev Phytopathol 36:453–483
Vespermann A, Kai M, Piechulla B (2007) Rhizobacterial volatiles affect the growth of fungi and *Arabidopsis thaliana*. Appl Environ Microbiol 73:5639–5641
Wang X, Sun C, Gao S, Wang L, Shuokui H (2001) Validation of germination rate and root elongation as indicator to assess phytotoxicity with *Cucumis sativus*. Chemosphere 44:1711–1721
Wenke K, Kai M, Piechulla B (2010) Belowground volatiles facilitate interactions between plant roots and soil organisms. Planta 231:499–506
Wenke K, Weise T, Warnke R, Valverde C, Wanke D, Kai M, Piechulla B (2012) Bacterial Volatiles Mediating Information Between Bacteria and Plants. In: Witzany G, Baluska F (ed) Biocommunication of Plants. Springer-Verlag, Berlin, pp 327–347

4

Printed paper-based arrays as substrates for biofilm formation

Anni Määttänen[1*], Adyary Fallarero[2], Janni Kujala[2], Petri Ihalainen[1], Pia Vuorela[3] and Jouko Peltonen[1]

Abstract

The suitability of paper-based arrays for biofilm formation studies by *Staphylococcus aureus* is demonstrated. Laboratory-coated papers with different physicochemical properties were used as substrates. The array platform was fabricated by patterning the coated papers with vinyl-substituted polydimethylsiloxane (PDMS) -based ink. The affinity of bacteria onto the flexographically printed hydrophobic and smooth PDMS film was very low whereas bacterial adhesion and biofilm formation occurred preferentially on the unprinted areas, i.e. in the reaction arrays. The concentration of the attached bacteria was quantified by determining the viable colony forming unit (CFU/cm^2) numbers. The distribution and the extent of surface coverage of the biofilms were determined by atomic force microscopy. In static conditions, the highest bacterial concentration and most highly organized biofilms were observed on substrates with high polarity. On a rough paper surface with low polarity, the biofilm formation was most hindered. Biofilms were effectively removed from a polar substrate upon exposure to (+)-dehydroabietic acid, an anti-biofilm compound.

Keywords: Biofilm formation; *Staphylococcus aureus*; Polarity; Surface roughness; AFM; PDMS

Introduction

Bacteria can switch between two different life styles: single cells floating in a liquid medium (planktonic mode) and sessile cells (biofilm mode). Biofilms are surface-attached bacterial life forms which can be described as well-organized communities of cells that are surrounded by a self-produced layer of an extracellular polymeric substance (EPS). Biofilms cause serious threat to human health. They are responsible for a significant number of chronic antibiotic-resistant infections (Donlan and Costerton 2002). Their chemoresistance has been attributed to various factors. One mechanism is related to the presence of the EPS, which acts as a protective barrier against biocides and toxins and it sequesters nutrients from the environment, thus being an essential part of the strategy of bacteria for persistence under extreme, unfavourable conditions (McDougald et al. 2011). Other mechanisms include slower growth rate and the presence of resistant subpopulations (persister cells) (Proctor et al. 1994; Fux et al. 2005; Anderson and O'Toole 2008). It has also been shown that biofilms are more tolerant to environmental bacteriophages and phagocytic amoebae than planktonic bacteria (Higashi and Sullam 2006).

Staphylococcus aureus is a Gram-positive bacterium regarded as an etiologic agent of a wide range of diseases associated with significant morbidity and mortality and is a leading cause of nosocomial infections (Jabra-Rizk et al. 2006). In this study *S. aureus* was selected as a model biofilm-forming organism given its profound association with serious biofilm-mediated pathologies. For instance, it has been established that *Staphylococcus* spp. biofilms account for more than a half of infections associated with prosthetic devices (Fluckiger et al. 2005). In addition, *S. aureus* colonize diabetic, pressure and venous ulcers as well as burn wounds which can result in non-healing infection and may even lead to death. Additionally, *S. aureus* biofilms are involved in the first stage of cystic fibrosis (CF), as well as in chronic otitis media and osteomyelitis (Lindsay and von Holy 2006). Unfortunately, all the antibiotics that are currently in use today have been developed to act against dividing phase planktonic bacteria and there is a pressing need for compounds that can selectively act on staphylococcal biofilms (Elliot et al. 1982; Landini et al. 2010; Worthington et al. 2012; Blackledge et al. 2013; Ausbacher et al. 2014).

* Correspondence: anni.maattanen@abo.fi
[1]Laboratory of Physical Chemistry, Center of Excellence for Functional Materials, Abo Akademi University, Turku, Finland
Full list of author information is available at the end of the article

Substrate properties are important for biofilm growth since bacteria generally prefer to grow on available surfaces rather than in the surrounding aqueous phase (Katsikogianni and Missirlis 2004). Physicochemical surface properties such as roughness and surface energetics, together with the properties of the suspending medium (e.g. pH, surface tension and/or presence of proteins), have been shown to have an influence on the strength and direction (decrease/increase) of microbial adhesion and subsequent biofilm formation (Absolom et al. 1983; Fletcher and Pringle 1985; Litzler et al. 2007; Chung et al. 2007; Zmantar et al. 2011; Mosier and Cady 2011; Nill et al. 2011; Singh et al. 2012). In addition, electrostatic interactions (surface charge) and viscoelastic properties (elastic modulus) of the substrate have been shown to influence the bacterial adherence (Brady 1999; Rosenhahn et al. 2009).

Paper is a sustainable and recyclable material, and its physicochemical properties (topography, roughness, stiffness, surface energy, polarity, porosity and pore geometry) can be modified quite conveniently by various coating materials and methods and surface treatments (Ihalainen et al. 2012). Furthermore, the surface of paper can be functionalized by printing of not only graphical inks but also e.g. functional polymers and biomaterials (Siegel et al. 2009; Määttänen et al. 2011; Tian and Shen 2011; Li et al. 2012; Wang et al. 2012). By patterning paper for instance with a PDMS-based ink, printable paper-based alternatives for conventional plastic microtiter plates can be fabricated (Määttänen et al. 2011; Juvonen et al. 2013). The role of the low surface energy PDMS film can be to confine the liquids or, to direct biomaterial adhesion and growth of e.g. human ARPE-19 cells on the unprinted (PDMS-free) areas as recently demonstrated (Juvonen et al. 2013).

The goals of this work were to study the effect of certain physicochemical properties of different paper-based substrates on bacterial biofilm growth in static conditions, as well as to demonstrate the use of the printed platform in the investigation of biofilm susceptibility to an antimicrobial agent. The array was constructed through flexographic printing of a patterned PDMS layer on four different types of coated paper substrates with characteristic surface topography, roughness and surface energetics. The changes in bacterial biofilm morphology were followed by high resolution microscopic techniques. The susceptibility tests were conducted using (+)-dehydroabietic acid. The effects of this potent anti-biofilm molecule against *S. aureus* have been recently reported (Fallarero et al. 2013) and they were verified here by topographical means.

Materials and methods

Substrates

Four different types of reverse gravure coated papers including substrates coated with two latex blends (coded: Latex 1, Latex 2) and substrates coated with two mineral pigments (coded: Kaolin and PCC (precipitated calcium carbonate)) were fabricated, characterized and used for the biofilm formation studies. A 96-well polystyrene microplate and a PDMS-printed paper substrate were used as reference substrates. Additional file 1: Table S1 summarizes the raw materials used in the substrates.

The detailed fabrication procedures and properties of the multilayer pigment-coated paper substrates have been described in earlier publications (Bollström et al. 2009, 2010; Määttänen et al. 2010).

The aqueous latex blends consisted of film-forming (low glass transition temperature (T_g)) and non-film-forming (high T_g) components. The hard high-T_g latex (plastic pigment) particles (size 100 – 200 nm) provide blocking resistance, mechanical strength and integrity to the film, while the soft (low T_g) latex particles act as a film-forming component (Schuler et al. 2000; Ihalainen et al. 2007, 2010). All the paper substrates were dried by using an IR lamp and calendered before use, pigment-based substrates at a nip pressure and temperature of 50 bars and 70°C and latex-based substrates at 70 bars and 35°C, respectively. The latex coatings were washed with purified water and absolute ethanol and further IR cured for 1 min in order to thermally modify the topography to gain a bimodal structure with a large surface area.

The printed array platforms (Figure 1a and b) for biofilm formation experiments were fabricated by flexographic printing of a single layer of the PDMS–based hydrophobic, translucent and solvent-free ink on the studied paper substrates by using pre-designed flexographic printing plates with well diameters of 2 mm or 5 mm (a photograph shown in Additional file 1: Figure S1). A more detailed description of the procedure is given elsewhere (Määttänen et al. 2011). In addition, the back side of the substrates was covered with the PDMS ink in order to direct the biofilm formation into the spherical non-printed areas. The arrays were disinfected with ethanol and sterile water prior to use in the biofilm studies and surface characterization.

Atomic force microscopy (AFM)

An NTEGRA Prima (NT-MDT, Russia) atomic force microscope (AFM) was used for analyzing the topography of the substrates and the biofilms. The microscope was placed on an active vibration isolation table (TS-150, Table Stable Ltd., Switzerland), which was further placed on a stone table to eliminate external vibrational noise. Topography imaging was carried out in intermittent-contact mode under ambient conditions (relative humidity (RH) = 30 ± 10%, room temperature (RT) = 26 ± 2°C) using uncoated rectangular silicon cantilevers (MikroMasch, model DP16/GP/AlBS (typical resonance frequency: 170 kHz, typical spring constant: 40

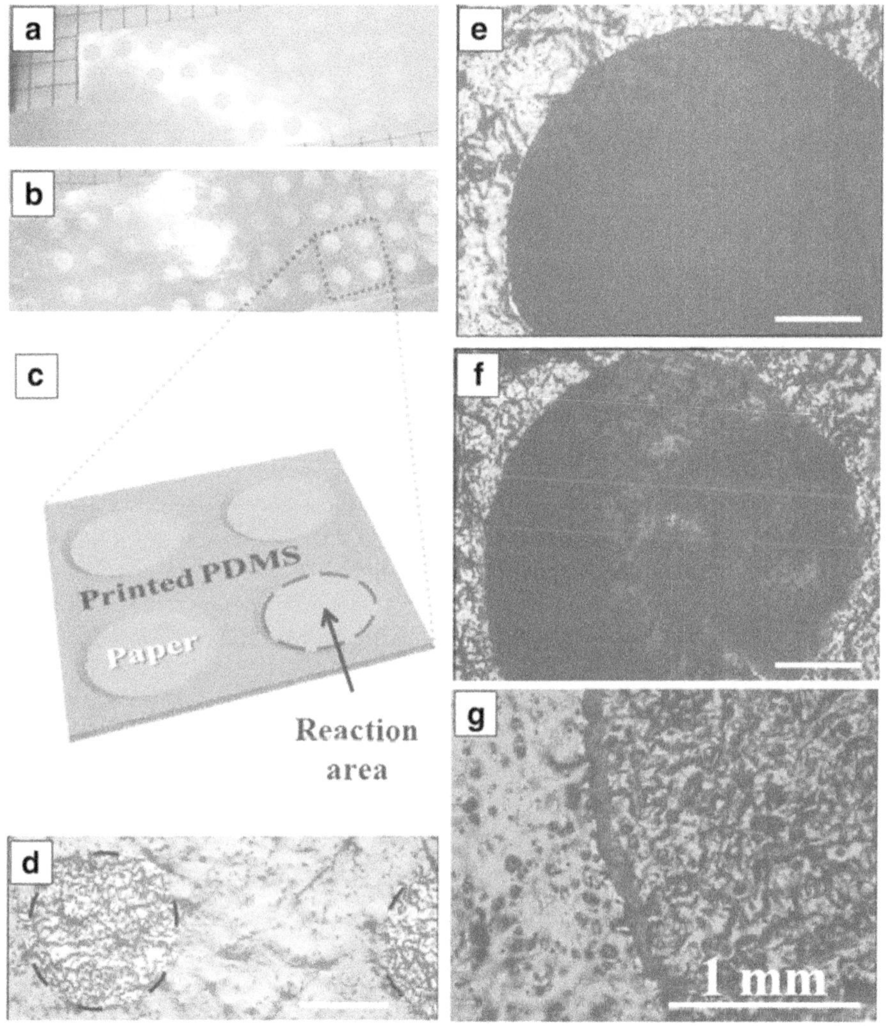

Figure 1 Paper-based platforms for biofilm studies. A digital photograph of the arrays fabricated on **a)** Kaolin coated paper and **b)** latex coated paper. A schematic illustration of the reaction array platform **c)**. Optical micrograph of an unexposed reaction area on Latex 2 substrate **d)**. Optical micrograph of the reaction array fabricated on Latex 2 after exposure to **e)** *S. aureus* (18 h), **f)** *S. aureus* for 5 min followed by incubation in TSB for 18 hours, and **g)** *S. aureus* (18 h) followed by exposure to K2 (24 h). Diameters of the reaction areas are 2 mm in d, and 5 mm in others. The scale bar is 1 mm.

N/m) and NT-MDT, models NSG10 (resonance frequency: 240 kHz, spring constant: 11.8 N/m) and NSG30 (resonance frequency: 240 kHz, spring constant: 40 N/m)). The images were recorded in the repulsive regime using a damping ratio of 0.6–0.7 and a scan speed of 0.2–0.4 Hz. The AFM images were processed and analyzed with the Scanning Probe Image Processor software (SPIP™, Image Metrology, Denmark) and the following roughness parameters were calculated for describing the surfaces: root mean square (RMS) roughness (S_q), root mean square gradient (S_{dq}) and the surface area ratio (S_{dr}). S_q expresses the standard deviation of the height values within the image, S_{dq} the root mean square average of the local surface slopes within the sampling area and S_{dr} the increment of the interfacial surface area relative to the area of the projected (flat) x-y plane (Stout et al. 1993). Also the height difference between the highest peak and the lowest valley (S_z or Z-range) is reported.

Contact angle and surface energy determination

A CAM 200 contact angle goniometer (KSV Instruments Ltd) was used for the measurements of static contact angles of water, diiodomethane (DIM) and ethylene glycol (EG) on the different substrates in ambient conditions (RH = 15 ± 5%, RT = 24 ± 1°C). Small droplets (1–2 μL) were gently deposited on the samples and the contact angles were recorded as a function of time using the software supplied with the instrument. Apparent contact angle values θ_a (average of three measurements) were obtained at a point of time where the drop diameter, contact angle and volume had stabilized to a constant level. The Owens-Wendt method, where the total surface energy (γ_s) is determined as the sum of the polar (γ_s^P) and dispersive (γ_s^d) components was used for the surface energy determination (Owens and Wendt 1969). The θ_a values (apparent values) were corrected for roughness by the Wenzel's equation $\cos \theta_a = r \cos\theta_r$, where θ_r is the contact angle for a smooth surface, and r is a roughness factor (Wenzel 1936). The r value was calculated by using the roughness parameter S_{dr} (Table 1); $r = 1 + S_{dr}/100$ (Peltonen et al. 2004). The surface energy values were determined by using the θ_r values. The values of surface tension components suggested by van Oss et al. were used for the probe liquids (van Oss et al. 1987, 1988).

Biofilm studies

Biofilm formation

Staphylococcus aureus (ATCC 25923) was used as a model for biofilm-forming bacteria in all the experiments. Typically, biofilms are prepared *in vitro* using 96-well microplates made of polystyrene (Nunclon Δ surface, Nunc, Roskilde, Denmark). Therefore, this plastic surface was used as a reference substrate to compare the biofilm formation with those observed for the paper substrates. Bacterial cultivation and biofilm forming experiments on 96-well microplates were conducted as previously described (Sandberg et al. 2008). Briefly, bacteria were cultured in 30 g/L Tryptic Soy Broth (TSB) (Fluka Biochemika, Switzerland) under aerobic conditions at 37°C for 4 hours and 200 rpm to reach exponential growth, up to a concentration of 10^8 CFU/mL. To promote biofilm formation, a bacterial suspension (10^6 CFU/mL, 200 μL/well, in TSB) was added into 96-well microplates and they were incubated for 18 h under aerobic conditions (37°C, 200 rpm) as described earlier (Sandberg et al. 2008). Paper-based substrates were used in a pattern of 4 well diameters (5 mm each). Biofilm formation in the paper substrates was monitored using a modified microtiter well plate assay. The substrates were placed in 6-well microplates, into which suspensions of exponentially grown bacteria (10^6 CFU/mL, 5 mL/well, in TSB) were added and biofilms were similarly allowed to be formed at 37°C, 200 rpm for 18 h. The total substrate area where biofilms could be formed in the printed substrates was calculated to be 0.784 cm^2. At the end of the incubation period, the suspensions were removed from the plates and the substrates were transferred to sterile 6-well plates for visualization with AFM. Parallel samples were rinsed in sterile water (by immersion for a few seconds) and biofilms were scraped off the substrates in 100 μL TSB using sterile plastic sticks and rinsed with additional 100 μL of TSB. To disperse the bacterial aggregates, samples were immersed in a high power ultrasonic bath (Bandelin Sonorex Digitec) using an in-house built-in device that allowed them to be in full contact with water without touching the bottom surface of the sonicator. The sonication time was kept short (5 min, 35 kHz) and performed at RT. The disaggregated biofilms were serially diluted, spread onto Tryptic Soy Agar (TSA) plates and incubated at 37°C overnight. The morphology of the resulting bacterial colonies was inspected and confirmed to be *S. aureus*. The inclusion of paper controls that were inoculated only with media, allowed the exclusion of potential contamination.

Viable bacterial cell densities were quantified and expressed as CFU/cm^2. The differences between the area distributions where biofilms were formed in the printed substrates and the microtiter well plates were taken into account, since in the wells biofilms were also formed on the sides of the wells, not only on the bottom.

Table 1 Determined surface roughness parameters for the studied substrates

Surface	S_z [nm]	S_q [nm]	S_{dq} [1/nm]	S_{dr} [%]
Latex 1	314.7	9.0	0.18	1.6
Latex 2	218.6	13.0	0.47	10.0
Kaolin	431.5	49.6	0.74	25.0
PCC	584.8	75.6	0.88	42.0
PS (96-well)	34.8	4.8	0.07	0.3
PDMS	15.5	1.5	0.04	0.1

Initial adhesion studies

To study the impact of the initial adhesion on the substrates, two of them (Latex 2 and PCC) were exposed to *S. aureus* suspensions, as described in *Biofilm formation* section. Instead of 18 h, the suspensions were added to the substrates only for 5 minutes to allow only the occurrence of initial attachment. After that time, suspensions were removed and fresh TSB was added, and the samples were incubated for 18 hours. AFM images of the samples were performed right after finishing this procedure.

Biofilm maintenance on the substrates

Comparative studies between sample Latex 2 and the 96-well polystyrene microplates were performed. Biofilms were formed as described earlier in *Biofilm formation* section. At the end of the incubation period (18 h), TSB was removed and samples were stored without culture media at +4°C for 1 or 2 weeks. At the end of the storage period, the presence of biofilms was confirmed by AFM. Also, the density of viable biofilm bacteria was measured by scraping bacteria off from the surface.

Effect of an anti-biofilm compound

After the 18 h biofilm formation period, 20 μL of (+)-dehydroabietic acid (coded K2, at a concentration of 400 μM) was added on top of the biofilms formed on the Latex 2 surface. Samples were then kept for 24 hours under aerobic conditions (37°C, 200 rpm). AFM images were taken at the end of the incubation period with K2.

Results

Characterization of the substrates

Figure 2 shows typical AFM topographical images (10 μm × 10 μm) of the studied substrates. Figure 3 depicts line profiles over the respective substrates showing their distinct morphologies with a gradually increasing surface peakedness and roughness. The determined roughness parameters for each substrate are listed in Table 1. The smoothest surface, PDMS, appeared featureless with clearly the smallest S_z, S_q, S_{dq} and S_{dr} values. The polystyrene 96-microwell was found to have the next smoothest surface, however, showing a surface texture generated by the molding process. The heat-treated latex-coated surfaces (Latex 1, Figure 2a and Latex 2, Figure 2b) consisted of very flat top areas with grooves and recesses abruptly interrupting the otherwise rather smooth surface. This resulted in a quite structured column-like surface morphology, especially in case of Latex 2. The Kaolin and PCC pigment-based coating surfaces consisted of randomly oriented pigment particles with their characteristic morphologies (platy-like Kaolin and rod-like PCC). The roughness values show that these surfaces were clearly the roughest.

Table 2 presents the roughness corrected contact angles (θ_r) as well as the polar (γ_s^p) and dispersive (γ_s^d) surface energy components and the total surface energy (γ_s) values for each substrate. Kaolin and Latex 2 coated samples were found to have the smallest water contact angles and highest polarities. On the contrary, the contact angles of the probe liquids on the PDMS surface are all over 90°, demonstrating the hydrophobic and non-wetting nature of the material.

Characterization of the biofilm growth

Figures 1a and b show two typical digital photographs of the paper-based arrays fabricated by printing a translucent PDMS pattern on a pigment coated paper and on a latex coated paper. Figure 1c shows a schematic illustration of the reaction array platform. As an example, Figure 1d shows an optical micrograph of the reaction areas (diameter 2 mm) on Latex 2 sample before exposure to the bacterial culture. Figures 1e and f show that culturing changed the optical contrast inside the reaction area from light to dark. When the Latex 2 sample that was first exposed to *S. aureus* for 18 h was treated with an anti-biofilm compound, the contrast changed back to light (Figure 1g). For Latex 1 and PCC similar clear contrast change was not observed (Additional file 1: Figure S2). AFM measurements were carried out to study in more detail the bacterial attachment and biofilm growth on the arrays fabricated on the different substrates (Figure 4).

Bacterial exposure to the different substrates (18 h) and biofilm formation

The arrays fabricated on the different substrates were exposed to *S. aureus* for 18 h. The total concentration of biofilm in the reaction areas was quantified by scraping the attached cells off and plating them in TSA (Table 3). Latex 2 and Kaolin surfaces were the most prone to the formation of *S. aureus* biofilm with the highest recovered counts (CFU/cm^2) of viable biofilms. The theoretical CFU/cm^2 value for a monolayer of hexagonally close-packed *S. aureus* cells is 1.14×10^8 assuming that the width of a single cell is 1 μm (Additional file 1: Figure S3). On Kaolin and Latex 2 samples the CFU/cm^2 value was higher than 1.14×10^8 indicating at least partial multilayer coverage. The multilayer formation is further confirmed by comparing the height of an individual cell (~560 nm, Additional file 1: Figure S3) to the S_z value (1168 nm) of the AFM image (Figure 4b). The AFM topographs show that adsorbed bacterial cells formed a relatively close-packed and fully covering biofilm on both samples (Figure 4b and c). This confirms that on Latex 2 the dark area in the optical micrograph corresponded to the bacterial biofilm and it supports the suitability of applying optical micrographic technique for visualizing biofilm growth on this type of non-conventional substrates.

Compared to Latex 2 and Kaolin samples, less biofilm was formed on the other studied substrates (Figures 4a, d-f). This is in accordance with the lower CFU/cm^2 values (Table 3). On Latex 1 sample the optical micrographs (Additional file 1: Figure S2) show unevenly distributed patches with a darker contrast. The AFM measurements confirmed that the dark areas corresponded to a biofilm (Figure 4a i) whereas the light areas contained no biofilm (Figure 4a ii). On the other hand, the surface of PCC sample contained both areas where no biofilm had formed (Figure 4f) and areas where loosely packed *S. aureus* capsules with a height of 400-700 nm partially covered the pigment coating (Additional file 1: Figure S3). Finally, very sporadically distributed bacterial cells were observed on the printed PDMS surfaces (Figure 4f).

Short bacterial exposure to the different substrates (5 mins) and further biofilm formation

Differences in the initial attachment of bacteria to the substrates were studied by using PCC (the substrate less prone to biofilm formation) and Latex 2 (one of the substrates more prone to biofilm formation). Figures 1f and 5a show that rapid bacterial attachment took place on

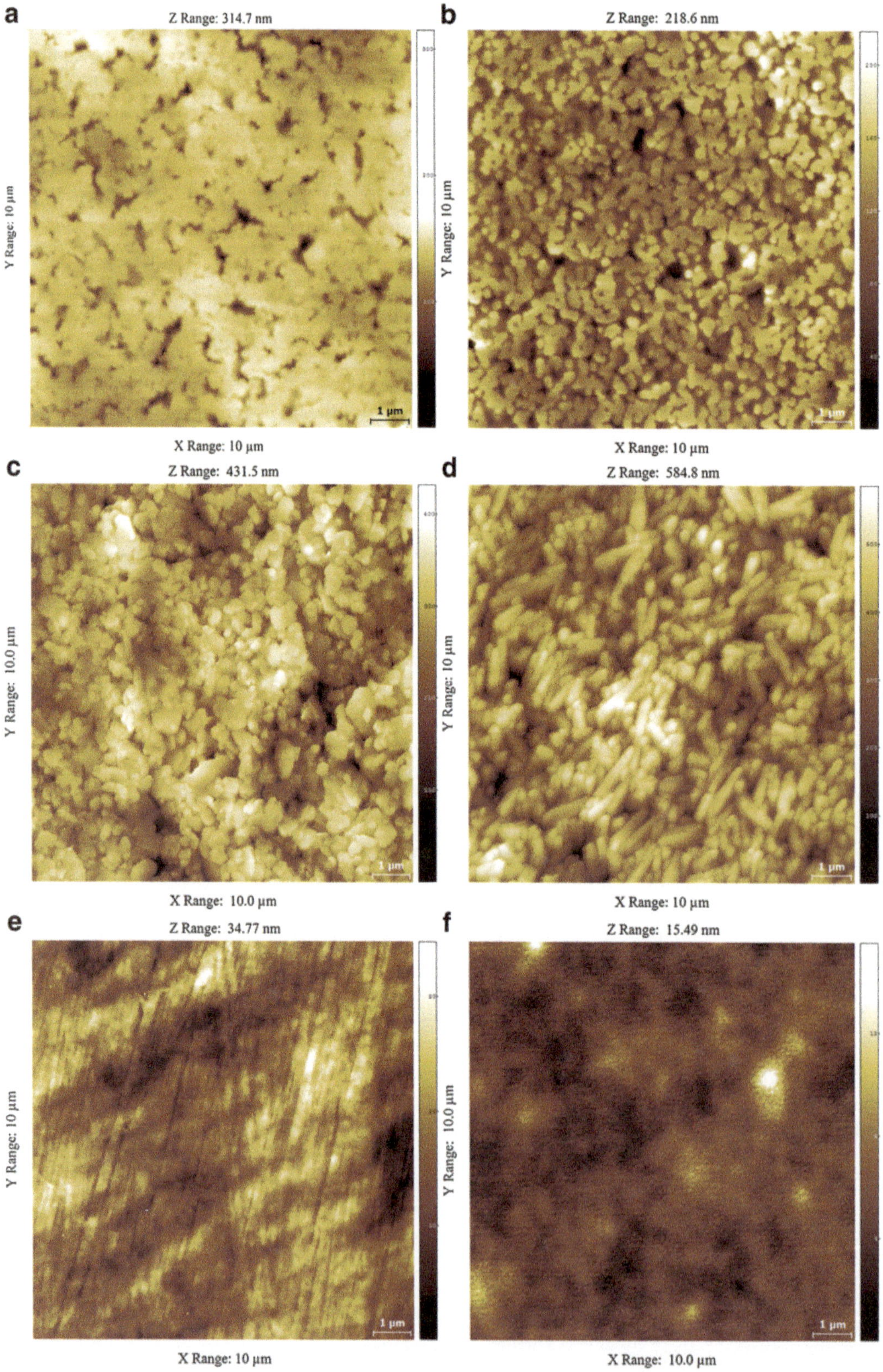

Figure 2 AFM characterization of the substrates. Topographical images (10 µm × 10 µm) of the paper substrates with coating **a)** Latex 1, **b)** Latex 2, **c)** Kaolin and **d)** PCC. Also included are topographs of the **e)** polystyrene 96-well microplate and **f)** the printed PDMS surface. The scale bar in each image is 1 µm.

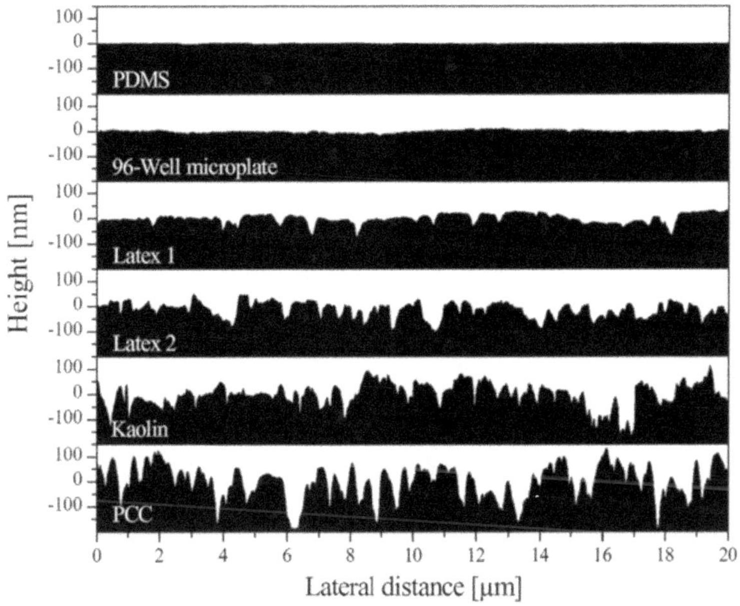

Figure 3 AFM line profiles of the studied substrates. Representative height profiles of the studied substrates showing the gradually increasing roughness.

Latex 2, and these initially attached cells were enough to allow biofilm formation during 18 hours. On the contrary to Latex 2, no indication of bacterial attachment was observed on the PCC sample as the rod-like pigment particles were still clearly visible after exposure to *S. aureus* solution (Figure 5b).

Maintenance of biofilms on a selected substrate

The stability of the *S. aureus* biofilm formed on Latex 2 sample was followed by storing freshly prepared biofilms for 1 and 2 weeks at +4°C. The influence of storage time on the density of the viable biofilm cells is shown in Table 3. The evolution of topography of the biofilm during storage was followed by AFM. Additional file 1: Figure S4 shows AFM topographs after 0, 1 and 2 weeks storage. Corresponding height profiles are shown in Additional file 1: Figure S5.

The effect of an anti-biofilm compound on the biofilm

The effect of the application of (+)-dehydroabietic acid (K2) on the pre-formed *S. aureus* biofilm (Figure 6a) was studied on the Latex 2 sample, as this substrate was found particularly prone to biofilm formation. From Figure 1g we already saw that the influence of K2 was significant as the dark optical contrast characteristic of the biofilm had nearly totally disappeared, indicating biofilm desorption. Only a narrow dark-appearing stripe was left at the perimeter of the reaction area. Detailed AFM characterization (Figure 6b) confirmed that the dark stripe corresponded to the remaining biofilm (the left hand side of Figure 6b). The average thickness of the biofilm (< 500 nm, see the line profile in Figure 6b) was somewhat reduced from the thickness of a fully functional biofilm and approximately 150 nm deep and 400-600 nm wide craters had been formed.

Discussion

Influence of substrate properties on biofilm growth

Roughness and topography

Previous studies have shown that initial bacterial attachment is directly dependent on the surface roughness of the substrate as increasing roughness usually leads to an increase in surface area accessible to bacteria (Kawai et al. 2000; Carlén et al. 2001). The PCC sample has clearly the highest S_q and S_{dr} values amongst the studied substrates (Table 1), and purely from the roughness perspective would be expected to have the highest initial bacterial attachment. However, the CFU counts and AFM measurements revealed that the *S. aureus* cells adhered better on the Latex 2 sample compared to the rougher PCC sample even after only 5 minutes of exposure.

While high roughness has been said to be beneficial for the initial adherence of bacteria (Kawai et al. 2000; Carlén et al. 2001), the size and shape of the grooves, scratches, depressions and other topographical features are also important. Bacteria have been shown to preferentially adhere to irregularities that conform to their size due to the increased bacteria-surface contact area (Katsikogianni and Missirlis 2004). Protruding topographical features on the other hand have been shown to decrease the biofilm

Table 2 Wetting and surface energy characterization

Surface	θ_r [°]			Surface energy [mN/m]		
	Water	DIM	EG	γ_s^p	γ_s^d	γ_s
Latex 1	84 ± 1	52 ± 1	73 ± 1	3.4	27.8	31.2
Latex 2	74 ± 3	47 ± 2	59 ± 2	6.6	31.3	37.9
Kaolin	64 ± 2	50 ± 1	48 ± 1	12.2	29.9	42.1
PCC	92 ± 3	52 ± 1	52 ± 2	1.4	34.7	36.1
PS (96-well)	80 ± 1[b]	37 ± 2[b]	53 ± 2[b]	3.2	38.2	41.4
PDMS[a]	114 ± 1[b]	92 ± 2[b]	96 ± 1[b]	0.6	11.7	12.3

[a]Not corrected for roughness, [b](Juvonen et al. 2013).

growth kinetics and enhance the aggregation of bacteria. Chung et al. for instance found out that an engineered PDMS surface with microtopographic features hindered the *S. aureus* biofilm formation compared to smooth PDMS surfaces (Chung et al. 2007). The protruded topographical features act as physical barriers to inhibit the expansion of small clusters of bacteria residing in surface cavities, physically disrupting further colonization and subsequent biofilm formation (Chung et al. 2007). Previously, the biofilm growth of *S. aureus* has been shown to be independent on surface morphology when the protruding height features were more than one order of

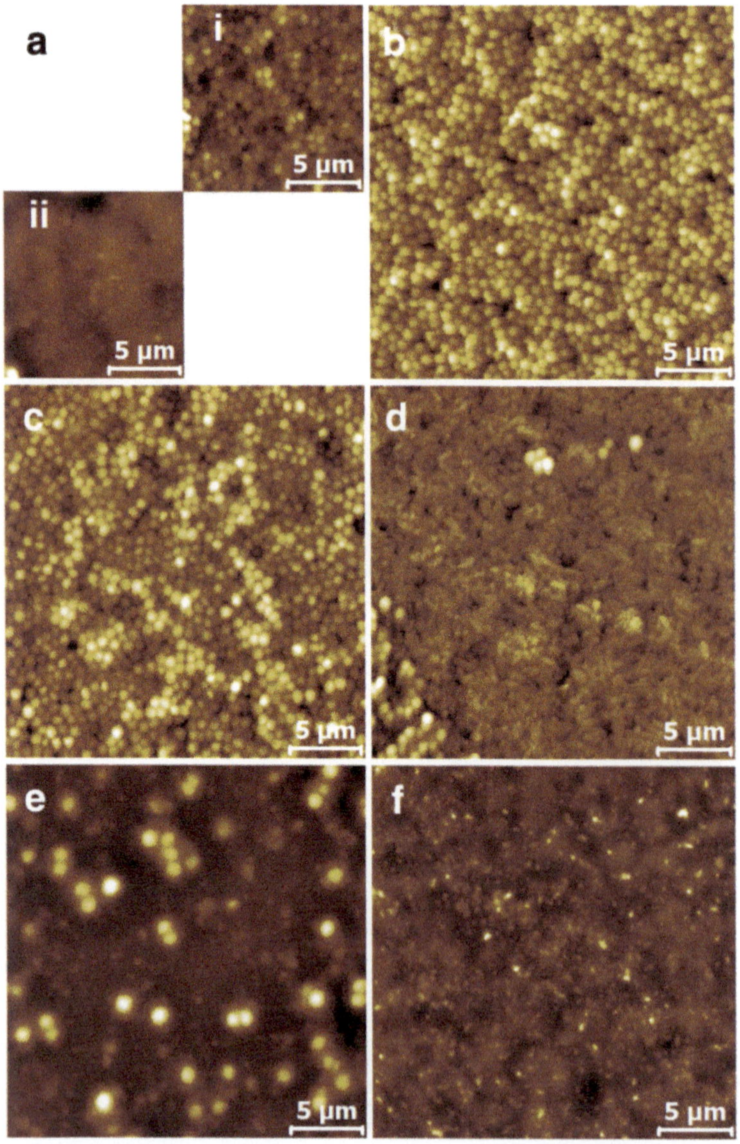

Figure 4 AFM characterization of the substrates exposed to *S. aureus* for 18 h. Topographical images (25 µm × 25 µm) of **a)** Latex 1, **b)** Latex 2, **c)** Kaolin, **d)** PCC, **e)** PS, and **f)** PDMS exposed to *S. aureus* for 18 h. Z-ranges are: **a i)** 768 nm, **ii)** 469 nm, **b)** 1168 nm, **c)** 749 nm, **d)** 907 nm, **e)** 780 nm and **f)** 158 nm. The scale bar in each image is 5 µm.

Table 3 Quantification of the bacterial biofilms attached to the substrates

Surface	Recovered biofilms		
	Right after the 18 h incubation period [CFU/cm^2]	After 1 week storage in + 4°C [CFU/cm^2]	After 2 weeks storage in + 4°C [CFU/cm^2]
Latex 1	1.8×10^7		
Latex 2	2.5×10^8	4.3×10^7	4.6×10^6
Kaolin	3.1×10^8		
PCC	1.4×10^7		
PS (96-well)	5.0×10^7	3.5×10^7	3.6×10^6
PDMS	1.0×10^6		

magnitude lower than the size of the bacterial cells (Litzler et al. 2007). Comparing the size of a bacterial capsule (height ~560 nm and width ~1 μm, Additional file 1: Figure S3) and the S_z values of the substrates, the morphology of PDMS and PS (96-well plate) substrates should not hinder the biofilm growth. The S_z values of the other substrates are considerably higher. Despite this, areas covered by a biofilm were seen on Latex 2 and Kaolin substrates, small local biofilm islands even on Latex 1. The reason might be that the spatial distance (spacing) between the protruding features is small compared to the size of the *S. aureus* cells. Therefore, from the perspective of the *S. aureus* cell capsules e.g. the Kaolin surface might appear not that much different from the Latex 2 surface. The PCC sample was the only coated paper substrate where the bacteria never formed a continuous fully covering biofilm. Since the surface energy, polarity and CFU's of the PCC and Latex 1 were very similar, the main remaining factor differentiating these two samples was roughness, i.e. the protruding topographical features obviously must play a role on the biofilm formation.

Surface energy

The initial adhesion of *S. aureus* to polymer substrates has been shown to depend on both the surface energy of the substrate and the surface tension of the suspending liquid medium (Absolom et al. 1983). When the surface tension of the liquid medium was near that of water (72.8 mN/m) the adhesion increased linearly with decreasing surface energy of the substrate (Absolom et al. 1983). However, when the surface tension of the suspending liquid was lower than that of *S. aureus* (~ 69 mN/m) (Absolom et al. 1983), the dependence was the opposite. With equal surface tension values no correlation was observed. Considering that the growth medium used here (TSB) had a lower surface tension (~44 mN/m) (Keller 1998) compared to *S. aureus*, it is expected that the substrates with the highest surface energies, i.e. PS (96-well), Kaolin and Latex 2 yield the largest initial bacterial adhesion. The results of the short term exposure test carried out for Latex 2 and PCC substrates were in agreement with this. However, no clear correlation between the total surface energy and CFU could be established (Tables 2 and 3). This indicates that the *S. aureus* biofilm growth was apparently independent on total surface energy at the conditions used in this work. The only significant observation is that clearly the lowest biofilm growth was seen on PDMS which has the lowest surface energy. The PDMS film features low surface energy due to the flexible polymer backbone of the polydimethylsiloxane and vinyl group terminated silicone polymers that readily adapt the lowest surface energy configuration (Duel and Owen 1983; Candries et al. 2001; Juvonen et al. 2013). PDMS elastomers have also gained attention in foul release systems on ship hulls (Brady 2000), as suitable polymer matrixes for incorporation of antimicrobials and thereby as materials that are highly resistant to microbial colonization (Pudleiner et al. 2006). It has been previously shown

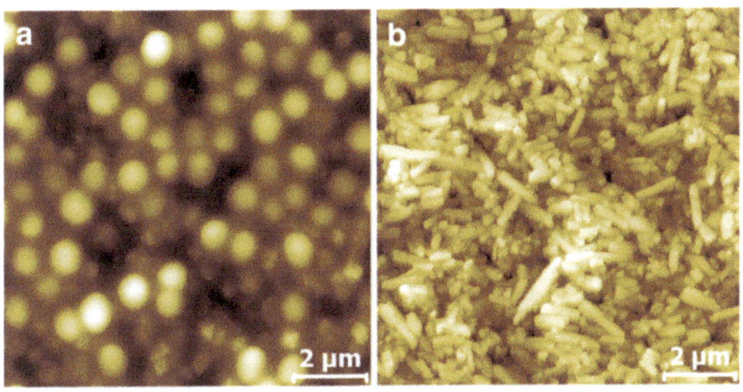

Figure 5 AFM characterization of Latex 2 and PCC substrates exposed to *S. aureus* for 5 minutes. Topographical images (10 μm × 10 μm) of **a)** Latex 2 and **b)** PCC sample exposed to *S. aureus* for 5 min followed by exposure to TSB for 18 h. The Z ranges are **a)** 799 nm and **b)** 1004 nm and scale bars 2 μm.

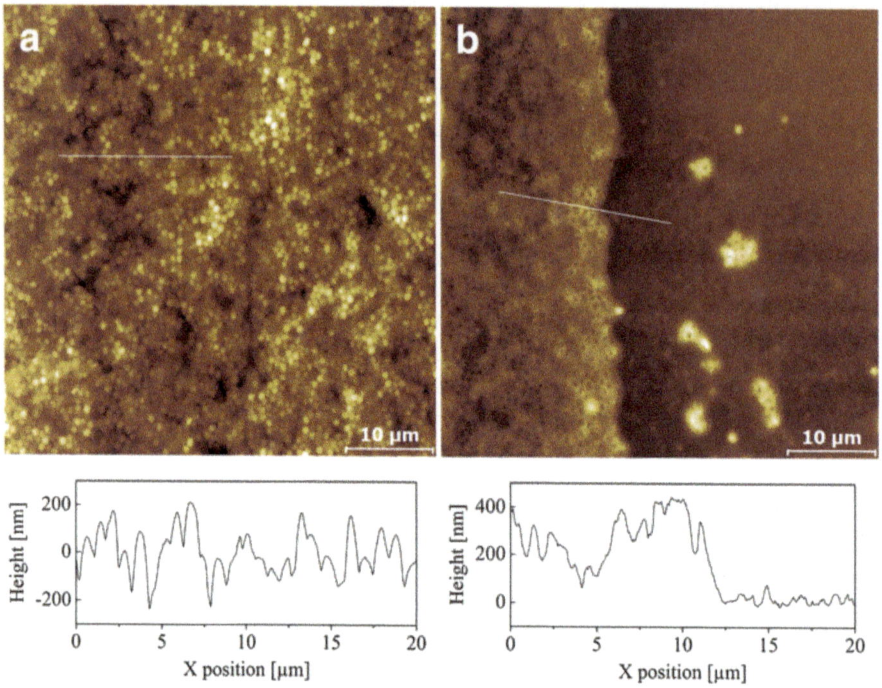

Figure 6 Influence of the antibiofilm compound on the biofilm. AFM topographical images (50 µm × 50 µm) of the **a)** untreated *S. aureus* biofilm on Latex 2 and **b)** preformed biofilm on Latex 2 treated with (+)-dehydroabietic acid (K2). The line profiles are shown below the corresponding topographs.

that biofilm growth of *S. aureus* was independent on the surface energy when the values were between 31 and 41 mN/m (Litzler et al. 2007). Here, the surface energies of the studies substrates (excluding PDMS) fall approximately between this range and hence might explain the poor correlation between the total surface energy and the CFU value. On the other hand, the CFU value shows a much clearer tendency to increase with increasing polarity (polar component value, γ_s^P) of the surface (Figure 7).

It has been observed that contact angle of water, which is closely related to the polarity of the substrate, relates better to bacterial film growth in aqueous systems compared to the total surface energy values (Fletcher and Pringle 1985). The amount of bacteria attached to the surface has been shown to increase with decreasing contact angle within the range from 110 to ca. 70-80 (Fletcher and Pringle 1985). Here, a similar trend was observed. While hydrophilic interactions increase with

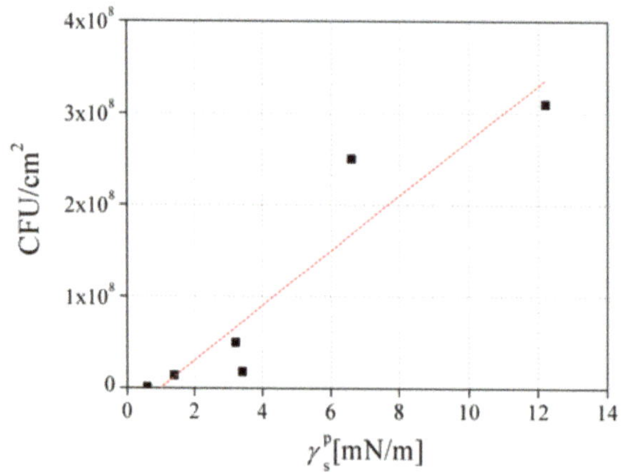

Figure 7 The trend between the CFU/cm^2 and polarity. Bacterial concentration values plotted as a function of the polar surface energy component. The line is not a theoretical fit but has been included in order to demonstrate the trend.

increasing polarity, similar dependence is shown here between the polarity and the viable colony counts.

Biofilm formation on paper-based arrays

S. aureus biofilms were more prone to form on the polar substrates as was clearly seen for Latex 2. Although the substrates used in this work were paper-based materials, it is important to note that they are furnished with a barrier layer which quite effectively prevents the absorption of water and inhibits the soaking effect commonly observed with the conventional paper grades, such as copy paper or filter paper. Soaking could have been one potential explanation for the higher bacterial attachment, but it is not the case with these substrates. In addition, the pigment coated papers have very small average pore size (50-100 nm) and thus the individual bacterial cells are not expected to fit into the pores and absorb deep into the coating. Finally, the hydrophobic PDMS further prevents the water soaking effect.

After a very short period of time (5 mins) *S. aureus* became effectively attached to Latex 2 and the cells present were enough to allow biofilm formation during 18 hours (as indicated by AFM results). The long-term stability and possible destruction of the biofilms on Latex 2 was studied by following the amount of biofilm during storage. The amount of bacteria on the surface dropped by approximately one order of magnitude per week during sample storage, being still on a level of 10^7 after one week storage (Table 3). This amount of bacteria is similar to that observed for the 96-well PS microplates which is still enough for conducting anti-biofilm studies. Similar decreasing CFU/cm^2 values as a function of time have been observed also on other substrates, e.g. polymer films with *S. aureus* and other bacteria (Aviv et al. 2007). To offer a proof-of-concept for the suitability of this substrate for anti-biofilm studies, Latex 2 was exposed to (+)-dehydroabietic acid, a molecule that has been recently shown to display high potency and efficacy against *S. aureus* biofilms, *in vitro* (Fallarero et al. 2013). It has been previously demonstrated (by fluorescence microscopy) that this compound significantly kills biofilm bacteria attached to 96-well microplates, causing more than 4-log reduction of the viable biofilm density, at 400 μM upon 24 hours exposure. The results obtained in this study using AFM imaging are in agreement with these previous findings. One advantage of using the planar paper-based samples is the fact that sample preparation and AFM analysis becomes more straightforward compared to the 96-well plate bottoms. Other future improvements and potential advantages related to the use of the planar and flexible paper-based platforms include e.g. the possibility to print potential antimicrobial agents on the wells prior to the biofilm formation experiments. Such experiments are currently underway. This concept is preliminary demonstrated in Additional file 1: Figure S6 which shows AFM images of *S. aureus* biofilms on Latex 2 pre-treated with printed penicillin.

Additional file

Additional file 1: Table S1. The main components of the paper coatings and the reference samples. **Figure S1.** A digital photograph showing the flexographic printing plates. **Figure S2.** High magnification optical micrographs of the dry and empty a) Latex 1 and b) PCC substrates before exposure to bacterial suspension and c, d) after the 18 h biofilm formation period. **Figure S3.** AFM topographical image (20 μm × 10 μm) of the PCC + *S. aureus* sample and corresponding height profiles. **Figure S4.** AFM topographical images (20 μm × 20 μm) of the formed biofilms stored at 4°C for a) 0, b) 1 and c) 2 weeks. **Figure S5.** AFM height profiles of the *S. aureus* biofilms stored for 0, 1 and 2 weeks. **Figure S6.** Preliminary antibiotic printing trials were carried out with two penicillin concentrations: 0.1 μM (0.0334 μg/mL) and 1 μM (0.334 μg/mL). The AFM topographs (20 μm × 20 μm) and the corresponding line profiles show that a much denser and thicker bacterial biofilm formed on the Latex 2 sample that was pre-treated with the lower penicillin concentration (a) compared to the higher concentration (b).

Competing interests
The authors declare that they have no competing interests.

Authors' contributions
AM participated in the design and coordination of the study, carried out AFM and CA measurements and contributed in the drafting of the manuscript. JK performed bacterial adhesion experiments. AF, PI, PV and JP participated in the design and coordination of the study and contributed in the drafting of the manuscript. All authors read and approved the final manuscript.

Acknowledgements
Technical contributions of Malena Skogman and Suvi Manner are greatly appreciated. Natalja Genina and Niklas Sandler are acknowledged for printing of penicillin. Laboratory of Paper Coating and Converting, Åbo Akademi University is acknowledged for preparing the pigment coated paper substrates. Funding from the Academy of Finland through the National Center of Excellence programme (FunMat CoE) is acknowledged (A.M., P. I., J. P.). Funding from Drug Discovery and Chemical Biology (Biocenter Finland) and Academy of Finland (decisions 128870, 264064 and 272266) are acknowledged (A.F., P.V.).

Author details
[1]Laboratory of Physical Chemistry, Center of Excellence for Functional Materials, Abo Akademi University, Turku, Finland. [2]Pharmaceutical Sciences, Department of Biosciences, Abo Akademi University, Turku, Finland. [3]Division of Pharmaceutical Biology, Faculty of Pharmacy, University of Helsinki, Helsinki, Finland.

References
Absolom DR, Lamberti FV, Policova Z, Zingg W, van Oss CJ, Neumann AW (1983) Surface thermodynamics of bacterial adhesion. Appl Environ Microbiol 46:90–97
Anderson GG, O'Toole GA (2008) Innate and induced resistance mechanisms of bacterial biofilms. Curr Top Microbiol Immunol 322:85–105
Ausbacher D, Fallarero A, Kujala J, Määttänen A, Peltonen J, Strøm MB, Vuorela PM (2014) *Staphylococcus aureus* biofilm susceptibility to small and potent $\beta^{2,2}$-amino acid derivatives. Biofouling 30:81–93
Aviv M, Berdicevsky I, Zilberman M (2007) Gentamicin-loaded bioresorbable films for prevention of bacterial infections associated with orthopedic implants. J Biomed Mater Res A 83:10–19
Blackledge MS, Worthington RJ, Melander C (2013) Biologically inspired strategies for combating bacterial biofilms. Curr Opin Pharmacol 13:699–706

Bollström R, Määttänen A, Tobjörk D, Ihalainen P, Kaihovirta N, Österbacka R, Peltonen J, Toivakka M (2009) A multilayer coated fiber-based substrate suitable for printed functionality. Org Electron 10:1020–1023

Bollström R, Määttänen A, Ihalainen P, Peltonen J, Toivakka M (2010) Method for creating a substrate for printed or coated functionality, substrate, functional device and its use. PCT/FI2010/050056, WO 2010/086511.

Brady RF Jr (1999) Properties which influence marine fouling resistance in polymers containing silicon and fluorine. Prog Org Coat 35:31–35

Brady RF (2000) Clean hulls without poisons: devising and testing nontoxic marine coatings. J Coat Technol 72:45–56

Candries M, Anderson CD, Atlar M (2001) Foul release systems and drag. Proceedings of the PCE 2001 Conference, Antwerp, pp 273–286

Carlén A, Nikdel K, Wennerberg A, Holmberg K, Olsson J (2001) Surface characteristics and in vitro biofilm formation on glass ionomer and composite resin. Biomaterials 22:481–487

Chung KK, Schumacher JF, Sampson EM, Burne RA, Antonelli PJ, Brennan AB (2007) Impact of engineered surface microtopography on biofilm formation of *Staphylococcus aureus*. Biointerphases 2:89–94

Donlan RM, Costerton JW (2002) Biofilms: survival mechanisms of clinically relevant microorganisms. Clin Microbiol Rev 15:167–193

Duel LA, Owen MJ (1983) ESCA studies of silicone release coatings. J Adhes 16:49–59

Elliott GR, Peterson PK, Verbrugh HA, Freiberg MR, Hoidal JR, Quie PG (1982) Influence of subinhibitory concentrations of penicillin, cephalothin, and clindamycin on *staphylococcus aureus* growth in human phagocytic cells. Antimicrob Agents Chemother 22:781–784

Fallarero A, Skogman M, Kujala J, Rajaratnam M, Moreira VM, Yli-Kauhaluoma J, Vuorela P (2013) (+)-Dehydroabietic acid, an abietane-type diterpene, inhibits *Staphylococcus aureus* biofilms in vitro. Int J Mol Sci 14:12054–12072

Fletcher M, Pringle JH (1985) The effect of surface free energy and medium surface tension on bacterial attachment to solid surfaces. J Colloid Interface Sci 104:5–14

Fluckiger U, Ulrich M, Steinhuber A, Döring G, Mack D, Landmann R, Goerke C, Wolz C (2005) Biofilm formation, icaADBC transcription, and polysaccharide intercellular adhesin synthesis by staphylococci in a device-related infection model. Infect Immun 73:1811–1819

Fux CA, Costerton JW, Stewart PS, Stoodley P (2005) Survival strategies of infectious biofilms. Trends Microbiol 13:34–40

Higashi J, Sullam P (2006) *Staphylococcus aureus* biofilms. In: Pace J, Rupp M, Finch R (eds) Biofilms, Infection, and Antimicrobial Therapy. CRC Press, Florida, pp 81–108

Ihalainen P, Backfolk K, Sirviö P, Peltonen J (2007) Thermal analysis and topographical characterization of latex films by scanning probe microscopy. J Appl Phys 101:043505

Ihalainen P, Backfolk K, Sirviö P, Peltonen J (2010) Topographical, chemical, thermal and electrostatic properties of latex films. Colloids Surf A Physicochem Eng Asp 354:320–330

Ihalainen P, Määttänen A, Järnström J, Tobjörk D, Österbacka R, Peltonen J (2012) Influence of surface properties of coated papers on printed electronics. Ind Eng Chem Res 51:6025–6036

Jabra-Rizk MA, Meiller TF, James CE, Shirtliff ME (2006) Effect of farnesol on *Staphylococcus aureus* biofilm formation and antimicrobial susceptibility. Antimicrob Agents Ch 50:1463–1469

Juvonen H, Määttänen A, Laurén IP, Urtti A, Yliperttula M, Peltonen J (2013) Biocompatibility of printed paper-based arrays for 2-D cell cultures. Acta Biomater 9:6704–6710

Katsikogianni M, Missirlis YF (2004) Concise review of mechanisms of bacterial adhesion to biomaterials and of techniques used in estimating bacteria-material interactions. Eur Cell Mater 8:37–57

Kawai K, Urano M, Ebisu S (2000) Effect of Surface Roughness of porcelain on adhesion of bacteria and their synthesizing glucans. J Prosthet Dent 83:664–667

Keller SW (1998) Determination of the Leak Size Critical to Package Sterility Maintenance. Virginia Polytechnic Institute and State University, Blacksburg, Virginia

Landini P, Antoniani D, Burgess J, Nijland R (2010) Molecular mechanisms of compounds affecting bacterial biofilm formation and dispersal. Appl Microbiol Biot 86:813–823

Li M, Tian J, Al-Tamimi M, Shen W (2012) Paper-based blood typing device that reports patient's blood type 'in writing'. Angew Chem Int Ed 51:5497–5501

Lindsay D, von Holy A (2006) Bacterial biofilms within the clinical setting: what healthcare professionals should know. J Hosp Infect 64:313–325

Litzler P-Y, Benard L, Barbier-Frebourg N, Vilain S, Jouenne T, Beucher E, Bunel C, Lemeland J-F, Bessou J-P (2007) Biofilm formation on pyrolytic carbon heart valves: influence of surface free energy, roughness, and bacterial species. J Thorac Cardiovasc Surg 134:1025–1032

Määttänen A, Ihalainen P, Bollström R, Toivakka M, Peltonen J (2010) Wetting and print quality study of an inkjet-printed poly(3-hexylthiophene) on pigment coated papers. Colloids Surf A 367:76–84

Määttänen A, Fors D, Wang S, Valtakari D, Ihalainen P, Peltonen J (2011) Paper-based planar reaction arrays for frinted diagnostics. Sens Actuators B 160:1404–1412

McDougald D, Rice SA, Barraud N, Steinberg PD, Kjelleberg S (2011) Should we stay or should we go: mechanisms and ecological consequences for biofilm dispersal. Nat Rev Microbiol 10:39–50

Mosier AP, Cady NC (2011) Analysis of bacterial surface interactions using microfluidic systems. Sci Prog 94:431–450

Nill P, Kern DP, Goehring N, Peschel A (2011) Influence of surface patterning on bacterial growth behavior. J Vac Sci Technol B Microelectron Nanometer Struct 29:06FA03

Owens DK, Wendt RC (1969) Estimation of the surface free energy of polymers. J Appl Polym Sci 13:1741–1747

Peltonen J, Järn M, Areva S, Linden M, Rosenholm JB (2004) Topographical parameters for specifying a three-dimensional surface. Langmuir 20:9428–9431

Proctor RA, Balwit JM, Vesga O (1994) Variant subpopulations of *Staphylococcus aureus* as cause of persistent and recurrent infections. Infect Agents Dis 3:302–312

Pudleiner H, Dujardin R, Albers R (2006) Active ingredient-containing silicone elastomers. US 11/718,587: PCT/EP2005/011365

Rosenhahn A, Finlay JA, Pettit ME, Ward A, Wirges W, Gerhard R, Callow ME, Grunze M, Callow JA (2009) Zeta potential of motile spores of the green alga Ulva linza and the influence of electrostatic interactions on spore settlement and adhesion strength. Biointerphases 4:7–11

Sandberg M, Määttänen A, Peltonen J, Vuorela PM, Fallarero A (2008) Automating a 96-well microtitre plate model for *Staphylococcus aureus* biofilms: an approach to screening of natural antimicrobial compounds. Int J Antimicrob Agents 32:233–240

Schuler B, Baumstark R, Kirsch S, Pfau A, Sandor M, Zosel A (2000) Structure and properties of multiphase particles and their impact on the performance of architectural coatings. Prog Org Coat 40:139–150

Siegel AC, Phillips ST, Wiley BJ, Whitesides GM (2009) Thin, lightweight, foldable thermochromic displays on paper. Lab Chip 9:2775–2781

Singh AV, Vyas V, Salve TS, Cortelli D, Dellasega D, Podestà A, Milani P, Gade WN (2012) Biofilm formation on nanostructured titanium oxide surfaces and a micro/nanofabrication-based preventive strategy using colloidal lithography. Biofabrication 4:025001

Stout KJ, Sullivan PJ, Dong WP, Mainsah E, Luo N, Mathia T, Zahouani H (1993) The development of methods for the characterization of roughness on three dimensions. Vol. Publication no. EUR 15178 EN of the Commission of the European Communities. Scanning Probe Image Processor, SPIP, ©Image Metrology A/S, Luxembourg

Tian J, Shen W (2011) Printing enzymatic reactions. Chem Commun 47:1583–1585

Van Oss CJ, Chaudhury MK, Good RJ (1987) Monopolar surfaces. Adv Colloid Interface Sci 28:35–64

Van Oss CJ, Chaudhury MK, Good RJ (1988) Interfacial Lifshitz-van Der Waals and polar interactions in macroscopic systems. Chem Rev 88:927–941

Wang J, Yiu B, Obermeyer J, Filipe CDM, Brennan JD, Pelton R (2012) Effects of temperature and relative humidity on the stability of paper-immobilized antibodies. Biomacromolecules 13:599–564

Wenzel R (1936) Resistance of solid surfaces to wetting by water. Industr Eng Chem 28:988–994

Worthington RJ, Richards JJ, Melander C (2012) Small molecule control of bacterial biofilms. Org Biomol Chem 10:7457–7474

Zmantar T, Bettaieb F, Chaieb K, Ezzili B, Mora-Ponsonnet L, Othmane A, Jaffrézic N, Bakhrouf A (2011) Atomic force microscopy and hydrodynamic characterization of the adhesion of *Staphylococcus aureus* to hydrophilic and hydrophobic substrata at different pH values. World J Microb Biot 27:887–896

Deletion of glucose oxidase changes the pattern of organic acid production in *Aspergillus carbonarius*

Lei Yang, Mette Lübeck and Peter S Lübeck*

Abstract

Aspergillus carbonarius has potential as a cell factory for the production of different organic acids. At pH 5.5, *A. carbonarius* accumulates high amounts of gluconic acid when it grows on glucose based medium whereas at low pH, it produces citric acid. The conversion of glucose to gluconic acid is carried out by secretion of the enzyme, glucose oxidase. In this work, the gene encoding glucose oxidase was identified and deleted from *A. carbonarius* with the aim of changing the carbon flux towards other organic acids. The effect of genetic engineering was examined by testing glucose oxidase deficient (Δgox) mutants for the production of different organic acids in a defined production medium. The results obtained showed that the gluconic acid accumulation was completely inhibited and increased amounts of citric acid, oxalic acid and malic acid were observed in the Δgox mutants.

Keywords: *Aspergillus carbonarius*; Citric acid; Glucose oxidase; Gluconic acid; Malic acid

Introduction

With depletion of crude oil and increased environmental concern, biologically based processes for producing organic acids that can be used as building blocks in the chemical industries have begun to raise attention in recent years (Holladay et al. 2007). It is widely accepted that a suitable industrial strain for organic acid production is one of the key factors affecting the feasibility of the entire production process. Filamentous fungi have been well studied as cell factories for different production of organic acids for decades because they exhibit excellent abilities to utilize a variety of carbon sources and naturally accumulate high amount of specific organic acids under stressed conditions, like citric acid production by *Aspergillus niger* and malic acid production by *Aspergillus flavus* (Battat et al. 1991; Papagianni et al. 1999). Although the mechanisms of organic acid accumulation by different fungi have not been fully understood, many successful attempts have been made to improve organic acid production in filamentous fungi by using genetic modification and production optimization. In this study, *Aspergillus carbonarius*, which has a close phylogenetic relationship to *Aspergillus niger* (Thom and Currie 1916) was selected to exploit its potential as a new cell factory for production of different organic acids. It resembles *A. niger* in many features, including morphology and high capacity of producing citric acid (Ghareib 1987; Joosten et al. 2001; Weyda et al. 2014), which creates the possibility to apply the knowledge accumulated for *A. niger* directly to metabolic engineering of *A. carbonarius* for production of organic acids.

In organic acid production using filamentous fungi, pH is an important parameter. It has been reported that low pH is necessary for obtaining high yields of citric acid using *A. niger*, and reversely pH around 5 ~ 6 is preferred for producing malic acid using *A. flavus* and fumaric acid using *Rhizopus oryzae* (Peleg et al. 1988a; Xu et al. 2012). The production pattern of organic acids can also be changed dramatically in filamentous fungi by adjusting pH during cultivation. In *A. niger*, production of citric acid was dramatically suppressed at near neutral pH as gluconic acid started accumulating in high amount (Bercovitz et al. 1990; Goldberg et al. 2006). In this study we investigated the effect of deleting the glucose oxidase in *Aspergillus carbonarius* for the purpose of organic acid production at pH 5–6. At this pH range, the fungus accumulates high amounts of gluconic acid, presumably due to secretion of glucose oxidase, whereby the fungus quickly converts glucose into gluconic acid outside the cell thus preventing further metabolism of glucose (Mischak et al. 1985). The aim of this study was

* Correspondence: psl@bio.aau.dk
Section for Sustainable Biotechnology, Aalborg University Copenhagen, A. C. Meyers Vænge 15, DK-2450 Copenhagen, SV, Denmark

therefore to eliminate the gluconic acid production in order to increase the carbon flux towards other organic acids. The hypothesis was: by eliminating gluconic acid production, *A. carbonarius* would increase 1,4-dicarboxylic acid production. Deletion of the gene encoding the glucose oxidase (*gox*) was conducted in *A. carbonarius* to suppress the conversion of glucose to gluconic acid. However, low frequency of homologous recombination in filamentous fungi often leads to very limited gene targeting efficiency in fungal transformation. Therefore, a Ku complex deficient strain of *A. carbonarius* (Gallo et al. 2012), which is supposed to dramatically increase the homologous recombination frequency due to the inactivation of Ku complex, was selected for this work.

Materials and methods

Strains and culture medium

A Ku deficient strain, KB1039 (*ΔkusA*, obtained from K. Bruno, PNNL, US), which is also uracil auxotrophic (*ΔpyrG)*, of the wild type *A. carbonarius* 5010 (ATCC® MYA-4641™) (Gallo et al. 2012), was used as the parental strain to construct the *Δgox* mutants. Both the *A. carbonarius* KB1039 (*ΔkusA*) and the wild type 5010 was cultured at PDA (potato dextrose agar) medium at 30°C and for the Ku deficient strain, the medium was supplemented with uracil and uridine, each at the final concentration of 2 mM.

Identification of the glucose oxidase gene in *A. carbonarius*

Due to lack of information about glucose oxidase (GOX) genes in *A. carbonarius*, the sequence of the *gox* gene in *A. niger* (accession no. X16061.1) was selected to identify the orthologous *gox* gene in *A. carbonarius* based on the close phylogenetic relationship between *A. niger* and *A. carbonarius.* One sequence with high identity was identified in the alignment hit as a putative *gox* gene in *A. carbonarius.* The sequence containing extension of 1000 bp from both 5' and 3' flanking regions of the *gox* encoding gene sequence was also identified with the purpose of using these 2 kb sequence flanking regions surrounding the putative *gox* gene for deletion of the gene. The sequence was submitted to GenBank with the accession no. KF741791.

Plasmid construction for gene knock-out

The genomic DNA was isolated from wild type *A. carbonarius* 5010 by using phenol-chloroform extraction (Andreou 2013) and used as template to amplify 5' and 3' flanking regions of the *gox* gene with primers containing uracil overhang (Table 1). Since the gene knock-out was carried out in a Ku deficient strain, it was not necessary to have very long fragments to increase the frequency of homologous recombination. It is reported that fragments about 1 kb both upstream and downstream to a target gene could provide efficient homologous recombination in *A. niger* (Meyer et al. 2007). Therefore, the size of fragments amplified from *A. carbonarius* was between 900 and 1000 bp. The PCR reaction was set up in 50 μL reaction volume: 5 μL 10X *pfu* turbox buffer; 1 μL 10μM dNTP; 2.5 μL 10 uM forward and reverse primer; 1 μL pfu turbo cx polymerase (Agilent); appropriate amount of DNA template and water added up to 50 μL. The PCR program was as follows: Initial denaturing step at 95°C for 3 min; 25–30 cycles of denaturing step at 94°C for 30s; annealing step at 55-65°C for 30s; elongation step at 72°C for specific amount of time calculated by the size of desired fragments (1 min/kb); final elongation step at 72°C for 5 minutes. The *pfu* turbocx polymerase was only used to amplify the DNA fragment with primers with uracil overhang in simpleUSER cloning (Hansen et al. 2014), whereas all other PCR reactions were carried out using RUN (taq) polymerase (A&A biotechnology). The plasmid pSB414 was designed and constructed for simpleUSER cloning (Hansen et al. 2014) and contains the following genetic elements: *gpdA* promoter, *trpC* terminator and *pyrG* gene, including a specific cassette facilitating simpleUSER cloning. The cassette was activated by the restriction enzyme *Pac*I and the nicking enzyme *Nb.BbvCI* to generate the complementary overhang to the target fragments. The target fragments were cloned into the plasmid through self-assembly followed by transformation of *E. coli* with the plasmid for further propagation using standard procedures.

Protoplast transformation

Protoplasts of *A. carbonarius* were made from young mycelium harvested after overnight growth in YPD medium. The cell walls were degraded by 60 mg/ml of the commercial product Vino Taste Pro (Novozymes A/S) in protoplasting buffer (1.2 M $MgSO_4$, 50 mM Phosphate Buffer, pH 5.0) for approx. 4 hours. Protoplasts were filtered and purified from the mixture, suspended in STC buffer (1.0 M sorbitol, 50 mM Tris, 50 mM $CaCl_2$ pH 8.0) and counted with appropriate dilution folds in a haemacytometer. The final concentration of protoplasts for aliquots was adjusted to 2×10^7/mL. Protoplast transformation was carried out by adding 5 μg plasmids in 100 μl protoplast suspension and incubated on ice for 15 minutes followed by incubation for 15 minutes at room temperature after adding 1 mL of 40% PEG. The mixture was transferred into 10 mL minimum medium (Gallo et al. 2012) with 1 M sorbitol at 30°C for 1 hour with agitation at 80 rpm in the incubator shaker (KS 4000 I control, IKA). Then the cells were concentrated by centrifugation for 5 minutes at 800 × g, re-suspended in minimal medium containing 1 M sorbitol and 0.8% agar and poured into petri-dishes. Next

Table 1 Primers used in this research

Name	Sequence (5′ → 3′)	Annotation
Gox upU Fw	*GGGTTTAAU*TCTCCTTGTGCTGACCAACCG	USER cloning of gene *gox* upstream
Gox upU Rv	*GGACTTAAU*GTTTACCAATCCCGCCGCGTC	USER cloning of gene *gox* upstream
Gox downU Fw	*GGCATTAAU*AGGTGAGATGGAGTTGTTG	USER cloning of gene *gox* downstream
Gox downU Rv	*GGTCTTAAU*TTGGGATGGGTAGGGTATT	USER cloning of gene *gox* downstream
Gox Fw1	GCCCTGCCACACTACATCCG	Amplify internal sequence of gene *gox*
Gox Rv1	TCGCCACAGCCGAGATCCTT	Amplify internal sequence of gene *gox*
Gox Fw2	GCTGCCAATCCTTCGGTCCA	Amplify internal sequence of gene *gox*
Gox Rv2	TAGTCGCCAAAGGTCTCGTT	Amplify internal sequence of gene *gox*
Gox Fw3	AACAACCTCACCCACCAGAG	Amplify the sequence containing gene *gox*
Gox Rv3	ACCATTGAAGTGGCAGGAAC	Amplify the sequence containing gene *gox*

day, a second layer of the same medium was poured on the top. The plates with potential transformants were incubated at 30°C for at least 3 days until transformants appeared.

Sporulating transformants were inoculated by streaking out on PDA medium and incubated at 30°C overnight. Single colonies were identified and picked out to verify the deletion of the target gene by extracting the genomic DNA from the transformants and amplifying the fragments with expected size in PCR. The transformants were further transferred to normal PDA plates and preserved for further steps.

Growth conditions

Spores of fungal transformants were harvested from PDA plates after 5–7 days of cultivation at 30°C, and collected through sterilized Miracloth (EMD Millipore, USA) in sterile 0.05 M phosphate buffer pH 6.8 in a 15 mL falcon tube. The spores were counted in a haemacytometer and then inoculated into 50 mL falcon tubes containing 10 mL pre-culture medium (3.6 g/L yeast extract and 10 g/L peptone). The final concentration of spores in the pre-culture medium was approximately 1×10^5/mL. Pre-cultivation was carried out at 30°C with agitation of 250 rpm for 2 days. Pellets formed in the pre-culture medium were then transferred into production medium by filtering the pre-culture medium through Miracloth, and all the pellets on the top were collected and transferred into the production medium, which was modified from the production medium C described by Peleg et al. 1988a, b: Glucose, 100 (g/L); $(NH_4)_2SO_4$, 2 (g/L); KH_2PO_4, 0.15 (g/L); K_2HPO_4, 0.15 (g/L); $MgSO_4$ $7H_2O$, 0.1 (g/L); $CaCl_2$ $2H_2O$, 0.1 (g/L); NaCl, 0.005 (g/L); $FeSO_4$ $7H_2O$, 0.005 (g/L), 0.1 g/L $ZnSO_4$ and $CaCO_3$, 60 (g/L) (Peleg et al. 1988a). Cultivation was carried out in 100 mL flasks containing 20 mL production medium at 30°C with agitation of 180 rpm. The cultivation time varied from 7 to 10 days. The pre-culture and acid production was carried out in triplicates and pH was kept at 5.5 for the entire procedure.

Analysis of extracellular metabolites

Samples taken from organic acid fermentation were acidified with 72% sulfuric acid to a final concentration of 5% in order to precipitate the calcium ion in form of calcium sulfate and exchange the organic acids back to liquid phase. The acidified samples were incubated at 80°C for at least 15 minutes to complete the reaction. After incubation, pH of the samples should be lower than 2 and was checked by pH indicator paper. The acidified samples were then centrifuged at 10,000 rpm for 1 minute and the supernatant was used for HPLC analysis. The analysis for sugar and organic acids were carried out in Aminex 87H column (Biorad) at 60°C by using HPLC mobile phase at a flow rate of 0.6 mL/minute. The HPLC samples were kept at 4°C in the machine during the analysis process and then stored at −20°C. The measurements of L-malic acid and D- gluconic acid in the samples were carried out respectively with L-malate (L-malic acid) kit and D-gluconate kit as described by the manufacturer (Megazyme).

Results

Protoplast transformation and deletion of the *gox* gene from *A. carbonarius*

Protoplast transformation was carried out with circular pSB414*gox* plasmids. Two *Δgox* transformants were obtained and checked for the deletion of the putative *gox* gene by 3 different pairs of primers (Table 1). Two pairs of primers (Gox Fw1-Rv1 and Fw2-Rv2) amplified the internal sequence of the putative *gox* gene, and the third pair of primers (Gox Fw3-Rv3) was designed to amplify the putative *gox* gene together with the flanking regions in order to check the replacement of the *gox* gene (Figure 1a). As shown in Figure 1b, no fragments could be amplified by the two pairs of internal primers in PCR, indicating that the *gox* gene was deleted. In PCR with external primers, a fragment at approx. 6.2 kb was amplified from both of the transformants, which indicated that the putative *gox* gene had been successfully replaced

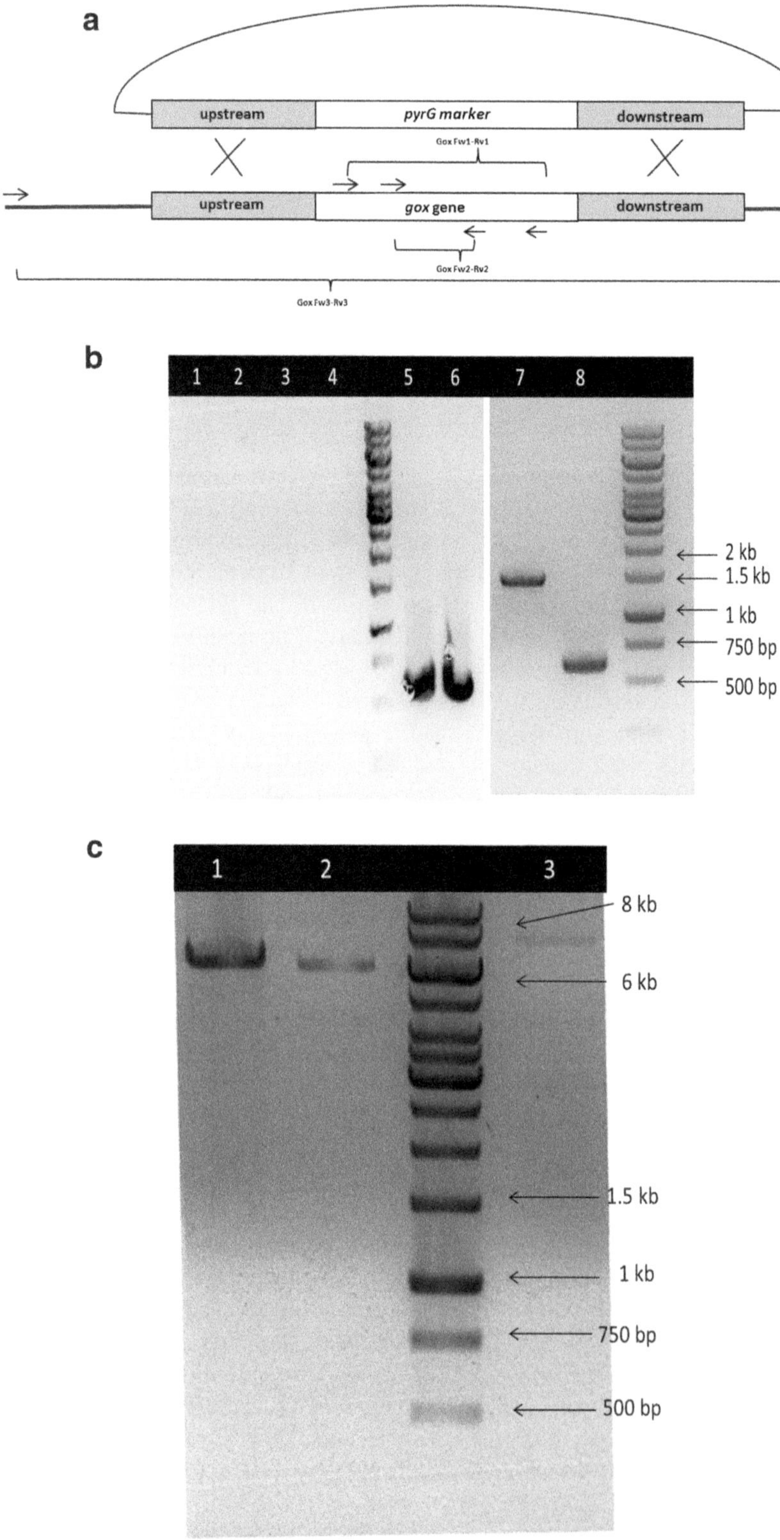

Figure 1 Verification of deletion of the *gox* gene in transformant 4 and 5 (a) Disruption of *gox* gene and primer binding sites (b) Amplification of internal sequence with primer Gox Fw1-Rv1 and Fw2-Rv2. Lane 1–2, Δgox transformant 4; lane 3–4, Δgox transformant 5; lane 5–6, ITS sequence amplified from Δgox transformant 4 and 5; lane 7–8, internal fragments of the gox gene from the wildtype strain (~1,5 kb and 0,6 kb) **(c)** Amplification of the *gox* gene containing region with the external primers Gox Fw3 and Rv3. Lane 1–2, amplified fragments from gox transformants 4 and 5. (~6.2 kb). Lane 3, amplified fragment from wild type (~7.7 kb).

by the marker gene, since the length of the original sequence containing the *gox* gene was approx.7.7 kb (Figure 1c). A PCR amplifying the ITS region (Bulat et al. 2000) confirmed the quality of the genomic DNA (Figure 1b). The size of all the amplified DNA fragments were estimated by comparing with 1 kb DNA ladder (Figure 1b).

Effect of deletion of the *gox* gene on gluconic acid production by *A. carbonarius*

The *gox* gene deleted from *A. carbonarius* is supposed to play a key role in the formation of gluconic acid. In order to evaluate the effect of this genetic modification on gluconic acid production, growth experiments were carried out with the *Δgox* mutants for 7 days at pH 5.5 by employing the *A. carbonarius* 5010 and KB1039 (*ΔkusA*) as control. As shown in Figure 2, the KB1039 (*ΔkusA*) and wild type strain (5010) produced high amount of gluconic acid (72 g/L and 53 g/L respectively) in the production medium after 7 days, and both of the *Δgox* transformants only produced 0.1 g/L gluconic acid by the end of the growth experiment. However, in order to confirm that the low gluconic acid production was not caused by different growth rates of the *Δgox* mutants, the biomass of the *Δgox* mutants was determined after the growth experiment. As shown in Figure 3, the *Δgox* mutants produced similar amount of biomass as the parent strain, which indicated that they grew equally well under the same conditions. The result confirmed that deletion of the *gox* gene can effectively inhibit the formation of gluconic acid by *A. carbonarius*.

Analysis of extracellular metabolites of the Δ*gox* mutants in the acid production media

Cultivation with the selected *Δgox* mutants was carried out in shaking flasks at pH 5.5 to investigate the effect of the *gox* gene deletion on production of organic acids in *A. carbonarius*. However, due to the accumulation and consumption of gluconic acid by wild type *A. carbonarius*, it is difficult to use glucose as the sole carbon source to compare the performance of wild-type strain with the *Δgox* mutant during growth. The evaluation of organic acid production was carried out based on the concentration of extracellular acid products in the cultivation broth. As shown in Figure 4a, the concentration of citric acid was dramatically increased after 7 days cultivation in the *Δgox* mutants compared with the wild type 5010 and the parent KB1039 (*ΔkusA*) strains. In addition, an accumulation of oxalic acid was also observed in the *Δgox* mutants during the cultivation (Figure 4b) and the production of malic acid also increased 2.4 and 1.8 folds, respectively, compared with the parent strain and the wild type strain (Figure 4c).

Discussion

In the present work, a glucose oxidase gene (*gox*) in *A. carbonarius* involved in gluconic acid production was identified and deleted. In order to achieve a high gene targeting efficiency to facilitate the process of deleting the *gox* gene, a Ku deficient strain KB1039 (*ΔkusA*) was selected to construct *Δgox* transformants. The Ku complex including Ku70 and Ku80 has been reported to play an essential role in the non-homologous end joining pathway (Dudášová et al. 2004). Deletion of the Ku

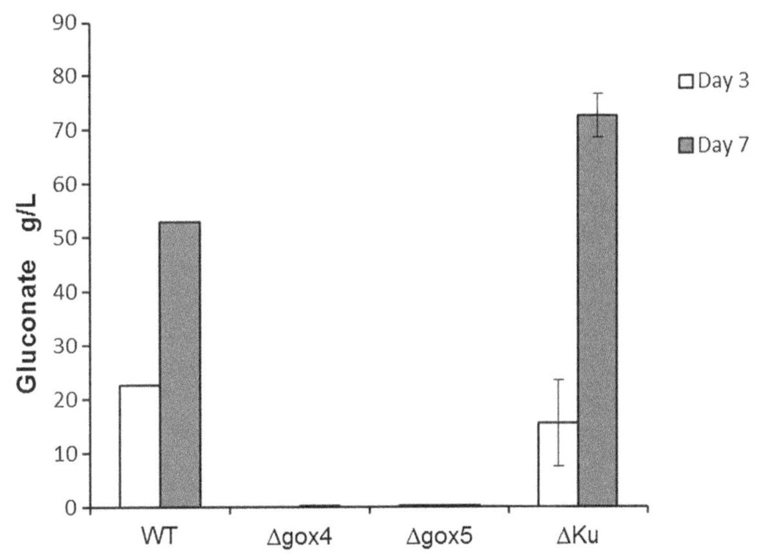

Figure 2 The results of fermentation test on Δ*gox* mutants in malic acid production medium. Only one sample from wild type strain was measured for gluconic acid production, whereas all others were tested in triplicates.

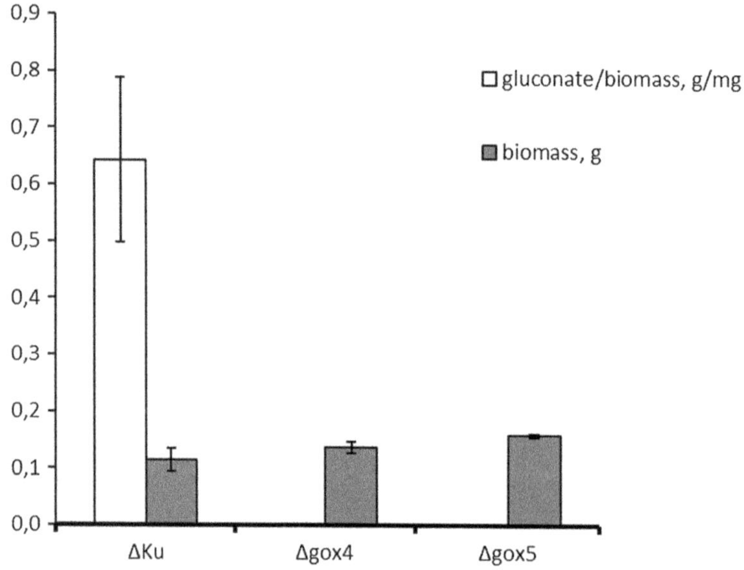

Figure 3 The yield of gluconic acid production based on the biomass growth. The yield was calculated based on the results from the samples taken on day 7.

encoding gene can inactivate the non-homologous recombination mechanism and increase the frequency of homologous recombination (Kooistra et al. 2004; Meyer et al. 2007; Ninomiya et al. 2004). In this study it was shown that the transformed DNA solely was integrated via homologous recombination in all the obtained transformants which proved that Ku deficient strain could be used as an efficient tool in gene targeting and study of gene function in *A. carbonarius.* By deleting the *gox* gene in a *Ku* deficient strain, it was also shown that the production of gluconic acid was dramatically reduced in *Δgox* mutants. Evaluation of growth experiments after 7 days of cultivation showed that the mutants produced increased amounts of malic acid and citric acid as well as an accumulation of oxalic acid (Figure 4).

Since the aim of this work was to investigate the effect of deleting glucose oxidase in *A.carbonarius* on the production of organic acids at a pH where the glucose oxidase is expressed, the cultivation was carried out at pH 5.5. Under this condition, *A. carbonarius* was capable of producing high amount of gluconic acid in media containing high concentration of glucose, resulting in a fast depletion of glucose, and it was shown that the elimination of glucose oxidase had an impact on production of other organic acids. Sugar concentration is considered as an important factor in organic acid production for filamentous fungi and has been used e.g. for citric acid production by *A. niger,* malic acid by *A. flavus,* and fumaric acid and lactic acid by *R. oryzae* (Alvarez-Vasquez et al. 2000; Peleg et al. 1988b; Ding et al. 2011). The effect of high sugar concentration on organic acid production has been investigated in *A. niger* for citric acid production. These studies indicate that high sugar concentration seems to repress the α-keto-glutarate dehydrogenase in *A. niger* and increases the intracellular concentration of fructose-2,6-bisphosphate which leads to high yield of citric acid production by activating phosphofructokinase (PFK1) (Kubicek-Pranz et al. 1990; Papagianni 2007). Reversely, the conversion of glucose to gluconic acid will decrease the sugar concentration very fast during cultivation and further eliminate the effect of high sugar concentration. This assumption was supported in this study by comparing the pattern of organic acid production in the *Δgox* mutants with the parental strain and the wild type strain after 3 days of cultivation. In the early phase of acid production, the glucose concentration still remained high although part of the glucose was already converted to gluconic acid. Therefore, the pattern of organic acids was similar among the *Δgox* mutants and the parental and wild type strains. However, when the glucose concentration in the media became much lower for the parental and wild type strains than for the *Δgox* mutants in the later phase of cultivation due to the accumulation of gluconic acid, the pattern of production of organic acid in the parental and wild type strains started varying from the *Δgox* mutants. The *Δgox* mutants produced higher amount of malic acid and citric acid compared with the parental and wild-type strains, and after 7 days of cultivation, production of oxalic acid was also observed in the *Δgox* mutants.

The enhanced production of citric acid and malic acid as well as accumulation of oxalic acid in the *Δgox* mutants must be a result of a higher carbon flux through the Glycolysis. Citric acid is solely produced through the

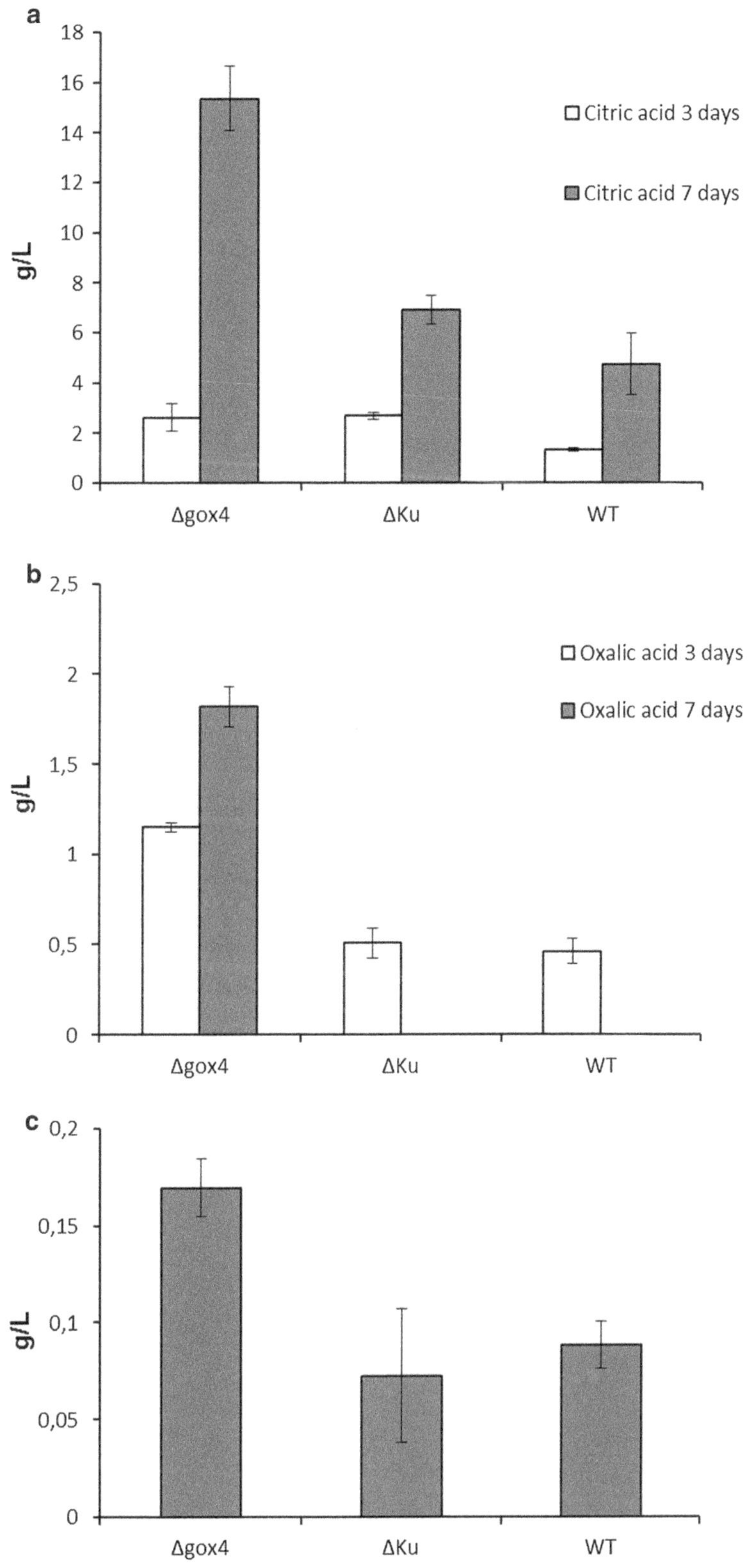

Figure 4 The effect of gene deletion on the production of organic acids by *A. carbonarius*. (a) The concentration of citric acid in the fermentation medium after day 3 and 7 **(b)** The concentration of oxalic acid in the fermentation medium after day 3 and 7 **(c)** The concentration of malic acid in the fermentation medium after day 7.

TCA cycle, and the enhanced production indicates an improvement of the carbon flux into relevant pathways in *A. carbonarius* including glycolysis and TCA cycle. Moreover, the accumulation of oxalic acid might imply an elevated carbon flux towards cytosolic oxaloacetate reduction pathway which converted oxaloacetate to malate in the cytosol. Although, the oxalic acid producing pathway has not been well studied in *A. carbonarius*, it has been shown in *A. niger*, that oxalic acid is mainly produced by hydrolyzing oxaloacetate into oxalate and acetate by the enzyme oxaloacetase located in the cytosol (Ruijter et al. 1999). Due to the close phylogenetic relationship between *A. niger* and *A. carbonarius*, it could be concluded that the accumulation of oxalic acid was most probably attributed to an increase of oxaloacetate in the cytosol, which was also in accordance with the increased production of malic acid during cultivation. Therefore, the high sugar concentration and improved carbon flux towards acid producing pathway might result in the different organic acid pattern in *Δ*gox mutant.

It is very likely that malic acid in *A. carbonarius*is transiently produced but not capable of accumulating at high concentration, especially in the *Δgox* mutant, the enhanced carbon flux towards the cytosolic oxaloacetate reduction pathway was supposed to result in a much higher concentration of malic acid during cultivation, but this was not achieved in our experiments. Malate, as an important intermediate of the TCA cycle, plays multiple roles in cell metabolism. A dramatically increase of intracellular malate concentration might influence the cell function and therefore result in immediately reduction via other pathways. The transiently produced malate may not be exported and accumulated during cultivation since *A. carbonarius* is able to produce high amount of citric acid and the malate instead may be directly used in intracellular anti-port of citrate and malate across the membrane of mitochondria as it is suggested for *A. niger* (de Jongh and Nielsen 2008). Therefore, it may suggest that a simple change in cultivation condition and single genetic modification may not be enough to completely reroute the carbon flux from citric acid production to other organic acids. A series of genetic modifications may be required in the future work on metabolic engineering of *A. carbonarius* for production of valuable organic acids.

Competing interests
The authors declare that they have no competing interests.

Authors' contributions
LY has made substantial contributions to conception and design, or acquisition of data, or analysis and interpretation of data; ML has been involved in discussing the results conducted and drafting the manuscript and revising it critically for important intellectual content; PL has been involved in all matters of the project (discussing the design of the experiments, the results and revision of the paper) and has given final approval of the version to be published. Each author has participated sufficiently in the work to take public responsibility for appropriate portions of the content. All authors read and approved the final manuscript.

Acknowledgements
Financial support from SUPRA-BIO EU grant 241640-2, BIOREF DSF grant 09-065165 and MycoFuelChem DSF grant 11-116803. We thank PhD-student Istvan Weyda for valuable discussion about growth conditions, and Gitte Hinz-Berg for HPLC analysis.

References
Alvarez-Vasquez F, González-Alcón C, Torres NV (2000) Metabolism of citric acid production by *Aspergillus niger*. Model definition, steady-state analysis and constrained optimization of citric acid production rate. Biotechnol Bioeng 70:82–108

Andreou L (2013) Preparation of genomic DNA from bacteria. Methods in Enzymology 529:143–151

Battat E, Peleg Y, Bercovitz A, Rokem JS, Goldberg I (1991) Optimization of L-malic acid production by *Aspergillus flavus* in a stirred fermentor. Biotechnol Bioeng 37:1108–1116

Bercovitz A, Peleg Y, Battat E, Rokem JS, Goldberg I (1990) Localization of pyruvate carboxylase in organic acid-producing *Aspergillus* strains. Appl Environ Microbiol 56:1594–1597

Bulat SA, Lubeck M, Alekhina IA, Jensen DF, Knudsen IMB, Lubeck PS (2000) Identification of a universally primed-PCR-derived sequence-characterized amplified region marker for an antagonistic strain of Clonostachys rosea and development of a strain-specific PCR detection assay. Appl Environ Microbiol 66:4758–4763

de Jongh WA, Nielsen J (2008) Enhanced citrate production through gene insertion in *Aspergillus niger*. Metab Eng 10:87–96

Ding Y, Li S, Dou C, Yu Y, He H (2011) Production of fumaric acid by *Rhizopus oryzae*: Role of carbon-nitrogen ratio. Appl Biochem Biotechnol 164:1461–1467

Dudášová Z, Dudáš A, Chovanec M (2004) Non-homologous end-joining factors of Saccharomyces cerevisiae. FEMS Microbiol Rev 28:581–601

Gallo A, Bruno KS, Solfrizzo M, Perrone G, Mulè G, Visconti A, Baker SE (2012) New insight into the ochratoxin a biosynthetic pathway through deletion of a nonribosomal peptide synthetase gene in *Aspergillus carbonarius*. Appl Environ Microbiol 78:8208–8218

Ghareib M (1987) Assimilation of galacturonate by *Aspergillus carbonarius*. Folia Microbiol (Praha) 32:211–215

Goldberg I, Rokem JS, Pines O (2006) Organic acids: old metabolites, new themes. J Chem Technol Biotechnol 81:1601–1611

Hansen NB, Lübeck M, Lübeck PS (2014) Advancing USER cloning into simpleUSER and nicking cloning. J Microbiol Methods 96:42–49

Holladay J, Bozell J, White J, Johnson D (2007) Top value-added chemicals from biomass. DOE Report PNNL-16983 website: http://chembioprocess.pnl.gov/staff/staff_info.asp

Joosten HMLJ, Goetz J, Pittet A, Schellenberg M, Bucheli P (2001) Production of ochratoxin A by *Aspergillus carbonarius* on coffee cherries. Int J Food Microbiol 65:39–44

Kooistra R, Hooykaas PJJ, Steensma HY (2004) Efficient gene targeting in *Kluyveromyces lactis*. Yeast 21:781–792

Kubicek-Pranz EM, Mozelt M, Rohr M, Kubicek CP (1990) Changes in the concentration of fructose 2,6-bisphosphate in *Aspergillus niger* during stimulation of acidogenesis by elevated sucrose concentration. Biochim Biophys Acta - General Subjects 1033:250–255

Meyer V, Arentshorst M, El-Ghezal A, Drews A, Kooistra R, van den Hondel CAMJJ, Ram AFJ (2007) Highly efficient gene targeting in the Aspergillus niger kusA mutant. J Biotechnol 128:770–775

Mischak H, Kubicek CP, Rohr M (1985) Formation and location of glucose oxidase in citric acid producing mycelia of *Aspergillus niger*. Appl Microbiol Biotechnol 21:27–31

Ninomiya Y, Suzuki K, Ishii C, Inoue H (2004) Highly efficient gene replacements in Neurospora strains deficient for nonhomologous end-joining. Proc Natl Acad Sci U S A 101:12248–12253

Papagianni M (2007) Advances in citric acid fermentation by *Aspergillus niger*: biochemical aspects, membrane transport and modeling. Biotechnol Adv 25:244–263

Papagianni M, Mattey M, Berovĭ M, Kristiansen B (1999) *Aspergillus niger* morphology and citric acid production in submerged batch fermentation: effects of culture pH, phosphate and manganese levels. Food Technol Biotechnol 37:165–171

Peleg Y, Rahamim E, Kessel M, Goldberg I (1988a) Malic acid accumulation by *Aspergillus flavus* - II. Crystals and hair-like processes formed by A. flavus in a l-malic acid production medium. Appl Microbiol Biotechnol 28:76–79

Peleg Y, Stieglitz B, Goldberg I (1988b) Malic acid accumulation by *Aspergillus flavus* - I. Biochemical aspects of acid biosynthesis. Appl Microbiol Biotechnol 28:69–75

Ruijter GJG, Van De Vondervoort PJI, Visser J (1999) Oxalic acid production by *Aspergillus niger*: An oxalate-non-producing mutant produces citric acid at pH 5 and in the presence of manganese. Microbiology 145:2569–2576

Thom C, Currie JN (1916) *Aspergillus niger* group. J Agr Res 7:1–15

Weyda I, Lübeck M, Ahring BK, Lübeck PS (2014) Point mutation of the xylose reductase (XR) gene reduces xylitol accumulation and increases citric acid production in *Aspergillus carbonarius*. J Ind Microbiol Biotechnol 41:733–739

Xu Q, Li S, Huang H, Wen J (2012) Key technologies for the industrial production of fumaric acid by fermentation. Biotechnol Adv 30:1685–1696

6

Contamination of water resources by pathogenic bacteria

Pramod K Pandey[1*], Philip H Kass[1], Michelle L Soupir[2], Sagor Biswas[1] and Vijay P Singh[3]

Abstract

Water-borne pathogen contamination in water resources and related diseases are a major water quality concern throughout the world. Increasing interest in controlling water-borne pathogens in water resources evidenced by a large number of recent publications clearly attests to the need for studies that synthesize knowledge from multiple fields covering comparative aspects of pathogen contamination, and unify them in a single place in order to present and address the problem as a whole. Providing a broader perceptive of pathogen contamination in freshwater (rivers, lakes, reservoirs, groundwater) and saline water (estuaries and coastal waters) resources, this review paper attempts to develop the first comprehensive single source of existing information on pathogen contamination in multiple types of water resources. In addition, a comprehensive discussion describes the challenges associated with using indicator organisms. Potential impacts of water resources development on pathogen contamination as well as challenges that lie ahead for addressing pathogen contamination are also discussed.

Keywords: Pathogens; Contamination; Water resources; Watershed; Pathogens transport

Introduction

Water-borne pathogen contamination in ambient water bodies and related diseases are a major water quality concern throughout the world. Pathogen contamination is a serious issue for almost all types of ambient water bodies, making its recognition and understanding essential (U.S. EPA 2012a). The United Nations identified improving water quality as one of the eight Millennium Development Goals (MDGs). Its target is to reduce the number of people without access to safe water by 50% by 2015 (WHO 2011). Because of the overwhelming scientific evidence for climate change (IPCC 2007), it is also important to understand how perturbations in weather patterns can potentially impact pathogen levels in water resources. To meet future demands of water for food, energy, and ecosystems, increasing water storage structures (i.e., dams) must be a component of long-range planning (World Bank 2010). However, such new structures can potentially degrade water quality and exacerbate public health risk.

* Correspondence: pkpandey@ucdavis.edu
[1]Department of Population Health and Reproduction, University of California, Davis, California, USA
Full list of author information is available at the end of the article

While several review papers are currently available (Bradford et al. 2013; Pachepsky and Shelton 2011; Pang 2009; Jin and Flury 2002; John and Rose 2005; Jamieson et al. 2004; Jamieson et al. 2002; Arnone and Walling 2007; Kay et al. 2007), there is a manifest need for additional transdisciplinary studies that assimilate knowledge gained from multi-research endeavors studying pathogen contamination, and provide a comprehensive synopsis in order to comprehend the entirety of the problem. Therefore, the goal of this review is to present a broad research scope assessment of pathogen contamination of water resources and the associated challenges it presents. We synthesize the potential health risks imposed by pathogens in water resources by providing existing knowledge that covers surface water, groundwater, fresh water, and saline water. Further, the impact of water resources development on pathogen contamination, future challenges, and recommendations are summarized. In addition, we provide a brief discussion describing water-borne pathogen footprints and potential challenges associated with the use of indicator organisms for assessing water quality.

Health risk

Water-borne diseases (i.e., diarrhea, gastrointestinal illness) caused by various bacteria, viruses, and protozoa have been the causes of many outbreaks (Craun et al. 2006). In developing countries, such as those in Africa, water-borne diseases infect millions (Fenwick 2006). According to World Health Organization (WHO), each year 3.4 million people, mostly children, die from water-related diseases (WHO 2014). According to United Nations Children's Fund (UNICEF) assessment, 4000 children die each day as a result of contaminated water (UNICEF 2014). WHO (2010) reports that over 2.6 billion people lack access to clean water, which is responsible for about 2.2 million deaths annually, of which 1.4 million are in children. Improving water quality can reduce the global disease burden by approximately 4% (WHO 2010).

Although water-associated diseases in developing countries are prevalent, they are also a serious challenge in developed countries. A study by Arnone and Walling (2007), who compiled data of outbreaks in the U.S. (1986 – 2000), reported 5,905 cases and 95 outbreaks associated with recreational water. Gastrointestinal Illness (GI) caused by variety of different microbes and germs, which causes symptoms, such as diarrhea, nausea, vomiting, fever, abdominal pain, was responsible for about 29.53% cases. More than 27% of cases were caused by *Shigella spp.* In addition, 10.99%, 10.08%, and 6.59% of the cases were caused by *Cryptosporidium parvum*, Adenovirus 3, and Leptospira, respectively. Nearly 23% and 21% of the outbreaks were caused by GI and *Shigella spp*, respectively. In addition, 16.84%, 12.63%, and 7.37% of the outbreaks were caused by *Naegleria fowleri*, *E. coli* 0157:H7, and *Schistosoma spp.*, respectively. Besides acute gastroenteritis, major etiological agents such as *Giardia*, *Cryptosporadium*, *E. coli* 0157:H7, *V. cholera*, and *Salmonella* were the agents responsible for many outbreaks (Craun et al. 2006). During the same period 437,082 cases and 48 outbreaks were caused by contaminated drinking water, of which about 95.89% of the cases were caused by *Cryptosporidium parvum*. Nearly 42% and 31% of the outbreaks were caused by *Giardia lamblia* and GI, respectively. Reporting statistics on water-borne outbreaks in the U.S., Craun et al. (2006) found that at least 1870 outbreaks (23 per year) occurred between 1920 and 2002. These reported outbreaks and their reported incidence of illnesses are likely to be an underestimation of actual numbers because of non-reported cases and missing exposure information. To protect public health, the U.S. EPA's National Primary Drinking Water Regulations (NPDWRs) contain standards describing the Maximum Contaminant Level (MCL) – the highest level of a contaminate allowable in drinking water. The U.S. EPA has defined the MCL of various microorganisms, such as *Cryptosporidium*, *Giardia lamblia*, *Legionella*, and Total Coliforms (including fecal coliform and *E. coli*), and viruses (U.S. EPA 2012b). The Maximum Contaminant Level Goal (MCLG) – the level of a contaminant in drinking water below which there is no known risk to public health, has also been proposed by the U.S. EPA. The MCLC levels for *Cryptosporidium*, *Giardia lamblia*, *Legionella*, and Total Coliforms are zero. The EPA requires 99% removal of *Cryptosporidium* in drinking water, and the removal percentages of *Giardia lamblia* and viruses are 99.9 and 99.99%, respectively. Although there is no limit for *Legionella*, EPA believes that if *Giardia lamblia* and viruses are removed/inactivated, then drinking water likely to be free of *Legionella*. The U.S. EPA requires routine sampling of drinking water for testing total coliform and *E. coli*, and if a routine sample is positive, then repeat samples are required. If, in any repeat sample, total coliform or *E. coli* is detected then the drinking water has an acute MCL violation. For a drinking water system that collects fewer than 40 routine samples per month, no more than one sample can be total coliform-positive per month. For a system that collects more than 40 routine samples, no more than 5% of samples total coliform-positive in a month is allowed (U.S. EPA 2012b).

Each year approximately 42,000 cases of salmonellosis are reported in the U.S. (CDC 2014). Schistosomiasis is not reported in the U.S. because it is not endemic; however, 200 million people are infected worldwide. In 2011, about 1,060 cases of Guinea worm disease, caused by the parasite *Dracunculus medinensis*, were reported in many remote parts of Africa that do not have safe drinking water. Malaria, a protozoal disease of the Genus Plasmodium transmitted by mosquitos breeding in contaminated water, affects 300–500 million people, and causes over one million deaths each year (more than 90% of deaths in Africa). Overall the morbidity and mortality caused by contaminated water are enormous and need to be controlled by improving the security of safe water (i.e., recreational as well as drinking water) in both developing and developed countries.

Historical perspective of water-borne diseases

Water contamination has a long presence in human history, with descriptions in the Sushruta Samshita about water-borne diseases resembling cholera in an Indian text written in Sanskrit as early as 500–400 B.C. (Colwell 1996). Although cholera infections have not been reported in recent years in developed countries mainly due to improved sanitation, millions of people each year continue to get infected by *Vibrio cholera* in developing countries (Nelson et al. 2009). The World Health Organization reports about 3–5 million cholera cases and 10,000 – 120,000 deaths, mainly in developing countries, due to cholera every year. Over time, cholera has caused millions of deaths in developing as well as developed countries (Colwell 1996; Okun 1996). For instance, a major outbreak of cholera was reported in

London in 1849. Dr. John Snow, a physician to Queen Victoria, showed a relationship between people infected by cholera and contaminated water (Snow 1854; Colwell 1996). Jordan et al. (1904), Ruediger (1911), Simons et al. (1922), and Rudolfs et al. (1950) provide excellent reviews on incidents during the early 19th century. Colwell (1996) reported that in the mid and late 18th century, cholera infected millions of people all over the world. The worst outbreak in recent memory occurred in Haiti following the devastating earthquake affecting the capital and surrounding regions, with almost a half a million cases, killing thousands of people (CDC 2011).

Water-borne pathogen footprints and challenges

Indicator organisms are commonly used to assess the levels of pathogens in water resources; i.e., water-borne pathogen footprints of water resources. Monitoring the levels of indicator organisms (such as fecal coliforms, *E. coli*) (Figure 1) is a common approach for quantifying the potential pathogen loads in ambient water bodies. For decades, public health officials/scientists have evaluated water quality by enumerating fecal coliforms and *E. coli* levels in rivers, lakes, estuaries, and coastal waters (Malakoff 2002; Pandey et al. 2012a; Pandey et al. 2012b; Pandey and Soupir 2013). There is, however, much debate regarding current indicator organisms and their ability to represent the potential presence of pathogenic bacteria. In addition, identifying the source of pathogens (e.g., human waste, animal waste, wildlife excreta, and waterfowl droppings) (Figure 2) is challenging (Malakoff 2002; Dickerson et al. 2007). There is potential to use a relatively new approach such as microbial source tracking (MST) to trace the origin of fecal coliform (Scott et al. 2002; Grave et al. 2007; Dickerson et al. 2007; Ibekwe et al. 2011; Ma et al. 2014). In the past, the MST method was exploited by antibiotic resistance analysis to assess the impact of cattle on water quality on a watershed scale (Grave et al. 2007). The authors suggested that host-origin libraries, based on a phenotypic method, are useful for tracking the pathogen sources. Many MST methods, however, rely on the assumption that some strains of bacteria are found only within a single kind or group of animals. This assumption can be debatable when it comes to the common fecal bacteria *E. coli* (Malakoff 2002). Therefore, caution is needed while using *E. coli* for source tracking (Gordon 2001). Further, the cost to develop libraries, implement extensive sampling programs needed for verifying the MST method, and calculate uncertainties associated with the method are legitimate issues, which requires attention before exploiting the MST method at watershed scale.

Currently, public health officials/scientists rely on exposure limits for assessing pathogen levels in water resources, which have been established to protect human

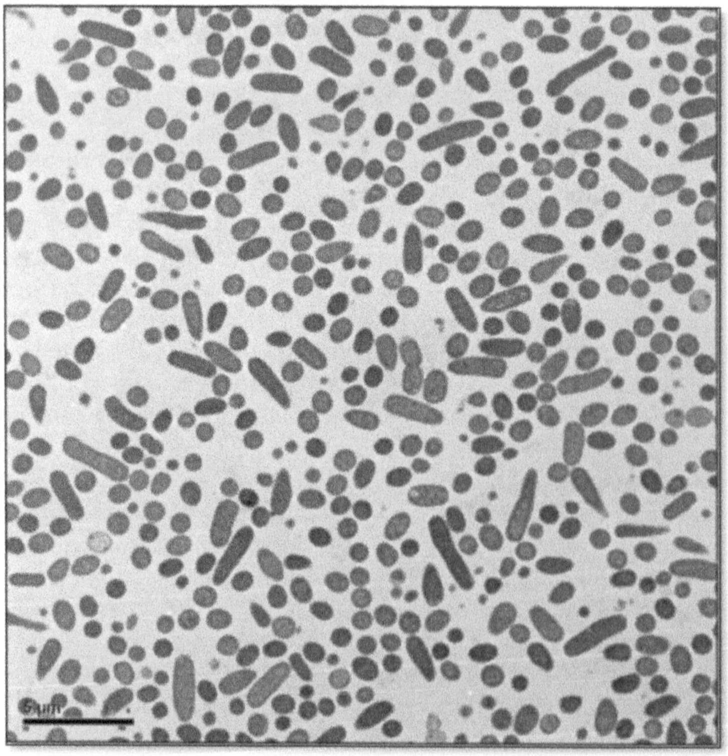

Figure 1 Transmission electron micrograph of *E. coli* (0157:H7; ATCC: 35150).

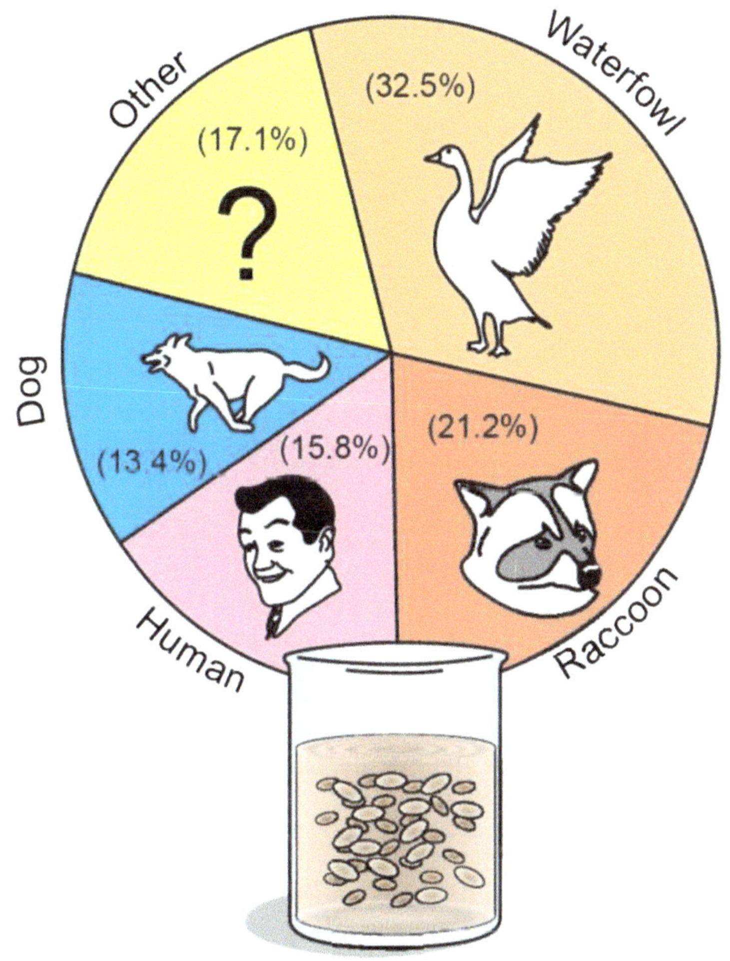

Figure 2 Challenges in identifying enteric pathogen sources (source: Malakoff, 2002).

health. The EPA defines acceptable recreational limits as those that will result in eight or fewer swimming-related gastrointestinal (GI) illnesses out of every 1,000 swimmers (U.S. EPA 1986). The current U.S. EPA fresh water quality criteria for *E. coli* is a geometric mean not exceeding 126 CFU/100 ml, or no samples exceeding a single sample maximum of 235 CFU/100 ml (U.S. EPA 2001). Criteria were developed based on the U.S. EPA measurements of total and Highly Credible Gastrointestinal Illnesses (HCGI), which correlated with *E. coli* densities (r = 0.804) in fresh recreational waters (Dufour 1984). Multiple studies have identified trends between indicator organisms in water and GI illness in humans, including vomiting, diarrhea, and fever (Cabelli 1983; Wade et al. 2006). Recent work by Edge et al. (2010) detected water-borne *E. coli* in 80% of water samples with *E. coli* levels of less than 100 CFU/100 ml. Another study by Wade et al. (2006) reported significant positive trends between increased GI illness and indicator organisms at the Lake Michigan beach, and a positive trend with indicators such as *E. coli* at a Lake Erie beach. Recently, the use of indicator organisms (e.g., fecal coliforms, *E. coli*) for assessing pathogen levels has been debated more often than ever; however, the use of indicator organisms is likely to continue for assessing pathogen levels in water resources potentially because of the lack of an alternative reliable solution.

Pathogen contamination in water resources

The U.S. EPA, which monitors water quality of various ambient water bodies, estimated that pathogens impair more than 480,000 km of rivers and shorelines and 2 million ha of lakes in the U.S. (U.S. Environmental Protection Agency 2010a). According to EPA estimates, pathogens are the leading cause of impairment for 303 (d) listed waters (i.e., list of impaired and threatened waters that the Clean Water Act requires all states to submit for EPA approval) (Figure 3) (U.S. EPA 2014a, 2014b, 2014c). A total of 71,917 causes of impairment have been reported, and the top five causes of impairment are shown in Figure 3. Pathogen contamination clearly dominates the causes of impairment (U.S. EPA 2014a, 2014b, 2014c).

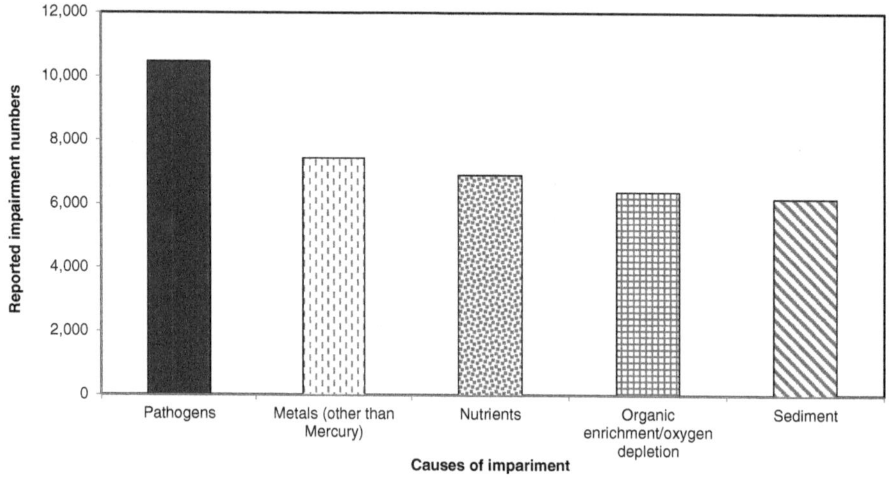

Figure 3 Causes of impairment in the U.S. (data source: U.S. EPA (2014a, 2014b, 2014c)).

Studies by Diffey (1991), Brookes et al. (2004), Jamieson et al. (2004), Gerba and Smith (2005), Gerba and McLeod (1976), Hipsey et al. (2008), Pachepsky and Shelton (2011) reviewed the current studies of water-borne pathogen transport, with particular reference to freshwater and estuarine sediments. In addition, many current reviews focus on specific aspects of water resources, for instance, John and Rose (2005) focused on groundwater, Brookes et al. (2004) focused on reservoirs and lakes, Jamieson et al. (2004) focused on agricultural watersheds, and Kay et al. (2007) reviewed catchment microbial dynamics. The review study presented here uses a relatively broader approach for understanding how water-borne pathogens can potentially impact public health and various ambient water bodies. In addition, existing challenges, while assessing pathogen levels in water resources are discussed.

Coastal and estuarine environments

In the U.S., elevated pathogen levels are a leading cause of impairments of coastal environments (U.S. EPA 2014a, 2014b, 2014c). Urban runoff and sewers have been identified as the primary source of coastal water impairments. Rippey (1994) reported about 400 outbreaks and 14,000 cases caused by pathogen contaminated coastal water since the late 1800s in the U.S. Impairments of coastal environments have major economic impacts on the U.S. For example, losses caused by pathogen contamination in Massachusetts are more than $75 million each year (Weiskel et al. 1996). The studies, which elaborate various pathogens in coastal environment and their survival mechanism, are summarized in Table 1.

The sources of coastal water contamination are: point discharges of treated and untreated sewage from shoreline outfalls, and non-point discharges. The non-point sources, such as runoff from naturally vegetated areas, discharge pathogens into coastal waters. Besides runoff from vegetated areas, the storm water runoff from urban, commercial, and industrial lands also discharges pathogens into coastal waters. In addition, other sources, such as malfunctioning or poorly sited septic systems, can also introduce significant amounts of pathogens (Sayler et al. 1975; Howe et al. 2002). Weiskel et al. (1996) reported that direct deposition of waterfowl feces was a considerable source of pathogens. Fayer and Trout (2005) summarized the transport of various pathogens, such as *Giardia*, *Toxoplasma*, and *Cryptosporidium* (zoonotic parasites) in the coastal environment. Moreover, the presence of sediment in seawater can also increase the survival chance of fecal coliforms, such as *E. coli* (Gerba nd McLeod 1976; Goyal et al. 1977). Solo-Gabriele et al. (2000) showed that the location and timing of storms of the coastal area in tropical and subtropical environments are also important factors that can potentially influence coastal water quality.

Previous studies have shown that the direct discharge of storm water runoff into coastal waters through storm drain systems can cause pathogen contamination, even where separate storm and sanitary sewer systems are in place. For instance, Weiskel et al. (1996) found that about 16% of the total fecal coliform inputs were caused by storm water entering Buttermilk Bay in Massachusetts. In addition, coastal rivers draining largely undeveloped watersheds with extensive riparian wetlands can be a natural source of fecal pathogens to coastal waters (Viau et al. 2011; Staley et al. 2014; Roberts et al. 2013; Liang et al. 2013; Wilkes et al. 2014). On-site septic systems can also contribute significant amounts of fecal pathogens to coastal waters in low-lying fine-grained geological settings where saturated soils enhance pathogen growth. Weiskel et al. (1996) reported that shoreline wrack deposits could act as a reservoir

Table 1 Studies describing pathogen contamination in saline water (coastal and estuary environments)

Author(s) year	Organisms	Results	Study remarks
Coastal water			
Gerba and McLeod (1976)	*E. coli*, fecal coliform	Longer survival in the sediment presence in seawater	Sediment influence on pathogen survival
Goyal et al. (1977)	Total coliform, fecal coliform	Sediments of shallow canal can act as a reservoir	Pathogens distribution in water and sediment
Kapuscinski and Mitchell (1983)	*E. coli*, bacteriophages	*E. coli* survives longer than bacteriophages	Survival of pathogens in sunlight
Rao et al. (1984)	Enteroviruses (Polio and Rota)	Abundance viruses were attached with sediment	Viruses distribution in water and sediment
Weiskel et al. (1996)	Fecal coliform	Waterfowl major source of fecal coliform inputs	Pathogen source and transport pathways
Sinton et al. (1999)	Bacteria and Fecal bacteriophages	Somatic coliphages shown prolonged survival	Sunlight influence on sewage-polluted seawater
Solo-Gabriele et al. (2000)	*E. coli*	Riverbanks as the primary pathogen source	Sources of pathogens on subtropical environment
Nasser et al. (2003)	*Cryptosporidium, viruses, E. coli*	*E. coli* die-off was faster than other pathogens	Comparative survival of various pathogens
Schriewer et al. (2010)	*Bacteroidales*, fecal indicator, protozoa, bacteria	*Bacteroidales* have shown higher predictive skill than fecal indicators	*Bacteroidales* as a predictor of pathogens in coastal water
Estuaries			
Ketchum et al. (1952)	Coliform, zooplankton	Sactericidal and predation caused coliform die-off	Processes responsible for pathogens decrease
Smith et al. (1978)	Echovirus 1, coxsackieviruses	Sediment prolonged viruses survival	Persistence of pathogen viruses
Rhodes and Kator (1990)	*E. coli*	Mortality rises in sunlight	Indigenous microbiota and sunlight influence
White et al. (1998)	*Perkinsus marinus*	Use of Kriging analysis for disease prevalence	Use of GIS in pathogen distribution analysis
Lipp et al. (2001)	*Vibrio vulnificus*	Salinity controls the distribution of pathogens	Distribution of human pathogens
Desmarais et al. (2002)	*E. coli, C. perfringens*	Pathogens re-grown with tides and sterile sediment	Influence of soil on fecal indicator in tidally influenced environment
Frias-Lopez et al. (2002)	*Clostridium, Campylobacter, Arobacter*	Pathogen partitioning was found in surface and overlying water	Bacterial communities partitioning between sea water, dead coral surface
Chandran and Hatha (2005)	*E. coli, Salmonella*	*E. coli* shows better survival capacity than *Salmonella*	Relative survival in microcosm studies
Baker-Austin et al. (2009)	*Vibrio vulnificus*	Prevalence of antibiotic resistance in a human pathogen	Multi-site analysis shows widespread antibiotic resistance in pathogens

of fecal bacteria, and the removal of wrack deposits from inter-tidal zones can improve the water quality of adjacent coastal waters.

Similar to the coastal environment, increasing water-borne pathogen levels in estuaries are a serious threat to public health. Human activities can impact estuary pathogen levels when they are adjacent to populated areas, and often provide a means of transportation and substantial recreation (Schriewer et al. 2010; Pachepsky and Shelton 2011). The most common pathogens, previously identified in estuaries by Rhodes and Kator (1990), were *Vibrio cholerae*, *Giardia*, *Cryptosporidium*, *Salmonella*, and *Campylobacter* spp. As shown in Table 1, the presence of various pathogens (e.g., *E. coli, C. perfringens, Clostridium, Salmonella*) has been reported in many previous studies. Municipal point sources are the primary cause of pathogen contamination in estuaries. Urban water disposed through combined sewer outflows is the cause of approximately 12% of estuary impairments in the U.S. (Arnone and Walling 2007). Pathogens, including *Vibrio vulnificus* which carries the highest fatality rate of any food-borne pathogen in the U.S., were detected in the Gulf of Mexico Estuary (Lipp et al. 2001; Baker-Austin et al. 2009). Several studies discovered that bed sediment plays a vital role (i.e., with the release of particle-attached pathogens from bed sediment to a water column through a resuspension process) for the persistence and transport of pathogens in the estuaries (Smith et al. 1978; Desmarais et al. 2002). Previous studies have shown that pathogen growth and decay are influenced by environmental conditions. For instance, a study by Chandran and Hatha (2005) revealed that sunlight is a major factor that influences survival of pathogens like *E. coli* and *S. typhimurium* in the estuarine water.

Groundwater

Groundwater is heavily used all over the world as the primary source of domestic drinking water supplies, and contaminated groundwater certainly enhances risk to public health. Nationally, 40% of the U.S. domestic water supply originates from groundwater, and over 40 million people use groundwater as their drinking water via private wells (Alley et al. 1999). Groundwater pathogen contamination has led to numerous disease outbreaks in the U.S.; for example, at least 46 outbreaks of disease occurred between 1992 and 1999, resulting in 2,739 cases of illness and several deaths (John and Rose 2005). These are reported cases; due to underdiagnosis and underreporting, the actual morbidity is almost certainly higher.

Several studies have shown that microbial pathogens, such as *Salmonella, E. coli, S. faecalis*, and enteroviruses are relatively stable in groundwater (Bittion et al. 1983; Schijven and Hassanizadeh 2000; Pang et al. 2004) (Table 2). Controlling groundwater pathogen contamination has recently been emphasized in many countries, as pathogens can survive up to 400 days depending on the soil temperature (Nevecherya et al. 2005; Filip and Demnerova 2009). For example, identifying sources of groundwater pathogen contamination has received significant attention in France (Grisey et al. 2010). Many studies reported that health risks caused by chlorine-resistant protozoans, such as *Cryptosporidium* spp. (Ferguson et al. 2003; Kay et al. 2007; Kay et al. 2008), are considerable. One of the major concerns is that wetlands without lining might cause pathogen contamination of groundwater (Kay et al. 2007). Similar concerns have been expressed in the United Kingdom by water regulators. The European Union (EU) has also emphasized protecting groundwater from pathogen contamination. Pathogen-contaminated groundwater can cause pollution in coastal environments. For example, a study of Buttermilk Bay has shown that groundwater is capable of transporting a large quantity of pathogens from surface to sub-surface water either by direct discharge or by discharge to rivers flowing into the bay (Moog 1987; Weiskel et al. 1996). The risk of contaminating groundwater particularly increases in areas where shallow aquifers exist. In these situations it is more likely that contaminated surface water or water from septic tanks can reach groundwater (Weiskel et al. 1996). Precipitation events are likely to increase groundwater pathogen contamination because of contaminated ground water recharge.

Pathogen contamination (e.g., bacteria, protozoa, and viruses) poses a serious risk in water resources. The transport of pathogens from surface water to groundwater increases the vulnerability of groundwater (Jin and Flury 2002). These authors reported that 70% of the water-borne microbial illness outbreaks in the United States have been associated with ground water. Pathogens such as viruses are much smaller than bacteria and protozoa, and many can potentially reach groundwater through porous soil matrices. Jin and Flury (2002) reviewed the fate and transport of viruses in porous media to understand mechanisms and modeling of virus sorption, and concluded that factors such as solution chemistry, virus properties, soil properties, temperature, and association with solid particle influences virus survival, transport, and sorption in porous media. Pang (2009) studied microbial removal rates in subsurface media, and reported that soil types considerably influence microbial removal rates. For instance, volcanic soils, pumice sand, fine sand, and highly weathered aquifer rocks showed high removal rates. The author found that microbial removal rates were inversely correlated with infiltration rates and transport velocity.

Considerable work recently has been done towards understanding pathogen transport in the vadose zone (Wang et al. 2014a; 2014b; Unc and Goss 2004; Darnault et al. 2004). Groundwater can be contaminated by seepage and percolation of contaminated water from the vadose zone (Darnault et al. 2004). The macropores of agricultural land are also known to play a considerable role in polluting groundwater, particularly from fields where manure is applied (Jamieson et al. 2002). Unc and Goss (2004) evaluated the influence of manure on the transport of bacteria from land receiving manure to water resources. These authors reported that manure application in the land influences pathogen transport in the vadose zone. The presence of straw and coarse organic matter influences the persistence of bacteria, and manure application changes the physical configuration of soil, the soil chemistry, and the properties of the microbial cells, which control the survival and persistence of bacteria in soils. Another recent study by Wang et al. (2014) assessed the transport of *E. coli* in soils with preferential flow. The authors reported that a decrease in macropore length resulted in a decreased apparent saturated hydraulic conductivity of the macropore and an increase in the mass transfer. Wang et al. (2014a; 2014b) concluded that macropore length has a considerable influence on preferential transport of *E. coli*.

Reservoirs and lakes

Previous studies have shown the presence of many pathogens in lakes and reservoirs (Table 2), and that these pathogens can pose risks to human health. In many countries surface reservoirs serve as the main source

Table 2 Pathogen contamination in freshwater environment (ground water, rivers and lakes and reservoirs)

Author(s) year	Organisms	Results	Study remarks
Groundwater			
Bitton et al. (1983)	*Salmonella, E. coli, S. faecalis, enteroviruses*	*S. faecalis* decay rate was similar to viruses	Survival of pathogens in groundwater
Schijven and Hassanizadeh (2000)	Viruses	Viruses attachment with soil was influenced by pH, and favorable sites	Soil passage impacts on virus removal at field-scale
Pang et al. (2004)	*E. coli* and F-RNA phages	Pathogens were sorbed in aquifer material	Transport of *E. coli* and F-RNA phages
Nevecherya et al. (2005)	*Salmonella*, viruses, *E. coli*, shigellos	Mathematical model was derived for temperature depends inactivation rate	Pathogenic bacteria and viruses survival in groundwater
Filip and Demnerova (2009)	*Bacillus megaterium* and *Staphylococcus*	Pathogens survived 10 to 100 days	Pathogens survival in groundwater; FT-IR characterization
Grisey et al. (2010)	Total coliforms, *E. coli, Enterococci, Pseudomonas aeruginosa, Salmonella and Staphylococcus aureus*	Bacterial density monitoring coupled with artificial trace experiments proved useful in locating pathogens sources	Pathogens survival in groundwater and landfill leachate
Rivers			
Chin (2010)	Fecal coliform	Summer and rainfall raises pathogens	Urban areas impacts on stream pathogens
Smith et al. (1973)	*Salmonella*, fecal coliforms, streptococci	*Salmonella* decline was close to fecal coliform	Usefulness of indicators organisms
Burton et al. (1987)	*Pseudomonas aeruginosa, Salmonella newport, E. coli*	Clay in sediments improves *E. coli* survival	Survival of pathogens in fresh water sediments
Smith et al. (1987)	*E. coli*, Fecal bacteria	U.S. rivers shows decline in fecal indicator	Water quality assessment from 1974 to 1981
Terzieva and McFeters (1991)	*E. coli, Campylobacter jejuni, Yersinia enterocolitica*	Temperate zone surface water serve as a rersisitent verhicle in transimsision of bacteria between animals and humans	Survival and injury study of bacteria in agricultural surface water
Terzieva and McFeters (1991)	*Yersinia enterocolitica* and *E. coli*	Experimental design and environmental play major role in pathogens survival	Survival in stream water, comparison between field and lab studies
Fraser et al. (1998)	Fecal coliform	Model predicted pathogens	Modelling non-point source pollution
Lakes and reservoirs			
Beaver and Crisman (1989)	Ciliates	Grazing habits of ciliates are discussed	Predators roles in fresh water
Rubentschik et al. (1936)	*Serratia marcescens*	Adsorption of bacteria could be questionable	Adsorption of bacteria in salt lakes
Mac Kenzie et al. (1994)	*Cryptosporidium*	*C. oocysts* study passes through the filtration system of water supply	Contaminated water from Milwaukee lake caused outbreak
Wcislo and Chrost (2000)	*E. coli*	Predators controlled pathogen levels	Survival of pathogens in man-made reservoir
Kistemann et al. (2002)	*E. coli*, coliform, fecal streptococcal, and *Clostridium perfringens*	Most of the pathogens increases during extreme runoff events	Microbial load in drinking water reservoir during rainfall events
Howe et al. (2002)	*Cryptosporidium oocysts*	Animal feces was a major source of pathogens	Water supply's oocysts caused outbreak in northern England
Ishii et al. (2006)	*E. coli*	*E. coli* survived longer in soil	Presence and growth of *E. coli* in Lake superior watershed

of drinking water, and these surface water bodies are often vulnerable to pathogen contamination (Kistemann et al. 2002). In the developed world, although there is increased awareness of water treatment for pathogen contamination and water quality, outbreaks of water-borne disease via public water supplies continue to be reported (Gibson et al. 1998; Howe et al. 2002; Brookes et al. 2004).

In the past, more than 403,000 residents of the greater Milwaukee, Wisconsin area experienced gastrointestinal illnesses due to infection with the parasite *Cryptosporidium parvum* following contamination of the city's water supply, which was associated with inadequate filtration of contaminated water from Lake Michigan (Mac Kenzie et al. 1994; Cicirello et al. 1997). In the 1990s, *Cryptosporidiosis* became the most common cause of outbreaks associated with public drinking water supplies in the United Kingdom (Howe et al. 2002). In developing countries, diseases such as diarrhea and cholera are the leading cause of morbidity (Nelson et al. 2009). Overall, diarrhea associated with drinking contaminated water is responsible for 2 to 2.5 million deaths annually (Fenwick 2006). In lakes and reservoirs, increased pathogens are often associated with storm events, and the stream inflow is considered to be the major source of pathogens during storm events. Elevated flows in rivers most likely agitate bed sediment, which causes enhancement of pathogen levels of the water column (Jamieson et al. 2005a; Jamieson et al. 2005b; Pandey and Soupir 2013; Bai and Lung 2005). During the rainy season, the influx of contaminated water from rivers to lakes and reservoirs can substantially increase pathogen levels (Kistemann et al. 2002). The quantity of pathogen influxes from tributaries of lakes and reservoirs during the rainy season is of particular importance in determining pathogen transport and distribution (Brookes et al. 2004).

Rivers

Pathogen contamination is a major cause of stream impairments. The sources of impairment and health risks induced by water-borne pathogens are extensively reported (Table 2). In the U.S. pathogen contamination is the leading cause of stream water pollution. The EPA's National Water Quality Inventory Report suggests that about 53% of the assessed rivers are impaired, and a majority of them are contaminated by pathogens (U.S. EPA 2012a). The cost to implement the total maximum daily load (TMDL) plans to improve stream water is estimated as $0.9 to $4.3 billion per year (U.S. EPA 2010b).

Pathogen influxes into rivers from agricultural lands (Figure 4) are the main cause of stream impairments (Chin 2010; U.S. EPA 2012a). A weak understanding of pathogen transport from agricultural lands to rivers is considered to be a major challenge in implementing and deriving suitable land management practices capable of improving stream water quality. For instance, despite common knowledge that agricultural land's non-point source pollution is a leading cause of stream impairment, it is difficult to identify points of origin of pathogens and the pathways by which they enter streams. As an example, pathogens are likely to enter rivers from many potential sources, including lateral inputs from pastures and riparian zones, influx of pathogen-contaminated groundwater, direct deposit of fecal matter from livestock and wildlife, discharge of contaminated sanitary sewer flows, and wastewater treatment plant effluents. In rainy events, pathogens in rivers are influenced by fresh input from watersheds as well as sub-surface flow. In addition, the resuspension of legacy pathogens from bed sediments can considerably increase pathogen levels (Cho et al. 2010; Droppo et al. 2009; Jamieson et al. 2005b; Kiefer et al. 2012; Nagels et al. 2002; Muirhead et al. 2004; Kim et al. 2010; Smith et al. 2008).

Controlling pathogen contamination from livestock/wildlife to streams is challenging (Terzieva and McFeters 1991). For example, it is doubtful that pathogen contamination can be prevented by fencing off riparian buffers, and even if buffers are useful in controlling stream water pathogens, it is not certain what their width must be (Nagels et al. 2002). There are review studies that elaborate on stream water pathogen contamination (Fraser et al. 1998; Jamieson et al. 2004; Pachepsky et al. 2006).

Many studies have emphasized the use of mathematical models to understand pathogen transport from agricultural land to rivers (Kim et al. 2010; Muirhead et al. 2004; Jamieson et al. 2005a; Jamieson et al. 2005b). Previous studies (Gerba and Smith 2005; Pandey et al. 2012b; Pandey and Soupir 2013; Pachepsky and Shelton 2011; Martinez et al. 2014) have emphasized the need to improve existing models for calculating the fate and transport of pathogens at the watershed scale. Currently, empirical as well as mechanistic models are being used to calculate microbial fate and transport (Muirhead and Monaghan 2012). Numerous studies are available for exploiting watershed scale models such as WATFLOOD (Dorner et al. 2006), the Soil and Water Assessment Tool (SWAT) (Neitsch et al. 2005), the Spatially Explicit Deliver Model (SEDMOD) (Fraser 1999), and KINEROS/STWIR (Guber et al. 2011) for predicting pathogen transport. Though numerous watershed scale models are available, which can be exploited for calculating pathogen transport at watershed scale, considerable difficulties exist while using the models (Pandey et al. 2012b; Pandey and Soupir 2013). For instance, correctly identifying the model input parameter values is a daunting task. Recently, considerable emphasis has been given to understand the sensitivity of the input parameters to the model output (Martinez et al. 2014; Parajuli et al. 2009).

When implementing watershed scale models for predicting pathogen transport, difficulties related to the selection of model input parameters are common (Martinez et al. 2014). A study by Parajuli et al. (2009) evaluated the sensitivity of fecal coliform bacteria loads modeled with SWAT, and the authors reported that many of the parameters (e.g., bacterial die-off rates, the temperature adjustment factor) were insensitive to the model output. Similarly, Coffey et al. (2010) reported that initial concentration of *E. coli* and the bacterial partition coefficient parameters of the SWAT model were overly

Figure 4 Simplified path of animal waste pathogen transport from agricultural land to rivers.

sensitive in affecting the model's output. Another recent study by Martinez et al. (2014) evaluated the KINEROS/STWIF model input parameter sensitivities to the output. These authors reported that environmental controls such as soil saturation, rainfall duration, and rainfall intensity had the most sensitivity, while parameters such as soil and manure properties were the least sensitive in affecting model output. Many previous studies (Pandey and Soupir 2012a, 2012b; Kim et al. 2010; Martinez et al. 2014; Parajuli et al. 2009) reported that identification and selection of model input parameters are major challenges when implementing the model for predicting fate and transport of pathogens at the watershed scale.

Blooming of cyanobacteria

While water-borne pathogens are a serious concern, excessive algal bloom (Figure 5) in water resources can potentially limit their uses for recreation activities as well as for drinking water. Cyanobacteria (blue-green algae) have unique roles in oxygenation of the atmosphere (Hofer 2013); however, their excessive growth or dense algal bloom in water resources diminishes the quality and quantity of light in the water column (U.S. EPA 2014b). When blooms are excessive, the risk of toxin contamination (released by harmful algal bloom (HABs)) is likely to be elevated. The HAB includes many types of algal taxa such as dinoflagellates, diatoms, and cyanobacteria. Eutrophication caused by excessive algal bloom can cause fish kills and reduce the diversity of aquatic life (U.S. EPA 2014b). In hypoxic water, dissolved oxygen levels can be less than 2 – 3 ppm (U.S. EPA 2014c). For example, in the hypoxic zone in the northern Gulf of Mexico, an area along the Louisiana-Texas coast, less than 2 ppm of dissolved oxygen concentration has been reported. This is believed to be caused by excess nutrients delivered from the Mississippi River, in combination with seasonal stratification of Gulf waters (U.S.G.S., 2014). The largest U.S. hypoxic zone, which occurred in 2002 in the Gulf of Mexico, was about 13,518 square kilometers (U.S. G.S. 2012).

Freshwater cyanobacterial blooms produce highly potent cyanotoxins and cyanobacterial HABs, which can affect the liver, nervous system, and skin (U.S. EPA 2014b). Cyanobacterial blooms can be potentially detrimental to

Figure 5 Algal bloom in Squaw Creek, Iowa, U.S.

human and animal health, aquatic habitats, and aquaculture industries (Kaloudis et al. 2013; Carmichael 2001; Falconer 2005; Codd et al. 1999). Previous studies (Mackintosh et al. 1990; Yoshizawa et al. 1990) have shown that microcystins (i.e., cynotoxins) are hepatotoxic and act as tumor promoters through the inhibition of protein phosphatases, which play a key role in cellular regulation. Eutrophic water conditions combined with warm surface water temperatures (15 – 30°C) can potentially enhance cyanobacterial blooms in water (Oikonomou et al. 2012; Vareli et al. 2009). Currently more than 400 hypoxic zones exist in the world affecting 245,000 square kilometers (Diaz and Rosenberg 2008). These dead zones in coastal waters have spread exponentially since the 1960s and have considerably impacted ecosystems. The increase in these dead zones is likely to be influenced by climate change. Various factors such as how climate change affects water-column stratification and how nutrient runoff affects organic matter production will determine further expansion of the dead zones (Diaz and Rosenberg 2008). For instance, climate predictions of the Mississippi River basin indicate a 20% increase in river discharge, which will enhance nutrient loading and result in expansion of the oxygen depleted area (IPCC 2007). Considering the importance of fresh water, estuarine, and marine environment to public health and aquatic life, identifying approaches capable of controlling excessive algal blooms is required before these dead zones spread globally.

Impact of water resources development

Water resources development involves altering the natural flow path of rivers and lakes, as well as designing irrigation schemes and dams. These activities have been alleged to be responsible for causing new diseases and enhancing health risks (Fenwick 2006; Steinmann et al. 2006).

The influence of water resources development in spreading diseases, such as schistosomiasis, a parasitic disease which is ranked second only to malaria with regard to the number of people infected, has been reported extensively; one estimate says that about 103 million out of 779 million infected people live in close proximity to large reservoirs and irrigation schemes (Steinmann et al. 2006).

Designing dams and irrigation schemes in tropical and subtropical climate zones has often resulted in disease outbreaks caused by water-borne pathogens. Consider, for example, the Sennar Dam on the Blue Nile River and Sudan's Gezira Scheme, the world's largest irrigation project. Because of the dam's commercial success, irrigation in the region has doubled from the 1940s and 1950s. After the 1950s, infections from malaria and schistosomiasis increased significantly, becoming the subject of the first integrated disease-control program, the Blue Nile Health Project, implemented from 1978 to 1990. The project failed to have any impact in controlling the prevalence of schistosomiasis (Eltoum et al. 1993; Fenwick 2006; Steinmann et al. 2006). Another example is China's Three Gorges Dam, built across the Yangtze River and completed in 2009, which created a 50,700 km^2 reservoir and submerged more than 220 counties. Hotez et al. (1997) reported that the reservoir would produce environmental changes that could lead to the transmission of schistosomiasis in the area served by the dam. A recent study by Schrader et al. (2013) found major high risk areas for schistosomiasis occurrence in the large lakes and flood plain regions of the Yangtze River. Another study by Gray et al. (2012) reported that the Three Gorges Dam will likely to impact the transmission of schistosomiasis in China.

In the U.S., because of increasing concern for produce safety, pathogen-free irrigation water is attracting considerable attention (Martinez et al. 2014). Painter et al. (2013) reported that produce accounted for nearly half of food-borne illnesses in the U.S. between 1998 and

2008. Growing concern about the safety of food and water will likely help in developing improved strategies while planning and designing large dams for irrigation purposes.

Challenges and recommendations

Infectious diseases caused by pathogens are the third leading cause of death in the United States, and the leading cause in the world (Binder et al. 1999). The past two decades have seen the emergence of many new pathogenic infectious diseases (Daszak et al. 2000). Many of these are caused by anthropogenic changes, such as water resources development, climate warming, and interactions between humans and animals, both domestic and wild (Krause 1994; Epstein 2001; Woolhouse 2002; Fenwick 2006; Schriewer et al. 2010).

Multidisciplinary knowledge about how ambient water bodies, wildlife, domestic animals, and human populations interact with and impact each other are crucial in dealing with future challenges. Generally, domestic animals, wildlife, and humans are considered to be major sources of water-borne pathogens; however, finding the specific culprit—the primary pathogen source—is challenging (Malakoff 2002). A watershed, for example, can have many pathogen sources, such as agricultural land, riparian areas, agricultural feeding operations, livestock, wildlife, and humans.

Developing models that are reliable in predicting pathogen survival and transport at the watershed scale can be helpful in implementing/evaluating the strategies for mitigating ambient water body pathogen levels. Evaluating the impacts of various environmental factors on pathogen survival in water resources is crucial. Various publications (i.e., Dorner et al. 2006; Kim et al. 2010; Rehmann and Soupir 2009; Droppo et al., 2011; Cho et al. 2010; Pandey et al. 2012b; Pandey and Soupir 2013) are available describing models capable of predicting pathogen contamination levels in ambient water bodies; however, further work is required to improve model predictions. The evaluation of predictions by existing models clearly demonstrates the need for improvement. Many pathogen transport models use only temperature-induced mortality and growth, and do not include interactions among other environmental factors (e.g., pH, nutrients, DO, solar radiation); future inclusion of these environmental factors will likely to improve the model predictions.

Typically most studies have relied on *E. coli* and other indicator bacteria to indicate pathogen levels in water. Although widely used in monitoring contamination levels, *E. coli* alone can lead to mercurial and misleading information (Gordon 2001). Schriewer et al. (2010) suggested that with improved pathogen detection technology (i.e., PCR-based detection) an indicator organism, such as *E. coli*, can be sufficiently accurate in most cases. Overall, improving technology to identify causative agents more accurately, creating standard epidemiological data for diseased populations, and enhancing the knowledge of disease dynamics can improve the understanding of risks caused by interactions among various populations (Harvell et al. 1999; Daszak et al. 2000; Harvell et al. 2002).

In the past, a number of studies on pathogen contamination have been conducted on a scale where the conditions of ambient water bodies were simulated in laboratories. These studies are helpful in understanding pathogen behavior only up to a point. For enhancing the understanding of pathogen interactions in the environment, more emphasis should be given to field-scale studies.

Conclusions

This review examines studies from various disciplines to understand pathogen contamination in ambient water bodies. The worldwide prevalence of pathogen contamination is a serious concern, and enhancing the understanding of major pathogen sources and their significant impacts on water resources is crucial. A considerable number of studies on pathogen contamination have been conducted on a laboratory-scale; more emphasis should be given to field-scale studies for enhancing the understanding of pathogen interactions in the environment. Developing new models, and improving existing modeling approaches commonly used for predicting water-borne pathogen levels will likely to help in assessing pathogen contamination at watershed-scale. Considering the limited ability of existing models to predict pathogen contamination, improvement and development of new models are needed so that pathogen levels can be predicted more accurately. Integrating knowledge from multiple fields (e.g., hydrology, microbiology, and ecology) would increase the understanding of pollution levels and potential causes of pollution, and can also help devise long-term strategies to improve water quality.

Competing interest

The authors declare that they have no competing interests.

Acknowledgment

The authors thank the Division of Agriculture and Natural Resources and Veterinary Medicine Extension, University of California, Davis, and National Science Foundation (award No. CBET-0967845) for supporting this work.

Author details

[1]Department of Population Health and Reproduction, University of California, Davis, California, USA. [2]Department of Agricultural and Biosystems Engineering, Iowa State University, Ames, Iowa, USA. [3]Department of Biological and Agricultural Engineering & Zachry Department of Civil Engineering, Texas A & M University, College Station, Texas, USA.

References

Alley WM, Reilly TE, Franke OL (1999) Sustainability of Groundwater Resources. U. S. Geological Survey (U.S.GS), Denver, CO

Arnone RD, Walling JP (2007) Waterborne pathogens in urban watersheds. J Water Health 5(1):149–162

Bai S, Lung WS (2005) Modeling sediment impact on the transport of fecal bacteria. Water Res 39:5232–5240

Baker-Austin C, McArthur J, Lindell A, Wright M, Tuckfield R, Gooch J, Warner L, Oliver J, Stepanauskas R (2009) Multi-site analysis reveals widespread antibiotic resistance in the marine pathogen: *Vibrio vulnificus*. Microb Ecol 57(1):151–159

Beaver JR, Crisman TL (1989) The role of ciliated protozoa in pelagic freshwater ecosystems. Microb Ecol 17(2):111–136

Binder S, Levitt AM, Sacks JJ, Hughes JM (1999) Emerging infectious diseases: public health issues for the 21st Century. Science 284(5418):1311–1313

Bitton G, Farrah SR, Ruskin RH, Butner J, Chou YJ (1983) Survival of pathogenic and indicator organisms. Groundwater 21(4):405–410

Bradford SA, Morales VL, Zhang W, Harvey RW, Packman AI, Mohanram A, Welty C (2013) Transport and fate of microbial pathogens in agricultural settings. Crit Rev Env Sci Tec 43:775–893

Brookes JD, Antenucci J, Hipsey M, Burch MD, Ashbolt NJ, Ferguson C (2004) Fate and transport of pathogens in lakes and reservoirs. Environ Int 30(5):741–759

Burton GA, Gunnison D, Lanza GR (1987) Survival of pathogenic bacteria in various freshwater sediments. Appl Environ Microbiol 53(4):633–638

Cabelli VJ (1983) Public health and water quality significance of viral diseases transmitted by drinking water and recreational water. Water Sci Technol 15(5):1–15

Carmichael WW (2001) Health effects of toxin-producing cyanobacteria: the CyanoHABs. Hum Ecol Risk Assess 7:1393–1407

Centers for Disease Control and Prevention (CDC) (2011) Cholera in Haiti: One Year Later., http://www.cdc.gov/haiticholera/haiti_cholera.htm (accessed on 5/23/2014)

Centers for Disease Control and Prevention (CDC) (2014) Salmonella., http://www.cdc.gov/salmonella/general/ (accessed on 4/28/2014)

Chandran A, Mohamed Hatha AAM (2005) Relative survival of Escherichia coli and Salmonella typhimurium in a tropical estuary. Water Res 39(7):1397–1403

Chin DA (2010) Linking pathogen sources to water quality in small urban streams. J Environ Eng 136(2):249–253

Cho KH, Pachepsky YA, Kim JH, Guber AK, Shelton DR, Rowland R (2010) Release of Escherichia coli from the bottom sediment in a first-order creek: Experiment and reach-specific modeling. J Hydrol 391(3–4):322–332

Cicirello HG, Kehl KS, Addiss DG, Chusid MJ, Glass RI, Davis JP, Havens PL (1997) Cryptosporidiosis in children during a massive waterborne outbreak in Milwaukee, Wisconsin: clinical, laboratory and epidemiologic findings. Epidemiol Infect 119(1):53–60

Codd G, Bell S, Kaya K, Ward C, Beattie K, Metcalf J (1999) Cyanobacterial toxins, exposure routes and human health. Eur J Phycol 34:405–415

Coffey R, Cummins EO, Flaherty V, Cormican M (2010) Pathogen sources estimation and scenario analysis using the Soil and Water Assessment Tool (SWAT). Hum Ecol Risk Assess 16(4):913–933

Colwell RR (1996) Global climate and infectious disease: The Cholera Paradigm. Science 274(5295):2025–2031

Craun GF, Fraun MF, Calderon RL, Beach MJ (2006) Waterborne outbreaks reported in the United States. J Water Health 4:19–30

Darnault CJD, Steenhuis TS, Garnier P, Kim YJ, Jenkins MB, Ghiorse WC, Baveye PC, Parlange JY (2004) Preferential flow and transport of *Cryptosporidium parvum* oocysts through the vadose zone: Experiments and modeling. Vadose Zone J 3(2):262–270

Daszak P, Cunningham AA, Hyatt AD (2000) Emerging infectious diseases of wildlife – threats to biodiversity and human health. Science 287:443–449

Desmarais TR, Solo-Gabriele HM, Palmer CJ (2002) Influence of soil on fecal indicator organisms in a tidally influenced subtropical environment. Appl Environ Microbiol 68(3):1165–1172

Diaz RJ, Rosenberg R (2008) Spreading dead zones and consequences for marine ecosystems. Science 321:926–929

Dickerson JW, Crozier JB, Hagedorn G, Hassall A (2007) Assessment of 16 s-23srDNA interagenic spacer regions in *Eterococcus* spp. for microbial source tracking. J Environ Qual 36:1661–1669

Diffey BL (1991) Solar ultraviolet-radiation effects on biological-systems. Phys Med Biol 36(3):299–328

Dorner SM, Anderson WB, Slawson RM, Kouwen N, Huck PM (2006) Hydrologic modeling of pathogen fate and transport. Environ Sci Technol 40(15):4746–4753

Droppo IG, Liss SN, Williams D, Nelson T, Jaskot C, Trapp B (2009) Dynamic existence of waterborne pathogens within river sediment compartments: Implications for water quality regulatory affairs. Environ Sci Technol 43(6):1737–1743

Droppo IG, Krishnappan BG, Liss SN, Marvin C, Biberhofer (2011) Modeling sediment-microbial dynamics in the South Nation River, Ontario, Canada: Towards the prediction of aquatic and human health risk. Water Res 45:3797–3809

Dufour A (1984) Health Effects Criteria for Fresh Recreational Waters. Toxicology and Microbiology Division, U.S. Environmental Protection Agency Report EPA- 600-1-84-2004. , Cincinnati, OH

Edge TA, Hill S, Seto P, Marsalek J (2010) Library-dependent and library independent microbial source tracking to identify spatial variation in faecal contamination sources along a Lake Ontario beach (Ontario, Canada). Water Sci Technol 62:719–727

Eltoum IA, Sulaiman SM, Elturabi H, Mahgoub E, Homeida MMA (1993) Infection with Schistosoma-Mansoni in 2 different endemic areas – a comparative population – based study in Elzidab and Gezira-Managil irrigation schemes, Sudan. J Trop Med Hyg 96(2):100–106

Epstein PR (2001) Climate change and emerging infectious diseases. Microbes Infect 3(9):747–754

Falconer IR (2005) Is there a human health hazard from microcystins in the drinking water supply? Acta Hydrochim Hydrobiol 33:64–71

Fayer R, Trout JM (2005) Zoonotic protists in the marine environment. In: Belkin S and Colwell RR (ed) Oceans and Health: Pathogens in the Marine Environment. Springer, New York, U.S, pp 143–163

Fenwick A (2006) Waterborne Infectious Diseases-Could they be consigned to History? Science 313:1077–1081

Ferguson C, Husman AMD, Altavilla N, Deere D, Ashbolt N (2003) Fate and transport of surface water pathogens in watersheds. Crit Rev Environ Sci Technol 33(3):299–361

Filip Z, Demnerova K (2009) Survival in groundwater and FT-IR characterization of some pathogenic and indicator bacteria. In: Jones JAA, Vardanian TG, Hakopian C (ed) Threats to Global Water Security. Springer, Dordrecht, The Netherlands, pp 117–122

Fraser RH (1999) SEDMOD: A GIS-based delivery model for diffuse source pollutants. PhD dissertation, School of Forestry and Environmental Studies, Yale University

Fraser RH, Barten PK, Pinney DAK (1998) Predicting stream pathogen loading from livestock using a geographical information system-based delivery model. J Environ Qual 27(4):935–945

Frias-Lopez J, Zerkle AL, Bonheyo GT, Fouke BW (2002) Partitioning of bacterial communities between seawater and healthy, black band diseased, and dead coral surfaces. Appl Environ Microbiol 68(5):2214–2228

Gerba CP, McLeod JS (1976) Effect of sediments on the survival of Escherichia coli in marine waters. Appl Environ Microbiol 32(1):114–120

Gerba CP, Smith J (2005) Sources of pathogenic microorganisms and their fate during land application of wastes. J Environ Qual 34:42–48

Gibson CJ, Haas CN, Rose JB (1998) Risk assessment of waterborne protozoa: current status and future trends. Parasitology 117:S205–S212

Gordon DM (2001) Geographical structure and host specificity in bacteria and the implications for tracing the source of coliform contamination. Microbiology 147(5):1079–1085

Goyal SM, Gerba CP, Melnick JL (1977) Occurrence and distribution of bacterial indicators and pathogens in canal communities along Texas Coast. Appl Environ Microbiol 34(2):139–149

Grave AK, Hagedorn C, Brooks A, Hagedorn RL, Martin E (2007) Microbial source tracking in a rural watershed dominated by cattle. Water Res 41:3729–3739

Gray DJ, Thrift AP, Williams GM, Zheng F, Li Y-S, Guo J, Chen H, Wang T, Xu XJ, Zhu R, Zhu H, Cao CL, Lin DD, Zhao ZY, Li RS, Davis GM, McManus DP (2012) Five-year longitudinal assessment of the downstream impact on schistosomiasis transmission following closure of the Three Gorges Dam. PLoS Negl Trop Dis 6(4):e 1588, doi:10.1371/journal.pntd.0001588

Grisey E, Belle E, Dat J, Mudry J, Aleya L (2010) Survival of pathogenic and indicator organisms in groundwater and landfill leachate through coupling bacterial enumeration with tracer tests. Desalination 261(1–2):162–168

Guber AK, Pachepsky YA, Yakirevich AM, Shelton DR, Sadeghi AM, Goodrich DC, Unkrich CL (2011) Uncertainty in modelling of faecal coliform overland transport associated with manure application in Maryland. Hydrolical Process 25(15):2393–2404

Harvell CD, Kim K, Burkholder JM, Colwell RR, Epstein PR, Grimes DJ, Hofmann EE, Lipp EK, Osterhaus AD, Overstreet RM, Porter JW, Smith GW, Vasta GR (1999) Emerging marine diseases-climate links and anthropogenic factors. Science 285(5433):1505–1510

Harvell CD, Mitchell CE, Ward JR, Altizer S, Dobson AP, Ostfeld RS, Samuel MD (2002) Climate warming and disease risks for terrestrial and marine biota. Science 296:2158–2162

Hipsey MR, Antenucci JP, Brookes JD (2008) A generic, process-based model of microbial pollution in aquatic systems. Water Resour Res 44(7), W07408

Hofer U (2013) Getting to the bottom of Cyanobacteria. Nat Rev Microbiol 11:818–819

Hotez PJ, Feng Z, Xu LQ, Chen MG, Xiao SH, Liu SX, Blair D, McManus DP, Davis GM (1997) Emerging and reemerging helminthiases and the public health of China. Emerg Infect Dis 3(3):303–310

Howe AD, Forster S, Morton S, Marshall R, Osborn KS, Wright P, Hunter PR (2002) *Cryptosporidium* oocysts in a water supply associated with a cryptosporidiosis outbreak. Emerg Infect Dis 8(6):619–624

Ibekwe AM, Murinda SE, Graves AK (2011) Genetic diversity and antimicrobial resistance of *Escherichia coli* from human and animal sources uncovers multiple resistances from human sources. PLoS One 6:e20819, doi:10.1371/journal.pone.0020819

Intergovernmental Panel for Climate Change (IPCC) (2007) IPCC Fourth Assessment Report, Summary for Policymakers, p 5

Ishii S, Ksoll WB, Hicks RE, Sadowsky MJ (2006) Presence and growth of naturalized *Escherichia coli* in temperate soils from Lake Superior watersheds. Appl Environ Microbiol 72(1):612–621

Jamieson R, Gordon R, Sharples K, Stratton G, Madani A (2002) Movement and persistence of fecal bacteria in agricultural soils and subsurface drainage water: A review. Can Biosyst Eng 44:1.1–1.90

Jamieson R, Gordon R, Joy D, Lee H (2004) Assessing microbial pollution of rural surface waters: A review of current watershed scale modeling approaches. Agric Water Manag 70(1):1–17

Jamieson R, Joy DM, Lee H, Kostaschuk R, Gordon R (2005a) Transport and deposition of sediment-associated Escherichia coli in natural streams. Water Res 39(12):2665–2675

Jamieson RC, Joy DM, Lee H, Kostaschuk R, Gordon RJ (2005b) Resuspension of sediment-associated Escherichia coli in a natural stream. J Environ Qual 34(2):581–589

Jin Y, Flury M (2002) Fate and transport of viruses in porous media. Adv Agron 77:39–102

John DE, Rose JB (2005) Review of factors affecting microbial survival in groundwater. Environ Sci Tech 39(19):7345–7356

Jordan EO, Russell HL, Zeit FR (1904) The longevity of the typhoid bacillus in water. J Infect Dis 1(4):641–689

Kaloudis T, Zervou S, Tsimeli K, Triantis TM, Fotiou T, Hiskia A (2013) Determination of microcystins and nodularin (cyanobacterial toxins) in water by LC-MS/MS. Monitoring of Lake Marathonas, a water reservoir of Athens, Greece. J Hazard Mater 263:105–115

Kapuscinski RB, Mitchell R (1983) Sunlight-induced mortality of viruses and Escherichia coli in coastal seawater. Environ Sci Technol 17(1):1–6

Kay D, Edwards AC, Ferrier RC, Francis C, Kay C, Rushby L, Watkins J, McDonald AT, Wyer M, Crowther J, Wilkinson J (2007) Catchment microbial dynamics: the emergence of a research agenda. Prog Phys Geogr 31(1):59–76

Kay D, Crowther J, Stampleton CM, Wyer MD, Fewtrell L, Anthony S, Bradford M, Edwards A, Francis CA, Hopkins M, Kay C, McDonald AT, Watkins J, Wilkinson J (2008) Faecal indicator organism concentrations and catchment export coefficients in the UK. Water Res 42:2649–2661

Ketchum BH, Ayers JC, Vaccaro RF (1952) Processes contributing to the decrease of coliform bacteria in a tidal estuary. Ecology 33(2):247–258

Kiefer LA, Shelton DR, Pachepsky Y, Blaustein R, Santin-Duran M (2012) Persistence of E. coli introduced into streabe dsedient with goose, deer and bovine animal waste. Lett Appl Microbiol 55(5):345–353

Kim JW, Pachepsky YA, Shelton DR, Coppock C (2010) Effect of streambed bacteria release on E. coli concentrations: Monitoring and modeling with the modified SWAT. Ecol Model 221(12):1592–1604

Kistemann T, Classen T, Koch C, Dangendorf F, Fischeder R, Gebel J, Vacata V, Exner M (2002) Microbial load of drinking water reservoir tributaries during extreme rainfall and runoff. Appl Environ Microbiol 68(5):2188–2197

Krause RM (1994) Dynamics of emergence. J Infect Dis 170(2):265–271

Liang Z, He Z, Zhou X, Powell C, Yang Y, He LM, Stoffella PJ (2013) Impact of mixed land-use practices on the microbial water quality in a subtropical coastal watershed. Sci Total Environ 449:426–433

Lipp EK, Rodriguez-Palacios C, Rose JB (2001) Occurrence and distribution of the human pathogen Vibrio vulnificus in a subtropical Gulf of Mexico estuary. Hydrobiologia 460(1):165–173

Ma LP, Li B, Zhang T (2014) Abundant rifampin resistance genes and significant correlations of antibiotic resistance genes and plasmids in various environments revealed by metagenomic analysis. Appl Microbiol Biotechnol 98:5195-5204. doi:10.1007/s00253-014-5511-3

Mac Kenzie WR, Hoxie NJ, Proctor ME, Gradus MS, Blair KA, Peterson DE, Kazmierczak JJ, Addiss DG, Fox KR, Rose JB, Davis JP (1994) A massive outbreak in Milwaukee of *Cryptosporidium* infection transmitted through the public water supply. N Engl J Med 331(3):161–167

MacKintosh C, Beattie KA, Klumpp S, Cohen P, Codd GA (1990) Cyanobacterial microcystin-LR is a potent and specific inhibitor of protein phosphatases 1 and 2A from both mammals and higher plants. FEBS Lett 264:187–192

Malakoff D (2002) Water quality: microbiologists on the trail of polluting bacteria. Science 295(5564):2352–2353

Martinez G, Pachepsky YA, Whelan G, Yakirevich AM, Guper A, Gish T (2014) Rainfall-induced fecal indicator organisms transport from manured fields: Model sensitivity analysis. Environ Int 63:121–129

Moog PL (1987) The hydrogeology and freshwater influx of Buttermilk Bay, Massachusetts, with regard to the circulation of coliforms and pollutants: a model study and development of methods for general application, M.S. Thesis. Boston University, Boston, MA, p 166

Muirhead RW, Monaghan RM (2012) A two reservoir model to predict Escherichia coli losses to water from pastures grazed by dairy cows. Environ Int 40:8–14

Muirhead RW, Davies-Colley RJ, Donnison AM, Nagels JW (2004) Faecal bacteria yields in artificial flood events: quantifying in-stream stores. Water Res 38(5):1215–1224

Nagels JW, Davies-Colley RJ, Donnison AM, Muirhead RW (2002) Faecal contamination over flood events in a pastoral agricultural stream in New Zealand. Water Sci Technol 45(12):45–52

Nasser AM, Zaruk N, Tenenbaum L, Netzan Y (2003) Comparative survival of *Cryptosporidium*, coxsackievirus A9 and *Escherichia coli* in stream, brackish- and sea waters. Water Sci Technol 47(3):91–96

Neitsch SL, Arnold JG, Kiniry JR, Williams JR, King KW (2005) Soil and water assessment tool: theoretical documentation, version 2005. Texas, U.S.A, Elsevier

Nelson EJ, Harris JB, Glenn Morris J, Calderwood SB, Camilli A (2009) Cholera transmission: the host, pathogen and bacteriophage dynamic. Nat Rev Microbiol 7(10):693–702

Nevecherya IK, Shestakov VM, Mazaev VT, Shlepnina TG (2005) Survival rate of pathogenic bacteria and viruses in groundwater. Water Resources 32(2):209–214

Oikonomou A, Katsiapi M, Karayanni H, Moustaka-Gouni M, Kormas KA (2012) Plankton microorganisms coinciding with two consecutive mass fish kills in a newly reconstructed lake. Sci World J 2012:1–14, doi:10.1100/2012/504135

Okun DA (1996) From cholera to cancer to cryptosporidiosis. J Environ Eng 122(6):453–458

Pachepsky YA, Shelton DR (2011) *Escherichia coli* and fecal coliforms in freshwater and estuarine sediments. Crit Rev Env Sci Tec 41(12):1067–1110

Pachepsky YA, Sadeghi AM, Bradford SA, Shelton DR, Guber AK, Dao T (2006) Transport and fate of manure-borne pathogens: Modeling perspective. Agric Water Manag 86(1–2):81–92

Painter JA, Hoekstra RM, Ayers T, Tauxe RV, Braden CR, Angulo FJ, Griffin PM (2013) Attribution of foodborne illnesses, hospitalizations, and deaths to food commodities by using outbreak data, United States, 1998–2008. Emerg Infect Dis 19(3):407–415

Pandey PK, Soupir ML (2012a) Non-point Source Pollution. Berkshire Encyclopedia of Sustainability: Ecosystem Management and Sustainability. Berkshire Publishing Group, LLC, Great Barrington, MA, U.S

Pandey PK, Soupir ML (2013) Assessing the impacts of *E. coli* laden streambed sediment on *E. coli* loads over a range of flows and sediment characteristics. J Am Water Resour Assoc 49(6):1261–1269

Pandey PK, Soupir ML, Rehmann CR (2012b) A model for predicting resuspension of *Escherichia coli* from streambed sediments. Water Res 46:115–126

Pang L (2009) Microbial removal rates in subsurface media estimated from published studies of field experiments and large intact soil cores. J Environ Qual 38:1531–1559, doi:10.2134/jeq2008.0379

Pang L, Close M, Goltz M, Sinton L, Davies H, Hall C, Stanton G (2004) Estimation of septic tank setback distances based on transport of *E. coli* and F-RNA phages. Environ Int 29(7):907–921

Parajuli PB, Douglas-Mankin KR, Barnes PL, Rossi CG (2009) Fecal bacteria source characterization and sensitivity analysis of SWAT 2005. Trans ASAE 52:1847–1858

Rao VC, Seidel KM, Goyal SM, Metcalf TG, Melnick JL (1984) Isolation of enteroviruses from water, suspended solids, and sediments from Galveston Bay: survival of poliovirus and rotavirus adsorbed to sediments. Appl Environ Microb 48(2):404–409

Rehmann CR, Soupir ML (2009) Importance of interactions between the water column and the sediment for microbial concentrations in streams. Water Res 43(18):4579–4589

Rhodes MW, Kator HI (1990) Effects of sunlight and autochthonous microbiota on Escherichia coli survival in an estuarine environment. Curr Microbiol 21(1):65–73

Rippey SR (1994) Infectious-diseases associated with molluscan shellfish consumption. Clin Microbiol Rev 7(4):419–425

Roberts MC, Soge OO, No D (2013) Comparison of multi-drug resistant environmental methicillin-resistant Staphylococcus aureus isolated from recreational beaches and high touch surfaces in built environment. Front Microbiol 4(74):1-8. doi:10.3389/fmicb.2013.00074

Rubentschik L, Roisin MB, Bieljansky FM (1936) Adsorption of bacteria in Salt Lakes. J Bacteriol 32(1):11–31

Rudolfs W, Falk LL, Ragotzkie RA (1950) Literature review on the occurrence and survival of enteric, pathogenic, and relative organisms in soil, water, sewage, and sludge, and on vegetation. Sewage Ind Wastes 22(10):1261–1281

Ruediger GF (1911) Studies on the self-purification of streams. Am J Public Health Assoc 1(6):411–415

Sayler GS, Nelson JD, Justice A, Colwell RR (1975) Distribution and significance of fecal indicator organisms in the Upper Chesapeake Bay. Appl Environ Microb 30(4):625–638

Schijven JF, Hassanizadeh SM (2000) Removal of viruses by soil passage: overview of modeling, processes, and parameters. Crit Rev Env Sci Tec 30(1):49–127

Schrader M, Hauffe T, Zhang Z, Davis GM, Jopp F, Remais JV, Wilke T (2013) Spatially explicit modeling of Schistosomiasis risk in eastern China based on a synthesis of epidemiological, environmental, and intermediate host genetic data. PLoS Negl Trop Dis 7(7):e2327

Schriewer A, Miller WA, Byrne BA, Miller MA, Oates S, Conrad PA, Hardin D, Yang HH, Chouicha N, Melli A, Jessup D, Dominik C, Wuertz S (2010) Presence of bacteroidales as a predictor of pathogens in surface waters of the Central California coast. Appl Environ Microbiol 76(17):5802–5814

Scott TW, Takken W, Knols BGJ, Boete C (2002) The ecology of genetically modified mosquitoes. Science 298(5591):117–119

Simons GW, Hilscher R, Ferguson HF, Gage SDM (1922) Report of the committee of bathing places. Am J Public Health 12(1):121–123

Sinton LW, Finlay RK, Lynch PA (1999) Sunlight inactivation of fecal bacteriophages and bacteria in sewage-polluted seawater. Appl Environ Microb 65(8):3605–3613

Smith RJ, Twedt RM, Flanigan LK (1973) Relationships of indicator and pathogenic bacteria in stream waters. Journal (Water Pollution Control Federation) 45(8):1736–1745

Smith EM, Gerba CP, Melnick JL (1978) Role of sediment in the persistence of enteroviruses in the estuarine environment. Appl Environ Microb 35(4):685–689

Smith RA, Alexander RB, Wolman MG (1987) Water-quality trends in the Nation's rivers. Science 235(4796):1607–1615

Smith J, Edwards J, Hilger H, Steck RTR (2008) Sediment can be a reservoir for coliform bacteria released into streams. J Gen Appl Microbiol 54:173–179

Snow J (1854) A Map of Cholera Deaths in London, 1840's. York University

Solo-Gabriele HM, Wolfert MA, Desmarais TR, Palmer CJ (2000) Sources of *Escherichia coli* in a coastal subtropical environment. Appl Environ Microb 66(1):230–237

Staley ZR, Rohr JR, Senkbeil JK, Harwood VJ (2014) Agrochemicals indirectly increase survival of E.coli O157:H7 and indicator bacteria by reducing ecosystem services. Ecol Appl, http://dx.doi.org/10.1890/13-1242.1

Steinmann P, Keiser J, Bos R, Tanner M, Utzinger J (2006) Schistosomiasis and water resources development: systematic review, meta-analysis, and estimates of people at risk. Lancet Infect Dis 6(7):411–425

Terzieva SI, McFeters GA (1991) Survival and injury of *E. coli*, *Campylobacter jejuni*, and *Yersinia Enterocolitica* in stream water. Can J Microbiol 37(10):785–790

U.S. Environmental Protection Agency (2010a) WATERS (Watershed Assessment, Tracking & Environmental Results). Washington, DC, http://water.epa.gov/scitech/datait/tools/waters/index.cfm. Accessed May 8 2014

U.S. Environmental Protection Agency (U.S. EPA) (1986) Ambient Water Quality Criteria for Bacteria. EPA 44015-84-002. Office of Regulations and Stand, Washington, DC, Accessed May 8 2014

U.S. Environmental Protection Agency (U.S. EPA) (2001) Protocol for Developing Pathogen TMDLs. National Summary of Impaired Waters and TMDL Information Washington, DC., Accessed May 8 2014

U.S. Environmental Protection Agency (U.S. EPA) (2010b) Protocol for Developing Pathogen TMDLs. National Summary of Impaired Waters and TMDL Information. Washington, DC, http://iaspub.epa.gov/waters10/attains_nation_cy.control?p_report_type=T. Accessed May 8 2014

U.S. Environmental Protection Agency (U.S. EPA) (2012a) Impaired Waters and Total Maximum Daily Loads., Accessed May 8 2014

U.S. Environmental Protection Agency (U.S. EPA) (2012b) Drinking Water Contaminants. National Primary Drinking Water Regulations., http://water.epa.gov/drink/contaminants/#one. Accessed May 18 2014

U.S. Environmental Protection Agency (U.S. EPA) (2014a) Protocol for Developing Pathogen TMDLs. National Summary of Impaired Waters and TMDL Information. Washington, DC, http://iaspub.epa.gov/waters10/attains_nation_cy.control?p_report_type=T. Accessed May 8 2014

U.S. Environmental Protection Agency (U.S. EPA) (2014b) Cyanobacterial Harmful Algal Blooms., http://www2.epa.gov/nutrient-policy-data/cyanobacterial-harmful-algal-blooms-cyanohabs. Accessed May 8 2014

U.S. Environmental Protection Agency (U.S. EPA) (2014c) Hypoxia., http://water.epa.gov/type/watersheds/named/msbasin/hypoxia101.cfm. Accessed May 8 2014

U.S. Geological Survey (U.S.G. S.) (2012) Dead Zone: the Source of the Gulf of Mexico's Hypoxia., http://www.usgs.gov/blogs/features/usgs_top_story/dead-zone-the-source-of-the-gulf-of-mexicos-hypoxia/. Accessed Apr 28 2014

U.S. Geological Survey (U.S.G. S.) (2014) The Gulf of Mexico Hypoxic Zone., http://toxics.usgs.gov/hypoxia/hypoxic_zone.html. Accessed May 8 2014

Unc A, Goss MJ (2004) Transport of bacteria from manure and protection of water resources. Appl Soil Ecol 25:1–18

United Nation Children's Fund (UNICEF) (2014) World Water Day 2025: 4,000 children die each day from a lack of safe water., http://www.unicef.org/wash/index_25637.html

Vareli K, Briasoulis E, Pilidis G, Sainis I (2009) Molecular confirmation of *Planktothrix rubescens* as the cause of intense, microcystin—synthesizing cyanobacterial bloom in Lake Ziros, Greece. Harmful Algae 8:447–453

Viau EJ, Goodwin KD, Yamahara KM, Layton BA, Sassoubre LM, Burns SL, Tong HI, Wong SHC, Lu Y, Boehm AB (2011) Bacterial pathogens in Hawaiian coastal streams — associations with fecal indicators, land cover, and water quality. Water Res 45:3279–3290

Wade TJ, Calderon RL, Sams E, Beach M, Brenner KP, Williams AH, Dufour AP (2006) Rapidly measured indicators of recreational water quality are predictive of swimming-associated gastrointestinal illness. Environ Health Perspect 114:24–28

Wang Y, Bradford SA, Simunek J (2014a) Estimation and upscaling of dual-permeability model parameters for the transport of Ecoli D21g in soils with preferential flow. J Contam Hydrol 159:57–66

Wang Y, Bradford SA, Simunek J (2014b) Physicochemical factors influencing the preferential transport of Escherichia coli in soils. Vadose Zone J 13(1) doi:10.2136/vzj2013.07.0120

Wcislo R, Chrost RJ (2000) Survival of *Escherichia coli* in freshwater. Pol J Environ Stud 9(3):215–222

Weiskel PK, Howes BL, Heufelder GR (1996) Coliform contamination of a coastal embayment: Sources and transport pathways. Environ Sci Technol 30(6):1872–1881

White DL, Bushek D, Porter DE, Edwards D (1998) Geographic information systems (GIS) and kriging: analysis of the spatial and temporal distributions of the oyster pathogen *Perkinsus marinus* in a developed and an undeveloped estuary. J Shellfish Res 17(5):1473–1476

WHO (2011) Millennium Development Goals: progress towards the health-related Millennium Development Goals., http://www.who.int/mediacentre/factsheets/fs290/en/index.html. Accessed 25 June 2012

Wilkes G, Brassard J, Edge TA, Gannon V, Jokinen CC, Jones TH, Lapen DR (2014) Coherence among different microbial source tracking markers in a small agricultural stream with or without livestock exclusion practices. Appl Environ Microbiol 79(20):6207–6219

Woolhouse MEJ (2002) Population biology of emerging and re-emerging pathogens. Trends Microbiol 10(10):s3–s7

World Bank (2010) Water Resource Management., Washington, D. C, http://www.worldbank.org/en/topic/waterresourcesmanagement. Accessed 26 Apr 2012

World Health Organization (WHO) (2010) Water Sanitation and Health., http://www.who.int/water_sanitation_health/diseases/en/. Accessed 20 Apr 2012

World Health Organization (WHO) (2014) Water Quality and Health. Drinking water chlorination – A review of disinfection practices and issues., http://www.waterandhealth.org/drinkingwater/wp.html. Accessed 28 Apr 2014

Yoshizawa S, Matsushima R, Watanabe MF, Harada K, Ichihara A, Carmichael WW, Fujiki H (1990) Inhibition of protein phosphatases by microcystis and nodularin associated with hepatotoxicity. J Cancer Res Clin Oncol 116:609–614

7

Production of aroma compounds from whey using *Wickerhamomyces pijperi*

Naoki Izawa*, Miyuki Kudo, Yukako Nakamura, Harumi Mizukoshi, Takahiro Kitada and Toshiro Sone

Abstract

The production of aroma compounds by the microbial fermentation of whey was studied. Seven strains of the yeast *Wickerhamomyces pijperi* were used for the fermentation of glucose-added whey (whey-g). Twelve aroma compounds (isobutanol, isoamyl alcohol, 2-phenylethanol, acetaldehyde, ethyl acetate, propyl acetate, isobutyl acetate, isoamyl acetate, ethyl butyrate, ethyl propionate, ethyl hexanoate and ethyl benzoate) were identified in the fermented broth using headspace gas chromatography mass spectrometry analysis. The major components were ethyl acetate (several tens to hundreds ppm), acetaldehyde (several tens ppm) and isoamyl alcohol (about 10 ppm). The strong fruity odor of ethyl benzoate (about 1 ppm) was detected in the broth of *W. pijperi* YIT 8095 and YIT 12779. The balance of aroma compounds produced was varied depending on the media used, and ethyl benzoate was only produced when using whey-g. The variation in the production of the aroma compounds over time using *W. pijperi* YIT 12779 at various culture temperatures (from 15–30°C) was also studied. From the results we propose that *W. pijperi* could be used as a novel microorganism for production of aroma compounds from whey.

Keywords: Whey; *Wickerhamomyces pijperi*; Aroma compound; Ethyl benzoate

Introduction

Whey is the liquid that remains after the precipitation and removal of milk casein during cheese-making. The dairy industry generates approximately 9 kg of whey for each kilogram of cheese, and its world production is estimated to be over 10^8 tons per year (Siso 1996; Grba et al. 2002). This whey by-product contains lactose (4.5–5% w/v), soluble proteins (0.6–0.8% w/v), lipids (0.4–0.5% w/v) and mineral salts (8–10% of dried extract). Cheese whey represents an important environmental problem because of its high organic matter content, with a BOD (Biochemical Oxygen Demand) = 30000–50000 ppm and a COD (Chemical Oxygen Demand) = 60000–80000 ppm (Siso 1996). Because of its rich nutrient content, whey has been used for the production of different chemicals such as ethanol fermentation. *Kluyveromyces marxianus*, a yeast capable of fermenting lactose to ethanol directly, has been studied in this respect. It has been used in ethanol production from crude whey (Zafar and Owais 2006), batch fermentation (Barba et al. 2001), fed-batch fermentation (Barba et al. 2001; Grba et al. 2002) and continuous fermentation (Christensen et al. 2011; Ozmihci and Kargi 2007). The production of other useful materials such as lactate, single-cell protein, enzymes and polysaccharides have also been studied using various microorganisms (Panesar and Kennedy 2012).

We are interested in studying the production of aroma compounds using whey. Aroma compounds have wide application in the food, cosmetic, chemical and pharmaceutical fields. Many aroma compounds on the market are produced via chemical synthesis or extraction from plant and animal sources. However, extraction from plants and animals is very expensive, and chemical synthesis often involves an environmentally unfriendly production process, sometimes affording undesirable racemic mixtures. Furthermore, consumers have a negative image of synthetic compounds (Vandamme and Soetaert 2002), and so (micro) biological origin aroma compounds have been the focus of much research. Many aroma compounds such as vanillin (Kaur and Chakraborty 2013; Priefert et al. 2001), 2-phenylethanol (2-PE, rose-like) (Etschmann et al. 2002), benzaldehyde (bitter almond, cherry) and 4-(*R*)-decanolide (fruity-fatty) (Krings and Berger 1998) have been observed via biotechnological production, but there are only a few studies that produce aroma compounds using whey as broth. Acetaldehyde, ethyl acetate, 1-propanol,

* Correspondence: naoki-izawa@yakult.co.jp
Yakult Central Institute, 5-11 Izumi, Kunitachi-shi, 186-8650 Tokyo, Japan

2-methyl-1-propanol, 2-methyl-1-butanol and 3-methyl-1-butanol were identified in whey fermented with *Kluyveromyces fragilis* (Parrondo et al. 2000). This group investigated the effects of temperature and agitation on the concentration of compounds in batch cultures, and went on to study certain compounds in continuous culture. In another study, forty aroma compounds were identified in distilled continuous whey fermentation using *K. marxianus* (Dragone et al. 2009). Whey utilization have been investigated using *Kluyveromyces* spp. and focused mainly on the ethanol fermentation (Guimarães et al. 2010). A possibility of the whey utilization would be expanded by using other microorganisms.

The aim of our study is to propose a novel method of reusing whey and adding new value through microbial fermentation, particularly with respect to the production of aroma compounds. In preliminary experiments, we noticed that a fruity odor was given off from whey fermented by the yeast *Wickerhamomyces pijperi* which were isolated from buttermilk, flower, partially decayed leaf and tree (http://www.nite.go.jp/en/nbrc/index.html). *W. pijperi* is phylogenetically closely related to *Candida solani*, and its morphologic and phylogenetic studies were examined by Imanishi et al. (2009). Only a few information is available about this yeast, to the best of our knowledge, there is only one report detecting this microorganism in the fermented liquid, Colombian Kumis (Chaves-López et al. 2012).

In this study, we investigated the *W. pijperi* fermentation of whey and analyzed the profiles of the aroma compounds produced when using different yeast strains, media and culture temperatures.

Materials and methods

Microorganisms and culture conditions

Seven strains of *W. pijperi* were used in this study. *W. pijperi* strain YIT (Yakult central Institute, Tokyo) 8095 (NBRC (Biological Resource Center, NITE) 1290), 12778 (NBRC 1791), 12779 (NBRC 1887), 12780 (NBRC 102058), 12781 (NBRC 102059), 12782 (NBRC 102060) and 12783 (NBRC 102061). Whey used in this study was the supernatant of 3% skimmed milk fermented using *Streptococcus thermophilus*, containing about 11 g/L lactose, 4 g/L galactose and 4 g/L lactate (no glucose present), 80 ng/L protein and had a pH of 4.0.

Twenty microliters of a 20% glycerol stock of *W. pijperi* was inoculated into 2 mL of YM (Yeast Mold) medium (Becton, Dickinson and company, Sparks, MD, USA) and cells were grown overnight at 30°C with 160 rpm reciprocal shaking (pre-preculture). Unless stated, all cultures in this study were carried out this way. For the screening of the strains, 20 μL of the pre-preculture broth was inoculated into 2 mL of whey with 10 g/L of glucose (whey-g), as *W. pijperi* cannot utilize lactose and galactose (Kurtzman and Fell 1998). Cells were further cultured overnight (preculture), then 50 μL of preculture broth was inoculated into 5 mL of whey-g in a test tube and cultured for 24 h.

For the comparison of media, 20 μL of pre-preculture was inoculated into 2 mL each of whey-g, YM medium (Becton, Dickinson and company), YPD (Yeast Extract Peptone Dextrose) medium (1% yeast extract, 2% peptone and 2% glucose) and 10% skim milk (Becton, Dickinson and company) and cultured overnight. Then, 50 μL of each culture broth was inoculated into 5 mL of the same media respectively, and cultured for 24 h.

For time-course analysis, 1 mL of whey-g preculture broth was inoculated into 100 mL of whey-g in a 200 mL Erlenmeyer flask and cultured at 15°C, 20°C, 25°C, 30°C and 35°C with 160 rpm rotary shaking. The culture broth was sampled at 0, 8, 24, 32, 48 and 56 h. All the samples were frozen at −30°C until use. Each datum point represents the mean of values from duplicate experiments.

Headspace gas chromatography mass spectrometry (HS-GC-MS)

An HS-GC-MS and an HS-GC- flame ionization detection (FID) protocols were modified from Dragone et al. (2009) and Saerens et al. (2008a) An HS-GC-MS system was used to identify the aroma compounds. This system consists of a 7000C GC/MS triple quadrupole mass spectrometer coupled to a GC 7890B (Agilent technologies, Palo Alto, CA, USA) and multi-purpose sampler MPS2-xt (Gerstel, Mühlheim/Ruhr, Germany). The samples were centrifuged at 8,000 rpm for 5 min and 2 mL of supernatant were collected in 20 mL glass tubes. Samples were heated for 15 min at 50°C with gentle shaking in the HS autosampler. One mL of sample was delivered into InertCap PureWAX column (30 m × 0.25 mm i.d., 0.25 μm, GL sciences, Tokyo, Japan). The flow rate of the carrier gas (He) was 1.0 mL/min, the injector temperature was 250°C and the transfer line temperature was 230°C. A split injection with a ratio of 1:10 was used. The oven temperature was held at 40°C for 5 min and then programmed to increase to 250°C at 10°C/min with a final 5 min hold at 250°C. The mass spectrometer was operated as follows: ionization voltage, 70 eV; ion source, EI mode; ion source temperature, 230°C. The mass spectrometer scanned from 34 to 400 (m/z) in 300 ms. The identification was performed by comparing the obtained mass spectra to the National Institute of Standards and Technology (NIST) 2011 mass spectral library.

HS-GC-FID

An HS-GC-FID system was used to quantify the aroma compounds. The samples were prepared in the same way as for HS-GC-MS. The system consisted of 7697A and GC7890 apparatus (Agilent technologies) with an

FID detector. Samples were heated for 15 min at 50°C with gentle shaking in the head space autosampler. Five mL of sample was delivered to an InertCap PureWAX column (30 m × 0.25 mm i.d., 0.25 μm, GL sciences). The flow rate of the carrier gas (He) was 3.0 mL/min. The injector temperature was 250°C, and the loop and transfer line temperatures were 100°C and 115°C, respectively. A split injection with a ratio of 1:20 was used. The oven temperature program was the same as for the HS-GC-MS analysis. The compounds were detected using an FID detector at a temperature of 250°C. Each identified aroma compounds was purchased and used for standard. Quantification was carried out by comparing with the peak area of 50 ppm (200 ppm for ethyl acetate) of each aroma compound.

Other analysis

Cell growth was measured at 600 nm using a Cary 60 UV-Vis spectrometer (Agilent technologies). Lactose, glucose, galactose, lactate and ethanol concentrations were quantified using a high performance liquid chromatography (HPLC) system. An HPLC protocol was modified from Chohnan et al. (2011). The system consisted of an alliance e2695 separation module (Waters corporation, Milford, MA, USA) equipped with a 2414 RI detector (Waters corporation). Ten micro liters of supernatant of culture broth was injected into a SUGAR SH1011 column (8.0 mm i.d. ×300 mm, Showa Denko, Tokyo, Japan) with 5 mM H_2SO_4 as eluent at 0.5 mL/min. The column and the detector temperatures were at 30°C and 37°C, respectively.

Results

Aroma compounds production of *W. pijperi*

Eleven aroma compounds were identified in whey-g fermented using *W. pijperi* (Table 1). Almost all of these compounds have a fruity aroma impression. The composition of aroma compounds varied depending on the strain of yeast used. High concentrations of acetaldehyde and ethyl acetate were observed, followed by isobutanol and isoamyl alcohol, and then isoamyl acetate in all culture broths. High production of ethyl acetate was observed in *W. pijperi* YIT 12779, 12780 and 12783. Ethyl benzoate was detected in the culture broth of *W. pijperi* YIT 8095 and 12779, which gave off an especially strong fruity odor. The odor of these two culture broths were recreated artificially using commercially available reagents at the detected concentration, and found that ethyl benzoate is important constituents of the fruity odor.

Influence of culture media in aroma compounds composition

The effects of the culture media on the aroma compounds composition were investigated. Whey-g, skim milk, YM and YPD were fermented using *W. pijperi* YIT 8095 and 12779 as they produced a strong odor in the initial whey-g experiments.

The composition of aroma compounds produced varied depending on the media. High production of ethyl benzoate, which has a characteristic fruity odor, was observed in whey-g with both strains. Isobutanol and isoamyl alcohol were highly produced in skim

Table 1 Aroma compounds produced by *W. pijperi* strains grown in whey-g

Compound (ppm)	8095	12778	12779	12780	12781	12782	12783	Aroma impression
Alcohols								
Isobutanol	8.15	7.25	7.41	9.58	6.94	5.33	9.37	Solvent[a]
Isoamyl alcohol	10.07	10.85	10.79	11.67	8.24	6.78	11.91	Alcoholic, banana[a]
Aldehyde								
Acetaldehyde	17.99	14.77	13.01	20.93	23.82	32.61	15.36	Fruity, solvent[b]
Acetate esters								
Ethyl acetate	42.72	80.17	198.38	174.55	63.89	75.91	159.20	Fruity, solvent[a]
Propyl acetate	0.00	0.13	0.29	1.22	0.00	0.00	0.24	Pineapple[b]
Isobutyl acetate	0.00	0.00	0.11	0.08	0.00	0.07	0.11	Fruity aroma[a]
Isoamyl acetate	0.04	0.17	0.30	0.21	0.28	0.08	0.24	Banana[a]
Ethyl esters								
Ethyl propionate	0.00	0.25	0.34	0.27	0.00	0.00	0.25	Fruity, rum[b]
Ethyl butyrate	0.60	0.14	0.61	0.13	0.00	0.00	0.14	Apple, banana[b]
Ethyl hexanoate	0.05	0.04	0.05	0.04	0.00	0.00	0.05	Apple, fruity[a]
Ethyl benzoate	1.14	0.00	1.24	0.00	0.00	0.00	0.00	fruity[c]

a: Pires et al. (2014).
b: Liu et al. (2004).
c: Niu et al. (2011).

Table 2 Difference in aroma compounds composition in different media

Compound (ppm)	12779				8095			
	Whey-g	Skim milk	YM	YPD	Whey-g	Skim milk	YM	YPD
Alcohols								
Isobutanol	15.17	33.80	17.04	27.53	14.79	25.97	15.86	13.34
Isoamyl alcohol	7.53	39.61	98.31	235.70	17.07	28.10	70.17	242.14
2-PE	0.00	0.00	27.81	28.49	0.00	0.00	14.73	30.60
Aldehyde								
Acetaldehyde	1.66	9.90	18.03	46.30	11.02	9.08	20.95	39.90
Acetate esters								
Ethyl acetate	305.73	7.18	102.18	234.76	69.05	3.52	28.43	291.34
Propyl acetate	0.98	0.00	0.00	0.67	0.20	0.00	0.00	0.69
Isobutyl acetate	1.25	0.07	0.10	0.00	0.08	0.00	0.00	0.00
Isoamyl acetate	5.07	0.24	1.56	3.09	0.58	0.00	0.38	2.07
Ethyl esters								
Ethyl propionate	1.01	0.18	1.17	2.12	0.57	0.00	0.86	2.37
Ethyl butyrate	0.57	0.56	1.08	0.80	0.55	0.27	1.00	3.25
Ethyl hexanoate	0.04	0.35	0.05	0.12	0.08	0.32	0.15	0.24
Ethyl benzoate	2.78	0.18	0.11	0.00	2.69	0.14	0.09	0.10

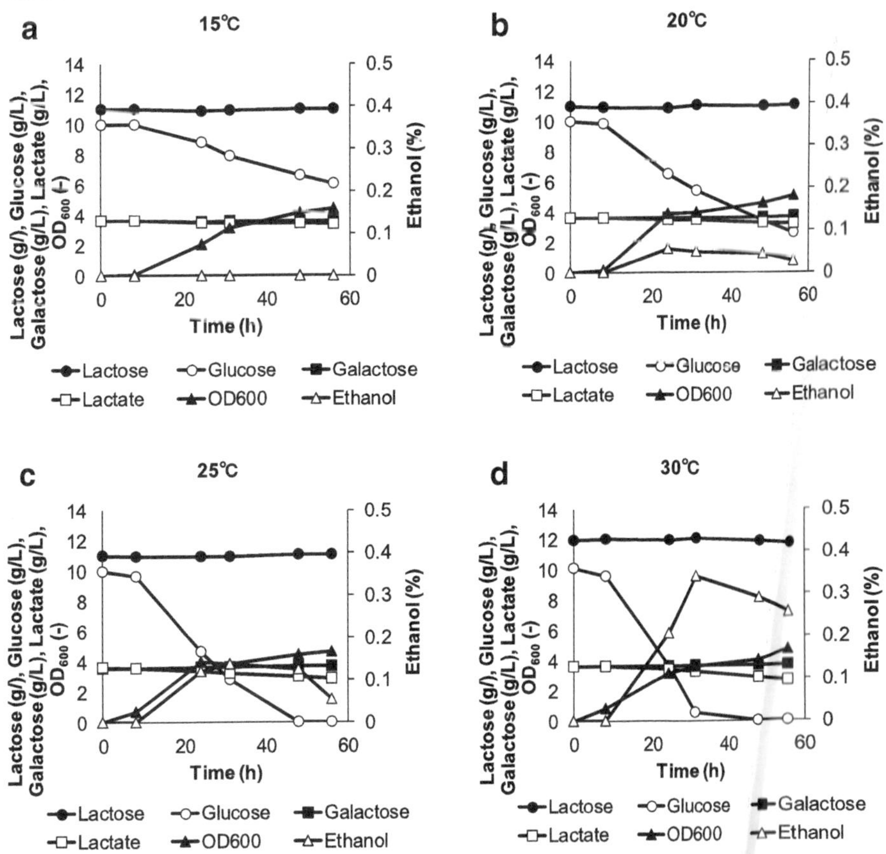

Figure 1 Time courses of growth, sugar and lactate consumption and ethanol production. *W. pijperi* YIT 12779 was cultured at **(a)** 15°C, **(b)** 20°C, **(c)** 25°C and **(d)** 30°C in the whey-g. Growth (*black triangles*), glucose (*white circles*), galactose (*black squares*), lactose (*black circles*), lactate (*white squares*) and ethanol (*white triangles*).

milk and YPD, respectively. 2-PE was only produced in YM and YPD (Table 2).

Influence of fermentation temperature on *W. pijperi* YIT 12779 growth and carbon source consumption

W. pijperi YIT 12779 was chosen for further research because of the distinctive fruity odor of its culture broth. The time courses for growth and carbon source consumption at various temperatures in whey-g are given in Figure 1 (this strain cannot grow at 35°C). Glucose was consumed first, followed by lactate at 25°C and 30°C, and the fastest glucose consumption was observed at 30°C. Compared with other temperatures, slow growth was observed at 15°C (Figure 1). Lactose and galactose were not consumed during the fermentation. Higher ethanol production was observed at higher temperatures, reaching 0.3% after 32 h at 30°C.

Influence of fermentation temperature on *W. pijperi* production of aroma compounds

Figures 2, 3, 4 show the effect of temperature on the production of aroma compounds. *W. pijperi* YIT 12779 was grown in whey-g at 15, 20, 25 and 30°C. Eleven aroma compounds were grouped as follows: alcohols (isobutanol and isoamyl alcohol) and aldehydes (acetaldehyde) (Figure 2); acetate esters (ethyl acetate, propyl acetate, isobutyl acetate and isoamyl acetate) (Figure 3); ethyl esters (ethyl propionate, ethyl butyrate, ethyl hexanoate and ethyl benzoate) (Figure 4). There were profile differences within the same group, for example, 3 acetate esters showed high production at 20°C and 25°C, but isobutyl acetate was also significantly produced at 15°C (Figure 3). In addition, the concentration of ethyl propionate kept increasing over the timeframe of the experiment, but levels of the other 3 ethyl esters decreased during the stationary phase (Figure 4).

Discussion

In this study, we found that whey-g fermentation using *W. pijperi* produces aroma compounds (Table 1). Twelve aroma compounds were identified in whey-g, skim milk, YM and YPD media (Table 2). Several tens ppm of isobutanol and isoamyl alcohol were detected in the whey fermented by *K. fragilis* (Parrondo et al. 2000). Additionally, several hundreds ppm of these compounds and ppb order of minor compounds were detected in the distilled *K. marxianus* fermented whey spirits (Dragone et al. 2009). So the level of isobutanol and

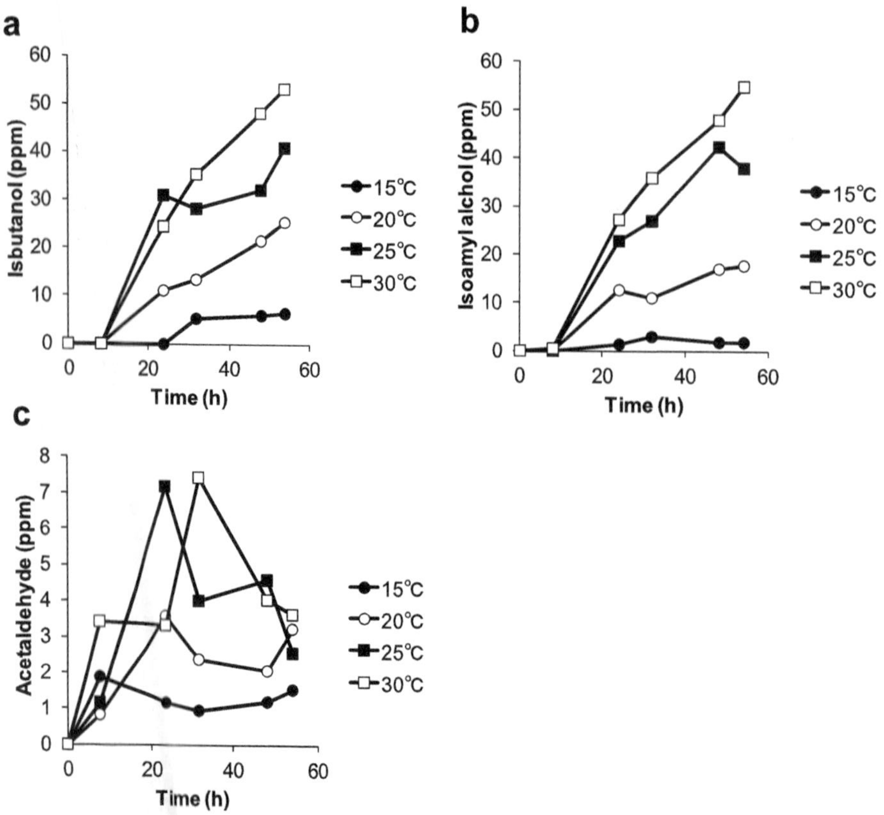

Figure 2 Time courses of alcohols and acetaldehyde concentrations at each temperature. *W. pijperi* YIT 12779 was cultured in the whey-g. The concentrations of **(a)** isobutanol, **(b)** isoamyl alcohol and **(c)** acetaldehyde. *Black circles*: 15°C, *white circles*: 20°C, *black squares*: 25°C and *white squares*: 30°C.

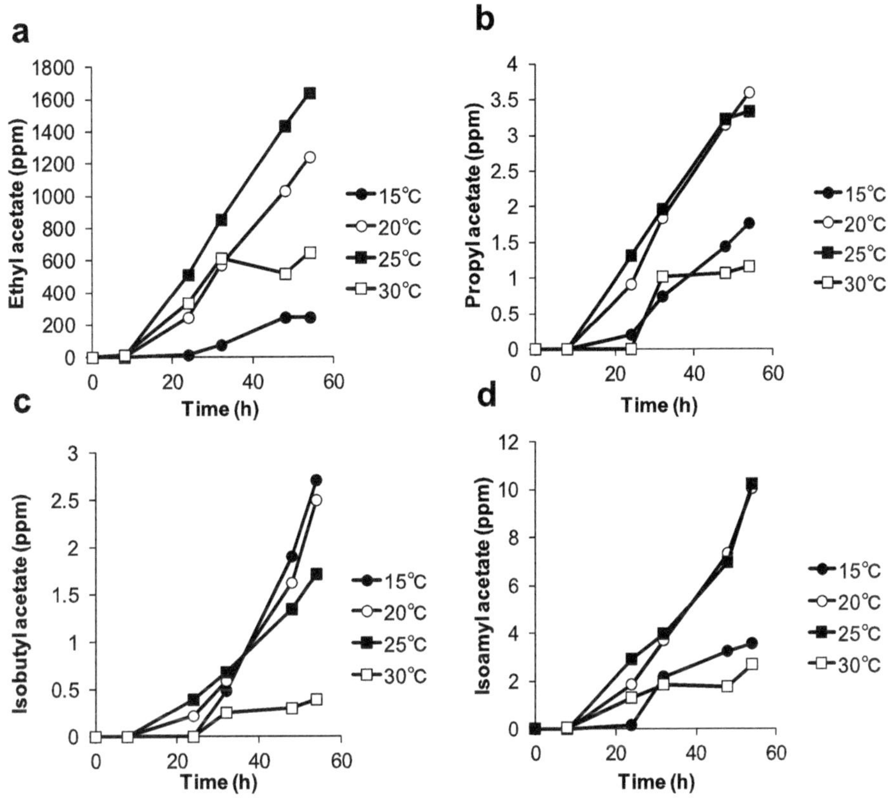

Figure 3 Time courses of acetate esters concentrations at each temperature. *W. pijperi* YIT 12779 was cultured in the whey-g. The concentrations of **(a)** ethyl acetate, **(b)** propyl acetate, **(c)** isobutyl acetate and **(d)** isoamyl acetate. *Black circles*: 15°C, *white circles*: 20°C, *black squares*: 25°C and *white squares*: 30°C.

isoamyl alcohol concentrations of *W. pijperi* fermented whey-g were low and the other minor aroma compounds were the same level as the *Kluyveromyces* spp. fermented whey. All of the aroma compounds had already been reported in *Saccharomyces* and *Kluyveromyces* fermentation, except for ethyl benzoate (Pires et al. 2014; Medeiros et al. 2000; Molina et al. 2007). It is novel that ethyl benzoate was produced from whey and *W. pijperi* possessed the ability to make it. The strong fruity odor was identified in *W. pijperi* YIT 8095 and 12779 fermented broth, these strains were collected from butter milk (Kurtzman et al. 2008) and partially decayed leaf, respectively (Mitaka and Banno 1979). Imanishi et al. (2009) showed that *W. pijperi* YIT 8095, 12778, 12779 and 12783 can be classified as neighbors by phylogenetic analysis based on three different nucleotide sequences (D1/D2 domain of 26S rDNA, the actin gene *ACT1* and the elongation factor 2 gene *FE2*). *W. pijperi* YIT 8095 and YIT 12779 are especially closely related according to D1/D2, and these 2 strains produced ethyl benzoate with levels up to 3 ppm (Tables 1 and 2). It is interesting that the ethyl benzoate strongly contributes to the fruity odor, considering that the absolute concentration of ethyl benzoate was very low compared with ethyl acetate, acetaldehyde, isoamyl alcohol and isobutanol. The phenotype that produces the strong fruity odor from YIT 8095 and YIT 12779 is extremely specific because YIT 12778 and YIT 12783 do not produce a fruity odor, even though these four strains are phylogenetically related.

The composition of the aroma compounds produced varied depending on the culture media (Table 2). Saerens et al. (2008a) noted that the precursor level present in the media can affect ester production while researching *Saccharomyces cerevisiae*. Likewise, a different composition of media could also affect the *W. pijperi* aroma compound production. Isoamyl alcohol was highly produced in YPD, and no production of 2-PE was observed in whey-g and skim milk. Considering that leucine and phenylalanine are precursors of isoamyl alcohol and 2-PE, respectively, this demonstrates that differences in the amino acid composition of the media affects the composition of the aroma compounds produced (Dickinson et al. 1997; Etschmann et al. 2002).

It is interesting that ethyl benzoate can be produced from whey-g but not from skim milk. Lactic acid bacteria, such as *S. thermophilus*, that are involved in the fermentation of skim milk have cell-envelope proteinase and peptide transport systems that use exogenous protein

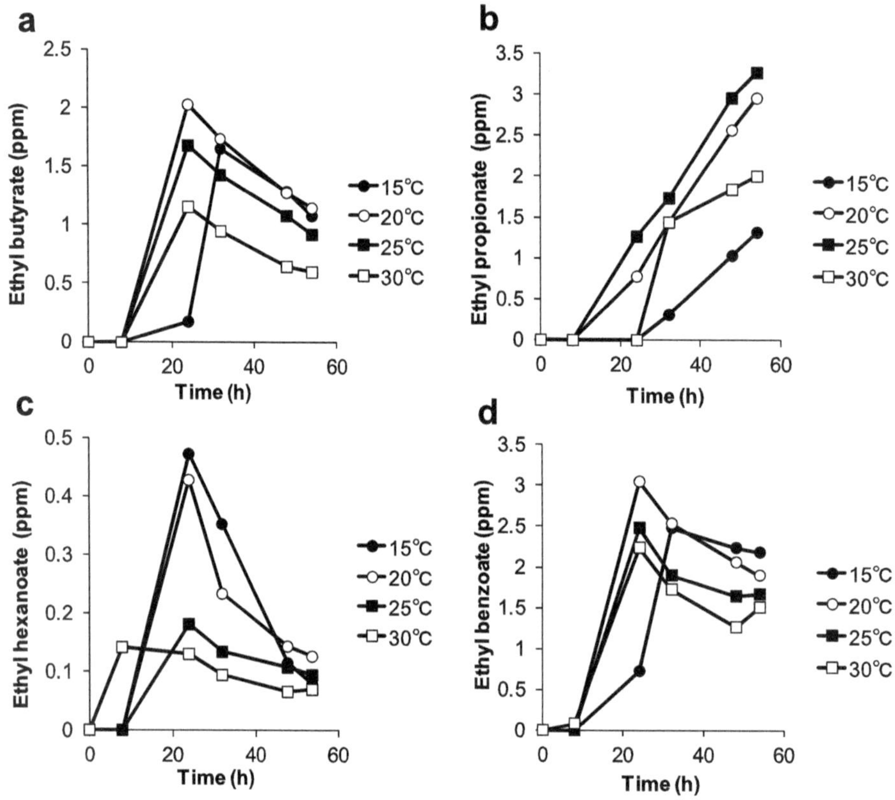

Figure 4 Time courses of ethyl esters concentrations at each temperature. *W. pijperi* YIT 12779 was cultured in the whey-g. The concentrations of **(a)** ethyl butyrate, **(b)** ethyl propionate, **(c)** ethyl hexanoate and **(d)** ethyl benzoate. *Black circles*: 15°C, *white circles*: 20°C, *black squares*: 25°C and *white squares*: 30°C.

(Savijoki et al. 2006). Extracellular proteinases can also release peptides from proteins in the culture medium (Yamamoto et al. 1994). This goes some way to explaining how *S. thermophilus* can convert milk protein into peptides, which can be utilized by *W. pijperi* to produce ethyl benzoate.

In the time-course experiments, glucose was consumed first, then lactate consumption began after the glucose levels had depleted when *W. pijperi* YIT 12779 was studied at temperatures up to 30°C (Figure 1). We confirmed *W. pijperi* YIT 8095 also did not use lactose and galactose experimentally. Referred to the NCYC (National Collection of Yeast Cultures) website (http://www.ncyc.co.uk/), *W. pijperi* do not use these sugars generally. The next challenge will be lactose and galactose utilization by *W. pijperi* in whey. This would be a great advantage environmentally, and gene manipulation is likely to be the best strategy to accomplish this. Although it has not yet been reported for *W. pijperi*, there are some reports demonstrating similar transformations in assortative *Pichia pastoris* (Morton and Potter 2000; O'Callaghan et al. 2002; Yuan et al. 2008).

Time-course studies of the production of aroma compounds were shown to be affected by temperature (Figure 2, 3, 4). Saerens et al. (2008a; 2008b) also reported that the production profile of esters by *S. cerevisiae* was affected differently by temperature, depending on the type of ester. High concentration of alcohols (ethanol, isobutanol and isoamyl alcohol) were obtained at high temperature, and levels of isobutanol and isoamyl alcohol tend to increase as time went on (Figures 1 and 2). The differences in the production profiles may be due to the differences in the biosynthetic pathways. Ethanol is produced at the end of the glycolytic pathway, but isoamyl alcohol and isobutanol are produced during leucine and valine synthesis, respectively (Dickinson et al. 1997; Matsuda et al. 2012).

Acetaldehyde was detected at higher levels at 25°C and 30°C, and the concentration decreased during the stationary phase (Figure 2c). Acetaldehyde is synthesized from pyruvate and metabolized to ethanol during ethanol fermentation. The profile for acetaldehyde may appear different to the other compounds as synthesis and consumption occur simultaneously. The observed decrease may also be due to evaporation during fermentation and sample preparation as the boiling point is very low (20.2°C).

Ester formation has been well studied in many organisms (Park et al. 2009), and the alcohol o-acyltransferases (ATF) are one class of enzymes involved in this process. In *S. cerevisiae*, acetate esters are produced from acetyl-CoA and alcohol using alcohol acetyltransferases, and ethyl esters are produced by a condensation reaction between an acyl-CoA unit and ethanol using acyl-CoA/ethanol o-acyltransferases. Although these enzymes have not been specifically observed in *W. pijperi*, it may work in a similar way. It is interesting that there was a difference in production between isoamyl acetate and isobutyl acetate, which are formed by the combination of branched fatty acids and acetic acid (Figure 3). It has been shown that ATFs have preferences for substrates (Rodriguez et al. 2014), so this difference may be due to the ATF in *W. pijperi* favoring isobutanol over isoamyl alcohol.

Ethyl propionate was shown to increase over time, but the levels of the 3 other ethyl esters decreased in the stationary phase (after 20 hours) (Figure 4). It is unlikely that these compounds could be metabolized to other substances. Although propionate is supplied directly via isoleucine metabolism, there is no direct amino acid direct supplier for hexanoate, benzoate and butyrate (KEGG: Kyoto Encyclopedia of Genes and Genomes, http://www.genome.jp/kegg/). It is thought that the different precursor supplementation pathways may cause the observed differences in the profiles, and that evaporation exceeded synthesis for the 3 esters that showed a decrease in the stationary phase.

We found that ethyl benzoate was important components of the fruity odor by sniffing artificially prepared samples. It is very important aroma compound used for fragrance, flavor and solvents. However, only a few papers discuss ethyl benzoate being found in fermentation products (Chen and Xu 2010; Zea et al. 2001) as it is mainly detected in fruit (Bartley and Schwede 1989; Thuaytong and Anprung 2011). To the best of our knowledge, fermentative production of ethyl benzoate has not yet been reported.

The biotechnological production of aroma compounds is very attractive for the reasons outlined in the introduction. Furthermore, as it is independent from agriculture and any potential disruptions to production, it can be carried out on an industrial scale using engineered pathways to produce what are regarded as natural products (Krings and Berger 1998). The results of this study will help to develop a novel bio production of aroma compounds. Further studies will be necessary to increase aroma compound production by investigating detailed fermentation conditions and developing genetically engineered yeast strains.

Abbreviations

YIT: Yakult Central Institute, Tokyo; NBRC: Biological Resource Center, NITE; YM: Yeast mold; YPD: Yeast extract peptone dextrose; HS-GC-MS: Headspace gas chromatography mass spectrometry; HS-GC-FID: Headspace gas chromatography flame ionization detection; HPLC: High performance liquid chromatography; NCYC: National collection of yeast cultures; ATF: Alcohol o-acyltransferases; KEGG: Kyoto encyclopedia of genes and genomes.

Competing interests

The authors declare that they have no competing interests.

Authors' contribution

NI designed the study and carried out the fermentation and the HS-GC-FID and drafted the manuscript. MK participated in the HS-GC-FID. YN, HM and TK carried out the HS-GC-MS. TS participated in the design and coordination of the study. All authors read and approved the final manuscript.

Acknowledgements

The authors thank Yasuyuki Takahashi, Nobutaka Takahashi, Akemi Imaoka, Yuki Ishii and Daisuke Niwa for helpful discussions.

References

Barba D, Beolchini F, Del Re G, Di Giacomo G, Veglió F (2001) Kinetic analysis of *Kluyveromyces lactis* fermentation on whey: batch and fed-batch operations. Process Biochem 36:531–6, doi:10.1016/S0032-9592(00)00242-9

Bartley JP, Schwede AM (1989) Production of volatile compounds in ripening kiwi fruit (*Actinidia chinensis*). J Agric Food Chem 37:1023–5, doi:10.1021/jf00088a046

Chaves-López C, Tofalo R, Serio A, Paparella A, Sacchetti G, Suzzi G (2012) Yeasts from Colombian kumis as source of peptides with angiotensin I converting enzyme (ACE) inhibitory activity in milk. Int J Food Microbiol 159:39–46, doi:10.1016/j.ijfoodmicro.2012.07.028

Chen S, Xu Y (2010) The influence of yeast strains on the volatile flavor compounds of Chinese rice wine. J Inst Brew 116:190–6, doi:10.1002/j.2050-0416.2010.tb00417.x

Christensen AD, Kádár Z, Oleskowicz-Popiel P, Thomsen MH (2011) Production of bioethanol from organic whey using *Kluyveromyces marxianus*. J Ind Microbiol Biotechnol 38:283–9, doi:10.1007/s10295-010-0771-0

Chohnan S, Nakane M, Rahman H, Nitta Y, Yoshiura T, Ohta H, Kurusu Y (2011) Fuel ethanol production from sweet sorghum using repeated-batch fermentation. J Biosci Bioeng 111:433–6, doi:10.1016/j.jbiosc.2010.12.014

Dickinson JR, Lanterman MM, Danner DJ, Pearson BM, Sanz P, Harrison SJ, Hewlins MJE (1997) A 13C nuclear magnetic resonance investigation of the metabolism of leucine to isoamyl alcohol in *Saccharomyces cerevisiae*. J Biol Chem 272:26871–8, doi:10.1074/jbc.272.43.26871

Dragone G, Mussatto SI, José M, Oliveira JM, José A, Teixeira JA (2009) Characterisation of volatile compounds in an alcoholic beverage produced by whey fermentation. Food Chem 112:929–35, doi:10.1016/j.foodchem.2008.07.005

Etschmann MMW, Bluemke W, Sell D, Schrader J (2002) Biotechnological production of 2-phenylethanol. Appl Microbiol Biotechnol 59:1–8, doi:10.1007/s00253-002-0992-x

Grba S, Stehlik-Tomas V, Stanzer D, Vahčić N, Škrlin A (2002) Selection of yeast strain *Kluyveromyces marxianus* for alcohol and biomass production on whey. Chem Biochem Eng Q 16:13–6

Guimarães PMR, Teixeira JA, Domingues L (2010) Fermentation of lactose to bio-ethanol by yeasts as part of integrated solutions for the valorisation of cheese whey. Biotechnol Adv 28:375–84

Imanishi Y, Jindamorakot S, Limtong S, Nakase T (2009) Mode of vegetative reproduction of the bipolar budding yeast species *Wickerhamomyces pijperi* and related strains. Microbiology 155:3142–8, doi:10.1099/mic. 0.029231-0

Kaur B, Chakraborty D (2013) Biotechnological and molecular approaches for vanillin production: a review. Appl Biochem Biotechnol 169:1353–72, doi:10.1007/s12010-012-0066-1

Krings U, Berger RG (1998) Biotechnological production of flavours and fragrances. Appl Microbiol Biotechnol 49:1–8, doi:10.1007/s002530051129

Kurtzman C, Fell JW (1998) The yeasts. A Taxonomic Study, 4th edn. Elsevier Science, Amsterdam, The Netherlands

Kurtzman CP, Robnett CJ, Basehoar-Powers E (2008) Phylogenetic relationships among species of *Pichia*, *Issatchenkia* and *Williopsis* determined from multigene sequence analysis, and the proposal of *Barnettozyma* gen. nov., *Lindnera* gen. nov. and *Wickerhamomyces* gen. nov. FEMS Yeast Res 8:939–54, doi:10.1111/j.1567-1364.2008.00419.x

Liu SQ, Holland R, Crow VL (2004) Esters and their biosynthesis in fermented dairy products: a review. Int Dairy J 14:923–45, doi:10.1016/j.idairyj.2004.02.010

Matsuda F, Kondo T, Ida K, Tezuka H, Ishii J, Kondo A (2012) Construction of an artificial pathway for isobutanol biosynthesis in the cytosol of *Saccharomyces cerevisiae*. Biosci Biotechnol Biochem 76:2139–41, doi:10.1271/bbb.120420
Medeiros ABP, Pandey A, Freitas RJS, Christen P, Soccol CR (2000) Optimization of the production of aroma compounds by *Kluyveromyces marxianus* in solid-state fermentation using factorial design and response surface methodology. Biochem Eng J 6:33–9, doi:10.1023/A:1013596330389
Mitaka K, Banno I (1979) Descriptive catalogue of IFO yeast collection II. IFO Res Comm 9:65–70
Molina AM, Swiegers JH, Varela C, Pretorius IS, Agosin E (2007) Influence of wine fermentation temperature on the synthesis of yeast-derived volatile aroma compounds. Appl Microbiol Biotech 77:675–87, doi:10.1007/s00253-007-1194-3
Morton CL, Potter PM (2000) Comparison of *Escherichia coli, Saccharomyces cerevisiae, Pichia pastoris, Spodoptera frugiperda*, and COS7 cells for recombinant gene expression. Mol Biotechnol 16:193–202, doi:10.1385/MB:16:3:193
Niu Y, Zhang X, Xiao Z, Song S, Eric K, Jia C, Yu H, Zhu J (2011) Characterization of odor-active compounds of various cherry wines by gas chromatography-mass spectrometry, gas chromatography-olfactometry and their correlation with sensory attributes. J Chromatogr B 879:2287–93, doi:10.1016/j.jchromb.2011.06.015
O'Callaghan J, O'Brien MM, McClean K, Dobson ADW (2002) Optimisation of the expression of a *Trametes versicolor* laccase gene in *Pichia pastoris*. J Ind Microbiol Biotechnol 29:55–9, doi:10.1038/sj.jim.7000268
Ozmihci S, Kargi F (2007) Continuous ethanol fermentation of cheese whey powder solution: effects of hydraulic residence time. Bioprocess Biosyst Eng 30:79–86, doi:10.1007/s00449-006-0101-0
Panesar PS, Kennedy JF (2012) Biotechnological approaches for the value addition of whey. Crit Rev Biotechnol 32:327–48, doi:10.3109/07388551.2011.640624
Park YC, Shaffer CEH, Bennett GN (2009) Microbial formation of esters. Appl Microbiol Biotechnol 85:13–25, doi:10.1007/s00253-009-2170-x
Parrondo J, Garcia LA, Diaz M (2000) Production of an alcoholic beverage by fermentation of whey permeate with *Kluyveromyces fragilis* II: aroma composition. J Inst Brew 106:377–82, doi:10.1002/j.2050-0416.2000.tb00528.x
Pires EJ, Teixeira JA, Brányik T, Vicente AA (2014) Yeast: the soul of beer's aroma—a review of flavour-active esters and higher alcohols produced by the brewing yeast. Appl Microbiol Biotech 98:1937–49, doi:10.1007/s00253-013-5470-0
Priefert H, Rabenhorst J, Steinbüchel A (2001) Biotechnological production of vanillin. Appl Microbiol Biotechnol 56:296–314, doi:10.1007/s002530100687
Rodriguez GM, Tashiro Y, Atsumi S (2014) Expanding ester biosynthesis in *Escherichia coli*. Nat Chem Biol 10:259–65, doi:10.1038/nchembio.1476
Saerens SMG, Delvaux F, Verstrepen KJ, Van Dijck P, Thevelein JM, Delvaux FR (2008a) Parameters affecting ethyl ester production by *Saccharomyces cerevisiae* during fermentation. Appl Environ Microbiol 74:454–61, doi:10.1128/AEM. 01616-07
Saerens SMG, Verbelen PJ, Vanbeneden N, Thevelein JM, Delvaux FR (2008b) Monitoring the influence of high-gravity brewing and fermentation temperature on flavour formation by analysis of gene expression levels in brewing yeast. Appl Microbiol Biotechnol 80:1039–51, doi:10.1007/s00253-008-1645-5
Savijoki K, Ingmer H, Varmanen P (2006) Proteolytic systems of lactic acid bacteria. Appl Microbiol Biotechnol 71:394–406, doi:10.1007/s00253-006-0427-1
Siso MIG (1996) The biotechnological utilization of cheese whey: A review. Bioresour Technol 57:1–11, doi:10.1016/0960-8524(96)00036-3
Thuaytong W, Anprung P (2011) Bioactive compounds and prebiotic activity in Thailand-grown red and white guava fruit. (*Psidium guajava L.*). Food Sci Technol Int 17:205–12, doi:10.1177/1082013210382066
Vandamme EJ, Soetaert W (2002) Bioflavours and fragrances via fermentation and biocatalysis. J Chem Technol Biotechnol 77:1323–32, doi:10.1002/jctb.722
Yamamoto N, Akino A, Takano T (1994) Antihypertensive effects of different kinds of fermented milk in spontaneously hypertensive rats. Biosci Biotech Biochem 58:776–8, doi:10.1271/bbb.58.776
Yuan T, Yang P, Wang Y, Meng K, Luo H, Zhang W, Wu N, Fan Y, Yao B (2008) Heterologous expression of a gene encoding a thermostable β-galactosidase from Alicyclobacillus acidocaldarius. 30:343-348. doi:10.1007/s10529-007-9551-y.
Zafar S, Owais M (2006) Ethanol production from crude whey by *Kluyveromyces marxianus*. Biochem Eng J 27:295–8, doi:10.1016/j.bej.2005.05.009
Zea L, Moyano L, Moreno J, Cortes B, Medina M (2001) Discrimination of the aroma fraction of Sherry wines obtained by oxidative and biological ageing. Food Chem 75:79–84, doi:10.1016/S0308-8146(01)00190-X

Phenotypic and metabolic traits of commercial *Saccharomyces cerevisiae* yeasts

Catarina Barbosa, Patrícia Lage, Alice Vilela, Arlete Mendes-Faia and Ana Mendes-Ferreira*

Abstract

Currently, pursuing yeast strains that display both a high potential fitness for alcoholic fermentation and a favorable impact on quality is a major goal in the alcoholic beverage industry. This considerable industrial interest has led to many studies characterizing the phenotypic and metabolic traits of commercial yeast populations. In this study, 20 *Saccharomyces cerevisiae* strains from different geographical origins exhibited high phenotypic diversity when their response to nine biotechnologically relevant conditions was examined. Next, the fermentation fitness and metabolic traits of eight selected strains with a unique phenotypic profile were evaluated in a high-sugar synthetic medium under two nitrogen regimes. Although the strains exhibited significant differences in nitrogen requirements and utilization rates, a direct relationship between nitrogen consumption, specific growth rate, cell biomass, cell viability, acetic acid and glycerol formation was only observed under high-nitrogen conditions. In contrast, the strains produced more succinic acid under the low-nitrogen regime, and a direct relationship with the final cell biomass was established. Glucose and fructose utilization patterns depended on both yeast strain and nitrogen availability. For low-nitrogen fermentation, three strains did not fully degrade the fructose.

This study validates phenotypic and metabolic diversity among commercial wine yeasts and contributes new findings on the relationship between nitrogen availability, yeast cell growth and sugar utilization. We suggest that measuring nitrogen during the stationary growth phase is important because yeast cells fermentative activity is not exclusively related to population size, as previously assumed, but it is also related to the quantity of nitrogen consumed during this growth phase.

Keywords: Yeast; Nitrogen requirements; Stress resistance; Alcoholic fermentation

Introduction

In most wine-producing countries, inoculating grape-juice with active-dried selected yeasts is a routine practice that yields reliable fermentations and reproducible product as well as predictable sensorial properties and quality. Modern winemaking is based on a monoculture inoculation technology that began in Germany with Müller-Thurgau in 1896 and was widely established in the second half of the twentieth century. Alternative yeast species combined with inoculation strategies yield wines with very different chemical and flavor profiles, such as a greater flavor complexity and diversity as well as enhanced mouth feel and flavor persistence (Ugliano and Henschke 2009). However, in certain types of wines, the yeast should minimally impact the varietal aroma derived from the grapes. The commercial yeast strains are natural isolates from vineyards or derived from breeding programs and are selected based on certain phenotypic characteristics, including tolerance to temperature variation, ethanol and SO_2, low foaming production, completion of alcoholic fermentation to dryness (residual sugars <2 g l^{-1}), no off-flavor production, including sulfides, and production of interesting aroma compounds (Boulton et al. 1996). During alcoholic fermentation, yeasts encounter a combination of several stress conditions. Yeasts are exposed to a number of factors due to the composition of grape-juice, such as high sugar levels, variable nitrogen concentration, low pH and potential inhibitors, including SO_2 and temperature variation. In addition, there are several stress conditions established by the yeast, including increased ethanol and inhibitor production. Yeast performance during alcoholic fermentation greatly depends on its ability to adjust to continuous

* Correspondence: anamf@utad.pt
Institute for Biotechnology and Bioengeneering – Centre of Genomics and Biotechnology, Universidade de Trás-os-Montes e Alto Douro, Vila Real, Portugal

environmental changes. The active dry yeast performance, including fermentation capacity and flavor release, greatly depends on the genetic background and fermentation conditions (Pretorius 2000). Estimates indicate that the market includes over 200 available yeast strains that are used globally for winemaking, and these are primarily *Saccharomyces cerevisiae* strains. In recent years, large-scale systematic studies on the genetic and phenotypic diversity of *S. cerevisiae* populations have emerged (Kvitek et al. 2008; Liti et al. 2009; Camarasa et al. 2011, Dunn et al. 2012; Richter et al. 2013). These studies used a high-throughput stress resistance analysis or adaptation to diverse environments (carbon and nitrogen sources, presence of toxins and nutrient limitations) to characterize the phenotypes of the yeast strains of interest, such as strains related to wine-making or fermentation of other alcoholic beverages (Camarasa et al. 2011; Nikolaou et al. 2006; Zuzuarregui and del Olmo 2004; Kvitek et al. 2008). In this work, we aimed to (i) study the phenotypic diversity of a small subset of commercial wine yeasts; (ii) assess their nitrogen requirements for growth and fermentation; and (iii) analyze their growth kinetics and metabolic traits. This study provides additional potential criteria for inclusion in a yeast selection program.

Material and methods

Yeasts and maintenance conditions

Twenty *S. cerevisiae* strains, all wine strains except K7 which is a sake yeast strain, from geographically diverse environments (Table 1) were characterized in this study. The strains were sourced from different yeast culture collections, such as the Oenology Culture Collection of the Department of Viticulture and Oenology, University of California, Davis, USA and the National Research Institute of Brewing (NRIB), Japan; the additional cultures were acquired from the market.

The strains were collected aseptically from active dried commercial preparations, re-hydrated in sterile water with 50 g l^{-1} of glucose in accordance with the manufacturer's instructions (37°C, for 30 min) and inoculated into yeast peptone dextrose medium (YPD), which included, per liter, 20 g glucose, 10 g peptone and 5 g yeast extract. The cultures were then streaked onto Wallenstein agar plates (WL) and grown at 30°C to determine the purity. Multiple representative colonies were inoculated into YPD broth and were starter cultures for the stress resistance assays. Pure cultures were routinely maintained at 4°C on YPD slants, and the stocks were stored at −80°C with glycerol at a final concentration of 40% (v/v).

Table 1 Commercial yeasts used in this study and respective origin countries and suppliers

Yeast strain	Geographical origin	Supplier
T73	Spain	Lalvin
EC1118	France	Lalvin
FERMIVIN	France	DSM
ZYMAFLORE VL1	France	Laffort
QA23	Portugal	Lalvin
CEG	Germany	UVAFERM
VIN 13	South Africa	Anchor
NT116	South Africa	Anchor
BM45	Italy	UVAFERM
BRL97	Italy	UVAFERM
FERMICRU XL	Chile	DSM
FERMICRU LVCB	Chile	DSM
XLD	Chile	Yeasts Collection DBVPG, Chile
UCD522	USA	Maurivin
UCD595	USA	UC Davis Collection, California, USA
UCD505	USA	UC Davis Collection, California, USA
AWRI796	Australia	Maurivin
AWRIR2	Australia	Maurivin
W3	Japan	National Research Institute of Brewing (NRIB), Higashiroshima, Japan
K7	Japan	National Research Institute of Brewing (NRIB), Higashiroshima, Japan

S. cerevisiae physiological and metabolic characterization

Stress resistance assays

To systematically analyze the particular stresses that affect yeast during wine production stages, each yeast strain was grown under different conditions that represent six environments to induce various physiological responses. The cells were grown in YPD medium until the mid-exponential growth phase ($OD_{600nm} = 1$), and three 10-fold serial dilutions were generated. For each strain, the four culture suspensions were spotted onto YPD agar plates with the appropriate stress and onto YPD agar plates without a stress agent, as a control, followed by incubation for 2–3 days, at 30°C.

Temperature variation

The tolerance to temperature variations was evaluated by spotting 5 μl of the dilutions from each yeast culture onto YPD agar plates after incubation at 15, 37 or 40°C, for 1–3 days.

Osmotic and oxidative shocks

The osmotic and oxidative shock tolerances were evaluated by spotting 5 µl of the dilutions from each yeast culture onto YPD agar plates with 1, 1.5 or 2 M of NaCl (from a 5 M stock solution) or 1.5 or 2.5 mM of H_2O_2 (from a 10 mM stock solution), respectively.

Ethanol resistance

Ethanol tolerance was evaluated by spotting 5 µl of the dilutions from each yeast culture onto YPD agar plates with different ethanol concentrations (10, 12.5, and 15% [vol/vol]).

Sulfur dioxide (SO_2) resistance

The sulfur dioxide resistance was evaluated based on the cell growth under different SO_2 concentrations (2, 4 and 6 mM) through adding 5 ml of a stock solution (30, 60 and 90 mM) to the YPD medium at pH 3.5.

Acetic acid resistance

To evaluate the capacity of yeast to grow in the presence of acetic acid, 5 µl of the dilutions from each yeast culture were spotted onto a YPD medium containing 90, 110 or 150 mM acetic acid from a 5 M stock solution at pH 4.5.

Each yeast strain was tested for each stress at least twice at the aforementioned 2–3 doses for most stresses on three different occasions. Final resistance scores for the initial suspension (OD_{600nm} = 1) and the 3 serial 10-fold dilutions were summed, and the resulting scores were then averaged for the replicates and the stress doses, providing a single score ranging from 0 (no growth) to 8 (active growth) for each strain and each stress condition.

Identifying 5, 5′, 5″-trifluor-D, L-leucine (TFL)- and/or cerulenin-resistant strains

To detect strains that produced certain flavor compounds, the strains were tested for the capacity to grow in the presence of TFL and/or cerulenin. The screens were conducted using agar plates and minimal medium (YNB) supplemented with glucose (2% [wt/vol]) and TFL (1 mM) (Oliveira et al. 2008) or cerulenin (6 µM) (Franco-Duarte et al. 2009) followed by incubation at 30°C for 3 days. For the remaining resistance assays, the final resistance scores were calculated using the above-described method.

Evaluation of sulfite reductase activity

To examine the capacity to produce H_2S, the sulfite reductase activity was estimated using a commercially available bismuth-containing agar (Bacto Biggy agar, Difco, Sparks, MD, USA). The culture media was prepared according to the manufacturer's instructions. After 48 hours at 30°C, H_2S formation was evaluated based on the varying colony color intensities, which turn light brown to black or remain white depending on the level of production (Mendes-Ferreira et al. 2002). The yeast growth scores were based on the colony color, which ranged from white (scored 0) through brown to near-black (scored 8).

Inoculum preparation and fermentation conditions

Based on the results, eight out of twenty strains were selected for fermentation experiments under two nitrogen regimes. Each yeast strain was rehydrated in accordance with the manufacturer's instructions (in water with 50 g l^{-1} of glucose, at 37°C, for 30 min) and inoculated in synthetic grape-juice media (GJM) at an initial cell count of 10^6 CFU/ml. The GJM, similar in the composition to typical grape juice, was formulated as previously described by Henschke and Jiranek (1993), with minor modifications. An equimolecular mixture of glucose and fructose (100:100 g l^{-1}) was used as carbon and energy source, and the yeast assimilable nitrogen (YAN) was supplied as diammonium phosphate (DAP) for the sole nitrogen source to facilitate monitor the nitrogen consumption profile. To evaluate the yeast nitrogen demands for growth and fermentation, two initial nitrogen concentrations were used: 67 mg l^{-1}, which yields a nitrogen-limiting condition that produces sluggish fermentation (Mendes-Ferreira et al. 2009), or 10-fold higher (670 mg l^{-1}), under which nitrogen is in excess. The pH was adjusted to 3.7 with NaOH prior to sterilization. The fermentation experiments were conducted in 250 ml-flasks filled to 2/3 of their volume and fitted with a side-arm port sealed and a rubber septum for anaerobic sampling; the conditions were maintained at 20°C in an orbital shaker at 120 rpm. The experiments were monitored daily for weight loss, which estimates CO_2 production. Aseptic samples were collected daily to assess fermentation and growth parameters using a syringe-type system, in which the fermentation gasses escaped through a glass tube connected to a two-way valve through Teflon tubing connected to a fermentation lock and an H_2S trap, as previously described (Mendes-Ferreira et al. 2009). Every 24 h, the flasks were removed from the shaker, weighed, disconnected from the trap, and connected to a new trap. Prior to sampling, the flasks were stirred to ensure homogeneity. The experiments proceeded until no further weight was lost, and completion was confirmed by stable sugar levels (< 2 g l^{-1}) using Clinitest tablets (Roche, Amadora, Portugal). The cells were collected at 5 different time-points: 12, 24, 36 and 96 hours after inoculation and at the end of the alcoholic fermentation. The supernatants were stored at −20°C for further analysis. The maximum fermentation rate was determined using the slope from the linear dependence of the steepest decline in weight (g) at the corresponding time points (h). Each set of fermentation experiments was repeated at least three times, and the reported data are mean values of these replicated experiments.

Determination of yeast growth and viability during fermentation

The yeast growth was followed by periodic optical density measurement (640 nm) of the appropriately diluted culture samples in a Shimadzu UV-2101 spectrophotometer (Shimadzu, Kyoto, Japan) and by counting the colony-forming units in YPD plates after incubation at 30°C, for 48 h. Further, the yeast cell biomass was determined using 2 × 50 and 3×15 ml samples centrifuged in pre-weighed tubes for 10 min at 2300 × g, washed twice with sterile deionized water, dried for 24 h at 100°C, and stored in a desiccator before weighing.

To evaluate the viability of nitrogen-starved cells, a sample from each low-nitrogen fermentation was spread onto yeast carbon base (YCB, Difco) agar plates with 0.73 mg N l^{-1} as amino acids: histidine.HCl (1 mg l^{-1}), methionine (2 mg l^{-1}) and tryptophan (2 mg l^{-1}). YCB was used to test the ability of yeasts to assimilate nitrogen by adding diverse nitrogen sources. The histidine, methionine and tryptophan concentrations were reduced to 10% of their original concentration in the yeast nitrogen base. This experiment was expected to detect viable cells in solid media without a nitrogen source.

Analytical determinations

Nitrogen determination

The ammonium concentration was determined during fermentation using a continuous-flow analysis system as previously described (Mendes-Ferreira et al. 2004). The maximum nitrogen consumption rate was determined by the slope from the linear dependence of the steepest decline in nitrogen concentration (mg) at the corresponding time points (h). The nitrogen consumption rates during the stationary growth phase under the high nitrogen regime were calculated using the slope from the linear dependence between nitrogen consumed (mg) and the respective duration (h).

H_2S determination

The quantity of H_2S released by the cultures was determined colorimetrically following selective collection of fermentation gases using a modified fermentation lock and sulfide-trapping system, as described previously (Mendes-Ferreira et al. 2009). The sulfide concentration was calculated using a calibration curve established from known sulfide quantities (0–12 µg) and standard methods (Acree et al. 1971; Rees et al. 1971).

Fermentation metabolites determination

The glucose and fructose consumed, ethanol converted, and glycerol, succinic and acetic acids produced were analyzed for each sample throughout fermentation using high-performance liquid chromatography (HPLC) and a liquid chromatographic system equipped with the ion exclusion, cation exchange column Aminex HPX-87H (Bio-Rad Laboratories, Hercules, CA, USA) and a refractive index detector. The column was eluted using sulfuric acid (0.013 N) at 62°C and a 0.6 ml min^{-1} flow rate. The samples were filtered through a membrane (Millipore, 0.22 µm pore size) before injecting 6 µl. The components were identified through their relative retention times compared with the respective standards, using the EZChrom Elite (vers. 3.x) Software (Agilent Technologies, Inc., Santa Clara, CA USA).

Statistical analysis

The strains were clustered based on the phenotype scores obtained, using the Pearson correlation and UPGMA clustering (Spotfire 4.5 – TIBCO Software Inc., 2007–2012, Somerville, Massachusetts, USA). Factorial design experiments were used to analyze the influence of the two nitrogen concentrations and eight yeast strains on metabolic activity. The effects of nitrogen concentration on each strain's fermentation activity were examined through a two-way analysis of variance (ANOVA) and principal component analysis (PCA) using SAS JMP 7.0 (2007, Cary, NC, USA). For paired comparisons, the Tukey HSD (honestly significant difference) test was used. $p < 0.05$ was considered statistically significant.

Results

Phenotypic characterization

Considering the objectives of this study, the strains were characterized for relevant alcoholic fermentation features, such as tolerance to temperature variation, presence of oxidizing agents and osmotic/ionic stress as well as the ability to produce certain volatile compounds. The phenotypic characteristics data are depicted in Table 2 and the Additional file 1: Table S1 as final scores for each strain and stress condition, ranging between 0 (no growth) and 8 (active growth). Additional file 1: Figure S1 shows examples of yeast growth under the different stress conditions. The strains and conditions were organized through hierarchical clustering using the Pearson correlation and UPGMA clustering, which were used to separate the isolates into distinct groups, as shown in Figure 1.

Yeast response to different stress conditions

The data herein demonstrate that temperature largely affects the growth of the yeast strains; each displayed active growth with a maximum score at 37°C, and 70% of the strains grew at 40°C, but eight out of 20 did not grow at 15°C. The strains K7, UCD505, UCD595, CEG and NT116 did not grow at 40°C but did grow at 15°C. The strains K7 and UCD505 diverged due to their low tolerance for most stress agents (acetic acid, SO_2 and ethanol), particularly UCD505. The yeasts in cluster II (CEG and NT116) diverged from the cluster III yeasts

Table 2 Phenotypes mean scores and conditions used to cluster the commercial yeasts tested

Yeast strain	Cerulenin	TFL	Temperature	SO_2	H_2O_2	NaCl	Acetic acid	Ethanol	H_2S
K7	0.0	0.0	4.0	0.0	6.0	3.7	1.0	1.0	8.0
W3	0.0	2.0	6.3	5.7	5.3	4.0	3.0	1.2	6.4
UCD595	1.0	1.5	3.3	5.0	3.3	0.8	0.8	1.5	5.6
UCD505	0.0	0.0	3.3	0.7	2.5	2.7	0.5	1.5	7.2
XLD	1.0	0.0	6.0	5.7	5.8	5.0	3.0	1.5	5.6
T73	0.0	0.0	6.0	5.7	6.3	2.7	3.3	1.5	6.4
AWRI R2	2.0	0.0	6.0	8.0	5.8	0.3	4.8	1.0	5.6
LVCB	0.0	0.0	6.0	4.8	6.8	1.2	3.3	4.2	4.0
QA23	0.0	0.0	5.5	4.3	7.0	1.7	2.7	3.0	3.2
BRL97	2.0	0.0	4.7	8.0	6.3	0.7	4.7	0.7	5.6
EC1118	0.5	0.0	6.8	4.7	6.5	1.7	3.0	3.0	3.2
CEG	0.5	0.0	2.8	4.7	5.3	0.0	2.8	0.7	6.4
FERMIVIN	0.5	0.0	8.0	5.3	7.0	3.2	3.8	1.7	5.6
VIN13	0.0	0.0	7.7	5.3	7.3	3.0	3.7	1.3	5.6
BM45	1.0	0.0	6.2	8.0	7.0	0.0	4.5	1.7	3.2
NT116	1.0	0.0	4.3	5.8	7.3	0.3	4.3	3.0	5.6
XL	1.0	0.0	4.7	5.7	5.0	3.5	3.3	2.3	5.6
UCD522	0.0	0.0	5.2	6.2	4.5	3.0	3.2	2.3	6.4
VL1	0.3	0.0	6.0	6.0	7.0	3.0	3.3	2.3	4.8
AWRI796	0.5	0.0	6.0	3.0	7.0	3.5	3.3	1.5	2.4

(BRL97, BM45 and AWRI2) because the latter exhibited a strong resistance to acetic acid, SO_2, and H_2O_2. Cluster V is enriched with the yeasts that were more resistant to temperature variations and H_2O_2 but had an average resistance to NaCl and acetic acid. Seventy-five percent of the yeast strains showed tolerance for the temperature range from 15 to 40°C, whereas 90% tolerated an acetic acid concentration up to 110 mM. Two strains, AWRI R2 and BM45, were particularly resistant to ethanol and acetic acid environmental stresses. Ethanol and sulfur dioxide resistance also varied among the strains. However, for the concentrations used, the strains were more sensitive to ethanol than sulfite; 85% of the strains did not grow with 12.5% ethanol, whereas only one, LVCB, tolerated 15% ethanol. For sulfur dioxide, 50% of the strains tolerated this antiseptic over the range 2 to 6 mM, but three, AWRI R2, BRL97 and BM45, reached maximum growth score with 6 mM of SO_2. In sum, the presence of NaCl, H_2O_2 and acetic acid best discriminated the strains compared with the remaining stress conditions. In contrast, TFL was the least discriminating agent, because only 2 strains, UCD595 and W3, were resistant.

The ability of yeasts to produce volatile compounds

TFL and cerulenin resistance were examined to identify the yeasts' ability to produce flavoring compounds. The experiments were performed on agar plates with cerulenin and TFL added separately and scored for no growth/active growth after incubation for 3 days. The yeasts showed generally greater resistance to cerulenin compared with TFL. Sensitivity to TFL was a feature of most strains, except UCD595 and W3. Conversely, 60% of the strains presented resistance to cerulenin, even with low growth scores. Only one strain, UCD595, presented resistance to both cerulenin and TFL.

For H_2S, the strains K7 and UCD505 displayed the highest sulfite reductase activities on Biggy agar (cluster I). The remaining strains exhibited variable colony color intensities, which indicate different abilities to produce H_2S. Based on the color intensity of the colonies, the yeasts screened herein were classified as high (K7 and UCD595), medium (CEG, UCD522, W3, T73), low (QA23, EC1118, BM45) and no sulfide producers (AWRI796). The remaining strains displayed a low to moderate ability to produce sulfide.

Yeast growth kinetics and fermentation profiles

Eight representative strains from the distinct phenotypic characterization groups regarded as strains with high (VL1, BRL97, T73, NT116), medium (QA23, UCD522, XL) and low resistance to stress (CEG) and with different abilities to produce H_2S were used for fermentation experiments under two nitrogen regimes. The growth and fermentation parameters are in Table 3. Accordingly,

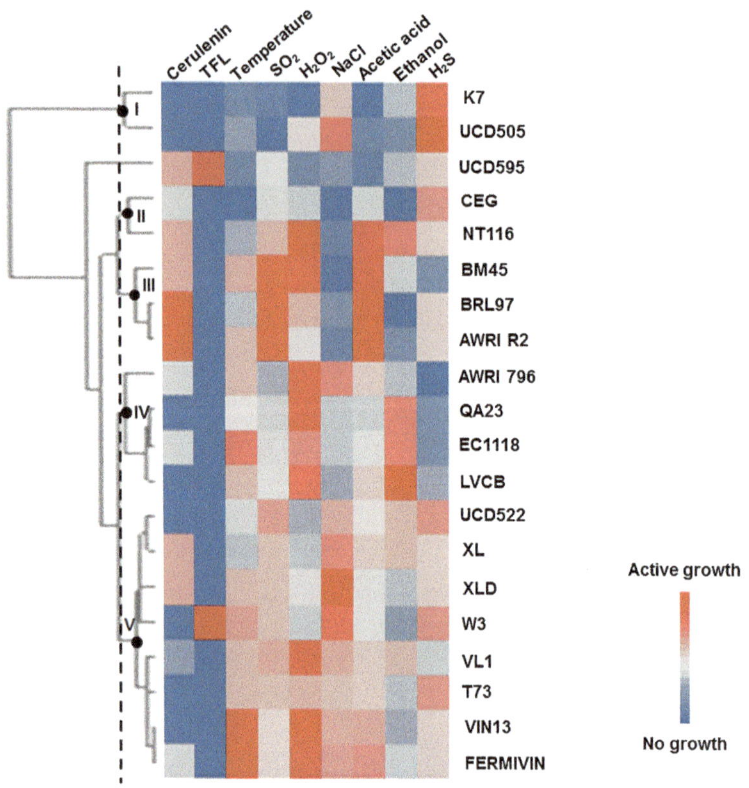

Figure 1 Phenotypic variation in the 20 commercial yeast strains. The strains were grown at least in duplicate on solid agar plates with 1–3 doses of each environmental variable, as described in Material and Methods. Each row on the plot represents a different strain, and each column indicates a particular environment. The colored boxes represent the average growth score of each strain in each environment, as in the legend shown at the lower right. For H_2S, the color gradient indicates the color intensity of the colonies on the BIGGY agar, low or absent (blue) and high (red).

vigorous fermentation was associated with high nitrogen levels and a high cell biomass. High nitrogen levels enhanced yeast growth and yielded higher fermentation rates with lower durations. With excess of nitrogen, the fermentation rates were much higher and the differences among the strains were great, whereas low nitrogen fermentations yielded relatively similar results among strains tested. Under the former conditions, 120 to 288 h were required for complete dryness depending on the yeast strain, whereas under the latter conditions, only five strains, QA23, UCD522, BRL97, VL1, and XL completed alcoholic fermentation after 672 to 696 h. The strains NT116, T73 and CEG did not fully degrade the sugars under low nitrogen. As the fermentation length per se does not fully define the performance of each individual strain, the pattern of sugars utilization was analyzed to better discriminate the yeast strains and nitrogen effects (Figures 2 and 3). Although fermentations started with equal quantities of glucose and fructose, the simultaneous but slower fructose use compared with glucose, yielded differences between their respective concentrations and the fermentation length depended on fructose use. Discrepancies in sugar consumption under high nitrogen levels were clearer for certain yeasts, such as VL1, T73, BRL97 and XL (Figure 3B, D-F). The strains QA23 and VL1 were, respectively, the fastest and slowest nitrogen-consumers (Figure 3A-B).

The yeast sugar consumption differed under low and high nitrogen. Sugar concentrations decreased slightly at the beginning of fermentation (0 to 2 days) (Figure 2A-H) through the fourth day; the sugar consumption rate was similar for each yeast strain but with lower decay curves than in the high nitrogen media. As shown in Figures 2 and 3, the area between the glucose and fructose consumption curves was clearly affected by both nitrogen availability and the yeast strain used. The fructose utilization kinetics influenced overall fermentation for the individual strains. The strains T73, CEG and NT116 yielded a larger area between the consumption curves for both sugars, which yielded an incomplete fermentation, likely due to an inability to entirely degrade the fructose in the medium.

The yeast cell viability (CFU's) and final biomass was significantly higher (2 to 3-fold) under high nitrogen, but each strain exhibited equal specific growth rates, irrespective of the initial nitrogen concentration used (Table 3).

Table 3 Overview of growth and fermentation parameters of the eight wine yeast strains grown in synthetic grape juice medium under low (67 mg l^{-1}) and high nitrogen (670 mg l^{-1}) concentrations

Strain	N level (mg l^{-1})	Specific growth rate μ (h^{-1})	Viable cells/ml	Final biomass (g l^{-1})	Maximum fermentation rate (g h^{-1})	$TN_{100\%}$ (h)	Maximum nitrogen consumption rate (mg h^{-1})	N_{EP} (mg)	N_{SP} (mg)	N_{SP} rate (mg h^{-1})	Total N consumed (mg)
T73	**67**	$0.185 \pm 0.00^{b,c,d,e,f}$	$3.28\ E+07 \pm 1.39\ E+07^{e,f}$	$2.6 \pm 0.06^{h,i}$	0.03 ± 0.01^{g}	26.11 ± 0.00^{e}	$4.0 \pm 0.01^{c,d,e}$	–	–	–	
	670	0.196 ± 0.008^{b}	$1.02\ E+08 \pm 0.03E+08^{d}$	$6.6 \pm 0.02^{b,c}$	$0.11 \pm 0.00^{e,f}$	–	5.0 ± 0.09^{b}	$240.9 \pm 4.69^{b,c}$	$84.8 \pm 14.56^{d,e}$	0.6 ± 0.10^{d}	$333.9 \pm 16.01^{c,d}$
UCD522	**67**	0.144 ± 0.01^{g}	$4.95E+07 \pm 0.07\ E+07^{e}$	$2.5 \pm 0.01^{h,i,j}$	0.04 ± 0.01^{g}	35.03 ± 0.00^{b}	$3.2 \pm 0.31^{e,f,g}$	–	–	–	
	670	$0.156 \pm 0.005^{d,e,f,g}$	$1.10\ E+08 \pm 0.01\ E+08^{c,d}$	$5.8 \pm 0.07^{d,e}$	0.19 ± 0.00^{a}	–	6.0 ± 0.09^{a}	288.4 ± 4.13^{a}	$120.7 \pm 29.97^{c,d}$	1.3 ± 0.31^{b}	$409.1 \pm 29.28^{a,b}$
BRL97	**67**	$0.152 \pm 0.01^{f,g}$	$3.56\ E+07 \pm 0.02E+07^{e,f}$	2.1 ± 0.04^{k}	0.03 ± 0.01^{g}	32.6 ± 0.00^{c}	$3.1 \pm 0.26^{f,g}$	–	–	–	
	670	$0.160 \pm 0.011^{d,e,f,g}$	$1.03\ E+08 \pm 0.03\ E+08^{d}$	5.6 ± 0.13^{e}	$0.14 \pm 0.00^{c,d}$	–	$4.4 \pm 0.33^{b,c,d}$	$211.3 \pm 15.68^{c,d}$	117.3 ± 10.33^{d}	$1.0 \pm 0.09^{b,c}$	328.6 ± 6.09^{d}
NT116	**67**	$0.186 \pm 0.01^{b,c,d,e}$	$4.45\ E+07 \pm 0.3\ E+07^{e}$	$2.4 \pm 0.03^{i,j}$	0.04 ± 0.01^{g}	24 ± 0.00^{f}	$4.4 \pm 0.30^{b,c,d}$	–	–	–	
	670	$0.184 \pm 0.005^{b,c,d,e,f}$	$1.50\ E+08 \pm 0.10\ E+08^{a,b}$	6.7 ± 0.07^{b}	$0.16 \pm 0.01^{b,c}$	–	$2.7 \pm 0.33^{g,h}$	131.2 ± 15.98^{e}	255.3 ± 22.75^{a}	2.7 ± 0.24^{a}	386.5 ± 6.88^{b}
QA23	**67**	$0.170 \pm 0.00^{c,d,e,f,g}$	$4.36\ E+07 \pm 0.21\ E+07^{e}$	2.3 ± 0.03^{j}	0.04 ± 0.01^{g}	22.34 ± 0.00^{g}	$3.7 \pm 0.07^{d,e,f}$	–	–	–	
	670	$0.173 \pm 0.007^{c,d,e,f,g}$	$1.71\ E+08 \pm 0.08E+08^{a}$	5.9 ± 0.07^{d}	$0.17 \pm 0.01^{a,b}$	–	5.0 ± 0.38^{b}	239.8 ± 18.47^{b}	177.5 ± 11.68^{b}	2.5 ± 0.16^{a}	417.3 ± 6.96^{a}
VL1	**67**	$0.210 \pm 0.01^{a,b}$	$4.87\ E+07 \pm 0.35\ E+07^{e}$	2.6 ± 0.04^{h}	0.04 ± 0.01^{g}	24 ± 0.00^{f}	$4.8 \pm 0.31^{b,c}$	–	–	–	
	670	0.220 ± 0.004^{a}	$1.15\ E+08 \pm 0.21\ E+08^{c,d}$	6.4 ± 0.05^{c}	0.08 ± 0.01^{f}	–	2.1 ± 0.12^{h}	100.9 ± 5.66^{f}	166.5 ± 8.87^{b}	$0.7 \pm 0.04^{c,d}$	267.4 ± 12.28^{d}
CEG	**67**	$0.153 \pm 0.00^{e,f,g}$	$1.49\ E+07 \pm 0.35\ E+07^{f}$	1.7 ± 0.06^{l}	0.04 ± 0.00^{g}	44.06 ± 0.00^{a}	2.1 ± 0.11^{h}	–	–	–	
	670	0.146 ± 0.002^{g}	$5.73\ E+07 \pm 0.09\ E+07^{e}$	5.3 ± 0.09^{f}	$0.13 \pm 0.01^{d,e}$	–	$2.9 \pm 0.12^{f,g,h}$	282.7 ± 13.24^{a}	55.3 ± 1.43^{e}	0.5 ± 0.01^{d}	$338.1 \pm 14.68^{c,d}$
XL	**67**	$0.192 \pm 0.02^{a,b,c,d}$	$3.80\ E+07 \pm 0.28\ E+07^{e,f}$	3.1 ± 0.12^{g}	0.04 ± 0.01^{g}	30.33 ± 0.00^{d}	$2.9 \pm 0.13^{f,g,h}$	–	–	–	
	670	$0.195 \pm 0.026^{a,b,c}$	$1.32\ E+08 \pm 0.07E+08^{b,c}$	7.5 ± 0.01^{a}	$0.14 \pm 0.01^{c,d}$	–	$4.2 \pm 0.61^{b,c,d}$	202.8 ± 29.23^{d}	$153.4 \pm 32.15^{b,c}$	1.3 ± 0.27^{b}	356.3 ± 4.61^{c}

Values in same column not connected by the same letter are significantly different, $p < 0.05$.
N_{EP} - Nitrogen consumed during exponential phase.
N_{SP} - Nitrogen consumed during stationary phase.
N_{SP} rate - Nitrogen consumption rate during stationary phase.
$TN_{100\%}$ - Time necessary to consume all nitrogen.

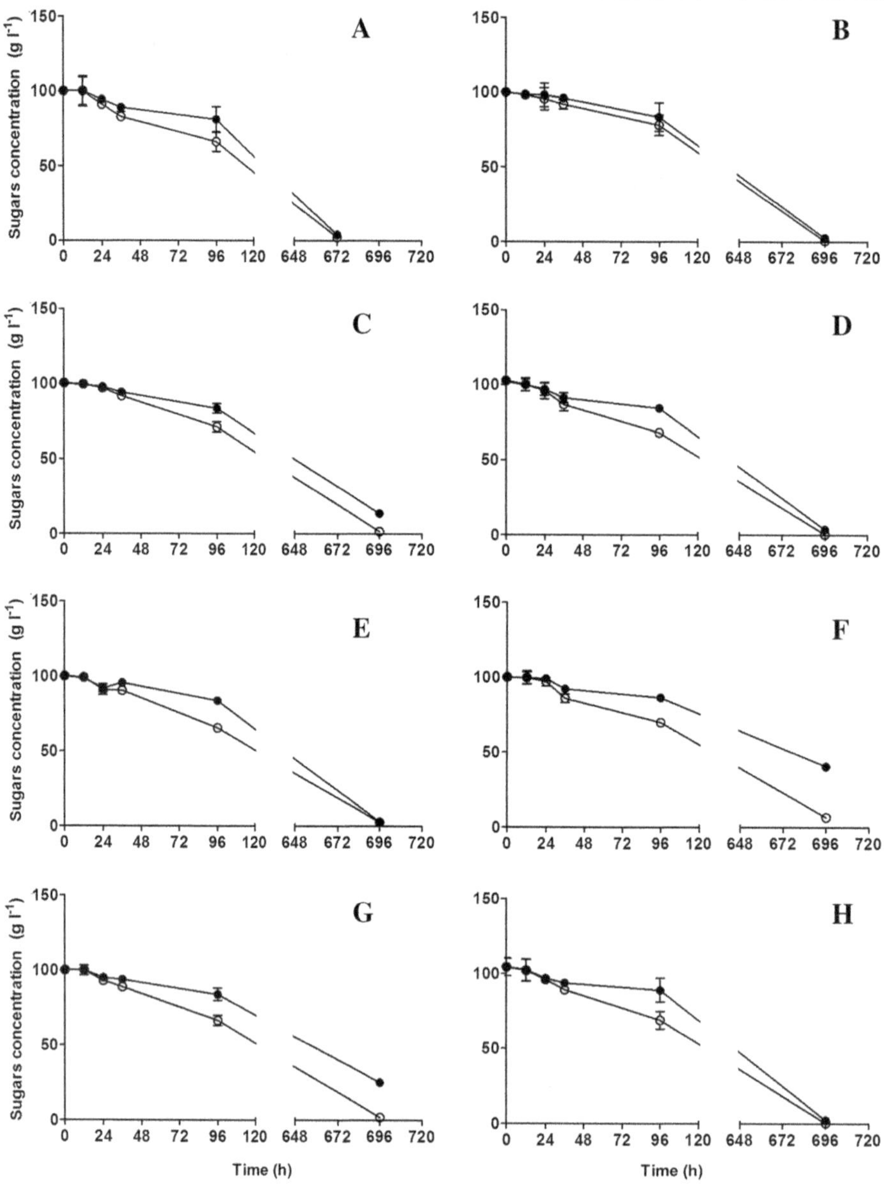

Figure 2 Glucose (○) and fructose (●) consumption profiles compared for the 8 yeast strains QA23 (A), VL1 (B), CEG (C), BRL97 (D), XL (E), T73 (F), NT116 (G) and UCD522 (H) in the media supplemented with low nitrogen (67 mg l^{-1}).

The yeast cells viability remained constant (either in YPD or YCB medium) under low nitrogen, and the yeast cells slowly continued to consume sugars for almost a month without extracellular nitrogen. As shown in Table 3, more nitrogen was used by the yeasts upon increasing the available nitrogen in the media; the nitrogen was fully consumed after 22 to 44 h (depending on the strain) in the media with low nitrogen. High nitrogen levels yielded different consumption rates during either the exponential or stationary growth phase. Both results demonstrate that nitrogen demands are highly strain-dependent and suggest that measuring nitrogen during the stationary growth phase is important. The quantity of nitrogen consumed during the stationary phase was positively correlated with a higher cell number and better fermentation competence, rather than the maximum nitrogen consumption rate. The strains with high fermentation activity, QA23, UCD522 and NT116, were high nitrogen-consumers with high nitrogen consumption rates, particularly during the stationary phase (Table 3). The strains NT116 and VL1 displayed high nitrogen consumption rates during low-nitrogen fermentations.

The effects of nitrogen concentration and yeast strain on H_2S release

H_2S release showed a strain-dependent variable response to nitrogen availability; despite this contrasting behavior,

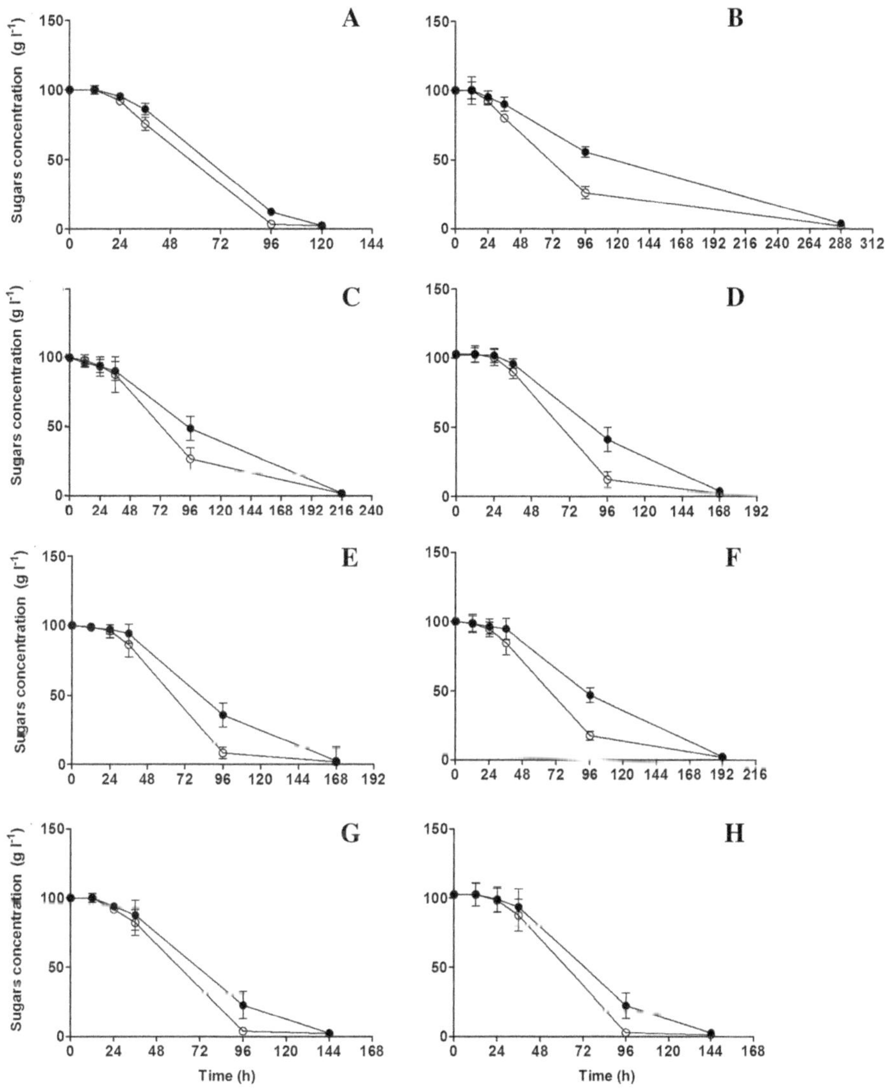

Figure 3 Glucose (○) and fructose (●) consumption profiles compared for the 8 yeast strains QA23 (A), VL1 (B), CEG (C), BRL97 (D), XL (E), T73 (F), NT116 (G) and UCD522 (H) in the media supplemented with high nitrogen (670 mg l^{-1}).

certain strains, QA23, NT116, CEG and BRL97, produced negligible quantities of H_2S under the two nitrogen regimes (Table 4). In contrast, the strains BRL97 and CEG showed a direct relationship between the initial levels of nitrogen in the media and the quantity of H_2S released. The quantity of H_2S released during alcoholic fermentation was positively correlated with the color intensity of the colonies on Biggy agar (Table 4 and Figure 1), particularly for the high and medium producers, UCD522, T73, XL and VL1.

The effects of nitrogen concentration and yeast strain on fermentation metabolites

At the end of fermentation, we quantified several metabolites, such as ethanol, glycerol, succinic and acetic acids and the residual sugars (Table 4). We observed differences in the final ethanol concentration, particularly under low nitrogen for the T73, NT116 and CEG strains, due to incomplete sugar degradation. The yeast strains differed in regard to the succinic acid concentration formed under high or low nitrogen, but each strain was more productive during low-nitrogen fermentations. Under the latter condition, the quantity of succinic acid produced by the yeasts can be directly correlated with the final biomass (Tables 3 and 4). We evaluated the glycerol concentration produced by each strain individually and observed that, even without significant differences, high glycerol concentrations were produced during high-nitrogen fermentations. The exception was the yeast strain CEG, which produced almost the same glycerol concentration under both nitrogen regimes (Table 4). With high nitrogen levels, glycerol production positively correlated with the final yeast biomass (Tables 3 and 4). Further, more acetic acid was produced during

Table 4 Overview of the levels of metabolites produced by the eight wine yeast strains grown in synthetic grape juice medium under low (67 mg l^{-1}) and high nitrogen (670 mg l^{-1}) concentrations

Strain	N level (mg l^{-1})	Time to reach dryness (h)	Residual glucose (g l^{-1})	Residual fructose (g l^{-1})	Succinic acid (g l^{-1})	Glycerol (g l^{-1})	Acetic acid (g l^{-1})	Ethanol (% v/v)	Total H_2S (μg l^{-1})
T73	**67**	>696	6.73 ± 1.65^{a}	40.48 ± 2.23^{a}	$1.91 \pm 0.02^{a,b,c}$	$6.74 \pm 1.49^{a,b}$	$0.32 \pm 0.06^{b,c,d}$	8.95 ± 0.31^{f}	214.4 ± 97.1^{a}
	670	192	2.09 ± 0.29^{b}	2.50 ± 0.13^{d}	0.54 ± 0.03^{e}	$8.63 \pm 0.29^{a,b}$	0.79 ± 0.07^{a}	$11.60 \pm 0.46^{a,b,c}$	$44.7 \pm 22.7^{b,c}$
UCD522	**67**	672	0.47 ± 0.04^{b}	2.42 ± 0.04^{d}	$2.00 \pm 0.12^{a,b}$	$7.24 \pm 0.54^{a,b}$	$0.22 \pm 0.05^{c,d}$	$12.00 \pm 0.00^{a,b}$	201.7 ± 19.3^{a}
	670	144	1.38 ± 0.32^{b}	2.77 ± 0.89^{d}	0.34 ± 0.03^{e}	$7.66 \pm 0.34^{a,b}$	$0.39 \pm 0.01^{b,c}$	12.30 ± 0.00^{a}	191.2 ± 14.3^{a}
BRL97	**67**	696	0.51 ± 0.01^{b}	3.52 ± 0.16^{d}	$1.68 \pm 0.04^{c,d}$	6.27 ± 0.21^{b}	0.44 ± 0.00^{b}	$11.80 \pm 0.20^{a,b}$	4.4 ± 3.8^{c}
	670	168	2.03 ± 0.35^{b}	3.69 ± 0.09^{d}	0.39 ± 0.00^{e}	8.79 ± 0.94^{a}	0.71 ± 0.01^{a}	$11.83 \pm 0.21^{a,b}$	8.9 ± 6.7^{c}
NT116	**67**	>696	1.85 ± 0.61^{b}	24.94 ± 3.62^{b}	$1.76 \pm 0.09^{a,b,c}$	$7.18 \pm 0.35^{a,b}$	$0.32 \pm 0.01^{b,c,d}$	10.2 ± 0.17^{e}	5.6 ± 3.0^{c}
	670	144	2.23 ± 0.63^{b}	$2.43 \pm 0.36^{c,d}$	0.38 ± 0.04^{e}	$8.00 \pm 0.30^{a,b}$	0.45 ± 0.01^{b}	$11.85 \pm 0.35^{a,b}$	2.1 ± 1.7^{c}
QA23	**67**	672	2.08 ± 1.43^{b}	3.89 ± 0.50^{d}	$1.69 \pm 0.04^{b,c,d}$	6.24 ± 0.16^{b}	0.16 ± 0.03^{d}	$11.50 \pm 0.14^{a,b,c,d}$	0.0 ± 0.0^{c}
	670	120	2.31 ± 0.76^{b}	2.38 ± 0.07^{d}	0.33 ± 0.01^{e}	$8.47 \pm 0.11^{a,b}$	0.43 ± 0.01^{b}	$11.90 \pm 0.10^{a,b}$	0.0 ± 0.0^{c}
VL1	**67**	696	0.62 ± 0.09^{b}	2.66 ± 1.09^{d}	$1.94 \pm 0.22^{a,b,c}$	6.34 ± 1.06^{b}	$0.34 \pm 0.04^{b,c,d}$	$11.38 \pm 0.34^{b,c,d}$	$45.1 \pm 14.1^{b,c}$
	670	288	1.70 ± 0.22^{b}	3.00 ± 0.31^{d}	0.47 ± 0.02^{e}	$8.32 \pm 0.42^{a,b}$	0.73 ± 0.04^{a}	$11.75 \pm 0.35^{a,b,c}$	4.4 ± 0.5^{c}
CEG	**67**	>696	1.49 ± 0.02^{b}	13.57 ± 0.61^{c}	1.46 ± 0.00^{d}	$7.05 \pm 0.63^{a,b}$	$0.33 \pm 0.11^{b,c,d}$	$10.87 \pm 0.13^{d,e}$	0.6 ± 0.0^{c}
	670	216	1.77 ± 0.03^{b}	1.99 ± 0.07^{d}	0.36 ± 0.04^{e}	$7.27 \pm 0.13^{a,b}$	0.69 ± 0.02^{a}	$11.37 \pm 0.21^{b,c,d}$	5.0 ± 7.0^{c}
XL	**67**	696	2.57 ± 0.00^{b}	2.94 ± 0.00^{d}	2.02 ± 0.00^{a}	$7.81 \pm 0.00^{a,b}$	$0.24 \pm 0.00^{b,c,d}$	$11.38 \pm 0.02^{a,b,c,d}$	$127.6 \pm 14.1^{a,b}$
	670	168	1.66 ± 0.15^{b}	2.16 ± 0.14^{d}	0.44 ± 0.05^{e}	$8.16 \pm 0.07^{a,b}$	0.77 ± 0.09^{a}	$11.63 \pm 0.4^{a,b,c}$	$35.6 \pm 23.7^{b,c}$

Values in same column not connected by the same letter are significantly different, $p < 0.05$.

high-nitrogen fermentations, being directly correlated with fermentation length.

For a general overview of the data, a principal component analysis (PCA) was performed using each strain and traits that characterized fermentations under the low- and high-nitrogen regimes (Figures 4 and 5). This approach was used to separate the strains based on growth parameters, nitrogen utilization and metabolites produced during fermentation. For fermentations under low nitrogen (Figure 4), PC1 and PC2 accounted for 42.8% and 25.4% of the variability, respectively. The strains were mainly discriminated by the time required to consume the available nitrogen (higher for CEG), nitrogen consumption rates and additional growth parameters. The center depicts

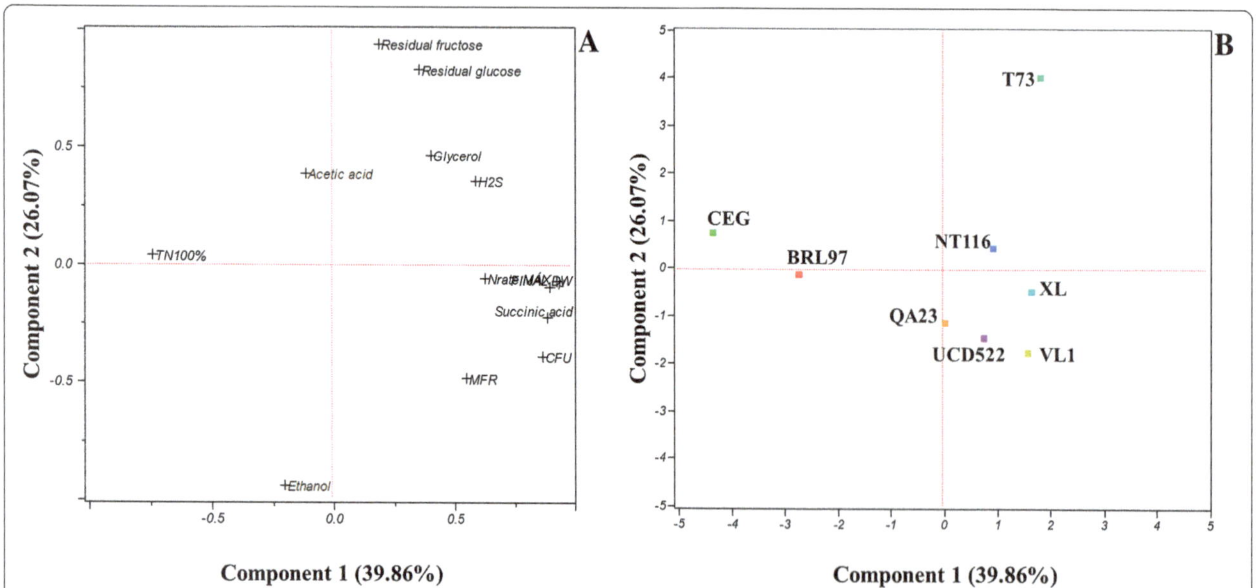

Figure 4 Principal component analysis (PCA) plot using the mean values of the metabolite concentrations as well as growth and fermentation parameters of the 8 strains under a low nitrogen regime (67 mg l^{-1}), the variable loadings (A) and scores (B) for the two first principal components.

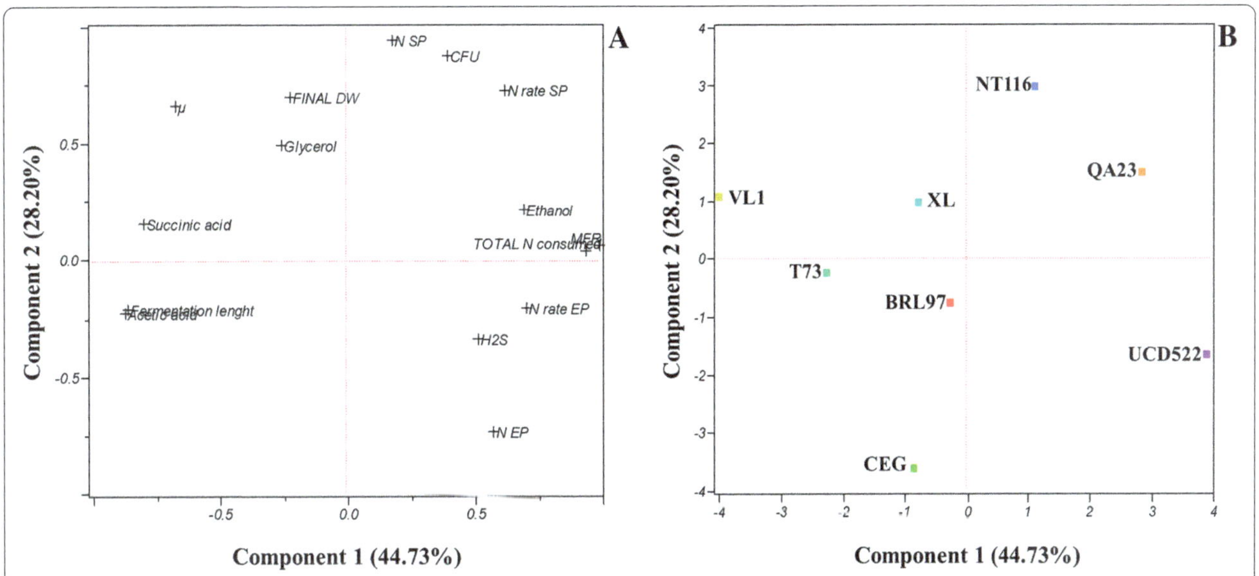

Figure 5 Principal component analysis (PCA) plot using the mean values of metabolite concentrations and growth and fermentation parameters of the 8 strains under a high nitrogen regime (670 mg l^{-1}), the variable loadings (A) and scores (B) for the two first principal components.

the most vigorous/more rapidly fermenting yeast strains, QA23, UCD522 and NT116, and the upper right quadrant only indicates strain T73 due to the residual sugar levels. Clearly, under low nitrogen, certain direct relationships between the parameters analyzed can be established; the strains that exhibit higher nitrogen consumption rates have a higher final biomass, higher specific growth and maximum fermentation rates, and high succinate concentrations. Thus, at low nitrogen levels, the nitrogen consumption rate defines the cell biomass yield; XL, T73 and VL1; NT116, QA23 and UCD522; and BRL97 and CEG are high, medium and low biomass producers, respectively. For high-nitrogen fermentation (Figure 5), PCA explains 72.9% of the total variability; PC1 accounted for 44.7% and PC2 accounted for an additional 28.2%. The high maximum fermentation rates are mainly due to the nitrogen consumed and its consumption rate during the stationary phase, which yielded a high cell population and fermentation velocity. Glycerol formation directly correlated with yeast cell biomass. H_2S liberation depended on the nitrogen consumption rate during the exponential growth phase. The strains were also discriminated by fermentation length; the more rapidly fermenting strains are located on the right quadrant (QA23, UCD522 and NT116), the medium in the center (BRL97 and XL), and the slowest on the left quadrant (CEG, T73 and VL1). Further, we observed a positive relationship between acetic acid formation and fermentation length.

Discussion

The yeast stress response is important for survival and proper processing during fermentation. Indeed, it is a main research interest both for fundamental knowledge of the *S. cerevisiae* wine strains and for the practical application thereof (Bisson et al. 2007). In this study, commercial yeast strains from different geographical origins were subjected to different environmental stresses that are similar to stresses encountered during alcoholic fermentation. Diversity among the strains was demonstrated through the unique phenotypic patterns of each strain, which is consistent with previous studies (Kvitek et al. 2008; Mendes et al. 2013). Overall, resistance to different NaCl, H_2O_2 and acetic acid concentrations was more strain-discriminating than the remaining stress conditions. *S. cerevisiae* yeast strains exhibit a higher natural predisposition for tolerating ethanol and sulfur dioxide (SO_2), due to adaptation and natural selection, allowing them to predominate in many fermentation environments. The wine *S. cerevisiae* strains are more resistant to SO_2, which has been a widely used preservative for centuries, due to higher *SSU1* gene expression, which encodes a plasma membrane sulfite pump that is required for efficient sulfite efflux and maintains sufficiently low sulfite levels to prevent toxicity (Park and Bakalinsky 2000). Except for the sake yeast strain K7 (sake is sulfite-free), which was highly sensitive to the range of SO_2 doses, all the strains tolerated the lowest concentration. Higher variability was observed among the yeast strains with the higher SO_2 concentrations; the AWRI R2 and BM45 strains displayed the highest scores even with very high SO_2 concentrations. For yeast tolerance to ethanol, we observed high diversity among the strains tested at 10%, which was diminished for the highest ethanol concentrations, wherein 80% of the strains did not grow. This is an intriguing result because a higher resistance phenotype was expected for the commercial wine strains. Additional studies have yielded similar results (Kvitek et al. 2008). The low growth scores for the highest ethanol levels observed may be due to use of yeast cells in the exponential growth phase, wherein the cells are significantly more sensitive to environmental stress (Werner-Washburne et al. 1993). An additional explanation may be that throughout alcoholic fermentation the yeast cells gradually adapt to the ethanol produced and increase their tolerance (Arroyo-López et al. 2010). High resistance to oxidative stress is an advantage for wine yeast because hypoxic fermentation in a high-sugar medium induces yeast cells to produce ROS (Reactive Oxygen Species) (Mendes-Ferreira et al. 2010; Landolfo et al. 2008). Therefore, the strains that better manage oxidative stress from either yeast respiratory metabolism or the presence of H_2O_2 produced during alcoholic fermentation (Bisson et al. 2007) will likely perform well during fermentation, such as for the QA23, FERMIVIN and VL1 strains herein. The high resistance to acetic acid demonstrated by the yeast strains AWRI R2, BRL97 and BM45, should be highlighted because this compound is known to affect yeast fermentative performance in winemaking. Additionally, it is important for yeast used in additional biotechnological processes, such as on bioethanol production, in which acetic acid is frequently the most dominant inhibitor in the plant-biomass hydrolysates (van Maris et al. 2006).

From this screen, eight strains, UCD522, XL, FERMIVIN, VIN13, W3, XLD, T73 and VL1, should be highlighted because displayed interesting industrial features, such as a high resistance to temperature variation and oxidative shock as well as, simultaneously, an average resistance to acetic acid.

Despite the wide variation in growth scores under the different conditions tested, hierarchical clustering showed phenotypic similarity between certain strains irrespective of geographical origin. This phenomenon was described by Dunn et al. (2012), who observed similar pairs or groups of strains, but could not find clusters based on geographic origin using array-Comparative Genomic Hybridization (aCGH) for 83 different *S. cerevisiae* strains. These observations support the hypothesis that genomic diversity driven by mutation or DNA sequence alteration and environmental adaptation may impact *S. cerevisiae* strain diversity more than geographical origin.

Considering the major role of nitrogen availability in yeast growth and fermentative performance (Bisson 1999; Salmon 1989; Manginot et al. 1998; Bell and Henschke 2005), alcoholic fermentation trials experiments were conducted under two different nitrogen regimes. For that purpose, eight representative strains from distinct phenotypic clusters were used to evaluate the rates and profiles of nitrogen consumption as well as analyze its impact on fermentation kinetics and the metabolites produced. For each strain, nitrogen concentration affected the fermentation length, maximum fermentation rate and final biomass, not specific growth rates, as proposed in literature (Mendes-Ferreira et al. 2004, 2009; Gardner et al. 2002; Varela et al. 2004). Consistent with a recent study (Gutiérrez et al. 2012) which demonstrated that the minimum nitrogen concentration required by yeasts to yield the μ_{max} is very low (5 to 18 mg N l^{-1}), we did not detect significant differences in the yeast growth rates under the two nitrogen regimes examined herein. However, the highest fermentation rates were observed under high nitrogen conditions due to the higher final yeast cell biomass (Varela et al. 2004; Albertin et al. 2011). Nevertheless, for each nitrogen level and comparing yeast strains, the same correlation between yeast cell final biomass and better fermentative performance was not established. For instance, the VL1 strain final biomass was significantly higher than for QA23, although taking more 168 h to fully degrade the sugars. On the other hand, a positive correlation was found between nitrogen consumption during the stationary growth phase and fermentative activity. The strains with a shorter fermentation length, QA23, UCD522 and NT116, exhibited the highest nitrogen consumption rates, particularly during the stationary phase. These findings suggest that nitrogen is not necessarily directed towards biomass formation, but it might be used to generate high fermentative vigor, which is associated with fructose utilization (Berthels et al. 2004). Fructose utilization and the time for its consumption impacted the fermentation length. The fructose consumption profile for the glucophilic wine yeast strains examined was dissimilar. The strains with a lower discrepancy between glucose and fructose consumption were assumed to perform better during fermentation (Berthels et al. 2004), as herein for the yeast strains QA23, UCD522 and NT116.

Fermentation behavior under excess of nitrogen was not a predictor for fermentation efficiency under limited nitrogen. For instance, the strains T73 and CEG as well as, more importantly, the highly efficient strain NT116 did not complete alcoholic fermentation under low nitrogen conditions, which reinforces the notion that nitrogen availability during the stationary-growth phase is crucial for fructose consumption, at least for certain strains.

The wine yeast strains fermentation by-products differed quantitatively but not qualitatively; this metabolic diversity yields the individual wine organoleptic characteristics. Glycerol, succinic acid and acetic acid are the main secondary metabolites derived from sugars fermentation. In this study, the levels of glycerol in both nitrogen regimes did not differ significantly, most likely because glycerol is primarily formed due to an osmotic stress response (Nevoigt and Stahl 1997), which occurred under both conditions. For acetic acid, although we observed quantitative differences among the strains, significantly higher levels were detected for high-nitrogen fermentations, as previously reported (Tromp 1984; Mendes-Ferreira et al. 2009; Barbosa et al. 2009), being positively correlated with biomass production. This direct relationship has also been reported for succinic acid (Albers et al. 1996). In contrast, herein succinic acid was formed at significantly higher levels under low nitrogen conditions, which is in agreement with Heerde and Radler (1978). Based on the results herein and considering that we did not observed significant differences in the levels of glycerol, the yeast cells likely use different mechanisms to balance the redox conditions under the two regimes. Under low nitrogen, the yeast cells likely produce succinic acid via the reductive branch of tricarboxylic acid cycle (TCA) to balance the $NADH/NAD^+$ ratio (Heerde and Radler 1978). Under high nitrogen the overproduction of NADPH produced by acetic acid formation is likely driven to yeast cell biomass and cellular redox balance is maintained without the need of succinic acid formation. Further investigation is required on the role succinic acid plays under varying fermentation conditions.

The cross-talk between phenotypic patterns, fermentation profiles and metabolic traits of commercial wine yeast strains allowed understanding their phenotypic relatedness. Moreover, this integrative study provides interesting data for inclusion in a yeast selection program and may be useful to the industry in designing better strategies to improve wine quality. This work contributes new findings on the relationships between nitrogen availability, yeast cell growth and sugars utilization.

Additional file

Additional file 1: Table S1. Raw phenotypes scores, conditions and stress doses used to characterize yeast strains. **Figure S1**: Growth variation of the 20 commercial yeast strains under control condition **(A)** and one dose of each stress agent tested: SO_2 – 6 mM **(B)**; NaCl 1M (C); Temperature 40°C **(D)**; H_2O_2 – 2.5 mM **(E)**; Acetic acid 90 mM **(F)**; Cerulenin 6 µM **(G)** and TFL 1 mM **(H)**. Cells were spotted at concentrations of (from left to the right): 10^7, 10^6, 10^5 and 10^4 cells ml^{-1}. Yeast strains were organized in three groups of four columns representative of the four cell suspensions. From the left to the right and top to the bottom, the first group of strains contains: K7, W3, UCD595, UCD505, XLD, T73, AWRI R2; the second: QA23, BRL97, EC1118, CEG, FERMIVIN, VIN13, BM45 and NT116; and the third group: XL, UCD522, VL1 and AWRI 796.

Competing interests
The authors have declared no conflict of interest.

Acknowledgments
We thank Dr. José Enrique Pérez-Ortín for his critical reading of the manuscript. The research presented in this paper was financially supported by FEDER through COMPETE (FCOMP-01-0124-FEDER-014043) and by national funds through FCT - Fundação para a Ciência e Tecnologia (PTDC/AGR-ALI/111224/2009 and PTDC/AGR-ALI/71460/2006). C.B. is a recipient of a Ph.D. granted by FCT (SFRH/BD/61881/2009).

References

Acree TE, Sonoff EP, Splittstoesser DF (1971) Determination of hydrogen sulfide in fermentation broths containing SO_2. Appl Microbiol 22:110–112
Albers E, Larson C, Lidén G, Niklasson C, Gustafsson L (1996) Influence of the nitrogen source on *Saccharomyces cerevisiae* anaerobic growth and product formation. Appl Environ Microbiol 62:3187–3195
Albertin W, Marullo P, Aigle M, Dillmann C, Vienne D, Bely M, Sicard D (2011) Population size drives industrial *Saccharomyces cerevisiae* alcoholic fermentation and is under genetic control. Appl Environ Microbiol 77:2772–2784
Arroyo-López FN, Salvadó Z, Tronchoni J, Guillamón JM, Barrio E, Querol A (2010) Susceptibility and resistance to etanol in *Saccharomyces* strains isolated from wild and fermentative environments. Yeast 27:1005–1015
Barbosa C, Falco V, Mendes-Faia A, Mendes-Ferreira A (2009) Nitrogen addition influences formation of aroma compounds, volatile acidity and ethanol in nitrogen deficient media fermented by *Saccharomyces cerevisiae* wine strains. J Biosc Bioeng 108:99–104
Bell SJ, Henschke PA (2005) Implications of nitrogen nutrition for grapes, fermentation and wine. Aust J Grape Wine Res 11:242–295
Berthels NJ, Cordero Otero RR, Bauer FF, Thevelein JM, Pretorius IS (2004) Discrepancy in glucose and fructose utilization during fermentation by *Saccharomyces cerevisiae* wine yeast strains. FEMS Yeast Res 4:683–689
Bisson LF (1999) Stuck and sluggish fermentations. Am J Enol Vitic 50:107–119
Bisson LF, Karpel JE, Ramakrishnan V, Joseph L (2007) Functional genomics of wine yeast *Saccharomyces cerevisiae*. Adv Food Nutr Res 53:65–121
Boulton B, Singleton VL, Bisson LF, Kunkee RE (1996) Yeast and biochemistry of ethanol fermentation. Principles and practices of winemaking. Chapman and Hall, New York, In, pp 139–172
Camarasa C, Sanchez I, Brial P, Bigey F, Dequin S (2011) Phenotypic landscape of *Saccharomyces cerevisiae* during wine fermentation: evidence for origin-dependent metabolic traits. PLoS ONE 6:e25147
Dunn B, Richter C, Kvitek DJ, Pugh T, Sherlock G (2012) Analysis of the *Saccharomyces cerevisiae* pan-genome reveals a pool of copy number variants distributed in diverse yeast strains from differing industrial environments. Genome Res 22:908–924
Franco-Duarte R, Umek L, Zupan B, Schuller D (2009) Computational approaches for the genetic and phenotypic characterization of a *Saccharomyces cerevisiae* wine yeast collection. Yeast 26:675–692
Gardner JM, Poole K, Jiranek V (2002) Practical significance of relative assimilable nitrogen requirements of yeast: a preliminary study of fermentation performance and liberation of H_2S. Aust J Grape Wine Res 8:175–179
Gutiérrez A, Chiva R, Sancho M, Beltran G, Arroyo-López FN, Guillamon JM (2012) Nitrogen requirements of commercial wine yeast strains during fermentation of a synthetic grape must. Food Microbiol 31:25–32
Heerde E, Radler F (1978) Metabolism of the anaerobic formation of succinic acid by *Saccharomyces cerevisiae*. Arch Microbiol 117:269–276
Henschke PA, Jiranek V (1993) Yeasts-metabolism of nitrogen compounds. In: Fleet GH (ed) Wine microbiology and biotechnology. Harwood Academic, Chur, Switzerland, pp 77–164
Kvitek DJ, Will JL, Gasch AP (2008) Variations in stress sensitivity and genomic expression in diverse *S. cerevisiae* isolates. PLoS Genet 4:e1000223
Landolfo S, Politi H, Angelozzi D, Mannazzu I (2008) ROS accumulation and oxidative damage to cell structures in *Saccharomyces cerevisiae* wine strains during fermentation of high-sugar-containing medium. Biochim Biophys Acta 1780:892–898
Liti G, Carter DM, Moses AM, Warringer J, Parts L (2009) Population genomics of domestic and wild yeasts. Nature 458:337–341
Manginot C, Roustan JL, Sablayrolles JM (1998) Nitrogen demand of different yeast strain during fermentation: Importance of the stationary phase. Enz Microbiol Technol 23:511–517
Mendes I, Franco-Duarte R, Umek L, Fonseca E, Drumonde-Neves J, Dequin S, Zupan B, Schuller D (2013) Computational models for prediction of yeast strain potential for winemaking from phenotypic profiles. PLoS ONE 8:e66523
Mendes-Ferreira A, Mendes-Faia A, Leão C (2002) Survey of hydrogen sulphide production by wine yeasts. J Food Prot 65:1033–1037

Mendes-Ferreira A, Mendes-Faia A, Leão C (2004) Growth and fermentation patterns of *Saccharomyces cerevisiae* under different ammonium concentrations and its implications in winemaking industry. J Appl Microbiol 97:540–545

Mendes-Ferreira A, Barbosa C, Falco V, Leão C, Mendes-Faia A (2009) Hydrogen sulphide and other aroma compounds production by wine strains of *Saccharomyces cerevisiae* in synthetic media with different nitrogen concentrations. J Ind Microbiol Biotechnol 36:571–583

Mendes-Ferreira A, Sampaio-Marques B, Barbosa C, Rodrigues F, Costa V, Mendes-Faia A, Ludovico P, Leao C (2010) Accumulation of non-superoxide aninon reactive oxygen species mediates nitrogen-limited alcoholic fermentation by *Saccharomyces cerevisiae*. Appl Environ Microbiol 76:7918–7924

Nevoigt E, Stahl U (1997) Osmoregulation and glycerol metabolism in the yeast *Saccharomyces cerevisiae*. FEMS Microbiol Rev 21:231–241

Nikolaou E, Soufleros EH, Oulompasi E, Tzanetakis N (2006) Selection of indigenous *Saccharomyces cerevisiae* strains according to their oenological characteristics and vinification results. Food Microbiol 23:205–211

Oliveira VA, Vicente MA, Fietto LG (2008) Biochemical and molecular characterization of *Saccharomyces cerevisiae* strains obtained from sugar-cane juice fermentations and their impact in cachaca production. Appl Environ Microbiol 74:693–701

Park H, Bakalinsky AT (2000) *SSU1* mediates sulphite efflux in *Saccharomyces cerevisiae*. Yeast 16:881–888

Pretorius IS (2000) Tailoring wine yeast for the new millennium: novel approaches to the ancient art of winemaking. Yeast 16:675–729

Rees TD, Gyllenspetz AB, Docherty AC (1971) The determination of trace amounts of sulphide in condensed steam with N, N-diethyl-p-phenylenediamine. Analyst 96:201–208

Richter CL, Dunn B, Sherlock G, Pugh T (2013) Comparative metabolic footprinting of a large number of commercial wine yeast strains in Chardonnay fermentations. FEMS Yeast Res 13:394–410

Salmon JM (1989) Effect of sugar transportation inactivation in *Saccharomyces cerevisiae* on sluggish and stuck enological fermentations. Appl Environ Microbiol 55:953–958

Tromp A (1984) The effect of yeast strain, grape solids, nitrogen and temperature on fermentation rate and wine quality. S Afr J Enol Vitic 5:1–6

Ugliano M, Henschke PA (2009) Yeasts and wine flavor. In: Moreno-Arribas MV, Polo MC (eds) Wine chemistry and biochemistry. Springer, New York, USA, pp 313–392

van Maris AJ, Abbott DA, Bellissimi E, van den Brink J, Kuyper M, Luttik MA, Wisselink HW, Scheffers WA, van Dijken JP, Pronk JT (2006) Alcoholic fermentation of carbon sources in biomass hydrolysates by *Saccharomyces cerevisiae*: current status. Antonie Van Leeuwenhoek 90:391–418

Varela C, Pizarro F, Agosin E (2004) Biomass content governs fermentation rate in nitrogen-deficient wine musts. Appl Environ Microbiol 70:3392–3400

Werner-Washburne M, Braun E, Johnston G, Singer R (1993) Stationary phase in *Saccharomyces cerevisiae*. Microbiol Rev 57:384–410

Zuzuarregui A, del Olmo M (2004) Analyses of stress resistance under laboratory conditions constitute a suitable criterion for wine yeast selection. Antonie Van Leeuwenhoek 85:271–280

9

Paracentrin 1, a synthetic antimicrobial peptide from the sea-urchin *Paracentrotus lividus*, interferes with staphylococcal and *Pseudomonas aeruginosa* biofilm formation

Domenico Schillaci[1*], Maria Grazia Cusimano[1], Angelo Spinello[1,2], Giampaolo Barone[1,2], Debora Russo[1,2], Maria Vitale[3], Daniela Parrinello[1] and Vincenzo Arizza[1,2]

Abstract

The rise of antibiotic-resistance as well as the reduction of investments by pharmaceutical companies in the development of new antibiotics have stimulated the investigation for alternative strategies to conventional antibiotics. Many antimicrobial peptides show a high specificity for prokaryotes and a low toxicity for eukaryotic cells and, due to their mode of action the development of resistance is considered unlikely. We recently characterized an antimicrobial peptide that was called Paracentrin 1 from the 5-kDa peptide fraction from the coelomocyte cytosol of the *Paracentrotus lividus*. In this study, the chemically synthesized Paracentrin 1, was tested for its antimicrobial and antibiofilm properties against reference strains of Gram positive and Gram negative. The Paracentrin 1 was active against planktonic form of staphylococcal strains (reference and isolates) and *Pseudomonas aeruginosa* ATCC 15442 at concentrations ranging from 12.5 to 6.2 mg/ml. The Paracentrin 1 was able to inhibit biofilm formation of staphylococcal and *Pseudomonas aeruginosa* strains at concentrations ranging from 3.1 to 0.75 mg/ml. We consider the tested peptide as a good starting molecule for novel synthetic derivatives with improved pharmaceutical potential.

Keywords: AMP (Antimicrobial peptides); Biofilm; Staphylococci; *Pseudomonas aeruginosa*; *Paracentrotus lividus*

Introduction

Many natural antimicrobial peptides show a high specificity for prokaryotes and a low toxicity for eukaryotic cells, and for their mode of action the development of resistance by pathogenic bacteria is considered unlikely (Hancock and Rozek 2002). At present, there has been an increase of interest in these molecules as potential new antimicrobials (Bax et al. 2000; Mor 2000).

In human medicine, chronic and persistent forms of some infectious diseases depend on the ability of pathogenic bacteria to develop bacterial communities called biofilms. Opportunistic pathogens, such as staphylococcal strains and *Pseudomonas aeruginosa* show a great ability to produce biofilms that preventing infected wounds to heal, render the treatment extremely challenging. In veterinary medicine, biofilms are believed to be involved in many diseases such as pneumonia, liver abscesses, enteritis, wound infections and mastitis, which is one of the most common diseases in dairy cattle (Clutterbuck et al. 2007). *Staphylococcus aureus,* a major pathogen of mastitis has good in vitro antimicrobial susceptibility, but the therapy used to treat the affected animals is often disappointing and results in chronic infections due to the growth of bacteria as biofilms (Melchior et al. 2006). The biofilm of *P. aeruginosa* is a severe complicating factor in bovine mastitis, which is often associated with contaminated udder washing water or contaminated intramammary dry-cow preparations (Melchior et al. 2009).

The treatment of biofilm-associated infections is complicated because microbial biofilms are typically highly resistant to conventional antibiotics (Gilbert et al. 2002). The discovery of anti-infective agents active not only against planktonic microorganisms but also against biofilms

* Correspondence: domenico.schillaci@unipa.it
[1]Dip. STEBICEF, Università degli Studi di Palermo, Via Archirafi, 32-90123 Palermo, Italy
Full list of author information is available at the end of the article

represents an important goal for an effective control of infections (Projan and Youngman 2002).

The antimicrobial defence system of marine invertebrates is an interesting source of new anti-infective agents (Arizza 2013). We focused, particularly, on the effector cells of the echinoderm immune system, the coelomocytes. In a recent study, the antimicrobial activity of a 5-kDa peptide fraction from coelomocyte cytosol (5-CC) of the *Paracentrotus lividus,* the sea-urchin from Mediterranean sea, was demonstrated in relation to a group of important human pathogens. The anti-biofilm activity of 5-CC was shown in *S. epidermidis* 1457, a clinical strain isolated from an infected central venous catheter, against reference staphylococcal biofilms and against *Candida albicans* and *Candida tropicalis* (Schillaci et al. 2010; Schillaci et al. 2012). We showed the presence of three principal peptides, in the 5-CC content, whose molecular weights were respectively 1251.7, 2088.1, and 2292.2. These peptides are the (9–19), (12–31), (24–41) fragments of a β-thymosin of *P. lividus.* We focused particularly on the smallest peptide, that we called paracentrin 1, 11 amino acids in length, because showed the common chemical-physical characteristics of an antimicrobial peptide (Wang and Wang 2004).

The present study was aimed at evaluating the antibacterial and anti-biofilm activity of a chemically synthesized paracentrin 1 (SP1) against a group of staphylococcal reference strains and isolates and against *P. aeruginosa* ATCC 15442.

Materials and methods

Synthetic peptide

The SP1 was purchased from CASLO, Lyngby, Denmark utilizing the following sequence EVASFDKSKLK derived from ESI-MS analysis. The peptide was synthesized using Fmoc solid phase technology and the peptide content and purity was determined by high performance liquid chromatography (HPLC) and mass spectrometry (MS) analysis.

Net molecular charge was calculated by using the following formula (1) and the pI was calculated by using a web tool (http://isoelectric.ovh.org/).

$$z = \sum_i N_i \frac{10^{pKa_i}}{10^{pH}} - \sum_j N_j \frac{10^{pH}}{10^{pH} + 10^{pKa_j}} \qquad (1)$$

Hydrophobic mean value was calculated by using the Liu-Deber hydrophobicity scale (Liu and Deber 1998). Secondary structure was evaluated utilizing the algorithm of Wang et al. (2009), present in antimicrobial peptide calculator web site (http://aps.unmc.edu/AP/prediction/prediction_main.php). The helical wheel projection was performed using the Gautier et al. (2008) algorithm present in the HeliQuest site (http://heliquest.ipmc.cnrs.fr) with a window size of 11 residues.

Microbial strains

The staphylococcal reference strains used were: *Staphylococcus aureus* ATCC 29213, *Staphylococcus aureus* ATCC 25923, *Staphylococcus aureus* ATCC 6538, and *Staphylococcus epidermidis* RP62A, known for its ability to form a biofilm (Schumacher-Perdreau et al. 1994). Four staphylococcal isolates from the Istituto Zooprofilattico Sperimentale, Sicily (IZS) bacterial collection including strain 657 isolated from a milk sample from an individual sheep affected by mastitis, strain 688 and 700 and strain 702 isolated from bulk milk samples from different sheep flocks. The isolates were selected on blood agar plates and on Mannitol Salt Agar (Difco, Sparks, MD). The colonies were typed by API Staph strip (bio-Mérieux) and tube coagulase test was performed (Canicatti and Roch 1989).

Pseudomonas aeruginosa ATCC 15442, the reference strain in official tests for antibacterial evaluation in vitro (UNI EN European Standard), was also used in this study.

Minimum inhibitory concentrations (MIC)

MICs were determined by a micro-method described previously (Schillaci et al. 2005). Briefly, a series of solutions of SP1 was prepared with a range of concentrations from 25 to 0.07 mg/ml(obtained by two-fold serial dilution). The serial dilutions were made in Tryptic Soy Broth (TSB) (Merck) in a 96-wells plate; to each well was added 10 μl of a bacterial suspension from a 24 h culture containing $\sim 1 \times 10^6$ CFU/ml. The plate was incubated at 37°C for 24 h; after this time, the MICs were determined by a microplate reader (ELX 800, Bio-Tek Instruments), and defined as the lowest concentration of compound whose O.D., read at 570 nm, was comparable with the negative control wells (broth only, without inoculum).

Evaluation of Biofilm formation

All the bacterial reference strains were tested for their ability to form biofilms. Briefly, bacteria were grown in Tryptic Soy Broth (TSB, Sigma) containing 2% glucose overnight at 37°C in a shaking bath and then diluted 1:200 to a suspension with optical density (OD) of about 0.040 at 570 nm corresponding to $\sim 10^6$ CFU/ml. Polystyrene 24-well tissue culture plates were filled with 1 ml of diluted suspension and incubated for 24-hours at 37°C. Then, the wells were washed three times with 1 ml of sterile phosphate-buffered saline (PBS) and stained with 1 ml of safranin 0.1% v/v for 1 min. The excess stain was removed by placing the plates under running tap water. Plates were dried overnight in inverted position at 37°C. Safranin stained adherent bacteria in each well were re-dissolved to homogeneity in 1 ml of 30% v/v glacial acetic acid, and the OD was read at 492 nm. Each assay was performed in triplicate and repeated at least twice.

Biofilm prevention assay

Procedure described above was used to evaluate the activity of SP1 in preventing biofilm formation. Polystyrene 24-well tissue culture plates were filled with 1 ml of diluted bacterial suspension (OD of about 0.040 at 570 nm), obtained and diluted as previously seen, and sub-MIC concentrations, ranging from 6.2 to 0.7 mg/ml, of SP1 were directly added to the bacterial suspension at time zero and incubated at 37°C for 24 hours. After that time the wells were washed and stained with safranin as seen in biofilm forming assay. By comparing the average optical density of the growth in control wells with that in the sample wells, the following formula was used to calculate the percentages of inhibition for each concentration of the sample:

$$\text{Inhibition } (\%) = \frac{OD_{570} \text{ growth control} - OD_{570} \text{ sample}}{OD_{570} \text{ growth control}} \times 100 \qquad (2)$$

Each assay was performed in triplicate and assays were repeated at least twice.

Scanning electron microscopy

The effects of SP1 on formation of bacterial biofilm were morphologically evaluated by scanning electron microscopy (SEM). Glass slides in the bottom of a polystyrene 24-well tissue culture plates, were filled with 1 ml of a *S. epidermidis* RP62A suspension, obtained and diluted as previously seen, and a sub-MIC concentration of 3.1 mg/ml of SP1 was directly added at time zero and incubated at 37°C for 24 hours. After that time, the glass slides were gently washed four times with PBS to remove non-adherent bacteria and fixed with 2.5% glutaraldehyde-2% paraformaldehyde in 0.1 M cacodilate buffer (pH 7.4) for 30 min at 4°C. The bacterial preparation were washed with phosphate saline buffer PBS and post-fixed in osmium tetroxide 1% for 30 min at 4°C, followed by an ethanol dehydration series: 15 minutes in 50:50 ethanol: H_2O, 15 minutes in 75:25 ethanol: H_2O, 15 minutes in 95:5 ethanol: H_2O, and 30 minutes in 100% ethanol than a critical point drying procedure was followed, and the preparations were mounted on aluminium stubs, and gold coated in a sputter coater. Imaging was conducted with a LEO 420 scanning electron microscope as previously reported (Arizza et al. 2011).

Haemolytic assay

The assay was performed as described (Arizza et al. 2013). In brief, freshly collected rabbit erythrocytes (RE) with heparin, kindly provided by the "Zooprophylaxis Institute of Sicily" (Palermo, Italy), were washed (400 g for 10 min at 4°C) to remove the buffy coat, and the erythrocytes obtained were washed three times with phosphate-buffered saline (PBS: 6 mM KH_2PO_4; 30 mM Na_2HPO_4; 0.11 M NaCl; pH 7.4) and suspended in 10 ml PBS to obtain an 80×10^6 cells/ml suspension. Aliquots of 200 μL of RE suspension were mixed with 200 μl of SP1 concentrations (1.5, 3.2, 6.2 and 50 mg/ml) prepared in PBS, two-fold sequentially diluted v/v with a PBS. After an 1 h incubation at 37°C, the reaction mixture was centrifuged at 800 g for 15 min at 4°C to remove debris and residual erythrocytes. The O.D. of haemoglobin release was measured spectrophotometrically at a 541 nm wavelength. Spontaneous haemoglobin release (0%) was estimated incubating the RE with PBS while the complete haemolysis (100%) was assessed incubating the erythrocytes in a solution of 0.1% Triton-X 100 in distilled water and the haemolysis percentage was calculated according to the following equation:

$$\text{Hemolysis } (\%) = \frac{OD_{541}\text{Hb release in the reaction mixture} - OD_{541}\text{spontaneous Hb release}}{OD_{541} \text{ complete Hb release}} \times 100 \qquad (3)$$

Molecular dynamics simulations

The molecular folding of the peptide in aqueous solution was investigated *in silico* by molecular dynamics (MD) simulations, following recently reported procedures (Lauria et al. 2014; Lentini et al. 2014). In details, a 400 ns of MD simulation was carried out at 300 K, in the explicit water solvent and in the presence of 150 mM Na^+ and Cl^- counterions, using the Amber99SB-ILDN force field (Lindorff-Larsen et al. 2010) implemented in the GROMACS 4.6.5 software package (Pronk et al. 2013).

Results

Molecular Dynamics of SP1

SP1 is an 11-residue-long cationic peptide mainly enriched by residues such as lysine, with a pI of 10.72 and a net charge of +1 at pH 7.0.

To investigate SP1 in physiological conditions, we performed a MD simulation using the model represented in Figure 1. The *in silico* study provided a comprehensive picture of the dynamic equilibrium existing among the possible backbone conformations of the small oligopeptide. In fact, the root mean square deviations (RMSD) in Figure 2 show that the peptide conformation continuously and consistently changes during the 400 ns of simulation. Essentially all possible backbone conformations are assumed. However, interestingly, between 150 and 250 ns the structure is more stable and preserved, before becoming again flexible and random. To analyse this time region quantitatively, we have reported the Ramachandran plot in Figure 3 of snapshots sampled every 10 ns. This plot shows that in this time range the different residues assume

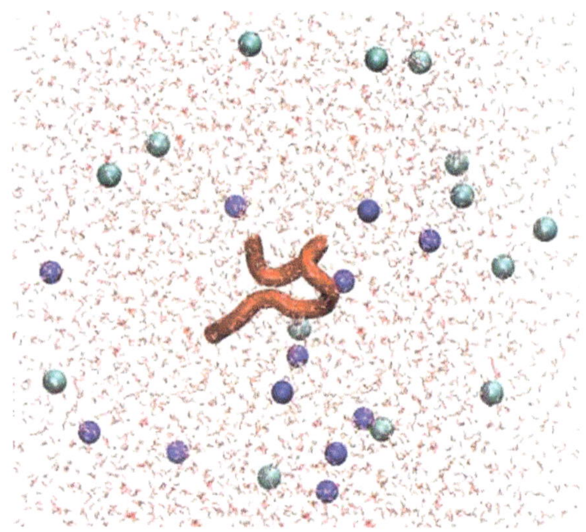

Figure 1 Example of MD simulation box, showing the SP1 molecule, the solution counterions, Na+ (in blue) and Cl− (in cyan), and the explicit water molecules (in red).

preferential local conformations. In particular, residues 2, 3, 4 and 8 are in the beta sheet conformation, residues 5, 6 and 9 are in the left-handed helix conformation, residue 7 is in the alpha helix conformation, while residue 10 is in the beta sheet conformation and only sporadically in the alpha helix conformation. The corresponding structure is depicted in Figure 4. The results obtained allow us to confirm that polar and apolar residues are not segregated on opposing surfaces through the long axis of the oligopeptide. In this peculiar conformation, the one that most frequently occurs in the investigated simulation time, the peptide possesses a hydrophobic non-amphipathic core constituted by V^2, A^3, S^4, F^5 with a hydrophobic region placed between E^1 and D^6. These features confer to SP1 properties similar to the AMPs with hydrophobic core.

Antibacterial activity of SP1

SP1 was tested at concentrations ranging from 25 to 0.07 mg/ml against a group of Gram positive and Gram negative reference strains. The antibacterial activity of SP1, expressed as minimum inhibitory concentrations (MICs) against planktonic cells of staphylococcal reference strains and isolates from veterinary origin and *P. aeruginosa* is listed in Table 1. All bacterial strains showed susceptibility to the antibacterial action of SP1. In particular the strain of *S. epidermidis* RP62A resulted to be the most susceptible, indeed, the MIC value was 6.2 mg/ml, while others attested to a value of 12.5 mg/ml. SP1 was active against all isolates of animal origin strains and in particular showed the highest activity against the strain #657 with a value of 6.2 mg/ml (Table 1).

Interference with biofilm formation

The interference with biofilm formation of SP1 against staphylococcal reference strains as *S. aureus* 25923, *S. aureus* 29213, *S. aureus* 6538, *S. epidermidis* RP62A and *P. aeruginosa* 15442 was observed. The inhibition was very evident to the highest concentrations of SP1 at 6.2 mg/ml, when the values, for all strains, reached approximately 80%. At the lowest concentrations of SP1 the degree of inhibition is reduced by following a dose dependence. *P. aeruginosa* strain 15442 was the most sensitive inhibitory activity of SP1, in fact, at a concentration of 3.1 mg/ml, a value of about 73% inhibition was observed, while for the other strains the inhibition values were of around 50% (Figure 5).

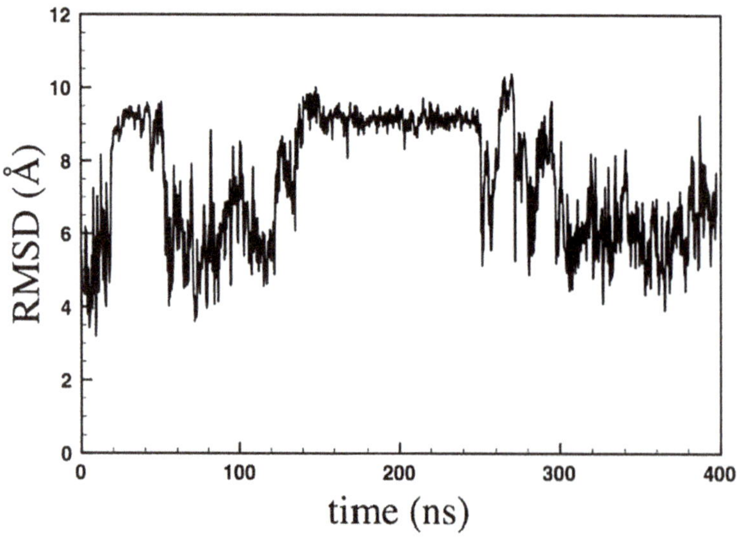

Figure 2 Plot of the RMSD obtained for SP1 up to 400 ns of MD simulation.

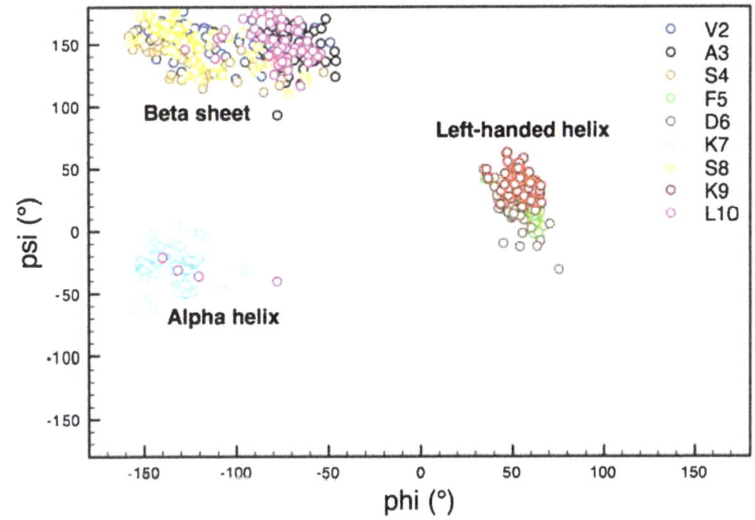

Figure 3 Ramachandran plot showing the values of psi and phi angles assumed by residues 2–10 of SP1 between 150 and 250 ns of the MD simulation.

SEM analysis was carried out to visualize the detailed architecture of biofilm and the antibiofilm effects of SP1 on bacterial biofilm of *S. epidermidis* RP62A (Figure 6). The control biofilm after 24 h was composed by multilayered conglomerated bacterial cells clusters that produced a dense biofilm matrix (Figure 6A). Sub MIC concentration (3.1 mg/ml) of SP1 was able to inhibit the formation of a multistratified structure, in fact in the sample treated with SP1 we found few monostratified bacterial cell clusters respect to the growth control (Figure 6B).

Haemolytic assay

The haemolytic activity of antimicrobial peptides against mammalian erythrocytes is often used as a preliminary evaluation of their selective toxicity and of the interactions of the cationic peptides with negatively charged membranes (Fischer et al. 2003; Wu et al. 2010) The performed hemolytic experiment to evaluate the interaction and the potential toxicity of the peptides SP1 did not show a measurable toxic effect against RBC (~1%) at MIC concentrations. A slight haemolytic activity, about 11.8%,

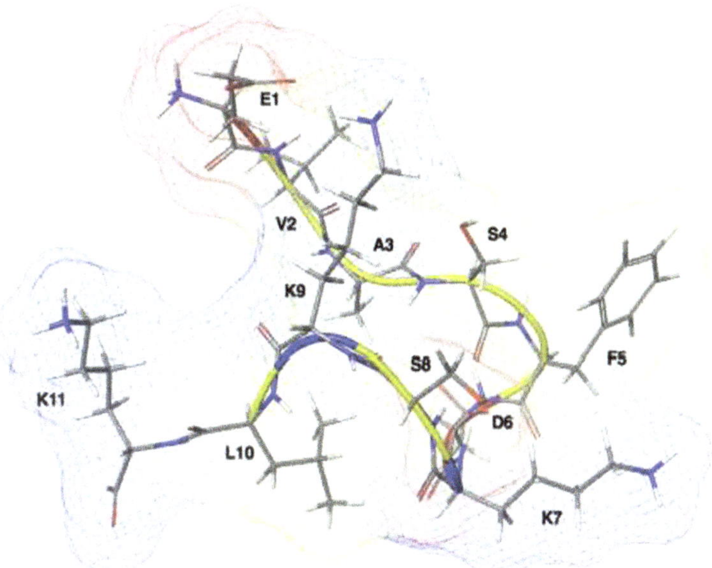

Figure 4 Molecular representation of SP1 at about 200 ns of MD showing the non amphipathic structure of the peptide. The potential surface is superimposed. Color code: acidic residues in red, basic residues in blue, and hydrophobic residues in yellow.

Table 1 MIC values of SP1 tested against bacterial strains

Bacterial strains	MIC (mg/ml)
S. aureus ATCC 25923	12.5
S. aureus ATCC 29213	12.5
S. aureus ATCC 6538	12.5
S. epidermidis RP62A	6.2
S. aureus 100	12.5
S. aureus 657	6.2
S. aureus 700	12.5
S. aureus 702	12.5
P. aeruginosa ATCC15442	12.5

MIC values for each strain are the means of three different assays performed in duplicate

was evident only at the highest concentration (50 mg/ml) (Figure 7).

Discussion

The echinoderms are considered a good source for AMPs and a variety of peptides with antimicrobial properties have been isolated from them. Antimicrobial activity has been reported in gonads of the asteroid *Marthasterias glacialis* (Stabili and Pagliara 1994), and *Paracentrotus lividus* contains, in the low molecular weight fraction (<5 kDa) of acid precipitate of their coelomocytes, peptides with antimicrobial activity against staphylococcal biofilms. In addition, we recently reported that immune mediators cells in the echinoderm *Holothuria tubulosa* is a source of novel AMPs with anti-staphylococcal biofim activity (Schillaci et al. 2013). Moreover, two cystein-rich AMPs, named centrocyns, have been characterized in the green sea-urchin *Strongylocentrotus droebachiensis* (Li et al. 2010).

The synthetic fragment SP1 of β-thymosin extracted from coelomocytes of the *P. lividus* shares structural characteristics of many antimicrobial peptides: it is a cationic peptide and possesses a significant proportion (~40%) of hydrophobic or non polar residues (Hancock and Lehrer 1998; Zasloff 2002). Although the α-helix is the most common secondary structure for the antimicrobial activity of AMPs (Mor and Nicolas 1994; Skerlavaj et al. 1996; Storici et al. 1994; Tossi et al. 1994) the *in silico* study suggests that the conformation assumed by SP1 (Figure 4) might be responsible for the observed antimicrobial activity. The helix structure of echinoderm AMPs shares a characteristic amphipathic structure with alternating hydrophobic and polar residues along the primary structure. According to the Shai-Matsuzaki-Huang model, the peptides bind to the membrane surface first and then with their amphipathic structure they enter into the membrane, breaking up the lipid chains and forming transient pore; this process can cause a collapse of the membrane at a critical peptide concentration (Hancock and Chapple 1999; Huang 2000; Matsuzaki 1998; Shai 2002). The SP1 peptide has a structure very different from other echinoderm AMPs (Schillaci et al. 2013; Schillaci et al. 2012). Moreover, such structure, unexpectedly, does not have an amphipathic nature with a hydrophobic face opposite to a hydrophilic one, because the polar charged residues and those hydrophobic are not arranged uniformly. The peptide contains a sequence of four hydrophobic/non polar residues which contribute to constitute a hydrophobic core, flanked at both ends by cationic and polar residues that can solubilize the peptides in aqueous solution providing a binding site for bacterial membranes. This particular conformation is different from other antimicrobial peptides, designed as transmembrane mimetic models and that spontaneously become inserted into the cell

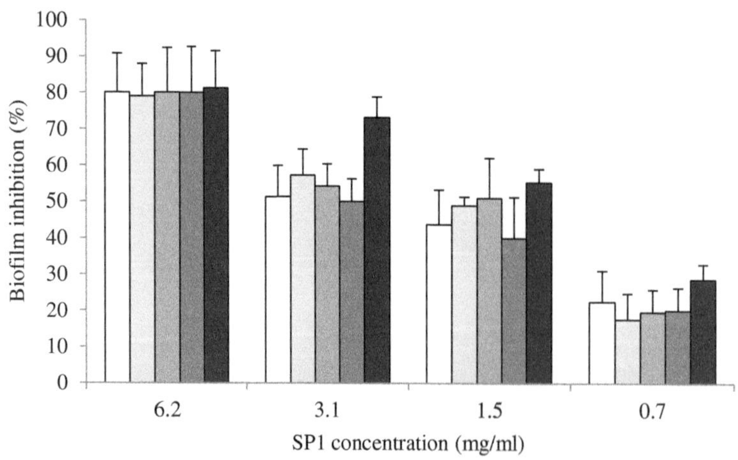

Figure 5 Interference with biofilm formation of SP1 against reference staphylococcal and *P. aeruginosa* strains. The percentage of inhibition were evaluated comparing the samples with not-treated 24 h old biofilms and staining with safranin. Bacterial strains (□) *S. aureus* 25923, (□) *S. aureus* 29213, (■) *S. aureus* 6538, (■) *S. epidermidis* RP62A, (■) *P. aeruginosa* 15442. Data for each strain are the mean of three distinct experiments ± S.D.

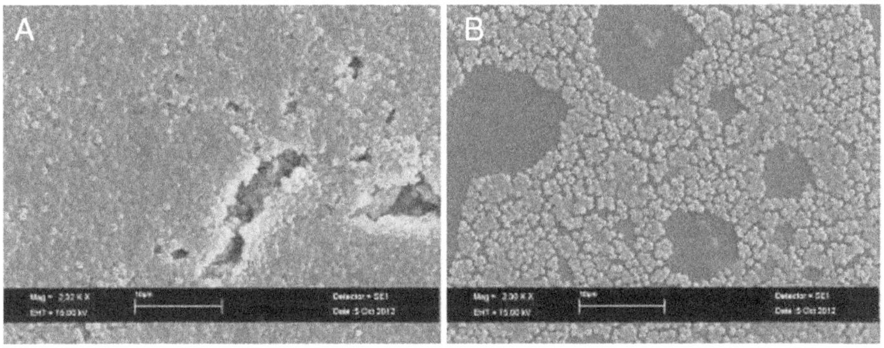

Figure 6 SEM micrography showing the effect of sub-MIC concentration of SP1 on *S. epidermidis* RP62A biofilm formation. A) Growth control (not treated with SP1); **B)** sample treated with a sub-MIC concentration (3.1 mg/ml) of SP1.

membranes (Chan et al. 2004; Liu and Deber 1998; Stark et al. 2002).

We found that the SP1 showed a broad antimicrobial activity against important pathogens such as *S. aureus* and *P. aeruginosa* but it acts at high concentration (12.5 or 6.2 mg/ml) against planktonic forms of these two microorganisms. Such weak activity is comparable to that reported for a different Echinoderm, *Holothuria tubulosa* (Schillaci et al. 2013) and to that of some described innate human defence protein like lactoferrin in vitro (de Andrade et al. 2014). MD results suggest that the observed weak activity could be due to the low stability of SP1, in fact, as showed in RMSD plot (Figure 2), the molecule is stable only in a limited time period and in this period it can present the active conformation of Figure 1. However, SP1 shows an interesting additional antimicrobial effect interfering with biofilm formation *in vitro* of the above cited pathogens, at lower concentrations than MIC evaluated against the planktonic forms. The prevention of biofilm formation – rather than its elimination – is the best strategy to contrast the growth, as a sessile community, of many pathogens.

We do not know the antimicrobial mechanisms of SP1, but we could speculate as for other cationic non-amphipathic microbial peptides that the positive charges carried by the peptide are essential for the membrane binding through electrostatic interaction between residues with anionic phospholipids. The peptide-membrane interaction could be responsible for membrane aggregation by peptides bridging simultaneously two membranes or for negative curvature of the membrane asymmetry that can form tubes (Lamaziere et al. 2007). The synthetic SP1 could act like Dermaseptin S9, a non-amphipathic antimicrobial peptides produced by the skin of the South American hylid frog, *Phyllomedusa sauvagei*, that contains, centrally located, a hydrophobic core that can insert the peptide in interior membrane (Lequin et al. 2006). Dermaseptin S9 exerts a microbicidal activity by perturbating both the membrane interface and the hydrophobic core of the bacterial membrane (Auvynet et al. 2008).

In a preliminary evaluation of selective toxicity, it was gratifying to note the general lack of hemolysis of rabbit erythrocytes by SP1, even at peptide concentrations up to

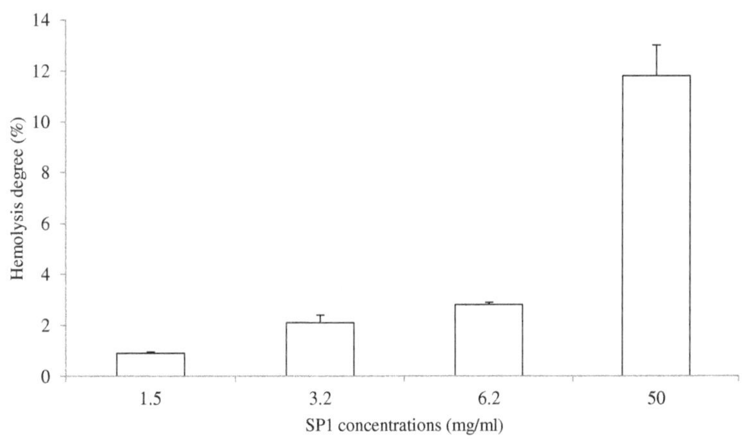

Figure 7 Hemolytic activity of SP1 peptides from *Paracentrotus lividus* hemocytes against rabbit blood cells at different concentrations: 1.5 mg/ml; 3.2 mg/ml; 6.2 mg/ml; 50 mg/ml. Data are the mean value of three separate experiments and expressed as percentage of hemolysis ± SD

50 mg/ml. Such selectivity of cationic non amphipathic antimicrobial peptides for bacterial membranes may be explicable, in part, by the differences in the compositions of eukaryotic and prokaryotic membranes (Matsuzaki 1999; Zasloff 2002). The outer leaflet of mammalian cells is predominantly composed of the zwitterionic phospholipids phosphatidylcholine and sphingomyelin (Verkleij et al. 1973), along with cholesterol (Turner and Rouser 1970), while bacterial membranes contain mainly anionic phospholipids and no cholesterol (Brock 1974). In addition, the outer surfaces of Gram-negative bacteria contain lipopolysaccharides, while those of Gram-positive bacteria contain teichoic acid, which in both cases add to the negative charge of the bacterial surface (Brock 1974). The cationic nature of native antimicrobial peptides clearly contributes to their preferential recognition by the negatively charged outer surfaces of bacterial membranes (Oren and Shai 1998; Shai 1999).

The extensive clinical use of classical antibiotics has led to the growing emergence of many medically relevant resistant strains of pathogens (Patel 2005). Therefore, the development of a new class of antimicrobials with a different mechanism of action than conventional antibiotics has become critical. The cationic antimicrobial peptides could represent such a new class (Andreu and Rivas 1998; Hancock 1997; Sitaram and Nagaraj 2002). The development of resistance to membrane active peptides whose sole target is the cytoplasmic membrane is not expected because this would require substantial changes in the lipid composition of cell membranes of microorganisms (Hancock 2001).

The anti-adhesion property showed by SP1 against *S. aureus* and *P. aeruginosa* strains is very interesting considering that the two opportunistic pathogens are able to form biofilms in open wounds, such as chronic diabetic foot ulcers, or infected wounds in clinical and veterinary medicine. New antimicrobial agents that are effective against staphylococci and *P. aeruginosa* to treat infected wounds are needed, and a potential topical application of SP1 could be supposed. Moreover, the two species are also involved in food spoilage and biofilm formation of food transmitted pathogens and an application of SP1 in the food processing is possible and closer, at this stage of our study, than application in clinical or veterinary health.

A chemotherapeutic approach combining conventional antibiotics and novel anti-biofilm agents could be a new strategy for the treatment of biofilm-associated infections like mastitis in veterinary field or the topical treatment of infected wounds in clinical and veterinary setting.

Finally, the tested synthetic peptide is a good starting point to design new synthetic derivatives with modified chemical-physical properties, with the aim to improve their antimicrobial activity against pathogens and their pharmaceutical potential (Brogden and Brogden 2011; Huang et al. 2010).

Competing interests
The authors declare that they have no competing interests.

Acknowledgements
Financial support from Fondi di Ateneo 2006, 2007 Università degli Studi di Palermo (Italy) and Ministero della Salute IZS is gratefully acknowledged.

Author details
[1]Dip. STEBICEF, Università degli Studi di Palermo, Via Archirafi, 32-90123 Palermo, Italy. [2]IEMEST, Istituto Euromediterraneo di Scienza e Tecnologia, Via Emerico Amari, 123-90139 Palermo, Italy. [3]Istituto Zooprofilattico Sperimentale della Sicilia "A. Mirri", Via G. Marinuzzi, 3-90129 Palermo, Italy.

References
Andreu D, Rivas L (1998) Animal antimicrobial peptides: an overview. Biopolymers 47(6):415–433

Arizza V (2013) Marine biodiversity as source of new drugs. Ital J Zool 80(3):317–318, doi:10.1080/11250003.2013.830370

Arizza V, Parrinello D, Cammarata M, Vazzana M, Vizzini A, Giaramita FT, Parrinello N (2011) A lytic mechanism based on soluble phospholypases A2 (sPLA2) and beta-galactoside specific lectins is exerted by *Ciona intestinalis* (ascidian) unilocular refractile hemocytes against K562 cell line and mammalian erythrocytes. Fish Shellfish Immun 30(4–5):1014–1023, doi:10.1016/j.fsi.2011.01.022

Arizza V, Vazzana M, Schillaci D, Russo D, Giaramita FT, Parrinello N (2013) Gender differences in the immune system activities of sea urchin *Paracentrotus lividus*. Comp Biochem Physiol A Mol Integr Physiol 164(3):447–455, doi:10.1016/j.cbpa.2012.11.021

Auvynet C, El Amri C, Lacombe C, Bruston F, Bourdais J, Nicolas P, Rosenstein Y (2008) Structural requirements for antimicrobial versus chemoattractant activities for dermaseptin S9. FEBS J 275(16):4134–4151, doi:10.1111/j.1742-4658.2008.06554.x

Bax R, Mullan N, Verhoef J (2000) The millennium bugs - the need for and development of new antibacterials. Int J Antimicrob Ag 16(1):51–59, doi:10.1016/S0924-8579(00)00189-8

Brock TD (1974) Biology of Microorganisms, 2nd edn. Prentice-Hall, Inc, Englewood Cliffs NJ

Brogden NK, Brogden KA (2011) Will new generations of modified antimicrobial peptides improve their potential as pharmaceuticals? Int J Antimicrob Ag 38(3):217–225, doi:10.1016/j.ijantimicag.2011.05.004

Canicatti C, Roch P (1989) Studies on Holothuria polii (Echinodermata) Antibacterial Proteins.1. Evidence for and Activity of a Coelomocyte Lysozyme. Experientia 45(8):756–759, doi:10.1007/Bf01974579

Chan C, Burrows LL, Deber CM (2004) Helix induction in antimicrobial peptides by alginate in biofilms. J Biol Chem 279(37):38749–38754, doi:10.1074/jbc.M406044200

Clutterbuck AL, Woods EJ, Knottenbelt DC, Clegg PD, Cochrane CA, Percival SL (2007) Biofilms and their relevance to veterinary medicine. Vet Microbiol 121(1–2):1–17, doi:10.1016/j.vetmic.2006.12.029

de Andrade FB, de Oliveira JC, Yoshie MT, Guimarães BM, Gonçalves RB, Schwarcz WD (2014) Antimicrobial activity and synergism of lactoferrin and lysozyme against cariogenic microorganisms. Braz Dent J 25:165–169

Fischer D, Li Y, Ahlemeyer B, Krieglstein J, Kissel T (2003) In vitro cytotoxicity testing of polycations: influence of polymer structure on cell viability and hemolysis. Biomaterials 24(7):1121–1131, doi:10.1016/S0142-9612(02)00445-3

Gautier R, Douguet D, Antonny B, Drin G (2008) HELIQUEST: a web server to screen sequences with specific α-helical properties. Bioinformatics 24(18):2101–2102, doi:10.1093/bioinformatics/btn392

Gilbert P, Allison DG, McBain AJ (2002) Biofilms in vitro and in vivo: do singular mechanisms imply cross-resistance? J Appl Microbiol 92:98s–110s, doi:10.1046/j.1365-2672.92.5s1.5.x

Hancock REW (1997) Peptide antibiotics. Lancet 349(9049):418–422, doi:10.1016/S0140-6736(97)80051-7

Hancock RE (2001) Cationic peptides: effectors in innate immunity and novel antimicrobials. Lancet Infect Dis 1(3):156–164, doi:10.1016/S1473-3099(01)00092-5

Hancock RE, Chapple DS (1999) Peptide antibiotics. Antimicrob Agents Chemother 43(6):1317–1323

Hancock RE, Lehrer R (1998) Cationic peptides: a new source of antibiotics. Trends Biotechnol 16(2):82–88

Hancock RE, Rozek A (2002) Role of membranes in the activities of antimicrobial cationic peptides. FEMS Microbiol Lett 206(2):143–149

Huang HW (2000) Action of antimicrobial peptides: Two-state model. Biochemistry 39(29):8347–8352, doi:10.1021/Bi0009461

Huang YB, Huang JF, Chen YX (2010) Alpha-helical cationic antimicrobial peptides: relationships of structure and function. Protein Cell 1(2):143–152, doi:10.1007/s13238-010-0004-3

Lamaziere A, Burlina F, Wolf C, Chassaing G, Trugnan G, Ayala-Sanmartin J (2007) Non-metabolic membrane tubulation and permeability induced by bioactive peptides. Plos One 2(2):e201, doi:10.1371/journal.pone.0000201

Lauria A, Bonsignore R, Terenzi A, Spinello A, Giannici F, Longo A, Almerico AM, Barone G (2014) Nickel(II), copper(II) and zinc(II) metallo-intercalators: structural details of the DNA-binding by a combined experimental and computational investigation. Dalton Trans 43:6108–6119, http://dx.doi.org/10.1039/c3dt53066c

Lentini L, Melfi R, Di Leonardo A, Spinello A, Barone G, Pace A, Palumbo Piccionello A, Pibiri I (2014) Towards a rationale for the PTC124 (Ataluren) promoted readthrough of premature stop codons: a computational approach and GFP-reporter cell-based assay. Mol Pharm 11:653–664, http://dx.doi.org/10.1021/mp400230s

Lequin O, Ladram A, Chabbert L, Bruston F, Convert O, Vanhoye D, Chassaing G, Nicolas P, Amiche M (2006) Dermaseptin S9, an alpha-helical antimicrobial peptide with a hydrophobic core and cationic termini. Biochemistry 45(2):468–480, doi:10.1021/bi051711i

Li C, Haug T, Moe MK, Styrvold OB, Stensvåg K (2010) Centrocins: Isolation and characterization of novel dimeric antimicrobial peptides from the green sea urchin, Strongylocentrotus droebachiensis. Dev Comp Immunol 34:959–968, doi: 10.1016/j.dci.2010.04.004

Lindorff-Larsen K, Piana S, Palmo K, Maragakis P, Klepeis JL, Dror RO, Shaw DE (2010) Improved side-chain torsion potentials for the Amber ff99SB protein force field. Proteins 78:1950–1958, http://dx.doi.org/10.1002/prot.22711

Liu LP, Deber CM (1998a) Guidelines for membrane protein engineering derived from de novo designed model peptides. Pept Sci 47(1):41–62, doi:10.1002/(SICI)1097-0282(1998)47:1<41::AID-BIP6>3.0.CO;2-X

Liu LP, Deber CM (1998b) Guidelines for membrane protein engineering derived from de novo designed model peptides. Biopolymers 47(1):41–62, doi:10.1002/(SICI)1097-0282(1998)47:1<41::AID-BIP6>3.0.CO;2-X

Matsuzaki K (1998) Magainins as paradigm for the mode of action of pore forming polypeptides. Biochim Biophys Acta Rev Biomembr 1376(3):391–400, doi:10.1016/S0304-4157(98)00014-8

Matsuzaki K (1999) Why and how are peptide–lipid interactions utilized for self-defense? Magainins and tachyplesins as archetypes. Biochim Biophys Acta Rev Biomembr 1462(1–2):1–10, http://dx.doi.org/10.1016/S0005-2736(99)00197-2

Melchior MB, Vaarkamp H, Fink-Gremmels J (2006) Biofilms: a role in recurrent mastitis infections? Vet J 171(3):398–407, doi:10.1016/j.tvjl.2005.01.006

Melchior MB, van Osch MHJ, Graat RM, van Duijkeren E, Mevius DJ, Nielen M, Gaastra W, Fink-Gremmels J (2009) Biofilm formation and genotyping of Staphylococcus aureus bovine mastitis isolates: Evidence for lack of penicillin-resistance in Agr-type II strains. Vet Microbiol 137(1–2):83–89, doi:10.1016/j.vetmic.2008.12.004

Mor A (2000) Peptide-based antibiotics: A potential answer to raging antimicrobial resistance. Drug Develop Res 50(3–4):440–447, doi:10.1002/1098-2299(200007/08)50:3/4<440::Aid-Ddr27>3.0.Co;2-4

Mor A, Nicolas P (1994) The Nh2-terminal alpha-helical domain 1–18 of Dermaseptin is responsible for antimicrobial activity. J Biol Chem 269(3):1934–1939

Oren Z, Shai Y (1998) Mode of action of linear amphipathic α-helical antimicrobial peptides. Pept Sci 47(6):451–463, doi:10.1002/(SICI)1097-0282(1998)47:6<451::AID-BIP4>3.0.CO;2-F

Patel R (2005) Biofilms and antimicrobial resistance. Clin Orthop Relat Res 437:41–47

Projan SJ, Youngman PJ (2002) Antimicrobials: new solutions badly needed. Curr Opin Microbiol 5(5):463–465

Pronk S, Pall S, Schulz R, Larsson P, Bjelkmar P, Apostolov R, Shirts MR, Smith JC, Kasson PM, van der Spoel D, Hess B, Lindahl E (2013) GROMACS 4.5: a high-throughput and highly parallel open source molecular simulation toolkit. Bioinformatics 29:845–854, http://dx.doi.org/10.1093/bioinformatics/btt055

Schillaci D, Petruso S, Sciortino V (2005) 3,4,5,3 ′,5 ′-Pentabromo-2-(2 ′-hydroxybenzoyl)pyrrole: a potential lead compound as anti-Gram-positive and anti-biofilm agent. Int J Antimicrob Ag 25(4):338–340, doi:10.1016/j.ijantimicag.2004.11.014

Schillaci D, Arizza V, Parrinello N, Di Stefano V, Fanara S, Muccilli V, Cunsolo V, Haagensen JJ, Molin S (2010) Antimicrobial and antistaphylococcal biofilm activity from the sea urchin *Paracentrotus lividus*. J Appl Microbiol 108(1):17–24, doi:JAM4394 [pii] 10.1111/j.1365-2672.2009.04394.x

Schillaci D, Vitale M, Cusimano MG, Arizza V (2012) Fragments of beta-thymosin from the sea urchin Paracentrotus lividus as potential antimicrobial peptides against staphylococcal biofilms. In: Goldstein AL, Garaci E (eds) Thymosins in Health and Disease Ii, vol 1270, Annals of the New York Academy of Sciences. Blackwell Science Publ, Oxford, pp 79–85

Schillaci D, Cusimano MG, Cunsolo V, Saletti R, Russo D, Vazzana M, Vitale M, Arizza V (2013) Immune mediators of sea-cucumber Holothuria tubulosa (Echinodermata) as source of novel antimicrobial and anti-staphylococcal biofilm agents. AMB Express 3(1):35, doi:10.1186/2191-0855-3-35

Schumacher-Perdreau F, Heilmann C, Peters G, Götz F, Pulverer G (1994) Comparative analysis of a biofilm-forming Staphylococcus epidermidis strain and its adhesion-positive, accumulation-negative mutant M7. FEMS Microbiol Lett 117:71–78

Shai Y (1999) Mechanism of the binding, insertion and destabilization of phospholipid bilayer membranes by α-helical antimicrobial and cell non-selective membrane-lytic peptides. Biochim Biophys Acta Rev Biomembr 1462(1–2):55–70, http://dx.doi.org/10.1016/S0005-2736(99)00200-X

Shai Y (2002) Mode of action of membrane active antimicrobial peptides. Biopolymers 66(4):236–248, doi:10.1002/Bip.10260

Sitaram N, Nagaraj R (2002) The therapeutic potential of host-defense antimicrobial peptides. Curr Drug Targets 3(3):259–267, doi:10.2174/1389450023347786

Skerlavaj B, Gennaro R, Bagella L, Merluzzi L, Risso A, Zanetti M (1996) Biological characterization of two novel cathelicidin-derived peptides and identification of structural requirements for their antimicrobial and cell lytic activities. J Biol Chem 271(45):28375–28381

Stabili L, Pagliara P (1994) Antibacterial protection in *Marthasterias glacialis* eggs: characterization of lysozyme-like activity. Comp Biochem Physiol B Biochem Mol Biol 109(4):709–713

Stark M, Liu LP, Deber CM (2002) Cationic hydrophobic peptides with antimicrobial activity. Antimicrob Agents 46(11):3585–3590

Storici P, Scocchi M, Tossi A, Gennaro R, Zanetti M (1994) Chemical synthesis and biological-activity of a novel antibacterial peptide deduced from a pig myeloid cDNA. FEBS Lett 337(3):303–307, doi:10.1016/0014-5793(94)80214-9

Tossi A, Scocchi M, Skerlavaj B, Gennaro R (1994) Identification and characterization of a primary antibacterial domain in Cap18, a lipopolysaccharide-binding protein from rabbit leukocytes. FEBS Lett 339(1–2):108–112, doi:10.1016/0014-5793(94)80395-1

Turner JD, Rouser G (1970) Precise quantitative determination of human blood lipids by thin-layer and triethylaminoethylcellulose column chromatography. I Erythrocyte lipids. Anal Biochem 38(2):423–436

Verkleij AJ, Zwaal RFA, Roelofsen B, Comfurius P, Kastelijn D, van Deenen LLM (1973) The asymmetric distribution of phospholipids in the human red cell membrane. A combined study using phospholipases and freeze-etch electron microscopy. Biochim Biophys Acta Rev Biomembr 323(2):178–193, http://dx.doi.org/10.1016/0005-2736(73)90143-0

Wang Z, Wang G (2004) APD: the Antimicrobial Peptide Database. Nucleic Acids Res 32(Database issue):D590–D592, doi:10.1093/nar/gkh025

Wang G, Li X, Wang Z (2009) APD2: the updated antimicrobial peptide database and its application in peptide design. Nucleic Acids Res 37(Database issue):D933–D937, doi:10.1093/nar/gkn823

Wu G, Wu H, Fan X, Zhao R, Li X, Wang S, Ma Y, Shen Z, Xi T (2010) Selective toxicity of antimicrobial peptide S-thanatin on bacteria. Peptides 31:1669–1673, doi:10.1016/j.peptides.2010.06.009

Zasloff M (2002) Antimicrobial peptides of multicellular organisms. Nature 415(6870):389–395, doi:10.1038/415389a

10

Diverse allosteric and catalytic functions of tetrameric D-lactate dehydrogenases from three Gram-negative bacteria

Nayuta Furukawa[1], Akimasa Miyanaga[2], Misato Togawa[1], Masahiro Nakajima[1] and Hayao Taguchi[1*]

Abstract

NAD-dependent D-lactate dehydrogenases (D-LDHs) reduce pyruvate into D-lactate with oxidation of NADH into NAD^+. Although non-allosteric D-LDHs from *Lactobacilli* have been extensively studied, the catalytic properties of allosteric D-LDHs from Gram-negative bacteria except for *Escherichia coli* remain unknown. We characterized the catalytic properties of D-LDHs from three Gram-negative bacteria, *Fusobacterium nucleatum* (FNLDH), *Pseudomonas aeruginosa* (PALDH), and *E. coli* (ECLDH) to gain an insight into allosteric mechanism of D-LDHs. While PALDH and ECLDH exhibited narrow substrate specificities toward pyruvate like usual D-LDHs, FNLDH exhibited a broad substrate specificity toward hydrophobic 2-ketoacids such as 2-ketobutyrate and 2-ketovalerate, the former of which gave a 2-fold higher $k_{cat}/S_{0.5}$ value than pyruvate. Whereas the three enzymes consistently showed hyperbolic shaped pyruvate saturation curves below pH 6.5, FNLDH and ECLDH, and PALDH showed marked positive and negative cooperativity, respectively, in the pyruvate saturation curves above pH 7.5. Oxamate inhibited the catalytic reactions of FNLDH competitively with pyruvate, and the PALDH reaction in a mixed manner at pH 7.0, but markedly enhanced the reactions of the two enzymes at low concentration through canceling of the apparent homotropic cooperativity at pH 8.0, although it constantly inhibited the ECLDH reaction. Fructose 1,6-bisphosphate and certain divalent metal ions such as Mg^{2+} also markedly enhanced the reactions of FNLDH and PALDH, but none of them enhanced the reaction of ECLDH. Thus, our study demonstrates that bacterial D-LDHs have highly divergent allosteric and catalytic properties.

Keywords: Allosteric regulation; NAD-dependent D-lactate dehydrogenase; Gram-negative bacteria; *Escherichia coli*; *Fusobacterium nucleatum*; *Pseudomonas aeruginosa*

Introduction

NAD-dependent D- and L-lactate dehydrogenases (D-LDH, EC 1.1.1.28; and L-LDH, EC 1.1.1.27) reduce pyruvate to D- and L-lactate, respectively, with oxidization of NADH into NAD^+, and act at the last step of the glycolytic pathway under anaerobic conditions (Holbrook et al. 1975). Although both types of enzymes catalyze identical reactions except for the distinct chirality of the lactate product, they are evolutionarily distinct from each other (Taguchi and Ohta 1991; Bernard et al. 1991; Kochhar et al. 1992). D-LDHs belong to a large D-2-hydroxyacid dehydrogenase (D-HydDH) superfamily, which comprises various kinds of D-isomer dehydrogenases such as the D-glycerate (Greenler et al. 1989; Goldberg et al. 1994), D-3-phosphoglycerate (Tobey and Grant 1986; Grant 1989), D-2-hydroxyglutarate (Buckel and Barker 1974; Martins et al. 2005), and D-2-hydroxyisocaproate (Lerch et al. 1989; Dengler et al. 1997) dehydrogenases, and even non- D-HydDHs such as the formate (Vinals et al. 1993; Lamzin et al. 1994; Popov and Lamzin 1994), phosphite (Costas et al. 2001; Woodyer et al. 2005), and L-alanine (Baker et al. 1998; Tripathi and Ramachandran 2008) dehydrogenases. Although these enzymes possess highly divergent primary sequences, their tertiary structures are very similar to one another.

In the case of L-LDHs, it is known that allosteric types of the enzyme are widely distributed in bacteria including *Lactobacilli* (Garvie 1980). These enzymes commonly require fructose 1,6-bisphosphate (FBP) to exhibit high catalytic activity, and usually exhibit sigmoidal shaped substrate saturation curves unless FBP is present. L-LDHs generally have tetrameric structures comprising

* Correspondence: htaguchi@rs.noda.tus.ac.jp
[1]Tokyo University of Science, 2641 Yamazaki, Noda, Chiba 278-8510, Japan
Full list of author information is available at the end of the article

four identical subunits, and allosteric L-LDHs undergo cooperative structural changes in the substrate and FBP binding sites through quaternary structural changes (Iwata et al. 1994; Arai et al. 2010). On the other hand, no allosteric property has been reported for *Lactobacillus* D-LDHs. The *Lactobacillus* D-LDHs consistently have a dimeric structure comprising identical subunits, and exhibit hyperbolic shaped saturation curves for substrate pyruvate. Interestingly, it was reported that D-LDH from *E. coli* exhibits a sigmoidal pyruvate saturation curve, which is changed to the hyperbolic one by 2-ketobutyrate (Tarmy and Kaplan 1968). The reported results of size-exclusion chromatography suggested that this enzyme has a homotetrameric structure, although the detailed catalytic properties and 3D structure of the enzyme remain unknown. Beside the D-LDH from *E. coli*, the D-3-phosphoglycerate dehydrogenases (D-PgDHs) from some bacteria, such as *E. coli* (Tobey and Grant 1986; Grant 1989; Schuller et al. 1995) and *Mycobacterium tuberculosis* (Dey et al. 2005), are only known as allosteric and homotetrameric enzymes in the D-HydDH superfamily.

In this paper, we describe the detailed biochemical analysis of allosteric D-LDHs from three Gram-negative bacteria; *Fusobacterium nucleatum* subsp. *nucleatum*, an obligate anaerobe; *Pseudomonas aeruginosa*, an obligate aerobe; and *E. coli*, a facultative anaerobe, demonstrating that bacterial D-LDHs have highly divergent catalytic properties.

Materials and methods

Cloning, expression, and purification

Genomic DNA of *F. nucleatum* subsp. *nucleatum* JCM14847 (=ATCC25586) and *P. aeruginosa* JCM8532 (=PAO1) was purchased from the Riken BioResource Center (Japan). Genomic DNA of *E. coli* BL21(DE3) was isolated by ISOGEN (NIPPON GENE, Tokyo, Japan). The genes encoding the FN0511 (FNLDH), PA0927 (PALDH), and ECD_01352 (ECLDH) proteins (Genbank accession numbers: AAL94707.1, AAG04316.1, and ACT43236.1, respectively) were amplified *via* PCR using genomic DNA as templates. The primer pairs used were, 5′-CGCTCGAGATGCAAAAAACTAAGATAATATTTTT TG-3′ and 5′-GGGATCCATTTATTGATTTTGTGGA ACTTC-3′ for the *fn0511* gene, 5′-CGCATATGCGCA TCCTGTTCTTCAGCAG-3′ and 5′- CGGGATCCTC AGGCCCGGACCCGATTG-3′ for the *pa0927* gene, and 5′-CGCATATGAAACTCGCCGTTTATAGC-3′ and 5′-CGGGATCCTTAAACCAGTTCGTTCGGGC-3′ for the *ecd_01352* gene. The amplified *fn0511*, *pa0927*, and *ecd_01352* genes were inserted into T-vector pMD20 (TaKaRa Bio, Shiga, Japan). After purification of the plasmids carrying these genes, the plasmids were digested with restriction enzymes and the genes were inserted into pCold I (TaKaRa Bio). *E. coli* Rosetta2 (DE3) (Merck Millipore, Darmstadt, Germany) was transformed using the constructed plasmids and cultured at 37°C in 2 × YT medium (1.6% tryptone, 1.0% yeast extract and 0.5% NaCl) containing 100 μg/ml ampicillin. After the optical density of the culture at 600 nm (OD_{600}) had reached 0.6, protein expression was induced using 0.5 mM isopropyl-β-1-D-thiogalactopyranoside at 15°C overnight. The harvested cells were lysed by sonication in 50 mM HEPES-NaOH buffer (pH 8.0) containing 150 mM NaCl and 10 mM imidazole on ice. After each sample had been centrifuged at 27,000 × *g*, the supernatant was applied to HisTrap FF crude column (5 ml) (GE Healthcare, Buckinghamshire, UK), and was then eluted with a linear gradient of 10–250 mM imidazole. After dialysis against 50 mM HEPES-NaOH buffer (pH 8.0), the protein solution was further purified using a UnoQ column (Bio-Rad Laboratories, California, USA). The target protein was eluted with a linear gradient of 0–300 mM NaCl. The buffer for FNLDH and ECLDH was changed to 5 mM HEPES-NaOH buffer (pH 8.0), and the buffer for PALDH to 5 mM sodium acetate buffer (pH 5.0) using Amicon Ultra 30,000 molecular weight cut-off (Merck Millipore). The concentrations of FNLDH, PALDH, and ECLDH were determined spectrophotometrically at 280 nm using theoretical extinction coefficients of 24,215, 19,285, and 17,140 M^{-1} cm^{-1}, respectively (Pace et al. 1995).

SDS-PAGE

The pufiried enzymes were separated on a 11% (w/v) SDS-polyacrylamide gel (Laemmli 1970). Approximately 5 μg of the enzymes were loaded as 5% (v/v) glycerol solutions containing 1% (w/v) SDS, 2.5% (v/v) β-mercaptoethanol and 0.05% bromophenol blue. Precision Plus Protein Unstained Standards (Bio-Rad Laboratories) was used as molecular weight markers. Proteins were stained with Coomassie Brilliant Blue.

Enzyme assay

Enzyme activity was determined by measuring the reduction of the absorbance at 340 nm derived from NADH for 1 minute. The reactions were performed at 30°C in various 50 mM buffers containing 0.1 mM NADH and a substrate. The data that showed significant cooperative effects of the substrate were interpreted using the Hill equation (Equation 1) (Dixon and Webb 1979).

$$(v - v_{min})/k_{cat} = [S]^{nH} / \left([S]^{nH} + S_{0.5}{}^{nH} \right) \quad (1)$$

where v is the reaction velocity, [S] the ligand concentration, such as pyruvate, FBP, or Mg^{2+}, k_{cat} the turnover rate of catalysis, v_{min} the reaction velocity with no ligand, $S_{0.5}$ the half-saturation concentration of a ligand, and n_H

the Hill coefficient. Kinetic parameters were obtained by curve fitting of the data with KaleidaGraph ver 3.51. The data that indicated significant substrate inhibition were interpreted using the equation for substrate inhibition (Equation 2) (Eszes et al. 1996).

$$v/k_{cat} = [S]/\left([S] + S_{0.5} + [S]^2/K_I\right) \tag{2}$$

where K_i is the inhibition constant. The kinetic parameters for oxamate at pH 7.0 were calculated by curve fitting of the data with GraFit version 7.0.3. The data that indicated significant competitive-type and mix-type inhibition were interpreted using the equation for competitive-type (Equation 3) and mix-type inhibition (Equation 4), respectively.

$$v/k_{cat} = [S]/(K_m(1 + [I]/K_I) + [S]) \tag{3}$$

$$v/k_{cat} = [S]/(K_m(1 + [I]/K_I) + [S](1 + [I]/K_I') \tag{4}$$

where K_m is the Michaelis constant, and K_i' the inhibition constant. The data of oxamate inhibition for FNLDH, PALDH, and ECLDH at pH 8.0 were interpreted using the equation for competitive-type (Equation 5), mix-type (Equation 6), and non-competitive-type (Equation 7) inhibition of allosteric enzyme, respectively.

$$v = [S]^{nH}/\left(K_m{}^{nH}(1 + [I]/K_I) + [S]^{nH}\right) \times \left(k_{cat} + k_{cat}'[I]^{nH'}/\left([I]^{nH'} + K_{act}{}^{nH'}\right)\right) \tag{5}$$

$$v = [S]^{nH}/\left(K_m{}^{nH}(1 + [I]/K_I) + [S]^{nH}(1 + [I]/K_I')\right) \times \left(k_{cat} + k_{cat}'[I]^{nH'}/\left([I]^{nH'} + K_{act}{}^{nH'}\right)\right) \tag{6}$$

$$v = k_{cat}[S]^{nH}/\left(\left(K_m{}^{nH} + [S]^{nH}\right)(1 + [I]/K_I)\right) \tag{7}$$

where k_{cat}' is the turnover rate of catalysis in the presence of oxamate, K_{act} the half-saturation concentration of oxamate for activation, and n_H' the hill coefficient for oxamate.

Temperature and pH stability

The pH-stability was determined by measuring the remaining activity after incubation of the enzymes (1 μM) in various 50 mM buffers at 30°C for 1 h. The thermostability was determined by measuring the remaining activity after incubation of FNLDH (1 μM) or ECLDH (1 μM) in 50 mM HEPES-NaOH buffer (pH 8.0), or of PALDH (1 μM) in 50 mM sodium acetate buffer (pH 5.0) at various temperatures for 30 min.

Size-exclusion chromatography

Each protein solution diluted to 1 mg/ml (1 ml) was applied to Superdex 200 (Hiload 16/60; GE Healthcare) equilibrated with 50 mM buffer containing 150 mM NaCl. HEPES-NaOH (pH 8.0) was used for FNLDH and ECLDH, and sodium acetate buffer (pH 5.0) for PALDH. Ovalbumin (44 kDa), conalbumin (75 kDa), aldolase (158 kDa), ferritin (440 kDa), and thyroglobulin (669 kDa) (GE Healthcare) were used as standard proteins. Blue dextran 2000 (2,000 kDa; GE Healthcare) was used to determine the void volume of the column.

Results

Basic properties

The recombinant FNLDH, PALDH and ECLDH consistently exhibited marked catalytic activity toward pyruvate, and were successively purified to homogeneous protein samples (Figure 1a). FNLDH, PALDH, and ECLDH were stable in the pH ranges of 5.0–8.5, 4.0–8.0, and 5.0–10.0, respectively, during treatment at 30°C for 1 h (Figure 2 a, b, c). FNLDH, PALDH, and ECLDH were stable up to 39, 57, and 49°C, respectively, under favorable pH conditions, i.e., pH 8.0 for FNLDH and ECLDH, and pH 5.0 for PALDH (Figure 2d).

The purified FNLDH and PALDH samples each gave a single protein peak on size-exclusion chromatography analysis, and exhibited apparent molecular weights of 160 and 120 kDa, respectively (Figure 1b). On the other hand, the ECLDH sample gave two peaks corresponding to molecular weights of 330 and 150 kDa. The theoretical molecular weights of FNLDH, PALDH, and ECLDH are 40.0, 37.8, and 38.5 kDa, suggesting that FNLDH and PALDH at that of homotetrameric and homotrimeric structures, respectively, and ECLDH was eluted at the position of homooctameric and homotetrameric structures. Since it has been reported that native D-LDH from *E. coli* cells gives only a peak that corresponds to the molecular weight of 130 kDa (Tarmy and Kaplan 1968), the apparently octameric form of ECLDH is likely an artificial product due to the overexpression, and exhibited slightly but significantly lower specific activity than the tetrameric form (data not shown). Therefore, only homotetrameric ECLDH was used for the detailed analysis below.

pH-Dependency of kinetic parameters

The kinetic parameters, k_{cat}, $S_{0.5}$, $k_{cat}/S_{0.5}$, and n_H values, were determined by pyruvate reduction assaying in the pH range of 4.5-9.0 (Figure 3). The three enzymes consistently showed virtually constant k_{cat} values independently of pH, and PALDH and ECLDH exhibited about 5-fold larger k_{cat} values than FNLDH. The three enzymes also consistently showed constant pyruvate $S_{0.5}$ values below pH 7.0, and PALDH and FNLDH exhibited approximately one order of magnitude smaller $S_{0.5}$ values than ECLDH. Above pH 8.0, the three enzymes exhibited increased $S_{0.5}$ values, depending on the pH, and their $k_{cat}/S_{0.5}$ values changed mostly according to

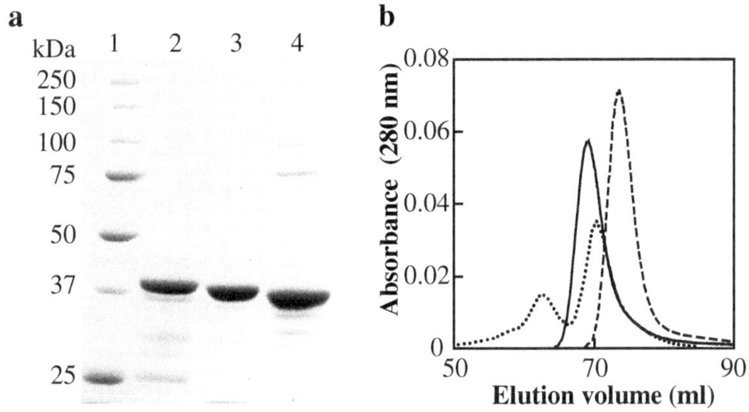

Figure 1 SDS-PAGE and size-exclusion chromatography of the purified enzymes. a) Lane 1, molecular mass standards (kDa); lane 2, FNLDH; lane 3, PALDH; lane 4, ECLDH. **b)** Elution of the enzymes from Superdex 200 with the buffers described in 'Materials and methods'. Solid, dashed, and dotted lines indicate FNLDH, PALDH, and ECLDH, respectively.

the changes in the $S_{0.5}$ values. ECLDH apparently showed lower pH-dependence of the $S_{0.5}$ value than the other two enzymes, of which PALDH showed slightly higher pH-dependence than FNLDH. Under acidic conditions below pH 6.0, the three enzymes consistently showed hyperbolic shaped pyruvate saturation curves, and no marked cooperativity in pyruvate binding. Under higher pH conditions, however, FNLDH and ECLDH showed markedly sigmoidal saturation curves for pyruvate, i.e., positive homotropic cooperativity, giving maximal n_H values of about

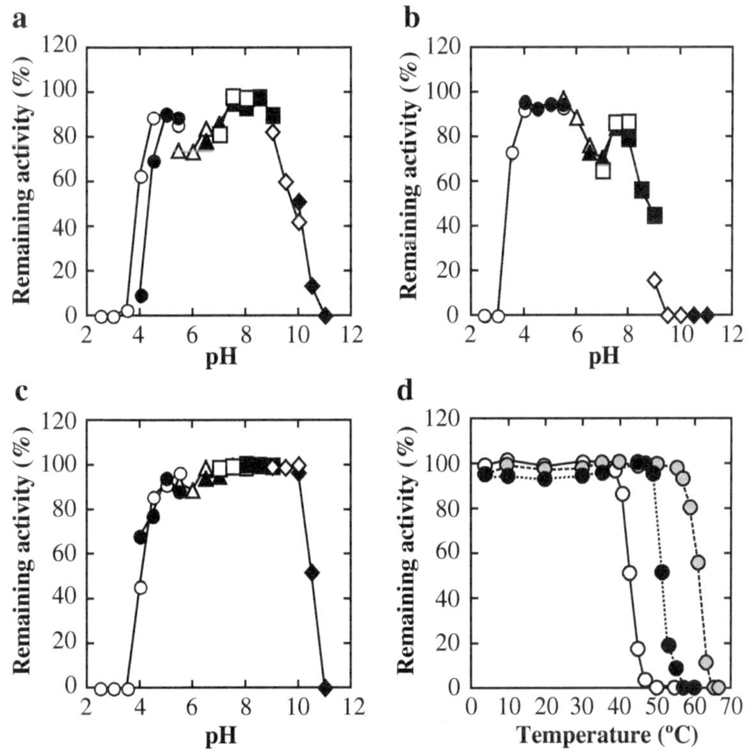

Figure 2 pH-stability of FNLDH (a), PALDH (b), and ECLDH (c), and heat-stability of the enzymes (d). a-c) Each enzyme was treated at 30°C for 1 h in sodium citrate (open circles), sodium acetate (closed circles), MES-NaOH (open triangles), MOPS-NaOH (closed triangles), HEPES-NaOH (open squares), Bicine-NaOH (*N,N*-Bis(2-hydroxyethyl)glycine) (closed squares), CHES-NaOH (*N*-cyclohexyl-2-aminoethanesulfonic acid) (open diamonds), and CAPS-NaOH (*N*-cyclohexyl-3-aminopropanesulfonic acid) (closed diamonds) buffers. **d)** FNLDH (white circles and solid lines), PALDH (grey circles and dashed lines), and ECLDH (black circles and dotted lines) were treated at various temperatures for 30 min in the buffers described in 'Materials and methods'.

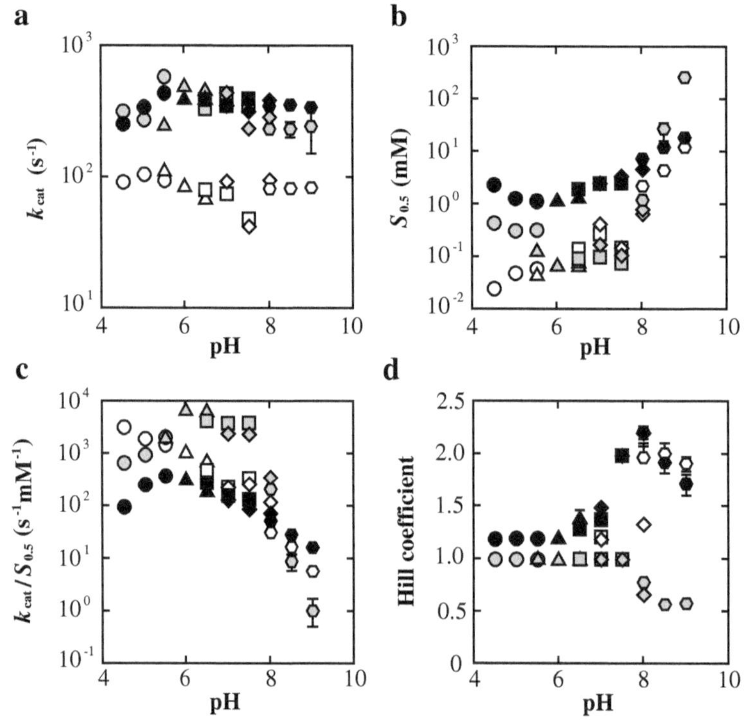

Figure 3 pH dependence of kinetic parameters on pyruvate reduction. White, grey, and black circles indicate FNLDH, PALDH, and ECLDH, respectively. The k_{cat} **(a)**, $S_{0.5}$ **(b)**, and $k_{cat}/S_{0.5}$ data **(c)** are plotted logarithmically, and the Hill coefficient **(d)** is plotted linearly. The buffers used for the assay were sodium acetate buffer (pH 4.5, 5.0, and 5.5; circles), MES-NaOH buffer (pH 5.5, 6.0, and 6.5; triangles), MOPS-NaOH buffer (pH 6.5, 7.0, and 7.5; squares), HEPES-NaOH buffer (pH 7.0, 7.5 and 8.0; diamonds), and Bicine-NaOH buffer (pH 8.0, 8.5, and 9.0; hexagons). We adjusted the pH of each buffer solution prior to the addition of substrate and cofactor, and confirmed that the pH value is not affected even in the presence of high concentrations of substrate.

2.0. In contrast, PALDH showed negative homotropic cooperativitiy above pH 8.0, giving a minimal n_H value of 0.5.

Substrate specificity

The kinetic parameters of the D-LDHs for various 2-ketoacids are summarized in Table 1. For pyruvate, PALDH exhibited the highest catalytic efficiency among the three enzymes, since it exhibited a small $S_{0.5}$ like FNLDH and a large k_{cat} like ECLDH. ECLDH showed a narrow substrate specificity toward pyruvate among hydrophobic 2-ketoacid substrates, although ECLDH consistently exhibited relatively high catalytic activity toward oxaloacetate, glyoxylate and hydroxypyruvate. PALDH showed a similar substrate specificity to that of ECLDH, exhibiting relatively high catalytic activity toward oxaloacetate, glyoxylate and hydroxypyruvate, although it generally exhibited higher catalytic efficiencies for these substrates than ECLDH. In contrast, FNLDH showed a broad substrate specificity toward relatively bulky hydrophobic 2-ketoacids, and exhibited particularly high catalytic activity toward 2-ketobutyrate and 2-ketovalerate, for which the catalytic efficiencies were comparable to that for pyruvate. FNLDH also showed relatively high activity to oxaloacetate and hydroxypyruvate, but poor activity toward glyoxylate.

Inhibition and activation by oxamate

Oxamate, an inert pyruvate analogue, inhibited the reactions of the three enzymes at pH 7.0, where the enzymes showed no marked homotropic cooperativity, in different manners, i.e. an apparently competitive manner with pyruvate for FNLDH, and a mixed manner for PALDH and ECLDH (Figure 4). This suggests that oxamate is bound not only to the catalytic site, but also to unknown allosteric sites in PALDH and ECLDH, whereas it is bound mostly to the catalytic site in FNLDH. Oxamate apparently enhanced the reactions of FNLDH and PALDH, and exhibited the highest activating effects at 20 mM and 2.5 mM, respectively (Figure 5a, b), whereas oxamate inhibited the ECLDH reaction (Figure 5c), at pH 8.0, where the three enzymes exhibited marked homotropic cooperativity. FNLDH and PALDH showed no significant homotropic (positive or negative) cooperativity in the presence of 20 mM and 2.5 mM oxamate, respectively (Figure 5d, e). In ECLDH, the inhibition occurred in a

Table 1 Kinetic parameters for various 2-ketoacids at pH 7.0[a]

	k_{cat} (s^{-1})	$S_{0.5}$ (mM)	$k_{cat}/S_{0.5}$ (s^{-1} mM^{-1})	K_i (mM)[b]	n_H[c]
Glyoxylate					
FNLDH	11 (0.1)	18 (0.4)	0.70 (0.007)		
PALDH	880 (60)	6.8 (0.6)	130 (5)	44 (8)	
ECLDH	100 (5)	20 (2)	5.3 (0.2)		1.5 (0.09)
Pyruvate					
FNLDH	80 (0.4)	0.34 (0.007)	230 (4)		
PALDH	400 (10)	0.10 (0.009)	4,000 (300)	1.8 (0.1)	
ECLDH	410 (8)	2.6 (0.09)	160 (2)		1.4 (0.05)
2-Ketobutyrate					
FNLDH	140 (0.7)	0.31 (0.006)	440 (7)		
PALDH	100 (0.6)	0.64 (0.01)	150 (3)		
ECLDH	31 (2)	30 (4)	1.0 (0.05)		1.4 (0.08)
2-Ketovalerate					
FNLDH	140 (0.5)	0.68 (0.01)	200 (3)		
PALDH	69 (0.5)	4.0 (0.07)	18 (3)		
ECLDH	N.D.[d]	N.D.	<0.01		
2-Ketoisovalerate					
FNLDH	120 (0.7)	3.5 (0.07)	35 (0.5)		
PALDH	N.D.	N.D.	<0.4		
ECLDH	N.D.	N.D.	<0.01		
2-Ketocaproate					
FNLDH	230 (1)	2.6 (0.06)	87 (1)		
PALDH	160 (1)	5.8 (0.1)	27 (0.4)		
ECLDH	N.D.	N.D.	<0.01		
2-Ketoisocaproate					
FNLDH	200 (2)	3.8 (0.1)	53 (1)		
PALDH	39 (0.3)	12 (0.2)	3.2 (0.03)		
ECLDH	N.D.	N.D.	<0.01		
Oxaloacetate					
FNLDH	80 (1)	2.5 (0.09)	31 (0.7)	140 (20)	
PALDH	410 (4)	1.7 (0.06)	230 (6)		
ECLDH	1,100 (200)	54 (10)	21 (0.6)	5.0 (1)	
Hydroxypyruvate					
FNLDH	160 (2)	18 (0.4)	8.7 (0.07)		
PALDH	420 (10)	3.0 (0.2)	140 (3)	19 (1)	
ECLDH	50 (0.5)	12 (0.3)	4.2 (0.06)		
Phenylpyruvate					
FNLDH	120 (7)	11 (0.9)	11 (0.3)	23 (3)	
PALDH	18 (0.2)	7.1 (0.2)	2.6 (0.04)		
ECLDH	N.D.	N.D.	<0.01		
Hydroxyphenylpyruvate					
FNLDH	N.D.	N.D.	<0.02		
PALDH	N.D.	N.D.	<0.4		
ECLDH	N.D.	N.D.	<0.01		

[a]Parameters were determined as described in the 'Materials and methods' section, and standard deviations are shown in parentheses. [b]The inhibition constants (K_i) and [c]Hill coefficients (n_H) are shown only for the cases in which significant substrate inhibition and cooperativity were observed, respectively. [d]N.D., not determined. The activities were too weak for determination of exact values.

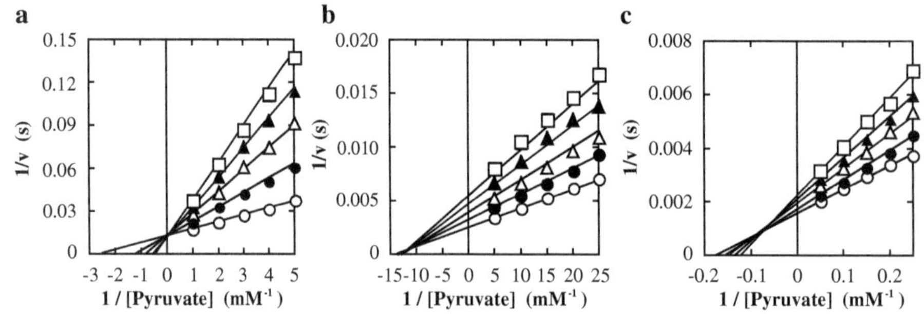

Figure 4 Effects of oxamate on the catalytic reactions at pH 7.0. The reaction velocities for FNLDH **(a)**, PALDH **(b)**, and ECLDH **(c)** were measured in 50 mM MOPS-NaOH buffer (pH 7.0) containing 0.1 mM NADH, the indicated concentrations of pyruvate and several concentrations of oxamate. The concentrations of oxamate for FNLDH were 0 mM (open circles), 10 mM (closed circles), 20 mM (open triangles), 30 mM (closed triangles), and 40 mM (open squares). The concentrations of oxamate for PALDH were 0 mM (open circles), 0.1 mM (closed circles), 0.2 mM (open triangles), 0.3 mM (closed triangles), and 0.4 mM (open squares). The concentrations of oxamate for ECLDH were 0 mM (open circles), 5 mM (closed circles), 10 mM (open triangles), 15 mM (closed triangles), and 20 mM (open squares). The reaction velocity and the concentration of pyruvate were plotted reciprocally. The data for PALDH and ECLDH were interpreted using the equation for mixed type inhibition, whereas the data for FNLDH were interpreted using the equation for competitive type inhibition. The kinetic parameters were as follows; FNLDH: $k_{cat} = 79 \pm 1$ (s^{-1}), $K_m = 0.39 \pm 0.01$ (mM), and $K_i = 9.4 \pm 0.2$ (mM). PALDH: $k_{cat} = 410 \pm 10$ (s^{-1}), $K_m = 0.074 \pm 0.006$ (mM), $K_i = 0.29 \pm 0.04$ (mM), and $K_i' = 0.33 \pm 0.04$ (mM). ECLDH: $k_{cat} = 670 \pm 10$ (s^{-1}), $K_m = 5.6 \pm 0.3$ (mM), $K_i = 20 \pm 2$ (mM), and $K_i' = 47 \pm 6$ (mM). GraFit ver 7.0.3 was used for non-linear regression and calculation of values. The lines were calculated with kinetic parameters.

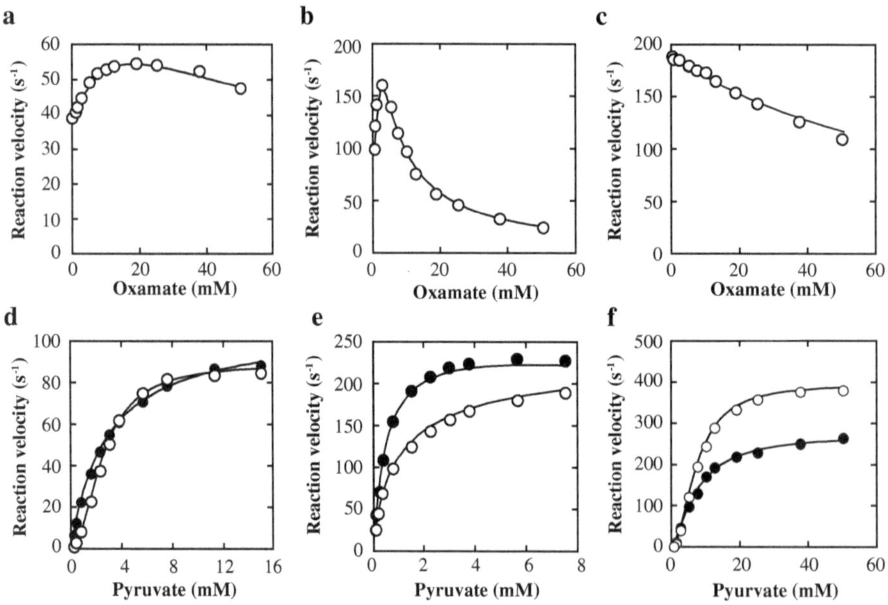

Figure 5 Effects of oxamate on the catalytic reactions at pH 8.0. a-c) The saturation curves for oxamate. The reaction velocities for FNLDH **(a)**, PALDH **(b)**, and ECLDH **(c)** were measured in 50 mM Bicine-NaOH buffer (pH 8.0) containing 0.1 mM NADH, 2.5 mM (for FNLDH), 1.2 mM (for PALDH), or 7.5 mM (for ECLDH) pyruvate, and the indicated concentration of oxamate. The data for FNLDH, PALDH, and ECLDH were interpreted using the equation for competitive-type, mix-type, and noncompetitive-type inhibition of allosteric enzyme. The kinetic parameters were as follows; FNLDH: $K_i = 51 \pm 10$ (mM), $k_{cat}' = 80 \pm 20$ (s^{-1}), $n_H' = 1.1 \pm 0.1$, and $K_{act} = 9.9 \pm 3$ (mM). PALDH: $K_i K_i' / (K_i + K_i') = 2.0 \pm 0.6$ (mM), $k_{cat}' = 470 \pm 100$ (s^{-1}), $n_H' = 1.7 \pm 0.2$, and $K_{act} = 1.7 \pm 0.4$ (mM). ECLDH: $K_i = 80 \pm 3$ (mM). The lines indicate the calculated saturation curves obtained with kinetic parameters. **d-f)** the saturation curves for pyruvate with or without oxamate. The reaction velocities for FNLDH **(d)**, PALDH **(e)**, and ECLDH **(f)** were measured in 50 mM Bicine-NaOH buffer (pH 8.0) in the presence of 0.1 mM NADH and the indicated concentrations of pyruvate with no effector (open circles), or 20 mM (for FNLDH), 2.5 mM (for PALDH), or 30 mM (for ECLDH) oxamate (closed circles). The lines indicate the calculated saturation curves obtained with kinetic parameters.

noncompetitive manner toward pyruvate (Figure 5f), suggesting that it is mostly bound to the allosteric site of ECLDH at pH 8.0.

Heterotropic activation of allosteric D-LDHs

Bacterial allosteric L-LDHs are commonly activated by FBP, which usually induces drastic improvement of the substrate $S_{0.5}$ values of the enzymes (Garvie 1980). In addition, *Lactobacillus casei* L-LDH requires some divalent metal ions (e.g., Mn^{2+}) (Arai et al. 2011), and the *Thermus caldophilus* enzyme is activated also by citrate under slightly acidic conditions (Taguchi et al. 1984). We therefore evaluated the effects of FBP, citrate and divalent metal ions at pH 8.0 in the presence of the $S_{0.5}$ pyruvate. FBP and citrate (1.0 mM) slightly enhanced the catalytic reactions of the three D-LDHs, and divalent metal ions such as Mg^{2+} and Mn^{2+} more markedly enhanced the reactions of FNLDH and PALDH (Figure 6a). In the case of FNLDH, Mg^{2+} and FBP showed significant activation effects, and Mg^{2+} gave a 2.8-fold smaller $S_{0.5}$ value (1.5 mM) than FBP (4.2 mM) for the enzyme activation (Figure 6 and Table 2). In the presence of 10 mM FBP or 5 mM Mg^{2+}, FNLDH exhibited hyperbolic pyruvate saturation curves, the pyruvate $S_{0.5}$ (K_m) value being reduced by approximately 1.4-fold or twice, respectively (Figure 6, and Table 3). For PALDH, Mg^{2+} and FBP also

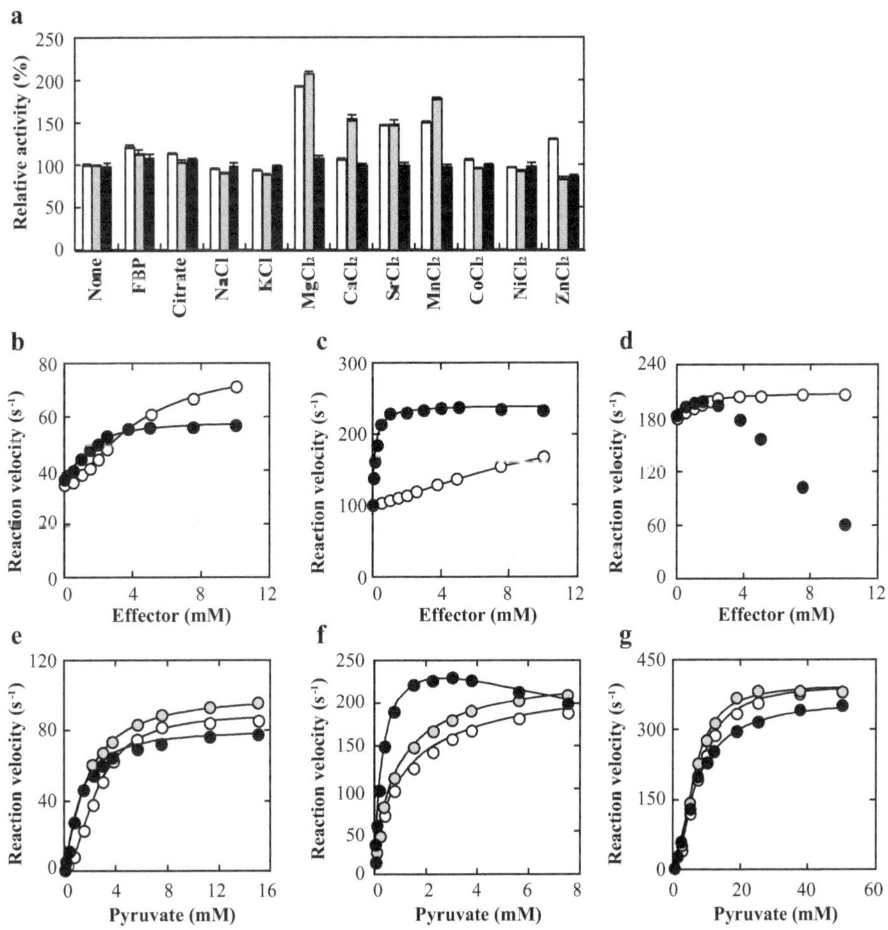

Figure 6 Effects of intermediary metabolites and ions at pH 8.0. a) Effects of intermediary metabolites and ions on FNLDH (white boxes), PALDH (grey boxes), and ECLDH (black boxes). The activities were measured in 50 mM Bicine-NaOH buffer (pH 8.0) containing 0.1 mM NADH, 2.5 mM (for FNLDH), 1.2 mM (for PALDH), or 7.5 mM (for ECLDH) pyruvate, and 1 mM indicated effectors. **b-d)** the saturation curves for FBP or $MgCl_2$. The reaction velocities for FNLDH **(b)**, PALDH **(c)**, and ECLDH **(d)** were measured in 50 mM Bicine-NaOH buffer (pH 8.0) containing 0.1 mM NADH, 2.5 mM (for FNLDH), 1.2 mM (for PALDH), or 7.5 mM (for ECLDH) pyruvate, and the indicated concentrations of FBP (open circles) or $MgCl_2$ (closed circles). The lines indicate the calculated saturation curves obtained with apparent kinetic parameters (Table 2). The data of $MgCl_2$ saturation curve of ECLDH was not fitted with any equations used in this study. **e-g)** the saturation curves for pyruvate with or without FBP or $MgCl_2$. The reaction velocities for FNLDH **(e)**, PALDH **(f)**, and ECLDH **(g)** were measured in 50 mM Bicine-NaOH buffer (pH 8.0) containing 0.1 mM NADH, the indicated concentrations of pyruvate with no effector (white circles), 10 mM (for FNLDH and PALDH) or 5 mM (for ECLDH) FBP (grey circles), and 5 mM (for FNLDH and PALDH) or 2.4 mM (for ECLDH) $MgCl_2$ (black circles). The lines indicate the calculated saturation curves obtained with kinetic parameters (Table 3).

Table 2 Apparent kinetic parameters for FBP and $MgCl_2$ at pH 8.0[a]

	k_{cat} (s^{-1})	v_{min} (s^{-1})[b]	$S_{0.5}$ (mM)	n_H[c]
FBP				
FNLDH	45 (1)	35 (0.6)	4.2 (0.2)	1.7 (0.05)
PALDH	150 (30)	100 (2)	12 (3)	1.3 (0.07)
ECLDH	25 (1)	180 (2)	1.3 (0.01)	1.5 (0.2)
Mg^{2+}				
FNLDH	22 (0.7)	37 (0.2)	1.5 (0.1)	1.3 (0.09)
PALDH	140 (2)	100 (1)	0.16 (0.008)	1.1 (0.08)
ECLDH	N.D.[d]	N.D.	N.D.	

[a]Parameters were determined as described in the 'Materials and methods' section, and standard deviations are shown in parentheses. [b]v_{min} is reaction velocity in the absence of FBP or Mg^{2+}. [c]Hill coefficients (n_H) are shown only for the cases that exhibited significant cooperativity. [d]N.D., not determined.

showed activation effects, and Mg^{2+} gave a 75-fold smaller $S_{0.5}$ value (0.16 mM) than FBP (12 mM) for the enzyme activation (Figure 6 and Table 2). In this case, FBP and Mg^{2+} also consistently reduced the pyruvate $S_{0.5}$ value significantly, and the negative cooperativity in pyruvate binding (Figure 6, and Table 3). In the case of ECLDH, FBP and Mg^{2+} exhibited only slight activation effects on the enzyme reaction, and the latter even markedly inhibited the reaction at high concentrations (Figure 6, and Tables 2 and 3).

Kinetic profiles for NADH

The apparent kinetic parameters for NADH were determined at the saturation concentration of pyruvate at pH 7.0 and 8.0 (Table 3). FNLDH and PALDH exhibited virtually the same apparent NADH $S_{0.5}$ value (about 0.035 mM), which was approximately twice smaller than that of ECLDH. Since the pyruvate $S_{0.5}$ values were determined at 0.1 mM NADH (Figure 3 and Table 3), these values might be larger than the real values because of an insufficient concentration of NADH, particularly in the case of ECLDH. PALDH and ECLDH exhibited hyperbolic NADH saturation curves independently of pH, FBP or Mg^{2+}. On the other hand, FNLDH exhibited a sigmoidal one at pH 8.0, and exhibited an n_H value of 1.6, which was slightly reduced by FBP or Mg^{2+}.

Discussion

The characteristics of FNLDH, PALDH and ECLDH demonstrate the high variety of the functions of bacterial D-LDHs. Of the three enzymes, PALDH exhibits a markedly smaller pyruvate $S_{0.5}$ value than ECLDH, and a larger k_{cat} value than FNLDH at pH 7.0 (Table 3). Although the crucial role of PALDH in the metabolism of *P. aeruginosa* remains uncertain, the enzyme is likely involved in regulation of the NAD^+/NADH ratio within the cells like the heart-type L-LDH isozymes of aerobic

Table 3 Kinetic parameters for pyruvate and NADH[a]

	k_{cat} (s^{-1})	$S_{0.5}$ (mM)	$k_{cat}/S_{0.5}$ (s^{-1} mM^{-1})	K_i (mM)[b]	n_H[c]
Pyruvate					
pH 7.0					
FNLDH	80 (0.4)	0.34 (0.007)	230 (4)		
PALDH	400 (10)	0.10 (0.009)	4,000 (300)	1.8 (0.1)	
ECLDH	410 (8)	2.6 (0.09)	160 (2)		1.4 (0.05)
pH 8.0					
FNLDH	87 (1)	2.5 (0.06)	35 (0.6)		2.0 (0.1)
PALDH	240 (6)	1.2 (0.1)	200 (10)		0.77 (0.03)
ECLDH	380 (5)	7.6 (0.1)	50 (0.4)		2.2 (0.06)
pH 8.0 + FBP[d]					
FNLDH	100 (1)	1.7 (0.06)	60 (1)		1.3 (0.04)
PALDH	240 (4)	0.85 (0.04)	280 (6)		0.9 (0.02)
ECLDH	400 (5)	6.5 (0.1)	61 (1)		2.1 (0.09)
pH 8.0 + Mg^{+2}[e]					
FNLDH	80 (1)	1.2 (0.04)	65 (1)		1.4 (0.05)
PALDH	290 (3)	0.36 (0.01)	820 (20)	20 (0.8)	
ECLDH	370 (6)	7.2 (0.2)	51 (1)		1.5 (0.07)
NADH					
pH 7.0					
FNLDH	98 (0.7)	0.022 (0.0005)	4,400 (80)		
PALDH	310 (2)	0.019 (0.0005)	17,000 (400)		
ECLDH	640 (20)	0.055 (0.004)	12,000 (500)		
pH 8.0					
FNLDH	90 (1)	0.036 (0.001)	2,500 (30)		1.6 (0.07)
PALDH	270 (2)	0.035 (0.0007)	7,700 (100)		
ECLDH	710 (8)	0.088 (0.002)	8,100 (100)		
pH 8.0 + FBP[d]					
FNLDH	80 (1)	0.028 (0.001)	2,800 (70)		1.5 (0.07)
PALDH	270 (2)	0.035 (0.001)	7,800 (100)		
ECLDH	730 (10)	0.092 (0.003)	8,000 (100)		
pH 8.0 + Mg^{+2}[e]					
FNLDH	110 (2)	0.031 (0.001)	3,700 (70)		1.3 (0.05)
PALDH	270 (2)	0.035 (0.0009)	7,700 (100)		
ECLDH	690 (9)	0.090 (0.002)	7,700 (100)		

[a]Parameters were determined as described in the 'Materials and methods' section, and standard deviations are shown in parentheses. [b]The inhibition constants (K_i) are shown only for the cases that exhibited significant substrate inhibition. [c]Hill coefficients (n_H) are shown only for the cases that exhibited significant cooperativity. [d]The concentrations of FBP for ECLDH, FNLDH, and PALDH were 5 mM, 10 mM, and 10 mM, respectively. [e]The concentrations of Mg^{2+} for ECLDH, FNLDH, and PALDH were 2.4 mM, 5 mM, and 5 mM, respectively.

tissues (Holbrook et al. 1975). On the other hand, ECLDH with the large $S_{0.5}$ is likely favorable for facultative anaerobes such as *E. coli*, in which the pyruvate level within the cells dynamically changes between anaerobic and aerobic conditions like the muscle-type L-LDH isozymes (Holbrook et al. 1975). FNLDH exhibits a small pyruvate $S_{0.5}$ value although *F. nucleatum* is obligate anaerobe. In *F. nucleatum* cells, pyruvate is likely consumed mostly through pathways other than lactate fermentation, such as butyrate fermentation (Kapatral et al. 2002), and therefore pyruvate within the cells may be maintained at low concentrations. It is particularly notable that FNLDH prefers relatively bulky hydrophobic 2-ketoacids, such as 2-ketobutyrate and 2-ketovalerate, which give apparently equivalent or even higher k_{cat}/K_m value than pyruvate, unlike in the cases of the other two enzymes (Table 1). This substrate preference suggests that FNLDH might be used for not only pyruvate reduction but also the reduction or oxidation of other bulky 2-ketoacid or 2-hydroxy acids, respectively, although the actual physiological substrates of the enzyme remains uncertain.

FNLDH apparently has a size-preference for 2-ketoacid substrates intermediate between those of conventional D-LDH and D-2-hydroxyisocaproate dehydrogenase (Hummel et al. 1985). Like usual D-LDHs, nevertheless, FNLDH has aromatic residues, Phe52 and Tyr299 (residue numbers are according to D-LDH from *Lactobacillus pentosus*), which are proposed to play key roles in the substrate recognition in the binding pocket for the substrate C3 side chains (Tokuda et al. 2003; Ishikura et al. 2005). It is also noteworthy that FNLDH exhibits relatively high activity toward 2-ketoisovalerate since little is known as to an enzyme that exhibits high catalytic activity toward C3-branched 2-ketoacid substrates in the D-LDH-related D-HydDH family. It is known that D-mandelate dehydrogenase from *Enterococcus faecalis* exhibits high catalytic activity toward hydrophobic C3-branched ketoacids such as 2-ketoisovalerate and benzoyl formate (Tamura et al. 2002). Nevertheless, this enzyme belongs to another D-HydDH family, the 2-ketopantate reductase-related family, which is evolutionally separate from the D-LDH-related family (Wada et al. 2008; Miyanaga et al. 2013). Hence, FNLDH is the first case of a D-LDH-related D-HydDH that exhibits relatively high catalytic activity toward C3-branched 2-ketoacid substrates and may be useful for enzymatic production of C3-branched D-2-hydroxyacids.

For FNLDH, oxamate inhibits the reaction competitively with pyruvate at pH 7.0, and substitutes for the role of pyruvate in the homotropic enzyme activation at pH 8.0, converting the sigmoidal saturation curve for pyruvate to a hyperbolic one (Figure 5d). These effects indicate that oxamate simply competes with pyruvate for a common binding site, i.e., the catalytic site in FNLDH. It is likely that PALDH binds oxamate at both the catalytic and allosteric sites, since oxamate inhibits the reaction in a mixed manner at pH 7.0 (Figure 4), as in the case of ECLDH, but stimulates the reaction at low concentration at pH 8.0, where oxamate likely exhibits an activation effect mostly through binding to the catalytic site and inhibitory effects mostly through binding to the unknown allosteric site (Figure 5b,e). In the case of ECLDH, it was suggested previously that oxamate is likely bound mostly to a binding site other than the catalytic site (Tarmy and Kaplan 1968), this being consistent with the results of this study (Figure 4c). These results indicate that the three enzymes differ in their allosteric behaviors, and specificities to allosteric effectors, although the actual allosteric binding sites or physiological effectors of these enzymes remain uncertain.

FNLDH and PALDH are the first cases of D-LDHs that are activated by FBP, which is a common activator for bacterial allosteric L-LDHs (Garvie 1980). It is nevertheless unlikely that FBP is really the specific activator of the two D-LDHs, since the two enzymes require apparently higher concentrations of FBP than the physiological one for their activation (Figure 6b,c and Table 2). On the other hand, the Mg^{2+} ion is more effective than FBP as to activation of the two D-LDHs, particularly PALDH (Figure 6, and Tables 2 and 3). It is, however, also unlikely that the Mg^{2+} ion is the specific activator of these enzymes, since some other divalent metal ions such as Sr^{2+} and Mn^{2+} also improve their reactions (Figure 6a). These two enzymes thus appear to be activated through the comprehensive effects of several activators with low specificities, physiologically, although it is possible that they have unknown activators with higher specificities. In the case of ECLDH, FBP and Mg^{2+} only slightly enhance the catalytic reaction, and the latter even markedly reduces the reaction at higher concentrations (Figure 6d,g and Table 3). We have evaluated other possible effectors, such as L-alanine, L-glutamate and L-aspartate, for this enzyme, but have not found out any potential activator for ECLDH (data not shown). This implies that ECLDH is only negatively controlled under alkaline conditions.

The allosteric and tetrameric D-PgDHs from *E. coli* and *M. tuberculosis* were extensively studied as to their structure-function relationship (Tobey and Grant 1986; Grant 1989; Schuller et al. 1995; Dey et al. 2005; Grant 2012). The D-PgDHs have *V*-type regulatory mechanisms, in which L-serine inhibits the enzyme reaction in a non-competitive manner, whereas the three D-LDHs apparently exhibit *K*-type rather than *V*-type regulation. These D-PgDHs have additional functional domain(s), ACT domain and ASB domain, which undergo not only interactions with allosteric effector, L-serine, but also inter-subunit interactions with each other for the construction of tetrameric structure. In contrast, D-LDHs have no additional protein domain. The three D-LDHs from Gram-negative bacteria thus greatly differ from these D-PgDHs in both regulatory mechanism and protein structure.

It is reported that bacterial allosteric L-LDHs consistently undergo the Monod-Wyman-Changeux-type (MWC-type: pre-existing type) (Monod et al. 1965; Iwata et al. 1994; Arai et al. 2010) allosteric transition, where the active (R) state with high affinity to pyruvate and FBP, and the inactive (T) state with low affinity. However, such drastic structural changes appear to be unnecessary for the three D-LDHs, which exhibit apparently much smaller changes in catalytic activity due to the allosteric transition than the L-LDHs. In fact, the MWC model is not available for the case of PALDH, which shows negative homotropic cooperativity. In addition, FNLDH shows significant positive homotropic cooperativity also for NADH binding, which is less affected by FBP or Mg^{2+} than pyruvate binding (Table 3), suggesting that FNLDH changes in its structure through two steps, the NADH binding and following pyruvate binding steps, in the allosteric transition. The *Lactobacillus* D-LDHs consistently have dimeric structure, whereas the three D-LDHs from Gram-negative bacteria have tetrameric structure. Thus, tetramerization might be correlated with the allostery of D-LDHs. It is highly desirable to determine the 3D-structures of these D-LDHs.

Abbreviations

D-LDH: NAD-dependent D-lactate dehydrogenase; L-LDH: L-lactate dehydrogenase; D-HydDH: D-2-hydroxyacid dehydrogenase; D-PgDH: D-3-phosphoglycerate dehydrogenase; ECLDH: D-LDH from *Escherichia coli*; FNLDH: D-LDH from *Fusobacterium nucleatum*; PALDH: D-LDH from *Pseudomonas aeruginosa*; FBP: Fructose 1,6-bisphosphate; MWC: Monod-Wyman-Changeux.

Competing interests

The authors declare that they have no competing interests.

Authors' contributions

NF designed the study, carried out most of the biochemical studies and drafted the manuscript. AM designed the study and helped to draft the manuscript. MT participated in the kinetics analysis of PALDH. MN participated in the design of the study and the kinetics analysis, and helped to draft the manuscript. HT conceived and designed the study, and helped to draft the manuscript. All authors read and approved the final manuscript.

Acknowledgements

This work was partially supported by a Grant-in-Aid (No. 23580120 to H.T.) from the Ministry of Education, Culture, Sports, Science, and Technology of Japan (MEXT).

Author details

[1]Tokyo University of Science, 2641 Yamazaki, Noda, Chiba 278-8510, Japan. [2]Tokyo Institute of Technology, 2-12-1 O-okayama, Meguro-ku, Tokyo 152-8551, Japan.

References

Arai K, Ishimitsu T, Fushinobu S, Uchikoba H, Matsuzawa H, Taguchi H (2010) Active and inactive state structures of unliganded *Lactobacillus casei* allosteric L-lactate dehydrogenase. Proteins 78:681–694, doi: 10.1002/prot.22597

Arai K, Ichikawa J, Nonaka S, Miyanaga A, Uchikoba H, Fushinobu S, Taguchi H (2011) A molecular design that stabilizes active state in bacterial allosteric L-lactate dehydrogenases. J Biochem 150:579–591, doi: 10.1093/jb/mvr100

Baker PJ, Sawa Y, Shibata H, Sedelnikova SE, Rice DW (1998) Analysis of the structure and substrate binding of *Phormidium lapideum* alanine dehydrogenase. Nat Struct Biol 5:561–567, doi: 10.1038/817

Bernard N, Ferain T, Garmyn D, Hols P, Delcour J (1991) Cloning of the D-lactate dehydrogenase gene from *Lactobacillus delbrueckii* subsp. *bulgaricus* by complementation in *Escherichia coli*. FEBS Lett 290:61–64, doi: 10.1016/0014-5793(91)81226-X

Buckel W, Barker HA (1974) Two pathways of glutamate fermentation by anaerobic bacteria. J Bacteriol 117:1248–1260

Costas AM, White AK, Metcalf WW (2001) Purification and characterization of a novel phosphorus-oxidizing enzyme from *Pseudomonas stutzeri* WM88. J Biol Chem 276:17429–17436, doi: 10.1074/jbc.M011764200

Dengler U, Niefind K, Kiess M, Schomburg D (1997) Crystal structure of a ternary complex of D-2-hydroxyisocaproate dehydrogenase from *Lactobacillus casei*, NAD^+ and 2-oxoisocaproate at 1.9 Å resolution. J Mol Biol 267:640–660, doi: 10.1006/jmbi.1996.0864

Dey S, Grant GA, Sacchettini JC (2005) Crystal structure of *Mycobacterium tuberculosis* D-3-phosphoglycerate dehydrogenase: extreme asymmetry in a tetramer of identical subunits. J Biol Chem 280:14892–14899, doi: 10.1074/jbc.M414489200

Dixon M, Webb EC (1979) Enzymes. Longman, London, pp 400–402

Eszes CM, Sessions RB, Clarke AR, Moreton KM, Holbrook JJ (1996) Removal of substrate inhibition in a lactate dehydrogenase from human muscle by a single residue change. FEBS Lett 399:193–197, doi: 10.1016/S0014-5793(96)01317-8

Garvie EI (1980) Bacterial lactate dehydrogenases. Microbiol Rev 44:106–139, doi: 10.1016/S0014-5793(96)01317-8

Goldberg JD, Yoshida T, Brick P (1994) Crystal structure of a NAD-dependent D-glycerate dehydrogenase at 2.4 Å resolution. J Mol Biol 236:1123–1140, doi: 10.1016/0022-2836(94)90016-7

Grant GA (1989) A new family of 2-hydroxyacid dehydrogenases. Biochem Biophys Res Commun 165:1371–1374, doi: 10.1016/0006-291X(89)92755-1

Grant GA (2012) Contrasting catalytic and allosteric for phosphoglycerate dehydrogenases. Arch Biochem Biophys 519:175–185, doi: 10.1016/j.abb.2011.10.005

Greenler JM, Sloan JS, Schwartz BW, Becker WM (1989) Isolation, characterization and sequence analysis of a full-length cDNA clone encoding NADH-dependent hydroxypyruvate reductase from cucumber. Plant Mol Biol 13:139–150, doi: 10.1007/BF00016133

Holbrook JJ, Liljas A, Steindel SJ, Rossmann MG (1975) Lactate dehydrogenase. The Enzymes 3rd ed (Boyer, PD, ed), Vol. 11. Academic Press, New York, pp 191–292

Hummel W, Schutte H, Kula M-R (1985) D-2-hydroxyisocaproate dehydrogenase from *Lactobacillus casei*. Appl Microbiol Biotechnol, doi: 10.1007/BF00252354

Ishikura Y, Tsuzuki S, Takahashi O, Tokuda C, Nakanishi R, Shinoda T, Taguchi H (2005) Recognition site for the side chain of 2-ketoacid substrate in D-lactate dehydrogenase. J Biochem 138:741–749, doi: 10.1093/jb/mvi170

Iwata S, Kamata K, Yoshida S, Minowa T, Ohta T (1994) T and R states in the crystals of bacterial L-lactate dehydrogenase reveal the mechanism for allosteric control. Nat Struct Biol 1:176–185, doi: 10.1038/nsb0394-176

Kapatral V, Anderson I, Ivanova N, Reznik G, Los T, Lykidis A, Bhattacharyya A, Bartman A, Gardner W, Grechkin G, Zhu L, Vasieva O, Chu L, Kogan Y, Chaga O, Goltsman E, Bernal A, Larsen N, D'Souza M, Walunas T, Pusch G, Haselkorn R, Fonstein M, Kyrpides N, Overbeek R (2002) Genome sequence and analysis of the oral bacterium *Fusobacterium nucleatum* strain ATCC 25586. J Bacteriol 184:2005–2018, doi: 10.1128/JB.184.7.2005-2018.2002

Kochhar S, Hunziker PE, Leong-Morgenthaler P, Hottinger H (1992) Primary structure, physicochemical properties, and chemical modification of NAD (+)-dependent D-lactate dehydrogenase: evidence for the presence of Arg-235, His-303, Tyr-101, and Trp-19 at or near the active site. J Biol Chem 267:8499–8513

Laemmli UK (1970) Cleavage of structural proteins during the assembly of the head of bacteriophage T4. Nature 227:680–685, doi: 10.1038/227680a0

Lamzin VS, Dauter Z, Popov VO, Harutyunyan EH, Wilson KS (1994) High resolution structures of holo and apo formate dehydrogenase. J Mol Biol 236:759–785, doi: 10.1006/jmbi.1994.1188

Lerch HP, Blöcker H, Kallwass H, Hoppe J, Tsai H, Collins J (1989) Cloning, sequencing and expression in *Escherichia coli* of the D-2-hydroxyisocaproate dehydrogenase gene of *Lactobacillus casei*. Gene 78:47–57, doi: 10.1016/0378-1119(89)90313-2

Martins BM, Macedo-Ribeiro S, Bresser J, Buckel W, Messerchmidt A (2005) Structural basis for stereo-specific catalysis in NAD(+)-dependent (R)-2-hydroxyglutarate dehydrogenase from *Acidaminococcus fermentans*. FEBS J 272:269–281, doi: 10.1111/j.1432-1033.2004.04417.x

Miyanaga A, Fujisawa S, Furukawa N, Arai K, Nakajima M, Taguchi H (2013) The crystal structure of D-mandelate dehydrogenase reveals its distinct substrate and coenzyme recognition mechanisms from those of 2-ketopantoate reductase. Biochem Biophys Res Commun 439:109–114, doi: 10.1016/j.bbrc.2013.08.019

Monod J, Wyman J, Changeux JP (1965) On the nature of allosteric transitions: a plausible model. J Mol Biol 12:88–118, doi: 10.1016/S0022-2836(65)80285-6

Pace CN, Vajdos F, Fee L, Grimsley G, Gray T (1995) How to measure and predict the molar absorption coefficient of a protein. Protein Sci 4:2411–2423, doi: 10.1002/pro.5560041120

Popov VO, Lamzin VS (1994) NAD(+)-dependent formate dehydrogenase. Biochem J 301:625–643

Schuller DJ, Grant GA, Banaszak LJ (1995) The allosteric ligand site in the V_{max}-type cooperative enzyme phosphoglycerate dehydrogenase. Nat Struct Biol 2:69–76, doi: 10.1038/nsb0195-69

Taguchi H, Ohta T (1991) D-lactate dehydrogenase is a member of the D-isomer-specific 2-hydroxyacid dehydrogenase family: cloning, sequencing, and expression in *Escherichia coli* of the D-lactate dehydrogenase gene of *Lactobacillus plantarum*. J Biol Chem 266:12588–12594

Taguchi H, Matsuzawa H, Ohta T (1984) L-Lactate dehydrogenase from *Thermus caldophilus* GK24, an extremely thermophilic bacterium: desensitization to fructose 1,6-bisphosphate in the activated state by arginine-specific chemical modification and the N-terminal amino acid sequence. Eur J Biochem 145:283–290, doi: 10.1111/j.1432-1033.1984.tb08550.x

Tamura Y, Ohkubo A, Iwai S, Wada Y, Shinoda T, Arai K, Mineki S, Iida M, Taguchi H (2002) Two forms of NAD-dependent D-mandelate dehydrogenase in *Enterococcus faecalis* IAM 10071. Appl Environ Microbiol 68:947–951, doi: 10.1128/AEM.68.2.947-951.2002

Tarmy EM, Kaplan NO (1968) Kinetics of *Escherichia coli* B D-lactate dehydrogenase and evidence for pyruvate-controlled change in conformation. J Biol Chem 243:2587–2596

Tobey KL, Grant GA (1986) The nucleotide sequence of the serA gene of *Escherichia coli* and the amino acid sequence of the encoded protein, D-3-phosphoglycerate dehydrogenase. J Biol Chem 261:12179–12183

Tokuda C, Ishikura Y, Shigematsu M, Mutoh H, Tsuzuki S, Nakahira Y, Tamura Y, Shinoda T, Arai K, Takahashi O, Taguchi H (2003) Conversion of *Lactobacillus pentosus* D-lactate dehydrogenase to a D-hydroxyisocaproate dehydrogenase through a single amino acid replacement. J Bacteriol 185:5023–5026, doi: 10.1128/JB.185.16.5023-5026.2003

Tripathi SM, Ramachandran R (2008) Crystal structures of the *Mycobacterium tuberculosis* secretory antigen alanine dehydrogenase (Rv2780) in apo and ternary complex forms captures "open" and "closed" enzyme conformations. Proteins 72:1089–1095, doi: 10.1002/prot.22101

Vinals C, Depiereux E, Feytmans E (1993) Prediction of structurally conserved regions of D-specific hydroxy acid dehydrogenases by multiple alignment with formate dehydrogenase. Biochem Biophys Res Commun 192:182–188, doi: 10.1006/bbrc.1993.1398

Wada Y, Iwai S, Tamura Y, Ando T, Shinoda T, Arai K, Taguchi H (2008) A new family of D-2-hydroxyacid dehydrogenases that comprises D-mandelate dehydrogenases and 2-ketopantoate reductases. Biosci Biotechnol Biochem 72:1087–1094, doi: 10.1271/bbb.70827

Woodyer R, Wheatley JL, Relyea HA, Rimkus S, van der Donk WA (2005) Site-directed mutagenesis of active site residues of phosphite dehydrogenase. Biochemistry 44:4765–4774, doi: 10.1021/bi047868c

11

Screening of liquid media and fermentation of an endophytic *Beauveria bassiana* strain in a bioreactor

Rieke Lohse, Desiree Jakobs-Schönwandt and Anant V Patel*

Abstract

A novel approach for biological control of insect pests could be the use of the endophytic entomopathogenic *Beauveria bassiana* isolate ATP-02. For the utilization of the endophyte as a commercial biocontrol agent, the fungus has to be mass-produced. *B. bassiana* was raised in shake flask cultures to produce high concentrations of total spores (TS), which include blastospores (BS) and submerged conidiospores (SCS). The highest concentration of 1.33×10^9 TS/mL and the highest yield of 5.32×10^{10} TS/g sucrose was obtained in the TKI broth with 5% sugar beet molasses which consists of 50% sucrose as a carbon source. In spite of the lower sugar concentration (2.5%) the amount of TS could be increased up to 11-times in contrast to the cultivation with 5% sucrose. The scale-up to a 2 L stirred tank reactor was carried out at 25°C, 200–600 rpm and 1 vvm at pH 5.5. A TS yield of 5.2×10^{10} TS/g sucrose corresponding to a SCS yield of 0.2×10^{10} SCS/g sucrose was obtained after 216 h. With regards to the culture medium the cost of 10^{12} TS amounts to 0.24 €. *Plutella xylostella* larvae, which were fed with oilseed rape leaves treated with spores from fermentation resulted in 77 ± 5% mortality. Moreover, spores from submerged cultivation were able to colonize oilseed rape leaves via leaf application. This is the first report of fermentation of an endophytic *B. bassiana* strain in a low-cost culture medium to very high yields of TS.

Keywords: Submerged culture; Fermentation; Beauveria bassiana; Endophyte; Blastospores; Submerged conidiospores; Biological control

Introduction

In the past decades, many microorganisms have been isolated and investigated for use as a biocontrol agent. Now, many promising strains are available for release into the environment and especially with the renewed interest in biocontrol await further exploitation for large-scale application in agriculture (Glare et al. 2012). The first step for commercialization of a biocontrol agent like *Beauveria bassiana* is the mass-production by fermentation (Burges 1998; Ravensberg 2011). *B. bassiana* strains that were applied to the insect and act on the outer surface of the plant show efficacy against a wide range of insect pests and have the potential of becoming a cost-effective biocontrol agent (Khachatourians 1986). However, the approved products of *B. bassiana* contain aerial conidia (AC), which are produced by either a solid-state or a diphasic fermentation. These processes are in classical biotechnology considered to be labour-intensive and unsuitable for conventional production of fungal biomass (Feng et al. 1994; Patel et al. 2011; Ravensberg 2011; Rombach et al. 1988). In contrast to these propagules, blastospores (BS) and submerged conidiospores (SCS) would be produced in submerged cultivations in a shorter time with higher yields and state-of-the-art process control. Furthermore, it was shown that BS and SCS of a *B. bassiana* isolate are as virulent to grasshoppers as the AC (Hegedus et al. 1992). Until today, no products with BS or SCS of *B. bassiana* are available. A few reports on growth requirements and shake flask culture of *B. bassiana* strains show the best growth and germination in complex media (Bidochka et al. 1987; Chong-Rodríguez et al. 2011; Hegedus et al. 1990; Humphreys et al. 1990; Pham et al. 2009; Rombach 1989; Safavi et al. 2007; Samsináková 1966; Thomas et al. 1987; Vega et al. 2003). However, only a few publications deal with the production of mycelium

* Correspondence: anant.patel@fh-bielefeld.de
Department of Engineering and Mathematics, University of Applied Sciences, Wilhelm-Bertelsmann-Str. 10, Bielefeld 33602, Germany

(Núñez-Ramírez et al. 2012) and the production of BS in complex (Humphreys et al. 1989, 1990) and mineral media (Lane et al. 1991) by submerged fermentation, respectively. Endophytic *B. bassiana* strains can exist asymptomatically in a variety of plants like banana (Akello et al. 2008), opium poppies (Quesada-Moraga et al. 2006), maize (Bing and Lewis 1992) and sorghum (Tefera and Vidal 2009). The recently isolated endophytic *B. bassiana* strain ATP-02 showed great potential for a novel plant control measure in a variety of crops (Tefera and Vidal 2009). However, it remained unknown if this strain can be mass-produced to high yields and if the spores from a submerged fermentation are able to colonize plants. That is why the objective of the present work was to produce spores of endophytic *B. bassiana* ATP-02 in a cost-effective culture medium on lab-scale and to scale-up the process to a 2 L stirred-tank reactor. Finally the virulence of the produced spores was checked in a bioassay with *Plutella xylostella* and their potential to colonize oilseed rape leaves via a leaf application was investigated.

Materials and methods

All materials used were purchased from Merck KGaA (Darmstadt, Germany), Carl Roth GmbH (Karlsruhe, Germany) or AppliChem GmbH (Darmstadt, Germany), if not mentioned otherwise. Sugar beet molasses with a dry matter content of 80% consisting of 50% sucrose was purchased from Suedzucker AG (Mannheim, Germany). All concentrations are given as (w/w).

Strain

B. bassiana isolate ATP-02, DSM 24665, was provided by Prof. Stefan Vidal, Georg-August-University, Department of Crop Sciences/Agricultural Entomology, Goettingen, Germany. The strain was raised at 25°C on SDA agar containing 1% casein peptone, 2% glucose and 1.5% agar-agar at pH 5.5. Temperature optimum was found at 25°C and pH optimum at 5.5 (data not shown).

Cultivation in shake flask culture

Different liquid media were used to cultivate *B. bassiana*: TKI medium with 5% carbon source (Thomas et al. 1987), Czapek-Dox medium (Kučera 1971), YPG medium and PG medium (Bidochka et al. 1987), Vogel's medium (Vogel 1956), SD medium (Odds 1991), CGM medium containing 1% glucose, 1% corn steep liquor, 0.5% NaCl, 0.1% $NaNO_3$, 1% $CaCO_3$ (Samsináková 1966), PWG medium containing 1% glucose, 8.75% whey powder, 0.25% peptone (Kassa et al. 2008), YG medium containing 1% glucose, 1% yeast extract (Leckie et al. 2008), YSM medium containing 2% sucrose, 0.5% yeast extract, 0.15% KH_2PO_4, 0.05% $MgSO_4{\cdot}7H_2O$, 0.001% $CaCl_2$, 0.000003% H_3BO_3, 0.000004% $MnSO_4{\cdot}4H_2O$, 0.0000025% $Na_2MoO_4{\cdot}2H_2O$, 0.000008% $CuSO_4{\cdot}5H_2O$, 0.00004% $ZnSO_4{\cdot}7H_2O$, 0.00005% $FeCl_3{\cdot}6H_2O$, 0.00004% $CoCl_2{\cdot}6H_2O$ (Rombach 1988, 1989) and YS medium containing 2.5% sucrose, 2.5% yeast extract (Rombach 1989). In each case 50 mL medium was placed in 250 mL DURAN® baffled flasks. The pH values of the media were adjusted to 5.5 with 0.5 M NaOH. As a starter inoculum AC from SDA agar (see above) were used. The AC were isolated by flooding the plates with 2 × 5 mL of sterile 0.1% Tween 80 and gently raking the plates with a sterile bristle brush. The shake flask cultures were inoculated with the spore suspension to give an initial spore density of $5.0{\times}10^4$ AC/mL. The flasks were incubated at 25°C on a rotary shaker at a speed of 150 rpm for 8–10 days. Every day, 1 mL samples were taken to check developmental stage and the concentration of the spores with a Thoma counting cell chamber under 400 × magnification (photomicroscope, Carl Zeiss AG, Oberkochen, Germany).

Fermentation

Batch fermentation was carried out in a 2 L BIOSTAT® Bplus stirred tank reactor (Sartorius Stedim System GmbH, Guxhagen, Germany) with a working volume of 1.5 L. The basal salts were dissolved in 1200 mL ddH_2O and were autoclaved in the bioreactor. Also, a few drops of the anti-foam agent Pluronic® PE 8100 (BASF SE, Ludwigshafen, Germany) were added before fermentation started. Likewise, 300 mL of a carbon source stock solution (75 g carbon source) were autoclaved separately and were inoculated with $7.5{\times}10^8$ aerial conidia ($5.0{\times}10^4$ spores/mL). To start the fermentation, the inoculum suspension was added to the bioreactor. Temperature was maintained at 25°C and fermentation time was between 8–10 days.

Analysis

The metabolic respiratory quotient (RQ) is an on-line parameter for the formation of biomass was calculated from the ratio of the generated carbon dioxide and the consumed oxygen, which were measured with an O_2 and CO_2 sensor (BlueSens GmbH, Herten, Germany) in the exhaust air. For the determination of fungal dry biomass 15 mL samples were centrifuged for 10 min at 20,000 *g*, washed two times with ddH_2O and centrifuged again. The pellets were suspended in 5–7 mL of ddH_2O. The cell suspensions were dried at 115°C to constant weight using a moisture analyzer (Sartorius AG, Goettingen, Germany). Each time, determination of fungal dry biomass was carried out in two replicates.

The colony forming units (CFU) of BS and SCS were determined by spreading 100 µL of diluted samples on SDA plates (Odds 1991) and incubating at 25°C for 4–6 days. To ensure that a sample will yield CFU in a range between 50 and 150 colonies requires several 10-fold dilutions of the sample with 0.9% NaCl. The CFU were determined on duplicate samples. The CFU/ml was calculated as follows:

$$\frac{CFU}{mL} = \frac{CFU}{plate} \cdot \frac{dilution\,factor}{0.1\;mL}$$

Insect virulence assay

Bioassays were conducted with 30 second instar larvae of *Plutella xylostella* L. (Yponomeutidae: Lepidoptera), which were provided by Prof. Stefan Vidal, Georg-August-University, Department of Crop Sciences/ Agricultural Entomology, Goettingen, Germany. The culture broth of the fermentation was centrifuged for 5 min at 20,000 *g*, washed two times with ddH_2O and centrifuged again. The washed spore mix, consisting of 95% BS and 5% SCS, as well as pure AC from a two-weeks-old SDA culture were suspended in 0.1% Triton-X114 to obtain a final concentration of 10^6 viable spores/mL. Aliquots of 1 mL of the suspensions were brushed on the adaxial side of secondary oilseed rape leaves with an area of 80 ± 10 cm^2. The control leaves were treated with 0.1% Triton-X114 only. High 500 mL beakers were filled with 100 mL water agar (1.0% agar-agar). In each case three stalks of the treated leaves were drilled into the solid water agar. The upper surface of the water agar was covered with sterile filter paper to prevent the larvae from getting stuck. Afterwards, ten larvae were transferred into each of the beakers with and without spores, respectively. The beakers were closed with silk gauze and incubated at room temperature.

After 14 days, the dead larvae were surface sterilized with 70% ethanol for 2 min, 5% sodium hypochlorite for 3 min and 70% ethanol for 2 min, rinsed twice in sterile distilled water, and then placed on sterile tissue paper in a laminar airflow cabinet. The larvae were placed on a modified *B. bassiana* selective medium, consisting of 1% casein peptone, 4% glucose, 0.1375% Syllit® (Spiess-Urania Chemicals GmbH, Hamburg, Germany), 0.0005% chlortetracycline, 0.0005% crystal violet and 1.5% agar-agar (Chase et al. 1986; Rangel et al. 2010), and were incubated at 25°C for 1 week. To evaluate the efficacy of the surface sterilization method the water used to rinse the tissues after surface sterilization was plated on selective medium and was incubated, too.

After re-isolation, the DNA of the mycelium was extracted. In each case 50 mg mycelium were cooled on ice, and 1 mL CTAB buffer (0.02 M Na-EDTA, 126 mM sorbitol, 36.8 mM n-laurylsarcosine, 22 mM CTAB, 90 mM polyvinylpyrrolidone, 10 mM Tris, 0.8 M NaCl at pH 8.0), 2 µL β-mercaptoethanol and 1 µL proteinase K (0.1 g/L) were added. The samples were incubated at 42°C for 10 min, at 65°C for 10 min, and then 800 µL chloroform: isoamylalcohol 24:1 were added, mixed, stored on ice for 10 min and centrifuged at 8,000 rpm for 10 min. The supernatants were carefully taken, mixed with 100 µL 5 M NaCl and 200 µL 30% PEG, stored for 10 min at room temperature and centrifuged at 14,000 rpm for 15 min. The pellets were washed twice with 600 µL 75% ethanol and dried in a thermoblock at 65°C. Afterwards, the DNA pellet was dissolved in 100 µL TE buffer (10 mM Tris at pH 8.0) and stored at −20°C. All samples were assessed by PCR using specific primers namely RD1-F (5′- TGGGT ATAGGCCGCAGCAC-3′) and RD1-R (5′- CTCTAAGG GTGACAGGGATAG-3′) which amplify a 208-bp region of the ITS1-5.8S-ITS2 region of *B. bassiana*. The primer sequences were compared with the NCBI nucleotide database using BLAST and were predicted to be 100% homologous to *B. bassiana* isolate DAOM210087 as well as ATP-02. All amplifications were performed in a TProfessional (Biometra GmbH, Goettingen, Germany) thermocycler. PCR reactions consisted of a reaction mix (final volume 10 µL) of 1 µl 10 x reaction buffer, 2.5 mM $MgCl_2$, 100 µM dNTPs, 0.3 µM RD1-F primer, 0.3 µM RD1-R primer, 0.3 U Taq polymerase (5 Prime GmbH, Hilden, Germany) and 1 µl template DNA (30 ng). The cycling program included an initial denaturation step of 5 min at 95°C, followed by 35 cycles of 1 min denaturation at 95°C, 1 min annealing at 59°C, and 1 min extension at 72°C. Amplification products were mixed with GelRed™ (Biotium Inc., Hayward, Canada), separated by electrophoresis in 1.5% agarose gels in 1× TBE buffer (89 mM Tris, 89 mM boric acid, 2 mM EDTA at pH 8.0) for 60 min at 100 V and visualized under UV radiation.

All tests were run for 14 days and each test was repeated thrice. The mortality data were analyzed statistically using one-way ANOVA test.

Penetration assay

Native oilseed rape (*Brassica napus* L.) cultivar "PULSAR" was obtained from the Deutsche Saatveredelung AG (Lippstadt, Germany). The seeds were planted singly in pots containing sterile soil/sand substrate (1:3 (v/v)) (Fruhstorfer Erde Type T25, HAWITA Group GmbH, Vechta, Germany) and were cultivated under greenhouse conditions, The plants were maintained at 18-22°C, 40–60% RH for 16 weeks and with a 12-h photoperiod (SON-T Agro 400 W, Philips, Amsterdam, Netherlands).

Formulation components consisting of 0.1% Triton X-114 as a wetter, 0.1% gelatine 280 Bloom (Gelita AG, Goeppingen, Germany) as a humectant, 1% sugar beet molasses as nutrient and 1% titanium dioxide as a UV protection agent were autoclaved for 20 min at 121°C. The spore suspension from a submerged fermentation was centrifuged for 5 min at 20,000 *g*, washed twice with ddH_2O and centrifuged again. Afterwards the spores were suspended in 0.9% NaCl and added to the formulation components up to a final concentration of 10^6 spores/mL. The control formulation was free of fungal biomass. Then, the formulations were brushed onto an area of approximately 3 cm of the tips of 9th secondary leaves oilseed rape plants. To increase the relative humidity up to 95%, the treated leaves were wrapped with plastic bags for the first 48 h. After 7 days the leaf tips were cut off and the untreated base of the leaves were harvested for the detection of endophytic colonization with *B. bassiana* by microscopy and PCR.

For microscopy, cross-sections of the leaf mid rip were stained with 0.5% rose bengal dissolved in 5% aqueous ethanol for 15 sec and were washed with ddH_2O (Saha et al. 1988). Growth of *B. bassiana* in the plant tissue was detected at 200-fold magnification with a light microscope. Afterwards, the leaves were surface sterilized as mentioned above and then placed on sterile tissue paper in a laminar airflow cabinet. In a preliminary test it was shown that this surface sterilization method kills all spores which were applied onto oilseed rape leaves.

For DNA extraction which was described above, approximately 400 mg plant tissue from the surface sterilized leaves and stems was taken. The plant tissue was crushed in a MM400 ball mill using a sterile 5 mm steel ball (Retsch GmbH, Haan, Germany) for 5 min at 30 Hz. To isolate *B. bassiana* DNA from a SDA culture, 50 mg fungal biomass were directly used for DNA extraction. All tissue samples of both treated and untreated oilseed rape plants were assessed by PCR.

Results

Screening of media in shake flask culture

The entomopathogenic and endophytic fungus *B. bassiana* isolate ATP-02 was cultivated in shake flasks. The different liquid media described above were used to study the effect of various nutrients, basal salts and other complex components on submerged spore formation. In Figure 1 the concentrations and yields of total spores (TS) with regard to the different culture media are illustrated. The most promising culture medium with regard to a maximum growth and optimum sporulation was the TKI medium with 5% sucrose as a carbon source: After a cultivation time of 168 h *B. bassiana* produced TS in a concentration of $1.10 \pm 0.01 \times 10^8$ TS/mL consisting of 100% BS. In this TKI medium a yield of $2.20 \pm 0.02 \times 10^9$ TS/g sucrose was obtained. Due to the lower substrate concentration of 2% sucrose and the also high concentration of $1.09 \pm 0.08 \times 10^8$ TS/mL in the YSM medium a yield of $5.43 \pm 0.38 \times 10^9$ TS/g sucrose was obtained. However, in contrast to the TKI medium, the YSM medium contains 0.5% yeast extract as a complex component. In all cases, the achieved concentration of SCS was lower than 11% of the TS. The highest, but still low SCS concentration of $0.29 \pm 0.04 \times 10^6$ SCS/mL was obtained in the YG medium.

Influence of different carbon sources on spore formation of *B. bassiana*

B. bassiana ATP-02 was cultured in TKI medium which was supplemented with 5% of different pure carbon sources according to Thomas et al. (1987). Furthermore, TKI medium was supplemented with 5% sugar beet molasses as a complex carbon source which consisted of 50% sucrose according to manufacturer specification (Suedzucker AG, Mannheim, Germany). In Figure 2 a and b the concentrations of TS and SCS with regard to

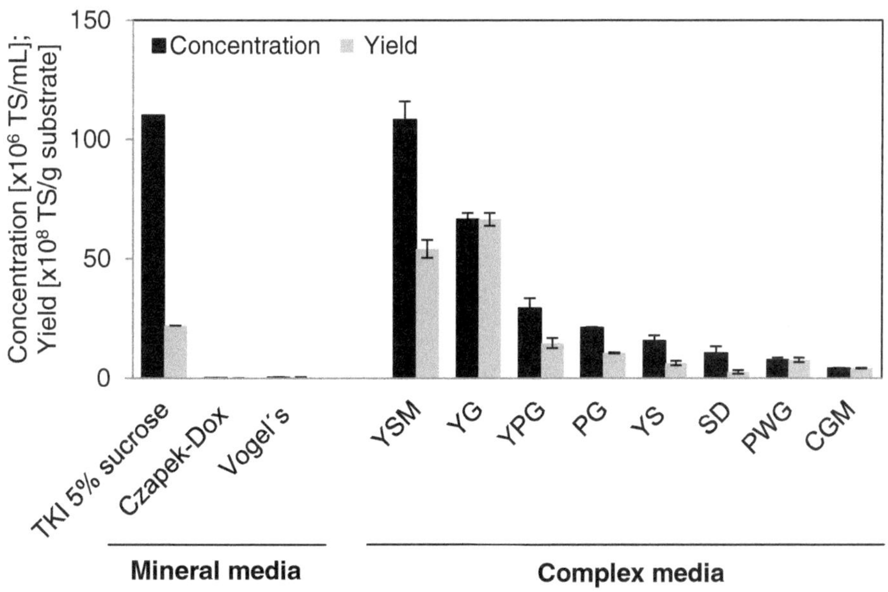

Figure 1 Influence of different culture media on spore formation. *B. bassiana* was cultivated in 250 mL shake flasks (n = 2). Mean (±SD) concentrations and mean (±SD) yields of TS 168 h after inoculation.

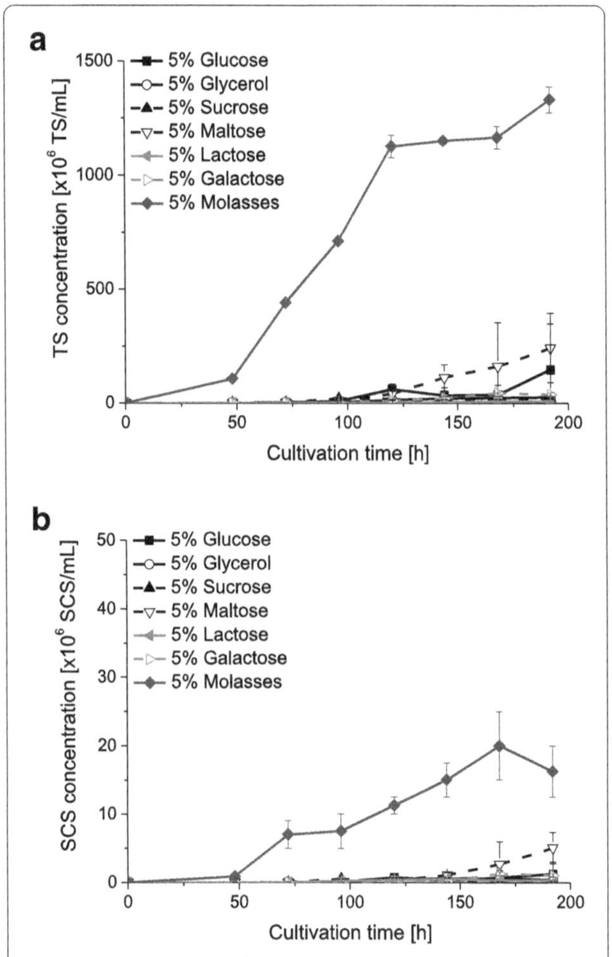

Figure 2 Influence of different carbon sources of the TKI medium on spore formation. *B. bassiana* was cultivated in 250 mL shake flasks (n = 2). **(a)** Mean (±SD) concentration of BS. **(b)** Mean (±SD) concentration of SCS.

the different carbon sources are illustrated. The highest concentration of $1.17 \pm 0.05\times10^9$ TS/mL corresponding to the highest yield of $4.68 \pm 0.20\times10^{10}$ TS/g sucrose was obtained in the TKI medium with 5% sugar beet molasses. Furthermore, in this medium *B. bassiana* also produced the highest concentration of $2.00 \pm 0.50\times10^7$ SCS/mL corresponding to a SCS yield of $8.00 \pm 2.00\times10^8$ SCS/mL 168 h after inoculation. However, the biomass consisted of more than 98% BS. In spite of the lower sugar concentration of the molasses (2.5%) the concentration of TS could be increased up to 11-times in contrast to the cultivation with 5% sucrose. Due to the lower sugar concentration the yield of TS could be even increased up to 21-times.

Interestingly, when TKI basal salts were omitted from the 5% molasses medium, TS concentration decreased by 96% compared to cultivation in the original TKI medium (data not shown).

Fermentation of *B. bassiana* ATP-02

B. bassiana ATP-02 was raised in a 2 L stirred tank reactor to produce high concentrations of TS. Based on the cultivations in shake flasks *B. bassiana* ATP-02 was cultivated in the mineral TKI medium with 5% sugar beet molasses and were inoculated with 5×10^4 AC/mL. The fermentation conditions and growth parameters are given in Table 1. During 48 h fermentation time, 12.6 g/L dry biomass was produced. Maximal specific growth rate μ_{max} was 0.14 h^{-1}. Average doubling time was 14.7 h. The minimal doubling time of 4.9 h was reached between 48 and 72 h after inoculation. The results of the fermentation could be verified in 7 sequentially performed runs, which are shown in Table 2.

The Figure 3a and b show the details of the fermentation no. 1. The fermentation process can be subdivided in two phases: a phase of mycelium formation and a following phase of spore formation.

At the beginning of the fermentation the amount of dry biomass increased because the fungus produced mycelium. After 62 h a sudden decrease of the RQ to 0.6 followed by a sharp decrease of pH to 4.7, a short recuperation of RQ and a decrease of pO_2 to 4% was observed. A sample taken shortly thereafter at 72 h still yielded 21 g biomass/L and the broth was still visibly viscous. Then, biomass dry weight decreased which was accompanied by a visible reduction of mycelium. Preliminary HPLC data indicated a formation of oxalate (data not shown). At this time the concentration of TS started to increase up to $1.29 \pm 0.04\times10^9$ TS/mL corresponding to a yield of $5.16 \pm 0.16\times10^{10}$ TS/g sucrose at the end of the fermentation. However, the biomass consisted of more than 95% BS. 96 h after inoculation the viability of TS started to decrease. The maximum concentration of 0.84×10^9 viable spores/mL corresponding to a yield of 3.36×10^{10} viable spores/g sucrose was obtained 168 h

Table 1 Cultivation of *B. bassiana* ATP-02 in a 2 L stirred tank reactor: fermentation conditions and growth parameters

Fermentation conditions	
Inoculum [aerial conidia/mL]	5×10^4
Start pH [–]	5.5
Stirrer speed [rpm]	600
Aeration rate [vvm]	1.0
Aeration rate [L/min]	1.5
Temperature [°C]	25
Growth parameters	
Biomass produced (after 48 h) [g/L]	12.6
Max. specific growth rate μ_{max} [h^{-1}]	0.14
Min. doubling time [h]	4.9
Mean doubling time [h]	14.7

Table 2 Cultivation of *B. bassiana* ATP-02 in 7 sequentially performed fermentation runs

No.	Total spores [x10^9 TS/mL]	Submerged conidiospores [x10^9 SCS/mL]
1	1.11 ± 0.00	0.04 ± 0.00
2	1.45 ± 0.01	0.10 ± 0.01
3	1.12 ± 0.01	0.06 ± 0.01
4	2.08 ± 0.02	0.12 ± 0.01
5	1.89 ± 0.01	0.09 ± 0.02
6	1.39 ± 0.04	0.08 ± 0.01
7	1.11 ± 0.09	0.04 ± 0.01

Achieved mean (±SD) concentrations of TS and SCS 168 h after inoculation. Standard deviations were calculated from two technical replicates.

after inoculation. During the further fermentation process the concentration of viable spores decreased to 0.78×10^9 TS/mL corresponding to a yield of 3.12×10^{10} TS/g sucrose at the end of the fermentation. Besides, the biomass dry weight decreased during the spore formation phase from 21 g/L to 12 g/L. Furthermore, 80 h after inoculation the RQ decreased continuously and the pO_2 reached a non-critical value of 20% and a clogging of the pO_2 electrode was noted.

Insect virulence assay

The spore mixture from a submerged fermentation, consisting of 95% BS and 5% SCS as well as pure AC harvested from a petri dish were applied in a virulence test against the diamondback moth, *P. xylostella*. After 14 days, in the control without fungal spores 93 ± 5% of the larvae developed into viable adult insects. However, 77 ± 5% of larvae fed with spore mix-treated leaves as well as 90 ± 8% of larvae fed with AC-treated leaves died within a week (Figure 4). The dead larvae were surface sterilized, placed on a *B. bassiana* selective medium and mycelium grew out of all larvae treated with fungal spores and the mycelium was identified as *B. bassiana* by PCR. The mycelium which grew out of dead larvae not treated with *B. bassiana* was clearly not *B. bassiana*,

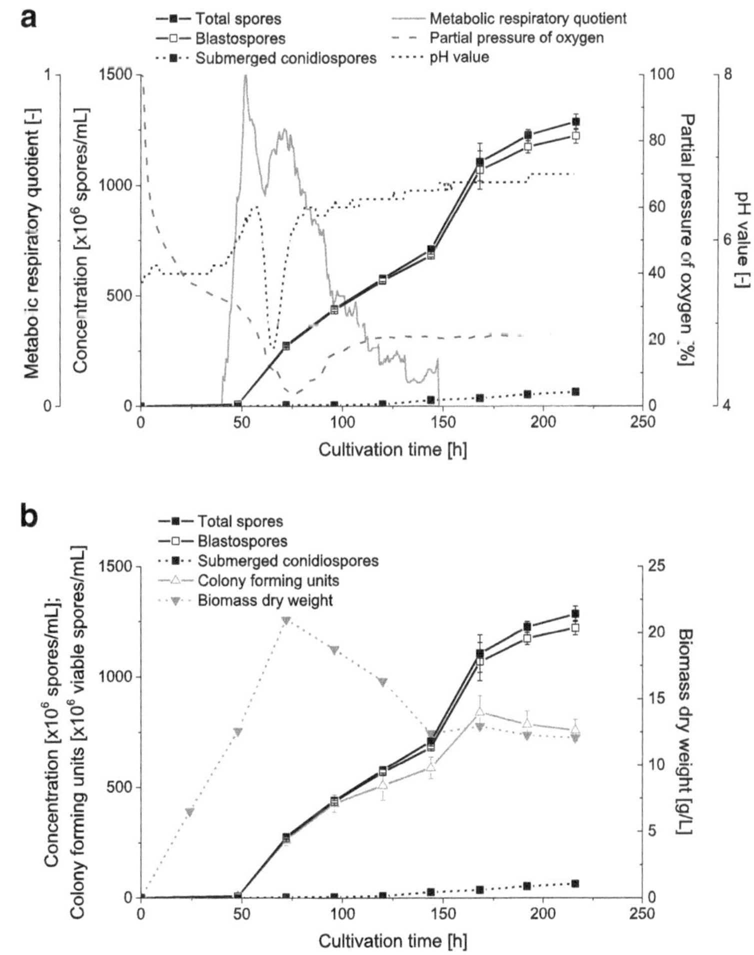

Figure 3 Cultivation of *B. bassiana* in a 2 L stirred tank reactor. The figure shows the mean (±SD) concentrations of TS, BS and SCS. **(a)** Process parameters. **(b)** Correlation of spore counts with biomass and mean (±SD) CFU. In each case, standard deviations resulted from two technical replicates.

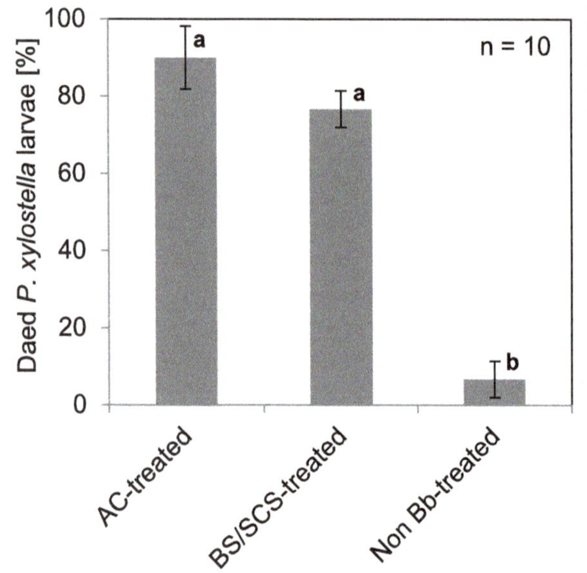

Figure 4 Virulence test with *P. xylostella* larvae. The larvae were fed with AC-treated as well as BS/SCS-treated (95% BS and 5% SCS) and non *B. bassiana* (*Bb*)-treated oilseed rape leaves. Means (±SD) followed by different letters are significantly different at P < 0.01 using one-way ANOVA test. In each case, standard deviations resulted from three replicates with 10 larvae.

so that these larvae did not die by *B. bassiana* induced mycosis. It was observed that the spores from submerged fermentation ($P < 0.01$; $F_{1,4} = 220.5$) as well as the pure AC ($P < 0.01$; $F_{1,4} = 156.3$) significantly affected the mortality of larvae. Besides, the number of dead larvae was not significantly affected by the type of spores.

Penetration assay

The influence of a formulation with *B. bassiana* spores (10^6 TS/mL) on the penetration of oilseed rape leaves was investigated. The formulation was brushed onto the 9th secondary leaf tips of seven oilseed rape plants and afterwards, endophytic *B. bassiana* was detected in the tissue of the untreated leaf base by PCR and microscopy. After 7 days, no hyphae growth was observed microscopically in control leaves treated without *B. bassiana* (n = 2). However, hyphae growth was observed in the mid rip cross-sections of 100% of leaves treated with the formulation. A randomly selected cross-section of these leaf mid rips is illustrated in Figure 5a. To verify that the microscopically detected mycelium was *B. bassiana*, a PCR was performed. *B. bassiana* was detected in all untreated areas of the leaves treated with the formulation by PCR and subsequent gel electrophoresis. The positive PCR signals of five randomly selected leaves are shown in Figure 5b and no PCR amplification was observed in control plants.

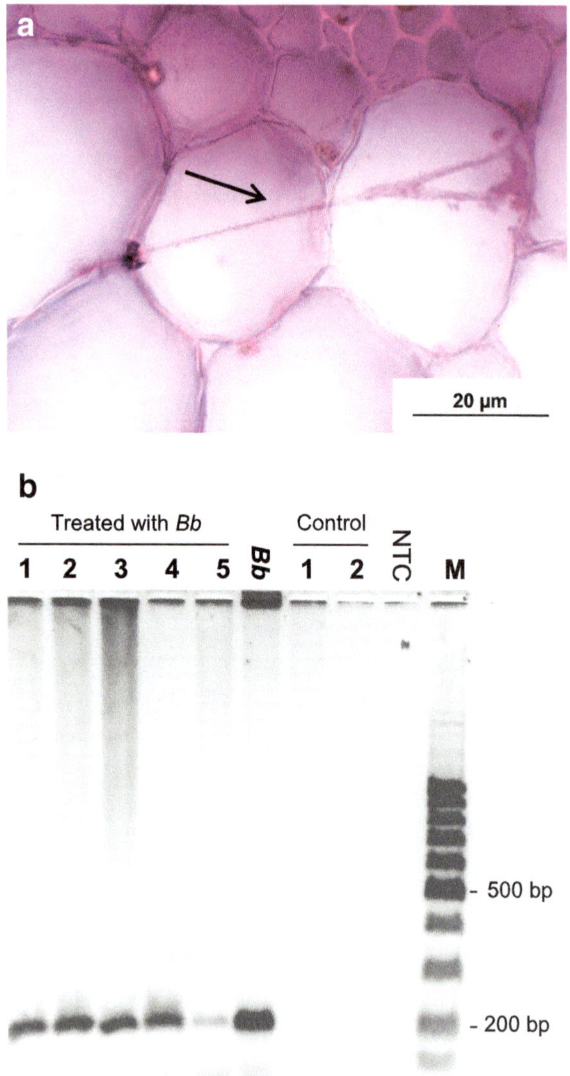

Figure 5 Penetration assay with *B. bassiana* ATP-02. Spray formulations with (n = 7) and without (n = 2) *B. bassiana* (*Bb*) were brushed onto the leaf tips. **(a)** Microscopic detection of *Bb* in a cross-section of a leaf mid rip from a *Bb*-treated leaf at 200-fold magnification. The arrow marks the hyphae in the leaf tissue. **(b)** *Bb* was detected in the untreated area of the *Bb*-treated leaves by PCR and subsequent gel electrophoresis. Lane 1–5: DNA from *Bb*-treated leaves; Lane 6: *Bb* DNA from pure SDA culture; Lane 7–8: DNA from control-treated leaves; Lane 9: no template control; Lane 10: 100 bp DNA ladder (AppliChem GmbH, Darmstadt, Germany).

Discussion

This study deals with the fermentation aspects of the endophytic *B. bassiana* ATP-02 which might prepare the way to exploit endophytes as commercial biocontrol agents.

Screening of media in shake flask culture

In the past, mass-production of *B. bassiana* has focused on AC, but the production through surface cultivation or a two-stage process in which the fungus is allowed to develop under submerged conditions and subsequently transferred to a solid media to sporulate requires long cultivation times, large amounts of space and can be labour-intensive (Hall and Papierok 1982). The obvious advantages of a submerged cultivation are that the fungus produces spores in a relatively short time with high yields under controlled sterile conditions as well as a simpler scale-up in contrast to solid-state fermentation (Feng et al. 1994; Hegedus et al. 1992; Patel et al. 2011). It was previously known that *B. bassiana* strains grow in a variety of liquid mineral and complex media, but that the conidiation of the fungi under submerged conditions may be strain-specific (Kassa et al. 2008) and needs to be investigated for each strain in detail. Furthermore, it was hypothesized that the endophytic *B. bassiana* strain ATP-02 does not grow in the same way as the established *B. bassiana* strains that are applied in classic biocontrol on the surface of plants or in the soil.

However, results on cultivation in shake flasks showed that *B. bassiana* ATP-02 was able to produce BS in the described culture media. Generally, in a submerged cultivation *B. bassiana* can produce two types of spores, namely BS and SCS. BS are relatively large, thin-walled and single-celled hyphal bodies (Bidochka et al. 1987). SCS, on the other hand, are small, spherical, more uniform in size and show a higher shelf life than BS (Thomas et al. 1987; Hegedus et al. 1992; Holder et al. 2007). They arise from the fungal mycelia or directly from BS in a process known as microcycle conidiation (Smith et al. 1981). Thomas et al. (1987) describe the direct formation of SCS from BS in the mineral TKI medium with 5% glucose after a cultivation time of 96 h. This phenomenon was not observed in the present work. From the biotechnological point of view the most important properties of the culture medium are high yields of TS and SCS, respectively. Although a higher TS yield was obtained in the YSM medium consisting of basal salts, 2% sucrose and 0.5% yeast extract a 2.5-fold, the mineral TKI medium without expensive complex components was further optimized because of cost-effectiveness.

Influence of different carbon sources on spore formation of *B. bassiana*

The highest concentrations and yields of TS and SCS were obtained in the TKI medium with 5% sugar beet molasses as a carbon source. It should be pointed out that the concentration of TS could be increased up to 11-times in contrast to the cultivation with 5% sucrose in spite of the lower sucrose concentration of 2.5% in the molasses. The utilized sugar beet molasses consisted of 50% sucrose and only traces of other sugars like glucose, fructose and raffinose as well as different proteins and basal salts according to manufacturer specification (Suedzucker AG, Mannheim, Germany). Furthermore, it could be shown that in addition to the present basal salts of the sugar beet molasses the TKI medium is necessary for optimal growth of *B. bassiana*. Sugar beet molasses is a residue of the agricultural industry and consequently, it is a low-cost source, which is a big advantage compared to other carbon sources. Therefore, the cost of 1 L TKI basal medium amended with 5% sugar beet molasses amounts to only 0.31 €.

Fermentation of *B. bassiana* ATP-02

After the optimized cultivation of *B. bassiana* ATP-02 in shake flasks the process was scaled-up to a 2 L stirred tank reactor. Based on the cultivations in shake flasks *B. bassiana* ATP-02 was cultivated in the TKI mineral medium with 5% sugar beet molasses. In a preliminary test the fermentation was inoculated with a 5-days old shake flask culture of *B. bassiana*, which contained only BS. During the fermentation the fungus produced only mycelium (data not shown). Since the objective of cultivation was a high concentration of TS, further fermentations were inoculated with 5.0×10^4 AC/mL. The achieved concentrations and yields of TS and SCS were comparable with the cultivation of *B. bassiana* ATP-02 in shake flasks and could be verified in 7 sequentially performed fermentation runs.

An unusual point during all fermentations is the sudden decrease of biomass dry weight in correlation with the low concentration of oxygen in the culture broth 72 h after inoculation. At this point it was observed that the finely dispersed mycelium lysed and the pH and the viscosity of the culture broth decreased. HPLC analysis indicated presence of oxalate. The reason for the rapid pH decrease in our fermentation is not clear, but it can be presumed that intracellular oxalate was suddenly released due to the lysis of mycelium. The following increase of pH suggests that oxalate was then metabolized by the growing spores. These presumptions are supported by other studies which also indicate that *B. bassiana* strains are able to produce, secrete and metabolize oxalate in vitro (Bidochka and Khachatourians 1993; Kirkland et al. 2005).

During the fermentation process the pH value was not regulated due to the fact that fluctuating pH values between 4.0 and 6.5 have no considerable impact on the growth of *B. bassiana* (Padmavathi et al. 2003; Thomas et al. 1987). The typical limitation of oxygen can be prevented by increase of the stirrer speed or agitation rate (Patel et al. 2011). But the primary objective of this fermentation process was not the production of finely dispersed mycelium but rather the mass-production of sprayable TS without any mycelium.

The recurring decrease of biomass dry weight in the spore formation phase cannot be explained in detail. It may be hypothesized that the mycelium is decreasing but the spore formation does not compensate the weight loss. It can be ruled out that spore biomass was lost during sample preparation as biomass was not filtered but centrifuged.

Furthermore, the achieved TS concentration and yield of the described fermentation process was higher than those obtained by other investigators in studies of liquid shake flask cultivations of epiphytic *B. bassiana* strains. For example, Thomas et al. (1987) reported a maximum concentration of 5.0×10^8 TS/mL corresponding to a yield of 1.00×10^{10} TS/g glucose, Rombach (1989) described that *B. bassiana* produced TS in a maximum concentration of 0.17×10^9 TS/mL corresponding to a yield of 0.85×10^{10} TS/g sucrose, Vega et al. (2003) obtained a maximum concentration of 1.24×10^9 BS/mL corresponding to a yield of 1.65×10^{10} BS/g glucose and Pham et al. (2009) reported a maximum concentration of 0.85×10^9 BS/mL. The highest described concentration of BS was reported by Chong-Rodríguez et al. (2011), who obtained an inconsistent concentration of $6.38\times10^9 \pm 3.63\times10^9$ BS/ml in a somewhat costly complex medium, which consisted of 5% sucrose, 2% corn steep liquor and basal salts. In comparison to other published data on cultivation of *B. bassiana* isolates in a solid-state or submerged cultivation it can be shown that the described fermentation process is very economical with regard to the achieved concentration and yield of TS. Furthermore, an obvious advantage of the fermentation process is that the cost of 10^{12} TS amounts to only 0.24 € with regard to the utilized culture medium. A further increase of the TS concentration can likely be realized by fed-batch fermentation.

Insect virulence assay

The mortality was 77 ± 5% for *P. xylostella* larvae, which were fed with oilseed rape leaves treated with BS and SCS, and 90 ± 5% for larvae fed with AC-treated leaves. These larvae mortalities are in accordance with those obtained by other investigators. Godonou et al. (2009) reported that AC of *B. bassiana* caused *P. xylostella* larvae mortality ranging from 20 to 94%. BS which were sprayed on *P. xylostella* larvae showed a mortality ranging from 95 to 100% (Fargues et al. 1983). In addition, Chong-Rodríguez et al. (2011) described that BS of *B. bassiana* maintained for six months at 4°C showed a mortality of more than 80% against third-instar *P. xylostella* larvae 8 days after application. Furthermore, Ortiz-Urquiza et al. 2010 observed that the composition of the culture medium affected the virulence of AC from *B. bassiana* because of an increased or decreased secretion of virulent proteins. But the influence of the culture media on the virulence of *B. bassiana* was not investigated in this work. Since no further mortality tests were conducted with pure BS and pure SCS, it can only be hypothesized that BS must have killed the larvae, because the spore mix consisted of 95% BS which are the preferred propagule of *B. bassiana* in the haemocoel of infected insects (Jackson et al. 2010; Shimizu et al. 1993; Sieglaff et al. 1997). Furthermore, BS are highly infective against a number of insect pests and have a lower LD_{50} when compared to AC or SCS (Hegedus et al. 1992). Finally, it was shown here that spores from a submerged cultivation are as virulent to *P. xylostella* larvae as AC.

Penetration assay

The simple penetration assay indicated that the spores from submerged fermentation show endophytic properties. This is in line with studies on AC that were applied to leaves and could to some extent colonize plants (Bing and Lewis 1991, 1992; Gurulingappa et al. 2010; Landa et al. 2013; Posada et al. 2007; Quesada-Moraga et al. 2006; Quesada-Moraga et al. 2009; Tefera and Vidal 2009; Wagner and Lewis 2000). Many questions about endophytism remain that are not within the scope of this fermentation study.

To the best of our knowledge, this is the first report of fermentation of an endophytic *B. bassiana* strain in a low-cost culture medium to very high yields of TS, which are able to penetrate oilseed rape leaves via a leaf application. This should further encourage the recent activities to exploit the biocontrol potential of endophytic entomopathogenic fungi. Besides, the evidence that the endophytic strain grows in simple cultivation conditions much like the classic biocontrol strains is further proof that in nature some microorganisms are facultative endophytes, which can optionally live inside plants and other habitats (Hardoim et al. 2008). The results also clearly suggest to further explore submerged cultivation for entomopathogenic fungi in general. Further studies are required to produce higher amounts of SCS of *B. bassiana*, which show a higher shelf life than thin-walled BS and may also persist longer when sprayed onto plant leaves.

Abbreviations

AC: Aerial conidia; BS: Blastospores; SCS: Submerged conidiospores; TS: Total spores; RQ: Respiratory quotient; CFU: Colony forming units.

Competing interest

The authors declare that they have no competing interests.

Authors' contributions

RL designed and carried out the cultivations, fermentations and mortality tests, analyzed the data and wrote the manuscript. DJS carried out the molecular biology studies. AVP conceived of the study, and participated in its design and coordination and helped to draft the manuscript. All authors read and approved the final manuscript.

Acknowledgements

We would like to thank Prof. S. Vidal (Georg-August-University of Goettingen, Department of Crop Sciences/Agricultural Entomology) for providing the endophytic *B. bassiana* strain ATP-02 and the *P. xylostella* larvae. The research performed in this study was funded by the German Federal Ministry of Education and Research (FKZ 17 N1510).

References

Akello J, Dubois T, Coyne D, Kyamanywa S (2008) Endophytic *Beauveria bassiana* in banana (*Musa* spp.) reduces banana weevil (*Cosmopolites sordidus*) fitness and damage. Crop Prot 27:1437–1441

Bidochka MJ, Khachatourians GG (1993) Oxalic acid hyperproduction in *Beauveria bassiana* mutants is related to a utilizable carbon source but not to virulence. J Invertebr Pathol 62:53–57

Bidochka MJ, Pfeifer TA, Khachatourians GG (1987) Development of the entomopathogenic fungus *Beauveria bassiana* in liquid cultures. Mycopathologia 99:77–83

Bing LA, Lewis LC (1991) Suppression of *Ostrinia nubilalis* (Hubner)(Lepidoptera: pyralidae) by endophytic *Beauveria bassiana* (Balsamo) Vuillemin. Environ Entomol 20:1207–1211

Bing LA, Lewis LC (1992) Endophytic *Beauveria bassiana* (Balsamo) Vuillemin in corn: the influence of the plant growth stage and *Ostrinia nubilalis* (Hübner). Biocontrol Sci Techn 1:29–47

Burges HD (1998) Formulation of microbial pesticides. Kluwer Academic Publishers, Dordrecht, The Netherlands

Chase AR, Osborne LS, Ferguson VM (1986) Selective isolation of the entomopathogenic fungi *Beauveria bassiana* and *Metarhizium anisopliae* from an artificial potting medium. Fla Entomol 69:285–292

Chong-Rodríguez MJ, Maldonado-Blanco MG, Hernández-Escareño JJ, Galán-Wong LJ, Sandoval-Coronado CF (2011) Study of *Beauveria bassiana* growth, blastospore yield, desiccation-tolerance, viability and toxic activity using different liquid media. Afr J Biotechnol 10:5736–5742

Fargues J, Reisinger O, Robert PH, Aubart C (1983) Biodegradation of entomopathogenic hyphomycetes: influence of clay coating on *Beauveria bassiana* blastospore survival in soil. J Invertebr Pathol 41:131–142

Feng MG, Poprawski TJ, Khachatourians GG (1994) Production, formulation and application of the entomopathogenic fungus *Beauveria bassiana* for insect control: current status. Biocontrol Sci Techn 4:3–34

Glare T, Caradus J, Gelernter W, Jackson T, Keyhani N, Köhl J, Marrone P, Morin L, Stewart A (2012) Have biopesticides come of age? Trends Biotechnol 30:250–258

Godonou I, James B, Atcha-Ahowé C, Vodouhè S, Kooyman C, Ahanchédé A, Korie S (2009) Potential of *Beauveria bassiana* and *Metarhizium anisopliae* isolates from Benin to control *Plutella xylostella* L. (Lepidoptera: plutellidae). Crop Prot 28:220–224

Gurulingappa P, Sword GA, Murdoch G, McGee PA (2010) Colonization of crop plants by fungal entomopathogens and their effects on two insect pests when in planta. Biol Control 55:34–41

Hall RA, Papierok B (1982) Fungi as biological control agents of arthropods of agricultural and medical importance. Parasitology 84:205–240

Hardoim PR, van Overbeek LS, van Elsas JD (2008) Properties of bacterial endophytes and their proposed role in plant growth. Trends Microbiol 16:463–471

Hegedus DD, Bidochka MJ, Khachatourians GG (1990) *Beauveria bassiana* submerged conidia production in a defined medium containing chitin, two hexosamines or glucose. Appl Microbiol Biotechnol 33:641–647

Hegedus DD, Bidochka MJ, Miranpuri GS, Khachatourians GG (1992) A comparison of the virulence, stability and cell-wall-surface characteristics of three spore types produced by the entomopathogenic fungus *Beauveria bassiana*. Appl Microbiol Biotechnol 36:785–789

Holder DJ, Kirkland BH, Lewis MW, Keyhani NO (2007) Surface characteristics of the entomopathogenic fungus *Beauveria* (*Cordyceps*) *bassiana*. Microbiology 153:3448–3457

Humphreys AM, Matewele P, Trinci APJ (1989) Effects of water activity on morphology, growth and blastospore production of *Metarhizium anisopliae*, *Beauveria bassiana* and *Paecilomyces farinosus* in batch and fed-batch culture. Mycol Res 92:257–264

Humphreys AM, Matewele P, Cunliffe B, Trinci APJ (1990) Comparison of sporulation of *Paecilomyces farinosus* and *Beauveria bassiana* in batch and fed-batch culture. Mycol Res 94:1046–1050

Jackson MA, Dunlap CA, Jaronski ST (2010) Ecological considerations in producing and formulating fungal entomopathogens for use in insect biocontrol. Biocontrol 55:129–145

Kassa A, Brownbridge M, Parker BL, Skinner M, Gouli S, Guo M, Lee F, Hata T (2008) Whey for mass production of *Beauveria bassiana* and *Metarhizium anisopliae*. Mycol Res 112:583–591

Khachatourians GG (1986) Production and use of biological pest control agents. Trends Biotechnol 4:120–124

Kirkland BH, Eisa A, Keyhani NO (2005) Oxalic acid as a fungal acaracidal virulence factor. J Med Entomol 42:346–351

Kučera M (1971) Toxins of the entomophagous fungus *Beauveria bassiana*: II effect of nitrogen sources on formation of the toxic protease in submerged culture. J Invertebr Pathol 17:211–215

Landa BB, Lopez-Diaz C, Jimenez-Fernandez D, Montes-Borrego M, Munoz-Ledesma FJ, Ortiz-Urquiza A, Quesada-Moraga E (2013) In-planta detection and monitorization of endophytic colonization by a *Beauveria bassiana* strain using a new-developed nested and quantitative PCR-based assay and confocal laser scanning microscopy. J Invertebr Pathol 114:128–138

Lane BS, Trinci APJ, Gillespie AT (1991) Endogenous reserves and survival of blastospores of *Beauveria bassiana* harvested from carbon- and nitrogen-limited batch cultures. Mycol Res 95:821–828

Leckie BM, Ownley BH, Pereira RM, Klingeman WE, Jones CJ, Gwinn KD (2008) Mycelia and spent fermentation broth of *Beauveria bassiana* incorporated into synthetic diets affect mortality, growth and development of larval *Helicoverpa zea* (Lepidoptera: noctuidae). Biocontrol Sci Techn 18:697–710

Núñez-Ramírez DM, Valencia-López JJ, Calderas F, Solís-Soto A, López-Miranda J, Medrano-Roldán H, Medina-Torres L (2012) Mixing analysis for a fermentation broth of the fungus *Beauveria bassiana* under different hydrodynamic conditions in a bioreactor. Chem Eng Technol 35:1954–1961

Odds FC (1991) Sabouraud('s) agar. J Med Vet Mycol 29:355–359

Ortiz-Urquiza A, Riveiro-Miranda L, Santiago-Álvarez C, Quesada-Moraga E (2010) Insect-toxic secreted proteins and virulence of the entomopathogenic fungus *Beauveria bassiana*. J Invertebr Pathol 105:270–278

Padmavathi J, Uma Devi K, Uma Maheswara Rao C (2003) The optimum and tolerance pH range is correlated to colonial morphology in isolates of the entomopathogenic fungus *Beauveria bassiana* – a potential biopesticide. World J Microb Biot 19:469–477

Patel AV, Jakobs-Schönwandt D, Rose T, Vorlop KD (2011) Fermentation and microencapsulation of the nematophagous fungus *Hirsutella rhossiliensis* in a novel type of hollow beads. Appl Microbiol Biotechnol 89:1751–1760

Pham TA, Kim JJ, Kim SG, Kim K (2009) Production of blastospore of entomopathogenic *Beauveria bassiana* in a submerged batch culture. Mycobiology 37:218–224

Posada F, Aime MC, Peterson SW, Rehner SA, Vega FE (2007) Inoculation of coffee plants with the fungal entomopathogen *Beauveria bassiana* (Ascomycota: hypocreales). Mycol Res 11:748–757

Quesada-Moraga E, Landa BB, Muñoz-Ledesma J, Jiménez-Diáz RM, Santiago-Álvarez C (2006) Endophytic colonisation of opium poppy, *Papaver somniferum*, by an entomopathogenic *Beauveria bassiana* strain. Mycopathologia 161:323–329

Quesada-Moraga E, Muñoz-Ledesma FJ, Santiago-Alvarez C (2009) Systemic protection of *Papaver somniferum* L. against *Iraella luteipes* (Hymenoptera: cynipidae) by an endophytic strain of *Beauveria bassiana* (Ascomycota: hypocreales). Environ Entomol 38:723–730

Rangel DEN, Dettenmaier SJ, Fernandes EKK, Roberts DW (2010) Susceptibility of *Metarhizium* spp. and other entomopathogenic fungi to dodine-based selective media. Biocontrol Sci Techn 20:375–389

Ravensberg WJ (2011) A roadmap to the successful development and commercialization of microbial pest control products for control of arthropods. Springer, Dordrecht, The Netherlands

Rombach MC (1989) Production of *Beauveria bassiana* [Deuteromycotina: hyphomycetes] sympoduloconidia in submerged culture. Entomophaga 34:45–52

Rombach MC, Aguda RM, Roberts DW (1988) Production of *Beauveria bassiana* [Deuteromycotina: hyphomycetes] in different liquid media and subsequent conidiation of dry mycelium. Entomophaga 33:315–324

Safavi SA, Shah FA, Pakdel AK, Rasoulian GR, Bandani AR, Butt TM (2007) Effect of nutrition on growth and virulence of the entomopathogenic fungus *Beauveria bassiana*. FEMS Microbiol Lett 270:116–123

Saha DC, Jackson MA, Johnsoncicalese JM (1988) A rapid staining method for detection of endophytic fungi in turf and forage grasses. Phytopathology 78:237–239

Samsináková A (1966) Growth and sporulation of submersed cultures of the fungus *Beauveria bassiana* in various media. J Invertebr Pathol 8:395–400
Shimizu S, Tsuchitani Y, Matsumoto T (1993) Production of an extracellular protease by *Beauveria bassiana* in the haemolymph of the silkworm, *Bombyx mori*. Lett Appl Microbiol 16:291–294
Sieglaff DH, Pereira RM, Capinera JL (1997) Pathogenicity of *Beauveria bassiana* and *Metarhizium flavoviride* (Deuteromycotina) to *Schistocerca americana* (Orthoptera : acrididae). J Econ Entomol 90:1539–1545
Smith JE, Anderson JG, Deans SG, Berry DR (1981) Biochemistry of microcylce conidiation. In: Cole GT, Kendrick B (eds) Biology of conidial fungi, vol 2. Academic Press, New York, NY, pp 329–353
Tefera T, Vidal S (2009) Effect of inoculation method and plant growth medium on endophytic colonization of sorghum by the entomopathogenic fungus *Beauveria bassiana*. Biocontrol 54:663–669
Thomas KC, Khachatourians GG, Ingledew WM (1987) Production and properties of *Beauveria bassiana* conidia cultivated in submerged culture. Can J Microbiol 33:12–20
Vega FE, Jackson MA, Mercadier G, Poprawski TJ (2003) The impact of nutrition on spore yields for various fungal entomopathogens in liquid culture. World J Microb Biot 19:363–368
Vogel HJ (1956) A conventional growth medium for *Neurospora* (Medium N). Microb Genet Bul 13:42–44
Wagner BL, Lewis LC (2000) Colonization of corn, *Zea mays*, by the entomopathogenic fungus *Beauveria bassiana*. Appl Environ Microbiol 66:3468–3473

12

Surface functionalization by covalent immobilization of an innovative carvacrol derivative to avoid fungal biofilm formation

Aïcha Gharbi[1], Thibaut Legigan[2], Vincent Humblot[3], Sébastien Papot[2] and Jean-Marc Berjeaud[1*]

Abstract

Carvacrol, an aromatic terpenic compound, known to be antimicrobial was grafted onto gold surfaces via two strategies based on newly-synthesized cross-linkers involving either an ester bond which can be cleaved by microbial esterases, or a covalent ether link. Surface functionalizations were characterized at each step by reflection absorption infrared spectroscopy (RAIRS). The two functionalized gold samples both led to a loss of culturability of the yeast *Candida albicans*, higher than 65%, indicating that the activity of the freshly-designed surfaces was probably due to still covalently immobilized carvacrol. On the contrary, when a phenyl group replaced the terpenic moiety, the yeast culturability increased by about 30%, highlighting the specific activity of carvacrol grafted on the surfaces. Confocal microscopy analyses showed that the mode of action of the functionalized surfaces with the ester or the ether of carvacrol was, in both cases, fungicidal and not anti-adhesive. Finally, this study shows that covalently immobilization of terpenic compounds can be used to design promising antimicrobial surfaces.

Keywords: Antibiofilm material; *Candida albicans*; Carvacrol derivatives; Surface functionalization

Introduction

In recent years, hospital-acquired infections have considerably increased, concomitantly to the enhanced use of temporarily or permanently inserted indwelling medical devices (IMDs) (Ramage et al. 2009), thus becoming a real problem of public health. Fungi as well as bacteria can cause IMD-related infections. Today, invasive fungal infections are mainly associated to the *Candida* (80%) and *Aspergillus* genera (Cowen 2008; Silva et al. 2010). *C. albicans* being by far the most frequently isolated fungal species (Miceli et al. 2011) in such a way that it became a model study organism for this kind of infection (Gow et al. 2002; McCullough et al. 1996).

It is now clearly recognized that more than 65% of all human infections without any distinction are linked to the growth of biofilms (Pemàn et al. 2008; Pierce et al. 2008; Ramage et al. 2009). Biofilm formation is a complex process that can be summed up in four steps (Seneviratne et al. 2008): (i) the early initial phase or adherence phase followed by (ii) the intermediate phase of the biofilm growth, then (iii) the maturation phase of the newly formed biofilm and finally (iv) a dispersion phase of microorganisms from the biofilm (Pemàn et al. 2008). These biofilms are a major problem as the required dose of biocide to eradicate them can be higher than the highest therapeutic levels that can be administered to the patient (Ramage et al. 2009). Resulting from the presence of biofilms, the diagnosis and treatment of systemic candidal infections is often difficult (Walker et al. 2010).

One promising solution to avoid such infections can be to prevent biofilm formation from the beginning, by inhibiting the adherence phase that initiates the biofilm development cycle (von Eiff et al. 2005). In practice, several strategies currently exist. (i) Hospital good practices can be adopted (Pearson 1996). (ii) The lock therapy that consists in filling the lumen of a catheter still in place in the patient, with a highly-concentrated antimicrobial solution, for six to twelve hours, can also be used (Andris et al. 1998). This therapeutic approach has still limitations, in particular in the case of fungal infections (Hall and Farr 2004). (iii) Antimicrobial materials can also be

* Correspondence: jean-marc.berjeaud@univ-poitiers.fr
[1]Ecologie & Biologie des Interactions - UMR CNRS 7267, Microbiologie de l'eau, Université de Poitiers, 1 Rue Georges Bonnet, TSA 51106, 86073 Poitiers Cedex 9, France
Full list of author information is available at the end of the article

developed. Indeed, non- functionalized biomaterials used to manufacture IMDs and catheters can be colonized by pathogenic microorganisms such as *C. albicans* (Costa et al. 2011). Modifying their surface, in order to make them unfavourable to microbial colonization, is an interesting and promising option (Tiller et al. 2001).

Material surface modifications can be performed in different ways in order to limit the establishment of microorganisms on the surface, to slow down their growth and thus to delay the biofilm formation (Roe et al. 2008). The coating approach by impregnation is today, the most developed one to design antimicrobial materials and the simplest one among all the technologies allowing a molecule release (Zhang et al. 2011). These surfaces suffer from a problem of antimicrobial activity depletion associated to the active substance release (Costa et al. 2011; Goddard and Hotchkiss 2007; Timofeeva and Kleshcheva 2011). To counteract this latter problem, the surface can also be functionalized by covalent grafting of the active molecule. Important achievements have been made during the last decade to covalently immobilized antimicrobial on surfaces (Costa et al. 2011; Goddard and Hotchkiss 2007; Timofeeva and Kleshcheva 2011). The inhibition of biofilm formation on a functionalized biomaterial may occur either through an anti-adhesive process (namely the reduction of microorganism adherence) or through a bacteriostatic action (namely avoiding cell multiplication on the surface in the presence of the antimicrobial agent) (Tenke et al. 2004) or finally through a bactericidal activity (namely by killing cells once in contact with the surface) (Aslam 2008). Different grafting strategies and types of antimicrobial agents have been more or less successfully tested such as heparin (Dwyer 2008), quaternary ammoniums (Abraham et al. 2002), or even antimicrobial peptides (*i.e.* Magainin I: (Humblot et al. 2009); Gramicidin A: (Yala et al. 2011)). In this field, one emergent trend consists in developing functionalized responsive materials from which the antimicrobial substance release is triggered by the microorganisms themselves, when approaching the surface. Indeed, as enzyme-sensitive linkages are introduced in such self-regulating systems, the active compound is only released in presence of the microorganisms that secrete the specific enzymes that can cleave the sensitive linkage in the construction (*e.g.* lipases (Komnatnyy et al. 2014)). Thus, the release of the active molecules can be tuned in function of the concentration of microorganisms surrounding the surface. This approach is supposed to increase the half-life of the biomolecule by avoiding its metabolization (Goddard and Hotchkiss 2007) or to limit the side-effects related to the molecule release that can lead to its accumulation in some vital tissues (*e.g.* brain, liver, spleen) (Costa et al. 2011).

Antimicrobial, antiseptic and antifungal and even anti-*Candida* properties of essential oils are often ascribed to their main volatile and aromatic terpenic components such as eugenol, thymol or carvacrol (CAR) (Braga et al. 2008; Pavel et al. 2010; Williams et al. 2011). Antifungal and anti-Candida biofilm properties of CAR have already been highlighted in a study from the laboratory (Dalleau et al. 2008). Regardless of the intensive research currently pursued on terpenes, their precise mechanisms of action remain slightly characterized (Gutierrez et al. 2010; Xu et al. 2008). In general, CAR has a fungicidal activity (namely by killing cells once in contact with the surface) on *C. albicans* (Ahmad et al. 2011; Braga et al. 2008).

In this work, two proofs of concept of innovative materials, functionalized with an original construction containing CAR covalently immobilized, was developed on gold surfaces. Indeed, CAR was attached to a specific spacer anchored on a gold surface through either an ester bound which could be cleaved by microorganisms' esterases or an ether one which cannot be split. Thus, in the "ester" case, the antibiofilm activity should result from the release of CAR triggered by microorganisms approaching the surface making the surface repellent to them, whereas in the "ether" case, this activity should be associated to CAR, still attached to the surface. The antibiofilm activity of the new surfaces was studied by CFU counting as well as microscopic observations to understand the mode of action of such functionalized surfaces. The relevance of these two systems was tested on *C. albicans* in the present work.

Materials and methods

Microorganisms and growth conditions

The used strain *C. albicans* ATCC 3153 was grown for 48 h at 37°C on Sabouraud Glucose agar containing Chloramphenicol (Fluka, Saint-Quentin Fallavier, France) to obtain a culture of synchronous stationary-phase yeast. A loopful of this 48 h-culture was transferred to 25 mL of Yeast Nitrogen Medium (YNB; BioChemika Sigma Aldrich, Saint-Quentin Fallavier, France) supplemented with 30 mM of glucose (Sigma Aldrich) (YNB-Glc 30 mM) and incubated for 16 h at 37°C without shaking. Blastospores were then harvested and washed twice in phosphate-buffered saline (PBS; pH 7.2) and adjusted to the desired concentration.

Carvacrol derivatives preparation

EstC-NH_2, Phe-NH_2 and EtC.-NH_2 (Figure 1A) were synthesized as described in the Additional file 1.

Functionalization of gold samples

Grafting steps

The surfaces cited below as gold surfaces were purchased from Arrandee (Werther, Germany). They were constituted of glass substrates (11 × 11 × 1 mm) and coated successively with a 50 Å thick layer of chromium

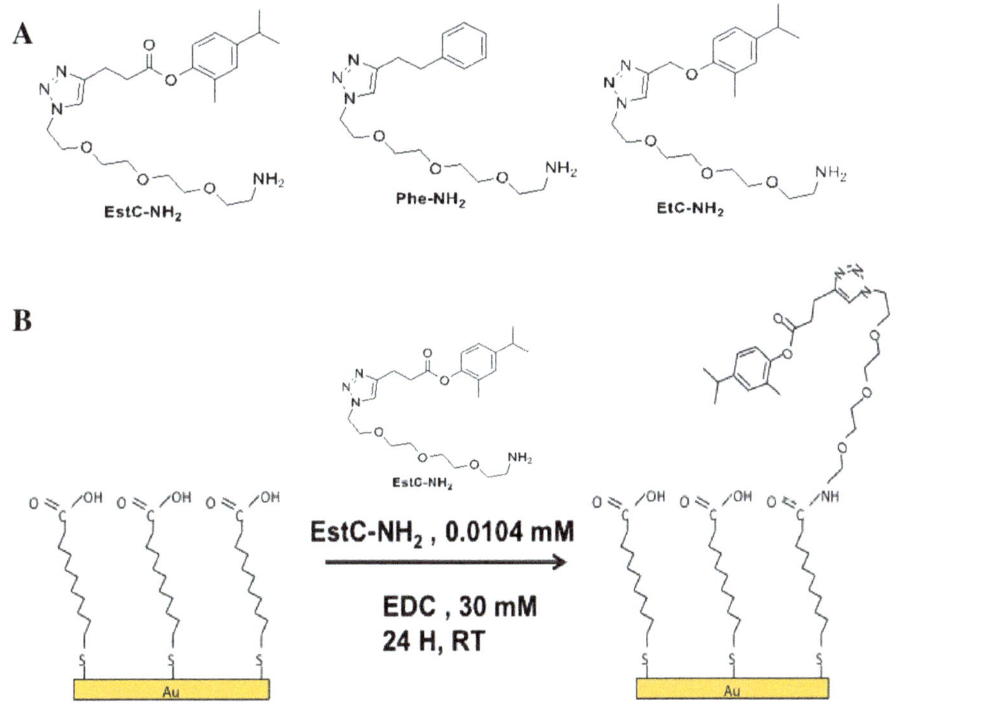

Figure 1 Structures of the carvacrol derivatives grafted on gold surfaces (A). Schematic representation of the two consecutive steps leading to the immobilization of EstC-NH_2 on Au surface, via esterification of the acidic functions of MUA by EDC **(B)**.

and a 200 nm thick layer of gold. The gold-coated substrates were annealed in a butane flame to ensure a good cristallinity of the topmost layers and rinsed first in a bath of absolute ethanol for 5 min and then twice in milliQ water for 5 min with shaking. The substrates were immersed in 10 mL of a 0.01 M solution of MUA in absolute ethanol for 3 h, to ensure an optimal homogeneity of the adlayer (Briand et al. 2006a; Briand et al. 2006b), and thoroughly rinsed in ethanol and water, and dried under a flow of dry nitrogen. Activation of acid groups into ester groups (EDC at 30 mM in water) and immobilization of EstC-NH_2, Phe-NH_2 or EtC.-NH_2 (1.04×10^{-2} mM in water respectively corresponding to 4.5 mg/L, 3.5 mg/L and 4 mg/L) on gold surfaces was carried out in one step, without any catalyst, by depositing a 150 μL drop on the MUA-Au-modified substrates at room temperature for 24 h. After the immobilization step, the surfaces were vigorously rinsed in water under shaking for 5 min and finally dried under a flow of dry nitrogen (see Figure 1B for an illustration of the two grafting steps).

PM-RAIRS measurements

The gold samples were placed in the external beam of an FT-IR instrument (Nicolet Nexus 5700 FT-IR spectrometer), and the reflected light was focused on a nitrogen-cooled Mercury-Cadmium-Telluride wide band detector. The infrared spectra were recorded at 8 cm^{-1} resolution, with co-addition of 128 scans. A ZnSe grid polarizer and a ZnSe photoelastic modulator to modulate the incident beam between p and s polarizations (HINDS Instruments, PEM90, modulation frequency = 36 kHz) are placed prior to the sample. The detector output was sent to a two-channel electronic device that generates the sum and difference interferograms. Those are processed and undergo Fourier transformation to produce the 'Polarization Modulation – Reflection Absorption Infrared Spectroscopy' (PM-RAIRS) signal $(\Delta R/R^0) = (Rp - Rs)/(Rp + Rs)$. Using a modulation of polarization enabled us to perform rapid analyses of the sample after treatment in various solutions without purging the atmosphere or requiring a reference spectrum.

XPS measurements

XPS analyses were performed using a PHOIBOS 100 X-ray photoelectron spectrometer from SPECS GmbH (Berlin, Germany) with a MgKa X-ray source (hn ¼ 1253.6 eV) operating at $P = 1 \times 10^{-10}$ Torr or less. Spectra were carried out with a 20 eV pass energy for the survey scan and 10 eV pass energy for the C1s, O1s, S2p and N1s regions. High-resolution XPS conditions have been fixed: "Fixed Analyser Transmission" analyses mode, a 7 × 20 mm entrance slit; leading to a resolution of 0.1 eV for the spectrometer, and an electron beam power of 150 W (12.5 kV and 12 mA). Such a low energy was used to keep the adsorbed layers as intact as possible. A takeoff angle of 90° from the surface was employed for each sample and binding energies were calibrated against the Au4f binding energy of the gold surface at 84.0 eV. Element peak intensities were corrected by Scofield factors (Scofield 1976). The spectra were fitted using the Casa XPS v.2.3.13 Software (Casa Software Ltd., UK) and applying a Gaussian/Lorentzian ratio, G/L equal to 70/30. In addition, the error in the peak fitting process was estimated at 5% or less with respect to the integration of the raw data.

Adhesion of yeasts on gold samples

Adhesion phase of yeast cells

A Petri dish was partially filled with approximately 7 mL of a kappa carrageenan gel (15 g L^{-1}; Sigma Aldrich) and cooled at room temperature. One gold sample (functionalized or control) was then carefully deposited per dish, the functionalized gold face upwards. A 100 μL drop containing 10^5 of freshly cultured yeast cells resuspended in PBS was deposited on the gold surface. After 3 h at 37°C, each sample was washed three times in physiological sterile solution (0.9% NaCl).

Enumeration of adherent yeast cells

After the 3 h adhesion phase and the washing step, the gold surfaces were transferred into a sterile tube containing 2 mL of physiological sterile solution (0.9% NaCl), then sonicated 3 min at 60 W. After the removing of the gold sample, the yeast cells were pelleted by centrifugation at 13 000 *g* for 5 min. Then, 1.8 mL of supernatant was removed cautiously, and the cells were resuspended by vortexing. The suspension was then diluted 50 and 100 times; 50 μL of each dilution were plated on Sabouraud agar plates, in duplicate for each dilution, using a spiral plater WASP (AES, France). The plates were incubated at 37°C for 24 h before counting the colonies.

Microscopic analysis

Viability of adherent yeasts to the surfaces was evaluated with the LIVE/DEAD® Bacterial Viability Kit (BacLight®). Syto9 (at 2.34 mM) and propidium iodide (at 20 mM). The two stains, were diluted 10 times, in physiological sterile solution (0.09% NaCl) and were kept as stock solutions at −20°C, in the dark. The final BacLight solution used for the microscopic observations was freshly prepared by mixing 1.5 μL of both stain stock solutions with 97 μL of distilled water. After the 3 h adhesion phase and the washing step described above, a 10 μL drop of this BacLight mixture was deposited on the surface of the sample, a slide was put on the drop and the whole set was incubated for 15 min in the dark, at room temperature. Samples were then examined with a confocal FV-1000 station installed on an inverted microscope IX-81 (Olympus, Tokyo, Japan). Images were acquired with an Olympus UplanSapo × 60 water, 1.2 NA, objective lens (800 × 800 pixels images with 0.13 mm per pixel corresponding to Nyquist criteria for optimal sampling). Multiple fluorescence signals were acquired sequentially to avoid cross talk between image channels. Fluorophores were excited with the 488 nm line of an argon laser (for Syto9) and the 543 nm line of an HeNe laser (for propidium iodide). The emitted fluorescences were detected through spectral detection channels between 500–530 and 555–655 nm, for green and red fluorescence, respectively. Several photographs (from 5 to 10), of different areas with same surfaces, were taken on the same sample. Adherent yeast cells and adherent permeabilized (red) cells were manually counted and average numbers were calculated.

Activity of already used surfaces

To investigate the persistence of the antifungal activity of already used Au-MUA-NH-EstC surfaces, microbial viability after adherence was assayed after 1 month's storage at 4°C by yeast numerations on plates and confocal microscopy analysis, using the same protocols described previously.

Results

EstC-NH_2 molecule grafting and surface characterization

PM-RAIRS analysis

Spectrum (a) on Figure 2, which corresponds to the grafting of MUA on the gold plate (Au-MUA), presents the typical infrared signature of MUA (Bain et al. 1989; Tielens et al. 2008). All usual IR features are present on the spectrum: a broad peak with two main frequencies at 1741 and 1714 cm^{-1} assigned to the νC = O of the carboxylic acid group of the acidic thiolate; The presence of two frequencies for the same group suggests either two different adsorption sites and/or a poor order of the SAM. This is confirmed by the average order defined by the νCH_2 at 2925 and 2854 cm^{-1} (highly ordered layer usually observed at 2919 and 2848 cm^{-1}). In addition, two bands at 1596 and 1424 cm^{-1}, assigned respectively to the anti-symmetric and symmetric νCOO−, suggest the presence of deprotonated carboxylate end groups (Bertilsson and Liedberg 1993; Briand et al. 2006a; Briand et al. 2006b). Thus, acidic functions are present under protonated and deprotonated chemical forms, both capable of reacting with EDC during the esterification process. In addition, the symmetric and antisymmetric stretching modes of the CH_2 groups of the back bone of the thiol molecule are clearly visible at 2925 and 2854 cm^{-1}, respectively. Eventually, PM-RAIRS spectra present bands related to the scissor mode of CH_2 groups at 1461 cm^{-1} and to the stretching of C–OH of protonated COOH moieties at 1243 cm^{-1}. Finally, there are some signs of residual contaminations in the MUA

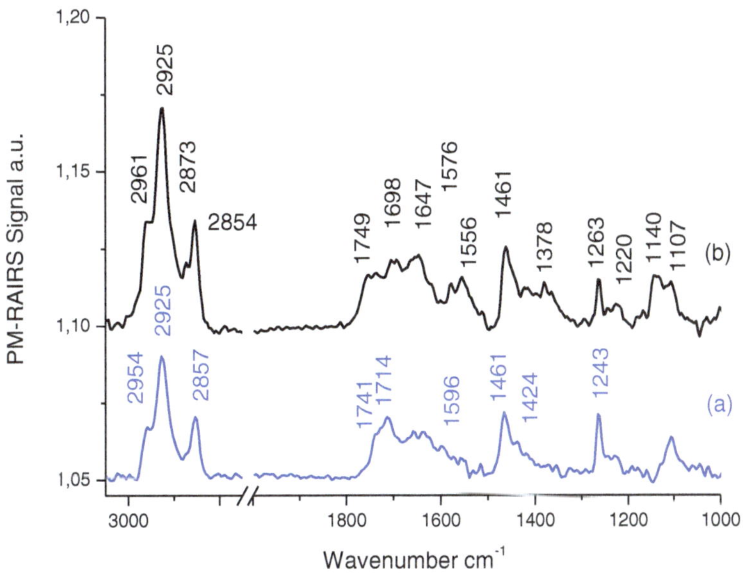

Figure 2 PM-RAIRS spectra of the two consecutive steps leading to the immobilization of EstC-NH_2: (a) Au–MUA; (b) covalent binding of EstC-NH_2 to MUA SAMs on gold surface.

self-assembled monolayers, with the presence of a weak band at 2954 cm^{-1} (usually assigned to CH_3 groups).

Spectrum (b), corresponding to the Au-MUA sample grafted with EstC-NH_2, exhibits differences with respect to the MUA layer, suggesting some reactions after 24 h of immersion in a solution containing EDC + EstC-NH_2. The binding of EstC-NH_2 is indicated by the presence of new IR features in the 1500 and 1750 cm^{-1} region. Intense peaks at 1647 and 1556 cm^{-1} are assigned to the amide I and amide II bands of the peptidic backbone formed between the activated COOH group and the NH_2 group of the EstC-NH_2 molecule. The grafting of EstC-NH_2 is also confirmed by the signature of the backbone of the molecules, mainly, the δOH at 1378 et 1263 cm^{-1}, ωCH2 at 1461 cm^{-1}, νC = O at 1698 and 1749 cm^{-1}, νCH2 and νCH3 symmetric and antisymmetric at 2961, 2925, 2873 et 2854 cm^{-1}, and probably the νC = C of the benzoic cycle à 1576 cm^{-1} also confirmed the successful grafting of the EstC-NH_2 onto the MUA-SAMs on gold surface.

XPS analysis

The general spectrum obtained by XPS spectroscopy as shown on Figure 3A for the gold surface functionalized with the EstC-NH_2, reveals a signal at ~400 eV corresponding to nitrogen (indicated on the figure as N1s) that is not present on the Au-MUA control (spectrum a), clearly suggesting the presence of the EstC-NH_2 molecules at the surface.

As shown on Table 1, the experimental percentages corresponding to the C, O, N and S elements are in good agreement with the estimated theoretical percentages. Sulphur is slightly under-estimated on the experimental gold surface, probably because of its position on the surface (see Figure 1). Nitrogen is also under-estimated for the benefit of carbon, maybe due to the molecular adsorption geometry undertaken by the EstC-NH_2 molecules when chemically grafted on the surface.

When looking at the high resolution XPS spectra of the mains elements, Core level C1s, O1s and N1s), the successful grafting of the molecules is confirmed. On the C1s XPS spectrum (Figure 3B (a)), all the expected components are found, in particular, the main peak shoulder at 287.0 eV, characteristic of C-O-C bindings that is quite represented in the EstC-NH_2 molecule. This signature also appears on the oxygen spectrum (Figure 3B (b)), with a key component at 533.2 eV. Finally, the nitrogen spectrum (Figure 3B (c)) is defined by two different components, one at 400.7 eV characteristics of both amide bindings (N-H) and the different nitrogen atoms on the unsaturated triazole cycle. The second contribution at 402.3 eV is not easily assignable and could be associated to a quaternary ammonium displaying a positive charge of the molecule.

The EstC-NH_2 grafting on the thiol MUA spacer through the formation of an amide binding was confirmed by the PM-RAIRS and XPS complementary experiments.

Antimicrobial activities of EstC-NH_2 functionalized gold samples

Culturability of Candida albicans after adherence on EstC functionalized gold samples

The culturability of adherent *C. albicans* ATCC 3153 on EstC-NH_2 functionalized gold surfaces was inhibited by

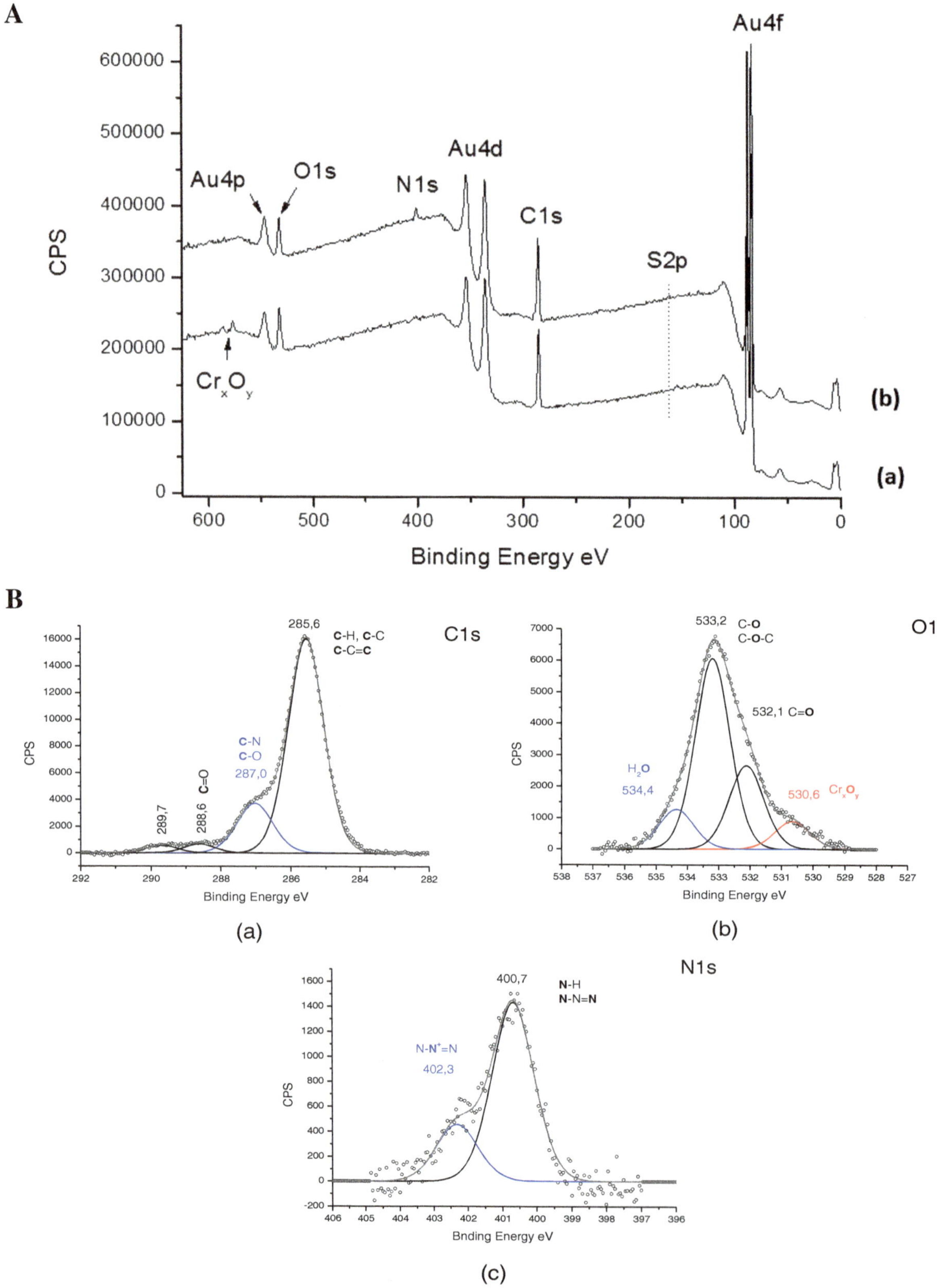

Figure 3 XPS general spectra of gold surfaces. A: XPS general spectra of gold surfaces functionalized (a) only with MUA (control) or (b) with the EstC-NH_2 molecule. **B:** XPS high-resolution spectra of gold EstC-NH_2 functionalized-surfaces for Carbon (a), for Oxygen (b) and for Nitrogen (c).

Table 1 Comparison between theoretical and experimental (XPS) percentages for C, O, N and S atoms, for the surface after both functionalization steps (Au_MUA-EstC-NH_2)

Atomic percentages	C	O	N	S
Theoretical	75.6%	13.3%	8.9%	2.2%
Experimental	81.0%	13.4%	4.3%	1.3%

76% ± 11% compared to the control Au-MUA (Figure 4A, n = 4). The experiment duration of 3 h is not long enough to permit biofilm formation.

Microscopic analysis

Confocal microscopy analyses were carried out using BacLight™ staining. Briefly, the cellular membrane of the yeasts which appeared red is damaged, whereas green cells have maintained their integrity and are considered as alive. Indeed, the red stain (PI) can only penetrate permeabilized cells, whereas the green one (Syto9) can cross over intact cellular membranes. Images obtained by confocal microscopy analyses are presented in Figure 5. Entire surfaces of three samples of each type of substrate were analysed. On the control surface, grafted with MUA only, *C. albicans* ATCC 3153 cells appeared as small green clusters corresponding to living yeasts (Figure 5A). As the yeast density differs according to the surface area that is observed under the microscope, several pictures were taken on the same surface and a mean density of each type of stained cells, red and green, was calculated for each surface after having manually counted both Syto9- and PI-stained cells. On the surface functionalized with

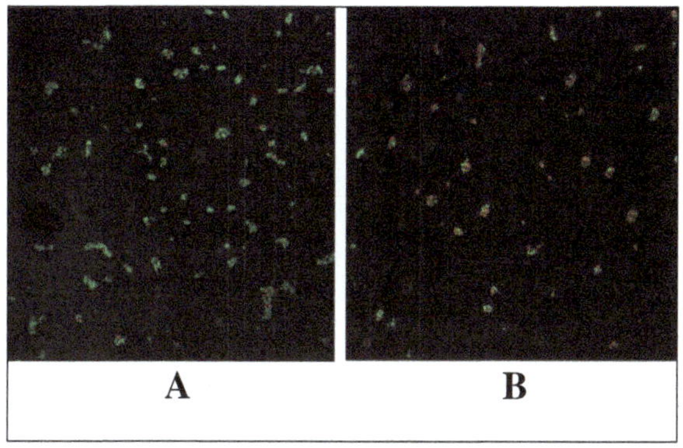

Figure 5 Viability of adherent *C. albicans* ATCC 3153, revealed by BacLight™ stain, attached to gold surfaces grafted with MUA (A) or with EstC-NH_2 (B) observed by confocal microscopy (magnification × 20). 10^6 Candida cells were deposited on the samples and allowed to adhere for 3 h.

EstC-NH_2, the number of total adherent cells was similar to the control Au-MUA but the number of red permeabilized and probably dead cells, increased by 60% comparing to the same control (Figure 4A and Figure 5B) (n = 2).

EtC.-NH_2 molecule grafting and surface characterization

PM-RAIRS analysis

Considering the spectrum b on Figure 6, EtC.-NH_2 was successfully grafted onto the MUA-SAMs on gold surface. Indeed the amid I and II bands (νC = O and δNH, respectively) are found at 1654 and 1545 cm^{-1}, showing the covalent bound formed between the activated COOH group and the NH_2 group of the EtC.-NH_2 molecule. On the same spectrum, one can observe the vibrations

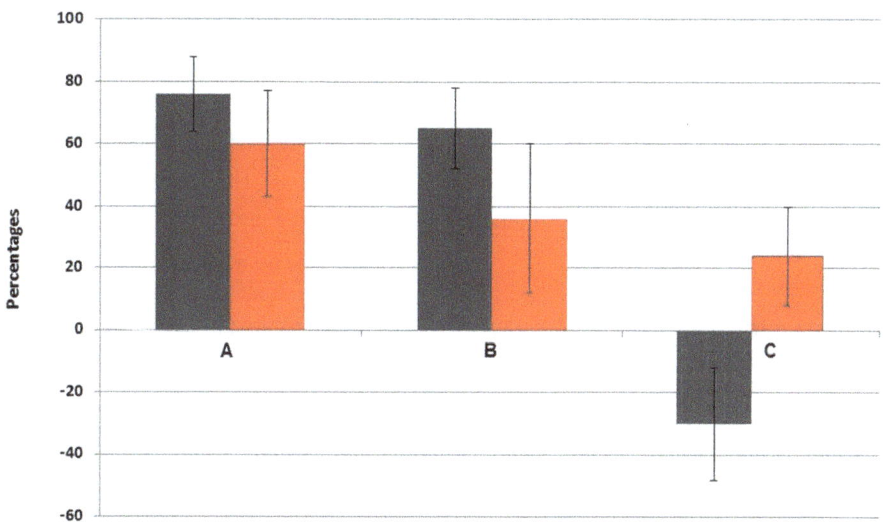

Figure 4 Percentages of inhibition of the culturability of *Candida albicans* after a 3 h of incubation on the different functionalized gold samples, A = EstC-NH_2, B = EtC.-NH_2 and C = Phe-NH_2 compared to the control MUA (grey) and percentages of the number of red stained cells compared to the control MUA (LIVE/DEAD® Bacterial Viability Kit (BacLight®) (red).

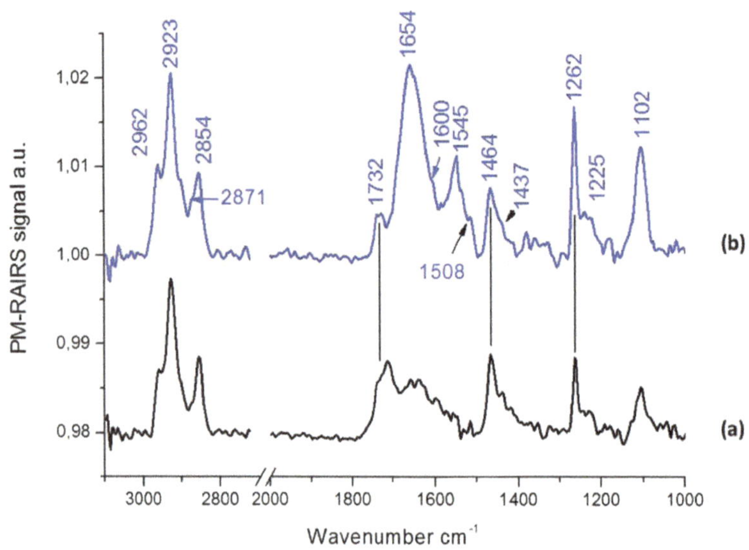

Figure 6 PM-RAIRS spectra of the two consecutive steps leading to the immobilization EtC.-NH_2. Spectrum (a) represents the control Au–MUA and spectrum (b) is associated to the covalent binding of the CAR derivative.

assigned to the Ether, with the νC = C and the νN = C-N respectively at 1600 cm^{-1} and 1508 cm^{-1} (also with a "shoulder" at 1545 cm^{-1}) and both νC-O-C symmetric and anti-symmetric stretching mode are found at 1102 cm^{-1} and at 1262 et 1225 cm^{-1}.

No further XPS experiments were performed at this stage as the comparison of PM-RAIRS results of EtC.-NH_2 with those of EstC-NH_2 was sufficient.

Antifungal activity and mode of action of EtC.- functionalized gold surfaces

The number of adhered culturable C. albicans ATCC 3153 was inhibited by 65% ± 13% on EtC.-functionalized gold surfaces compared to the control Au-MUA (Figure 4B; n = 4).

On the surface functionalized with EtC.-NH_2, the number of total adherent cells was similar to the control Au-MUA but the number of red permeabilized and probably dead cells increased by 36% comparing to the MUA control as shown on Figure 7 (n = 2) (See also Figure 4B).

Complementary functionalization of gold surfaces and its antifungal activity associated to the Phe-NH_2 molecule

PM-RAIRS

On spectrum b, on Figure 8, corresponding to the Phe-NH_2 molecule, the amid I and II bands (νC = O and δNH, respectively) are found at 1646 and 1553 cm^{-1} showing the covalent bound formed between the activated COOH group and the NH_2 group of the designed molecule. Moreover, the vibrations associated to the triazole cycle are observed, namely the νC = C at 1604 cm^{-1} and the weak bands νN = C-N at 1518 cm^{-1} (with a "shoulder" at 1533 cm^{-1}). Finally the characteristic νC-O-C symmetric and antisymmetric vibrations of the backbone of the molecules are found on one hand at 1105 and 1048 cm^{-1} and at 1262 and 1220 cm^{-1} on the other hand. PM-RAIRS measurements confirmed the grafting of Phe-NH_2 molecule.

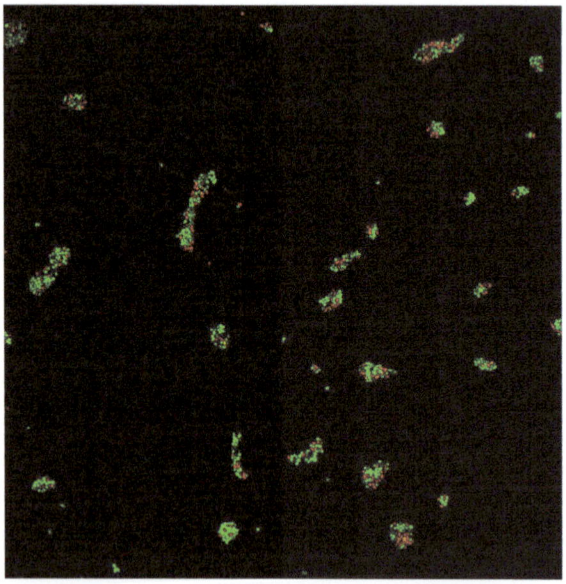

Figure 7 Viability of *C. albicans* ATCC 3153, revealed by BacLight™ stain, attached to gold surfaces grafted with EtC.-NH_2, observed by confocal microscopy (magnification × 20). 10^6 Candida cells were deposited on the samples and allowed to adhere for 3 h.

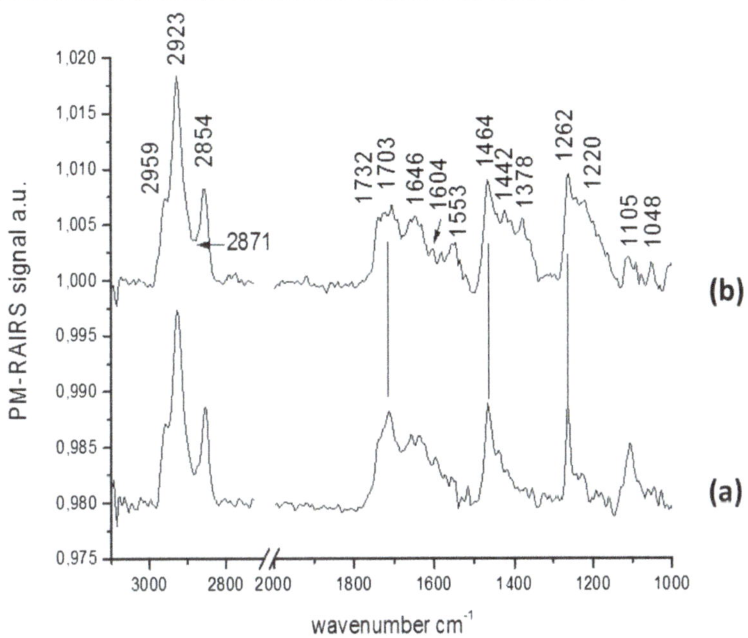

Figure 8 PM-RAIRS spectra of the two consecutive steps leading to the immobilization Phe-NH_2. Spectrum (a) represents the control Au-MUA and spectrum (b) is associated to the covalent binding of the CAR derivative.

Antifungal activity and mode of action of Phe - functionalized gold surfaces

The number of adhered culturable C. albicans ATCC 3153 was enhanced by a mean of 30% on Phe-functionalized gold surfaces (Figure 4C; n = 4) and not inhibited as in the cases of the two CAR derivatives.

On the surface functionalized with the Phe-NH_2 molecule, the number of adherent cells was similar to the control Au-MUA but the number of red permeabilized and probably dead cells increased by only 24% comparing to the MUA control as shown on Figure 9 (n = 2) (See also Figure 4C).

Activity of already used surfaces

To investigate the persistence of the antifungal activity of already used Au-MUA-NH-EstC surfaces, microbial viability after adherence was assayed after 1 month's storage at 4°C, using the same protocol as above. Samples were washed in pure water, dried under nitrogen, and stored at 4°C before the second assay. The culturability of adherent *C. albicans* ATCC 3153 on already-used EstC-NH_2 functionalized gold surfaces was still inhibited by 80% ± 13% compared to the control Au-MUA (10^6 *Candida* cells were deposited on the samples and allowed to adhere for 3 h; n = 2). On confocal microscopy pictures of already-used EstC-NH_2 surfaces, the number of total adherent cells was similar to the control Au-MUA but the number of red permeabilized and probably dead cells increased by 68% comparing to the same control (n = 2).

In order to estimate the amount of released CAR from the EstC-NH_2 functionalized surfaces, previously used surfaces were further analysed by PM-RAIRS, XPS and AFM. These assays revealed that there is nearly no loss of CAR on the surface after having been in contact with yeast cells (data not shown).

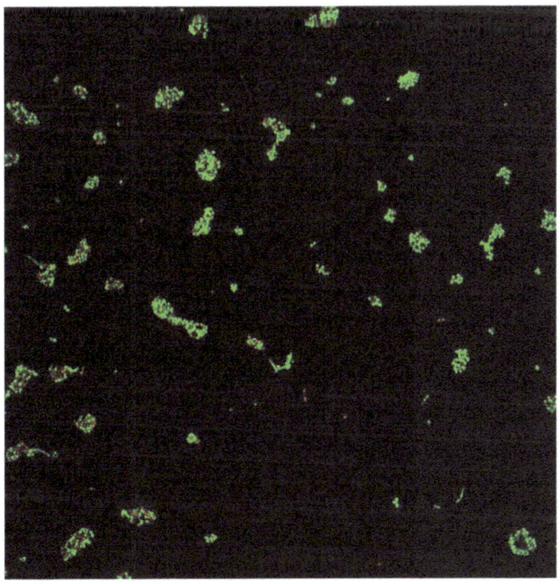

Figure 9 Viability of *C. albicans* ATCC 3153, revealed by BacLight™ stain, attached to gold surfaces grafted with Phe-NH_2, observed by confocal microscopy (magnification × 20). 10^6 Candida cells were deposited on the samples and allowed to adhere for 3 h.

Discussion

The idea behind this study was to anchor the volatile CAR on a surface in order to stabilize it while keeping its interesting antimicrobial/antifungal activity. Thus, two new molecules deriving from this terpene were specifically synthesized to be grafted on gold surfaces. They were named "Carvacrol Ester" or EstC-NH_2 and "Carvacrol Ether" or EtC.-NH_2. These two molecules both include a spacer which function is to take the active part of the molecule containing CAR away from the gold surface to increase its accessibility. A triazole cycle is present in the spacer structure. This triazole cycle results from the chemistry used to synthesize the EstC-NH_2 molecule (*i.e.* The Click Chemistry between an alkyne function and an azide function). The terminal part of the two constructs includes CAR which is linked to the rest of the molecule through an ester bond or an ether one. The first postulate of this work relied on the fact that the ester bond could be broken by esterase enzymes from microorganisms (*i.e.* yeasts or bacteria) arriving on the functionalized surface, thus allowing the liberation of CAR and its antifungal/antimicrobial activity. It is supposed to allow a customized release of CAR which should avoid a quick exhaustion of CAR on the surface or a loss of its activity often induced by a permanent binding. After CAR release, it could exert its antimicrobial activity on microorganisms, adherent or not, that are in its surrounding environment.

The other objective of this work was to design surfaces to which CAR is covalently attached and to study their antifungal activity. Once the functionalization of gold surface with EstC-NH_2 performed and checked by PM-RAIRS measurements, the inhibition of culturability of adherent *C. albicans* ATCC 3153 on these surfaces reached 76% ± 11% compared to the control Au-MUA. This decrease in the number of culturable cells attached to the functionalized surfaces can be explained by two processes. Either the grafting of EstC-NH_2 makes the surfaces antiadhesive thus preventing the yeasts from attaching onto the substrates, or it exerts an antifungal activity toward cells after their adhesion via a fungicidal or fungistatic mode of action. To answer this question, confocal microscopy analyses, using BacLight™ staining showed that on the surface functionalized with EstC-NH_2, the number of adherent cells was similar to the control Au-MUA but the number of red permeabilized cells, increased by more than 60%. The mode of action of these newly-developed EstC-NH_2 functionalized surfaces appears to be rather fungicidal but not antiadhesive.

In the same way, the loss of culturability on EtC.-NH_2 functionalized gold surfaces was 65% ± 13% and 36% more cells appeared red comparing to the MUA control. Thus, the mode of action of these EtC.-NH_2 functionalized surfaces appears to be fungicidal but less than the EstC-NH_2 functionalized one.

Consequently, the antifungal activity of both EstC-NH_2 and EtC.-NH_2 functionalized surfaces certainly results from the CAR molecule. This statement was reinforced by the experiments on Phe-NH_2 functionalized surfaces which did not impair the yeast culturability. Moreover, the activity seems to be mainly due to still fixed CAR that is not released by the action of esterases. As shown by PM-RAIRS, XPS and AFM experiments on previously used EstC-NH_2 functionalized surfaces, there is nearly no loss of CAR on the surface having been in contact with yeast cells (data not shown). As a consequence, it could be stated that CAR does not required to be released from the surface to exert its antifungal activity.

At this stage of the study, another question emerged concerning the process that governs the interesting fungicidal activity recorded for the two types of surfaces functionalized with CAR derivatives. Is this activity directly linked to the antifungal activity of CAR or is it due to the presence of the triazole cycle in the EstC and EtC. molecule backbones? Indeed, triazole moieties are essential components of the molecules that belong to the triazole class, a major group of systemic antifungals (Odds et al. 2003). However the non-activity of Phe-NH_2 functionalized surfaces demonstrated that the triazole cycle is not antifungal in this specific case.

Taking in consideration the methodology applied to prepare the surfaces and the possibility of having performed esterification and etherification reactions with a limited yield, it is possible that the grafting reaction does not cover the full treated surface. Therefore it could be considered that the obtained surfaces were not fully homogeneous, leading to domains free of CAR within the samples. Indeed, the confocal microscopy images displayed in Figures 5, 7 and 9 might indicate the formation of clusters of living (or dying) cells but without interconnection between colonies that could be in correlation with the heterogeneous distribution of CAR among the surface.

The persistence of the antifungal activity of already used Au-MUA-NH-EstC surfaces was verified after 1 month's storage at 4°C. Indeed, the culturability of adherent *C. albicans* was still inhibited by 80% ± 13% and confocal microscopy photographs, revealed that the number of red cells on already-used EstC-NH_2 surfaces increased by 68% comparing to the control. Thus, the antifungal and fungicidal activity of the newly-functionalized gold surfaces with the EstC-NH_2 molecule is conserved once used and sonicated even after being stored one month at 4°C which is a promising and desired property. Similar results were obtained previously for surfaces grafted with antimicrobial peptides Magainin and Gramicidin (Humblot 2009, Yala 2011) indicating that, as expected, the covalent grafting of antimicrobial molecules extends the shelf-life of antimicrobial materials.

To conclude, this proof of concept work demonstrated that gold surfaces grafted with carvacrol (CAR) integrated in specifically designed constructs displayed a promising antifungal activity. Confocal microscopy analyses highlighted that the newly-developed surfaces mode of action was fungicidal. It was shown, using a phenyl analogue replacing CAR as well as by replacing the ester bond by an ether linkage in the derivative structure, that the antifungal activity was directly linked to CAR. This result as well as the subsequent surface analysis of already-used functionalized gold samples indicated that the antifungal activity was finally not related to the cleavage of the ester bond and to the resulting release of CAR. Interestingly, the activity of the already-used grafted surfaces even persisted after the sonicating and washing steps. This study revealed that the immobilization of antimicrobial aromatic terpenic components, such as carvacrol, involved in an innovative chemical structure designed to overcome their high volatility and thus the resulting limits to their use, can considerably inhibit yeast adhesion and subsequently delay biofilm formation. The developed surfaces were first tested on *C. albicans*. As terpenes are known to exert both an antimicrobial and antifungal activity, one next step will be to further investigate the activity of these new functionalized samples on other *Candida* and yeast species but also on bacteria using experimental conditions mimicking real biological conditions. Prospects before making these new surfaces usable in clinical practice are many.

Additional file

Additional file 1: Carvacrol derivatives preparation.

Competing interests
None of the authors of this paper has a financial or personal relationship with other people or organisations that could inappropriately influence or bias the content of the paper.

Authors' contributions
AG carried out the microbiological tests, participated to the functionalization of surfaces and drafted the manuscript. TL carried out the chemical syntheses of the derivatives moieties. VH carried out the surfaces analysis and participated to the design surfaces modifications, SB designed the chemical synthesis of carvacrol derivatives, JMB conceived of the study, and participated in its design and coordination and helped to draft the manuscript. All authors read and approved the final manuscript.

Acknowledgements
We thank Anne Cantereau from ImageUP (Université de Poitiers) for confocal microscopy analyses.

Author details
[1]Ecologie & Biologie des Interactions - UMR CNRS 7267, Microbiologie de l'eau, Université de Poitiers, 1 Rue Georges Bonnet, TSA 51106, 86073 Poitiers Cedex 9, France. [2]UMR-CNRS 7285, Groupe Systèmes Moléculaires Programmés, Université de Poitiers, Poitiers, France. [3]UPMC Université Paris 06, UMR CNRS 7197, Laboratoire de Réactivité de Surface, Sorbonne Universités, Paris, France.

References
Abraham GA, de Queiroz AAA, San Roman J (2002) Immobilization of a nonsteroidal antiinflammatory drug onto commercial segmented polyurethane surface to improve haemocompatibility properties. Biomaterials 23(7):1625–1638, doi:10.1016/s0142-9612(01)00289-7
Ahmad A, Khan A, Akhtar F, Yousuf S, Xess I, Khan LA, Manzoor N (2011) Fungicidal activity of thymol and carvacrol by disrupting ergosterol biosynthesis and membrane integrity against Candida. Eur J Clin Microbiol Infect Dis 30(1):41–50, doi:10.1007/s10096-010-1050-8
Andris DA, Krzywda EA, Edmiston CE, Krepel CJ, Gohr CM (1998) Elimination of intraluminal colonization by antibiotic lock in silicone vascular catheters. Nutrition 14(5):427–432, doi:10.1016/s0899-9007(98)00013-6
Aslam S (2008) Effect of antibacterials on biofilms. Am J Infect Control 36(10):S175.e9-11
Bain CD, Troughton EB, Tao YT, Evall J, Whitesides GM, Nuzzo RG (1989) Formation of monolayer films by the spontaneous assembly of organic thiols from solution onto gold. J Am Chem Soc 111(1):321–335
Bertilsson L, Liedberg B (1993) Infrared study of thiol monolayer assemblies on gold: preparation, characterization, and functionalization of mixed monolayers. Langmuir 9(1):141–149, doi:10.1021/la00025a032
Braga PC, Culici M, Alfieri M, Dal Sasso M (2008) Thymol inhibits Candida albicans biofilm formation and mature biofilm. Int J Antimicrob Agents 31(5):472–477, doi:10.1016/j.ijantimicag.2007.12.013
Briand E, Salmain M, Compère C, Pradier C-M (2006a) Immobilization of Protein A on SAMs for the elaboration of immunosensors. Colloids and Surfaces B: Biointerfaces 53(2):215–224 http://dx.doi.org/10.1016/j.colsurfb.2006.09.010
Briand E, Salmain M, Herry J-M, Perrot H, Compère C, Pradier C-M (2006b) Building of an immunosensor: How can the composition and structure of the thiol attachment layer affect the immunosensor efficiency? Biosensors and Bioelectronics 22(3):440–448 http://dx.doi.org/10.1016/j.bios.2006.05.018
Costa F, Carvalho IF, Montelaro RC, Gomes P, Martins MCL (2011) Covalent immobilization of antimicrobial peptides (AMPs) onto biomaterial surfaces. Acta Biomater 7(4):1431–1440 doi:351433415,12,1
Cowen LE (2008) The evolution of fungal drug resistance: modulating the trajectory from genotype to phenotype. Nat Rev Microbiol 6(3):187–198, doi:10.1038/nrmicro1835
Dalleau S, Cateau E, Berges T, Berjeaud J-M, Imbert C (2008) *In vitro* activity of terpenes against *Candida* biofilms. Int J Antimicrob Agents 31(6):572–576 http://dx.doi.org/10.1016/j.ijantimicag.2008.01.028
Dwyer A (2008) Surface-Treated Catheters-A Review. Semin Dial 21(6):542–546, doi:10.1111/j.1525-139X.2008.00499.x
Goddard JM, Hotchkiss JH (2007) Polymer surface modification for the attachment of bioactive compounds. Prog Polym Sci 32(7):698–725, doi:10.1016/j.progpolymsci.2007.04.002
Gow NAR, Brown AJP, Odds FC (2002) Fungal morphogenesis and host invasion. Curr Opin Microbiol 5(4):366–371, doi:10.1016/s1369-5274(02)00338-7
Gutierrez L, Batlle R, Sanchez C, Nerin C (2010) New Approach to Study the Mechanism of Antimicrobial Protection of an Active Packaging. Foodborne Pathog Dis 7(9):1063–1069, doi:10.1089/fpd.2009.0516
Hall K, Farr B (2004) Diagnosis and management of long-term central venous catheter infections. J Vasc Interv Radiol 15(4):327–334, doi:10.1097/01.rvi.0000121405.46920.87
Humblot V, Yala JF, Thebault P, Boukerma K, Hequet A, Berjeaud JM, Pradier CM (2009) The antibacterial activity of Magainin I immobilized onto mixed thiols Self-Assembled Monolayers. Biomaterials 30(21):3503–3512, doi:10.1016/j.biomaterials.2009.03.025
Komnatnyy VV, Chiang W-C, Tolker-Nielsen T, Givskov M, Nielsen TE (2014) Bacteria-Triggered Release of Antimicrobial Agents. Angew Chem Int Ed Engl 53(2):439–41, doi:10.1002/anie.201307975
McCullough MJ, Ross BC, Reade PC (1996) Candida albicans: A review of its history, taxonomy, epidemiology, virulence attributes, and methods of strain differentiation. Int J Oral Maxillofac Surg 25(2):136–144, doi:10.1016/s0901-5027(96)80060-9
Miceli MH, Diaz JA, Lee SA (2011) Emerging opportunistic yeast infections. Lancet Infect Dis 11(2):142–151, doi:10.1016/s1473-3099(10)70218-8
Odds FC, Brown AJP, Gow NAR (2003) Antifungal agents: mechanisms of action. Trends Microbiol 11(6):272–279, doi:10.1016/s0966-842x(03)00117-3

Pavel M, Ristic M, Stevic T (2010) Essential oils of Thymus pulegioides and Thymus glabrescens from Romania: chemical composition and antimicrobial activity. J Serb Chem Soc 75(1):27–34, doi:10.2298/jsc1001027p

Pearson ML (1996) Guideline for prevention of intravascular-device-related infections. Infect Control Hosp Epidemiol 17(7):438–473

Pemàn J, Canton E, Valentin A (2008) Actividad de la anidulafungina sobre biopeliculas de *Candida*. Revista Iberoamericana de Micologia 25:124–128

Pierce CG, Uppuluri P, Tristan AR, Wormley FL, Mowat E, Ramage G, Lopez-Ribot JL (2008) A simple and reproducible 96-well plate-based method for the formation of fungal biofilms and its application to antifungal susceptibility testing. Nat Protoc 3(9):1494–1500, doi:10.1038/nprot.2008.141

Ramage G, Mowat E, Jones B, Williams C, Lopez-Ribot J (2009) Our Current Understanding of Fungal Biofilms. Crit Rev Microbiol 35(4):340–355, doi:10.3109/10408410903241436

Roe D, Karandikar B, Bonn-Savage N, Gibbins B, Roullet JB (2008) Antimicrobial surface functionalization of plastic catheters by silver nanoparticles. J Antimicrob Chemother 61(4):869–876, doi:10.1093/jac/dkn034

Scofield JH (1976) Hartree-slater subshell photoionization cross-sections at 1254 AND 1487EV. J Electron Spectrosc Relat Phenom 8(2):129–137, doi:10.1016/0368-2048(76)80015-1

Seneviratne CJ, Jin L, Samaranayake LP (2008) Biofilm lifestyle of Candida: a mini review. Oral Dis 14(7):582–590, doi:10.1111/j.1601-0825.2007.01424.x

Silva S, Negri M, Henriques M, Oliveira R, Williams D, Azeredo J (2010) Silicone colonization by non-Candida albicans Candida species in the presence of urine. J Med Microbiol 59(7):747–754, doi:10.1099/jmm. 0.017517-0

Tenke P, Jackel M, Nagy E (2004) Prevention and Treatment of Catheter-Associated Infections: Myth or Reality ? EAU Update Series 2:106–115

Tielens F, Costa D, Humblot V, Pradier C-M (2008) Characterization of ω-Functionalized Undecanethiol Mixed Self-Assembled Monolayers on Au(111): A Combined Polarization Modulation Infrared Reflection – Absorption Spectroscopy/X-ray Photoelectron Spectroscopy/Periodic Density Functional Theory Study. J Phys Chem C 112(1):182–190, doi:10.1021/jp074023c

Tiller JC, Liao CJ, Lewis K, Klibanov AM (2001) Designing surfaces that kill bacteria on contact. Proc Natl Acad Sci U S A 98(11):5981–5985, doi:10.1073/pnas.111143098

Timofeeva L, Kleshcheva N (2011) Antimicrobial polymers: mechanism of action, factors of activity, and applications. Appl Microbiol Biotechnol 89(3):475–492, doi:10.1007/s00253-010-2920-9

von Eiff C, Jansen B, Kohnen W, Becker K (2005) Infections associated with medical devices - Pathogenesis, management and prophylaxis. Drugs 65 (2):179–214, doi:10.2165/00003495-200565020-00003

Walker LA, Gow NAR, Munro CA (2010) Fungal echinocandin resistance. Fungal Genet Biol 47(2):117–126, doi:10.1016/j.fgb.2009.09.003

Williams DW, Kuriyama T, Silva S, Malic S, Lewis MA (2011) Candida biofilms and oral candidosis: treatment and prevention. Periodontol 2000 55:250–265, doi:10.1111/j.1600-0757.2009.00338.x

Xu J, Zhou F, Ji BP, Pei RS, Xu N (2008) The antibacterial mechanism of carvacrol and thymol against Escherichia coli. Lett Appl Microbiol 47(3):174–179, doi:10.1111/j.1472-765X.2008.02407.x

Yala JF, Thebault P, Hequet A, Humblot V, Pradier CM, Berjeaud JM (2011) Elaboration of antibiofilm materials by chemical grafting of an antimicrobial peptide. Appl Microbiol Biotechnol 89(3):623–634, doi:10.1007/s00253-010-2930-7

Zhang L, Gowardman J, Rickard CM (2011) Impact of microbial attachment on intravascular catheter-related infections. Int J Antimicrob Agents 38(1):9–15, doi:10.1016/j.ijantimicag.2011.01.020

13

Cultivation and detection of endophytic aerobic methanotrophs isolated from *Sphagnum* species as a perspective for environmental biotechnology

Zofia Stępniewska and Agnieszka Kuźniar*

Abstract

Enriched cultures of microorganisms are an essential step in the production of inoculum of these organisms for biotechnology and bioengineering. The potential application of methanotrophic microorganisms for removal of methane produced from landfills and coal mines as well as biodegradation of toxic compounds has been widely studied. Therefore, searching for new sources of methanotrophs can contribute to increasing the possibilities of biotechnology and bioengineering.

Enrichment cultures of endophytic methanotrophs from *Sphagnum* sp. were initiated in NMS medium, a most widely used medium for cultivation of methanotrophic bacteria from various environments proposed in 1970 by Whittenbury. Incubation was carried out at 10, 20, 30, and 37°C with vigorous shaking on a shaker (180 rpm). The source of carbon and energy for endophytes were methane at the concentration range between 1-20%.

It appeared that the consortium of endophytic bacteria grew only at the temperature of 20 and 30°C. During the culture of endophytes, the measurements of gas concentration showed a steady loss of methane and oxygen, as well as accumulation of carbon dioxide as a CH_4 oxidation product.

The use of FISH has made characterization of endophytic consortia possible. It turned out that the population of endophytes consists of type I and II methanotrophs as well as associated non-methanotrophic bacteria. Furthermore, we determined the potential of the examined bacteria for methane oxidation, which ranged up to 4,7 μMCH_4 per ml of the population of endophytes per day.

Keywords: Endophytic bacteria; Methanotrophs; Methane; *Sphagum* sp

Introduction

The history of the use of microorganisms by man is as old as the human civilization itself. Microorganisms have long served humans in industrial applications e.g. production of food, drug, and cosmetics. Recently, progress can be seen in the use of microorganisms for environmental biotechnology, namely removal of greenhouse gases from various sites (landfills, coal mines), biodegradation of toxic compounds, wastewater treatment etc. (Hamer 2010). Each species of microorganisms found in the earth has its own specific requirements of nutrients and growth conditions that are often difficult to mimic in the culture media in the laboratory. Therefore, one of the most up-to-date trends in environmental biotechnology is seeking for new sources of microorganisms with a wide range of growth conditions and, especially, the possibility of utilizing waste material for cultivation (Minamisawa 2006).

A unique and fascinating group of microorganisms are methanotrophic bacteria, which were discovered over a century ago and yet are still of great interest. Aerobic methanotrophic bacteria, which use methane as the sole source of carbon and energy, act as a major methane sink. Methanotrophic bacteria have been studied in soils, deserts, landfills, tundra, wetlands, rice paddies, sediments, lakes, and marine environments (Hanson and Hanson 1996), as well as in the atmosphere (Santl-Temkiv et al. 2013) and coal mines (Stępniewska et al. 2006).

Methanotrophs have significant potential for applied microbiology and biochemical engineering, including bioremediation of pollutants (e.g. halogenated hydrocarbons) via co-metabolism by MMOs, biotransformation of diverse

* Correspondence: agawoloszyn@kul.pl
Department of Biochemistry and Environmental Chemistry, The John Paul II Catholic University of Lublin, Konstantynow 1I, 20-708 Lublin, Poland

organic substrates (e.g., propylene to epoxypropane, production of chiral alcohols), assimilation of methane to mitigate greenhouse effects, and production of commercially relevant compounds (e.g., single cell protein, poly-_ hydroxybutyrate, astaxanthin). Therefore, engineering of methanotrophs is very important to their industrial applications. Recent years have seen significant progress in functional genomics and proteomics of methanotrophs (Hamer 2010; Jiang et al. 2010; Trotsenko and Murrell 2008). Understanding of natural microbially mediated processes has been severely retarded by the common requirement to study microorganisms only as pure monocultures under aseptic conditions, in spite of the fact that, in all real environments, microbial strains function in a community (Hamer 2010).

Based on basic research and deep knowledge of methanotrophs, there are more possibilities to make methanotrophs become important and universal industrial microorganisms (Jiang et al. 2010). Therefore, these studies focused on recognition of the conditions of culturing aerobic endophytic methanotrophs and their application in environmental biotechnology and bioengineering.

The aim of our present study was to investigate a new perspective of the use of *Sphagnum* endophytic methanotrophs in environmental biotechnology.

Material and methods

Plant material

Sphagnum sp. materials were selected from Moszne peatbog (the Poleski National Park, Eastern Poland). The samples of the crops were collected in September 2011 from four different sites (meadow peat soil, bog forest).

Isolation of endophytic methanotrophs

The endophytic bacteria were isolated from young *Sphagnum* sp. plants. Randomly selected plants were uprooted manually and washed in running tap water. Leaf sections of 1–2 cm length were excised using a flame-sterilized scalpel. The samples were blotted dry with filter paper; next, surface sterilization of shoot and root pieces was performed with the following immersion sequence: 70% ethanol for 1 min and 3% sodium hypochlorite for 5 min. Then, they were rinsed four times with sterile water and dried in laminar flow. The cut ends of surface sterilized segments were removed with a flame-sterilized scalpel and placed in appropriate NMS solid medium, i.e. a most widely used medium for cultivation of methanotrophic bacteria from various environments proposed in 1970 by Whittenbury, with the cut surface touching the medium. Due to the volatile nature of the essential metabolic substrates used by methanotrophs (CH_4, O_2), the cultures were grown in glass bottles with a capacity of 120 cm^3, equipped with a hermetically closed cover, allowing application and collection of gases. The glass bottles were incubated for six to ten days at 30°C. The growth of methanotrophic bacteria was stimulated by supplying CH_4 (10%v/v) for the cultivation. The proportion between the solid medium and gaseous culture was always 1:5. Subsequently, a single colony was transferred to 120 cm^3 bottles containing liquid NMS medium and the cultures were incubated at 10°C, 20°C, and 30°C with 180 rpm shaking. Meanwhile, the concentration of bacterial cells in the suspension was determined spectrophotometrically by absorbance at 600 nm. Concurrently, methane consumption in the headspace was measured with a gas chromatograph equipped with a flame ionization detector, a thermal conductivity detector, and an electron capture detector (SIMADZU, GC 2010). Nitrogen and helium were used a carrier gas (30 mL min−1) and the injector, oven, and detector temperatures were 250°C (FID) and 150°C (TCD). The flame gases including H_2 and compressed air were set at 20 and 30 mL min−1, respectively.

It has been demonstrated that in natural conditions methanotrophic bacteria cooperate with other microorganisms and their pure cultures are unstable for extended periods of time (Hoefman et al. 2010). In these studies, we investigated a consortium of whole microbial communities.

Morphology and cell shape

The isolated population of endophytic methanotrophs (actively growing) were collected by centrifugation, connected with 0,2% phosphotungstic acid in the ratio 1:1. Then, the mixtures were transferred onto a copper grid covered with a formvar film. After drying the grid, photographs of endophytic methanotrophs were taken with the use of electron microscopy techniques (LEO 912AB) with an electron energy filter. *Methylosinus trichosporium* and *Methylomonas methanica* originating from Russian Academy of Sciences, Institute of Biochemistry and Physiology of Microorganisms were used as a positive control.

Identification of the endophytic bacterial population

DNA extraction

Total DNA was extracted from enrichment cultures where visual turbidity developed using the method described previously by Sambrook and Russell (2001) with some modifications. Cells from 10 ml samples of late-exponential cultures were collected by centrifugation. The pellet was suspended in 250μl of TE buffer containing 50mM Tris–HCl (pH=8.0) and 50mM EDTA (pH=8.0). To achieve complete lysis of the cells, 1 ml of GES buffer (pH=8.0) containing 5M guanidine thiocyanate, 100mM EDTA, and 0.5% sarkosyl was added. The mixture was incubated at room temperature for 10 minutes and then "crude lysates" were cooled on ice. After addition of 125μl of ammonium acetate (7.5M), the samples were mixed and further incubated on ice. The DNA obtained was purified with 250µl of a chloroform-isoamyl alcohol (24:1) mixture, precipitated with isopropanol, washed with cold ethanol, and dissolved in 50µl of sterile distilled water.

PCR amplification

Specific primers for functional genes and 16S rRNA of the methanotrophic bacteria synthesized at GENOMED, Warsaw (Poland), were used. Polymerase chain reactions (PCR) were run in a programmable thermal cycler (MJmini, Bio-Rad). The reaction mixture (50µl) consisted of 1xPCR MIX: PCR amplification buffer, 0.05 U/µl Taq DNA polymerase, 0.4mM of each dNTP, 4mM $MgCl_2$ (FERMENTAS), forward and reverse primers at 0.1mM and 3.5 µl of template DNA. The reaction conditions consisted of initial denaturation at 96°C for 4 minutes, 30 cycles of 94°C for 2 min, primer annealing at 56°C for type I and 55°C for type II each for 1 min, and elongation at 72°C for 1 min. Final elongation was performed at 72°C for 3 minutes. The amplification products were analyzed by electrophoresis in 1% agarose gel and stained with ethidium bromide.

The sequencing processes were performed on the purified product immediately after the PCR reaction in the Laboratory of DNA Sequencing and Oligonucleotide Synthesis (GENOMED, Warsaw, Poland). The sequences obtained were compared to the closest relatives in the GenBank database using the BLAST program.

All sequences have been deposited in GenBank under accession numbers: KJ657737 - KJ657745 and KJ623261, KJ 1713769.

FISH

Fixation procedure

Cells growing in the logarithmic phase were harvested by centrifugation and resuspended in 0.5ml of phosphate-buffered saline (PBS). The suspensions were then mixed with 1ml of 4% (w/v) freshly prepared paraformaldehyde solution and fixed for 1 h at room temperature. The fixed cells were collected by centrifugation (6000•g for 1min) and washed twice with PBS to ensure removal of paraformaldehyde. The resulting pellet was resuspended in 0.3 ml of 99.9% (v/v) ethanol and stored at 20°C until use. For use in FISH, probe Mγ84: AGCCCGCGACTGCTCACC (type I), Mγ705: CTAGACTTCCTTGTGGTC (type I), and Mα450: CTATTACTGCCATGGACCTA (type II) was labeled with indocarbocyanine dye (Eller et al. 2001). The oligonucleotide probes were synthesized and labeled with fluorochromes (CY3, CY5) by MWG Biotech (Ebersberg, Germany).

Hybridization was done on 75% ethanol-rinsed and dried slides with eight wells for independent positioning of the samples. Approximately 4µl of the fixed cell suspension was spread on each well, air dried, and dehydrated by successive passages through an ethanol series (50, 75, and 99.9% (v/v)) for 3 min each. A 50ml polypropylene Falcon tube containing a slip of filter paper soaked in hybridization buffer was used as a hybridization chamber, as described by Stahl and Amann 1991. Hybridization buffer (10µl) containing 1M Tris–HCl (pH=8), 5M NaCl, 10% sodium dodecyl sulfate (SDS), and 20% formamide was placed on each spot of the fixed cells and then 1µl of fluorescent probe solution was added. The chamber was incubated for 60 min at the hybridization temperature. Then, the slides were washed at the hybridization temperature for 15 min in washing buffer (1M Tris–HCl, 0.5M EDTA, and 5M NaCl) and rinsed twice with distilled water. The slides were air dried, stained with 4,6- diamidino-2-phenylindole (DAPI; 2 M) for 10 min in the dark, rinsed again with distilled water, and finally air dried.

Microscopy

The slides were analyzed by fluorescent microscope techniques (Nikon Eclipse 80i). The pictures were taken with a Digital Sight camera (Nikon) and processed using software program.

Results

Culture growth

It appeared that the consortium of endophytic bacteria grew at the temperature of 20 and 30°C, but not at 10°C and 37°C (Figure 1). A steady loss of methane and oxygen was shown, as well as accumulation of carbon dioxide as a CH_4 oxidation product according to the equation:

$$CH_4 + 2O_2 \xrightarrow{MMO} CO_2 + H_2O + 780\ kJ\ mol^{-1} + \text{biomass}$$

During the initial phase, adaptation of endophytes ranged from 5 to 8 days at 30°C and up to 12 days at 20°C, which can be due to cellular metabolism at lower temperature (Figure 2). Then, there was rapid growth of the endophytic population (logarithmic phase), which was reflected by an increase in optical density in the range from 0.3 to 2.0 at 30°C and from 0.25 to 1.4 at 20°C, depending on the methane concentration (1-20%). After the logarithmic phase, lack of carbon/energy sources was observed and an increase in the concentration of waste products to a harmful level for methanotrophic consortia (about day 13, Figure 2). This time was referred to as the stationary phase of the enrichment cultures.

Methanotrophic activity of the endophytic population

The CH_4 consumption in the enrichment culture (at 20 and 30°C) from the four localizations of *Sphagnum* sp. showed a clear decrease in the CH_4 concentration tested in the headspace after 12–13 days of incubation (Figure 1).

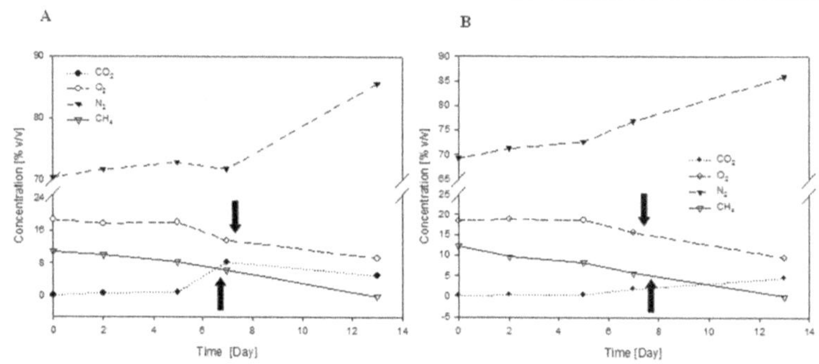

Figure 1 The dynamics of gases during the growth of the methanotrophic community, at 30oC (A) and 20oC (B) under 10% CH4 in the headspace.

The highest CH_4 consumption, below 20% of the initial CH_4 concentration, was found in the isolated population from *Sphagnum magellanicum* (4,7 μMCH_4 per ml liquid culture per day, Figure 3) at 30°C. No significant correlation in methanotrophic activity between the 20 and 30°C temperatures (UMW, $p=0.89$) was observed and no statistically significant differences between the activity of methanotrophic bacteria from the different species of *Sphagnum* sp. (UMW, $p=0.85$) were confirmed. A linear relationship with a high correlation coefficient ($R^2=0.99$) between the methanotrophic activity of *Sphagnum* sp. endophytes and the methane concentration (in the range of CH_4 from 0 to 10%, Table 1) was shown. The higher concentration of methane (over 10%) did not increase the methanotrophic activity of endophytes isolated from *Sphagnum* sp.. Furthermore, at a concentration of 20% CH_4 it was observed to be even slightly lower, as in the case of the *S. magellanicum* M2 endophytes. The results obtained suggest that at 10% of methane complete saturation of the methane monooxygenase active center took place.

Characterization of the endophytic isolates

Transmission electron microscopy (TEM) showed that the SM2 and SM4 populations consisted of single coccoid cells that showed a cell wall typical of Gram-negative bacteria. In addition, cells from all the tested endophyte populations had a rod shape. SM2 and SA3 isolates exhibited a motile character. TEM microscopy (negatively stained cells) indicated that the cells of all the tested populations had extracellular structures called nanopods with a length of ca. 480 nm (Figure 4, Table 2).

Fluorescence in situ hybridization

The endophytic consortia were characterized with the use of FISH (**F**luorescence **i**n **s**itu **H**ybridization) analysis. The cultured population of endophytic microorganisms consisted of

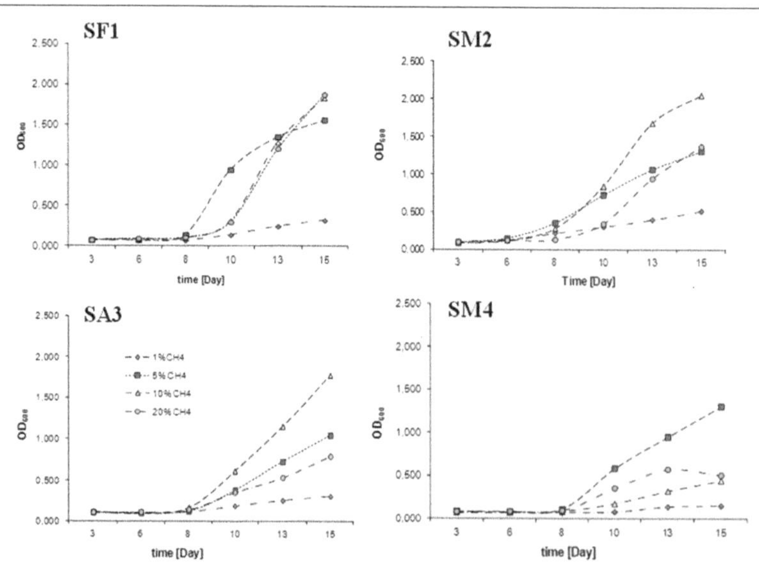

Figure 2 The growth curves of methanotrophic consortium cultures (at different 1-20% concentrations of CH4) isolated from leaves of *Sphagnum* sp. gametophytes: SF1 – *S. flexuosum* , SM2– *S. magellanicum* 2, SA3– *S. fallax* , SM4 – *S. magellanicum* 4.

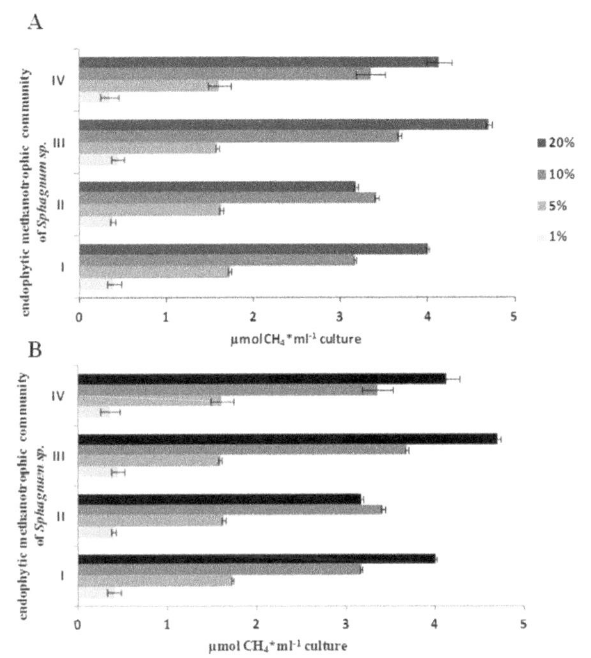

Figure 3 Methanotrophic activity (MA) of the endophytic metanotrophic population obtained from leaves of *Sphagnum* sp gametophyte determined at 20oC (A) and 30oC (B) under different concentrations of methane (1- 20%). I – *S. flexuosum* M1, II – *S. magellanicum* M2, III – *S. fallax* M3, IV – *S. magellanicum* M4.

type I, which were the dominant population (pink cells), and type II methanotrophs (green cells), as well as associated non-methanotrophic bacteria (dark blue cells, Figure 5).

Identification of the isolated endophytes

The methanotrophic communities isolated from *Sphagnum* mosses were investigated with use of general taxonomic markers like 16S rRNA, specific for type I and II methanotrophs. The 16S rRNA PCR products obtained (after purification, type I – 700 bp, type II – 530 bp) were sequenced using genomic DNA from *Sphagnum* mosses as a template. Then similarity of the tested sequences to nucleotide sequences from the GenBank database was assessed.

Table 1 Influence of the methane concentration (x) on the methanotrophic activity of endophytes (MA, y) determined at 20°C and 30°C (n = 8)

Population	AM (μMCH$_4$ ml^{-1} liquid culture day^{-1})		R^2	
	20°C	30°C	20°C	30°C
SF1	y=0.3063x+0.1316	y=0.3063x+0.1315	0.99	0.99
SM2	y=0.337x+0.0147	y=0.337x+0.0147	0.99	0.99
SA3	y=0.3618x+0.0268	y=0.3618x+0.0268	0.99	0.99
SM4	y=0.3336x+0.0088	y=0.3336x+0.0089	0.99	0.99

The identification results revealed the presence of both type I and type II methanotrophs. Among cultivable methanotrophs type I, there were different strains of the genus *Methylomonas*, whereas methanotrophs type II were represented by cultured strains belonging to the genera *Methylocystis* and *Methylosinus*. Molecular identification with the use of BLAST revealed high similarity of the methanotrophs cultured on the NMS medium to the genus *Methylocystis* (GenBank:. KJ623261, KJ657737, KJ657742) in the population isolated from *Sphagnum flexuosum* (99%), *Methylomonas* (GenBank: KJ657738, KJ657739, KJ657740, KJ657743) *Methylocystis* (GenBank: KJ 1713769) in the community isolated from *S. magellanicum* (94-100%), and to the genera *Methylocystis*, and *Methylosinus* (GenBank: KJ657744, KJ657745) in the population isolated from *S. fallax* (96-98%, Table 3).

Discussion

Enriched cultures of microorganisms are an essential step in the production of inoculum of different organisms for biotechnology and bioengineering. In our studies, we showed an ability of the endophytic bacterial community to be cultured only at 20 and 30°C (Figure 1, 3). Most of methanotrophs hitherto isolated were classified as mesophilic (Hanson and Hanson 1996). There were registered isolates belonging to the psychrophile group, whose optimum growth was reported at 3.5-10°C and a slight growth at 20°C (Omel'chenko et al. Omel'chenko et al. 1993; Trotsenko and Khmelenina 2005). Methanotrophic microorganisms belonging to the thermophilic/thermostable group, whose optimum growth was determined in the range of 40 - 65°C, have been identified (Trotsenko et al. 2009). It is worth noting that the psychrophilic methanotrophs do not occur in the temperate climate zone, subtropical and tropical zones.

The use of the enrichment culture of endophytes allowed determination of the length of the adaptation phase (Figure 2) and the rate of growth at different concentrations of methane, which may be a forecast in the use of these cultures for biotechnology and significantly reduce the time to adapt in successive cycles of life. Analysis of methanotrophic activity showed that optimal oxidation of methane was found below 10% v/v of the initial methane concentration and at 30°C. Next, the culture was run at higher availability of methane from 3.167 to 3.678 μM CH$_4$ per ml liquid culture and per day (Figure 3). If we apply the culture to pilot-scale bioreactors (200 dm^3), we need to provide only from 0.633 mM to 0.736 mM CH$_4$ per day for methanotrophic endophytes isolated from different species of *Sphagnum* sp. As a result, we believe that the culture of methanotrophic endophytes will be promising for environmental biotechnology. Besides, methane, which is the sole source of carbon and energy for these bacteria, is often

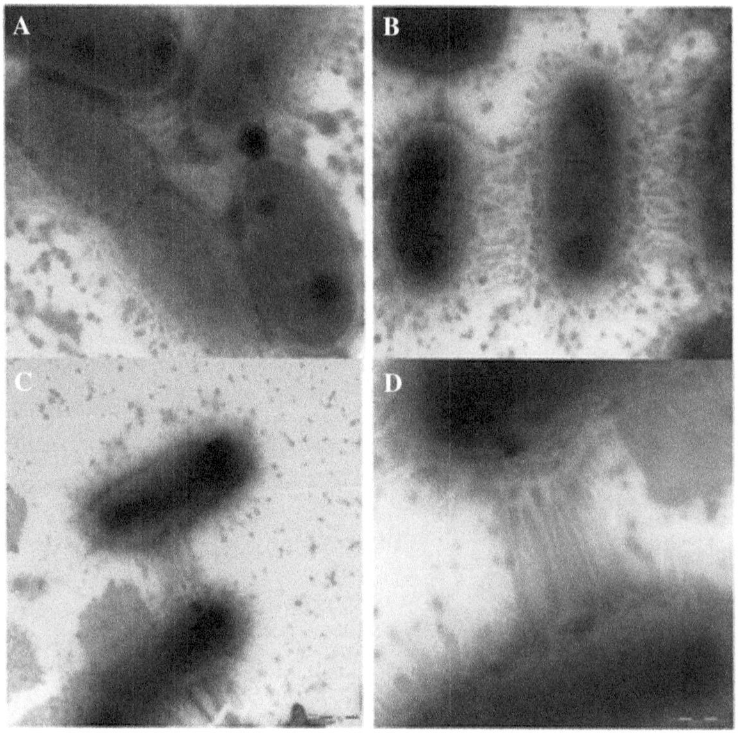

Figure 4 Transmission electron micrograph of the methanotrophic community grown on the NMS medium with visible nanopod structure connecting neighboring bacteria.

released as a waste product into the atmosphere; for instance, the use of methane recovered as a biogas from waste deposits for energy purposes could become unprofitable.

The composition of the bacterial community possible to culture at 30°C and 10% of CH_4 was identified (Table 3). The results show that the SF1 belongs to the genus *Methylocystis* and SM2 and SA3 to *Methylomonas*. The most diverse community is SA3, which consists

Table 2 Microscopic characterization of the methanotroph community isolated from different *Sphagnum* species

Name of isolates	Length/ Width [μm]	Cell shape		Nanopods	Flagella
		Cocco-bacilli	Rods		
SF1	1,89×0,32	-	+	+	-
SM2	1,99×0,52	+	+	+	+
	φ=0,94				
SFA3	1,89×1,01	-	+	+	+
	2,72×0,84				
	1,65×0,76				
SM4	3,69×2,22	+	+	+	-
	3,47×0,65				
	φ=1,09				

of cultured strains belonging to the genera: *Methylosinus, Methylocystis,* and *Methylomonas.* (Steenbergh et al. 2010) suggested different life strategies for these two groups of methanotrophic bacteria type I and II. In their view, type I methanotrophs have lower initial cell numbers in combination with the fast growth rate, which is in agreement with an R-type life strategy, investing in reproduction (Steenbergh et al. 2010) instantaneous upon the presence of favorable conditions. In contrast, type II methanotrophs have higher initial cell numbers and slower growth. This situation is referred to as a K-type life strategy, connected with investing in survival and longevity (Andrews and Harris 1986; Golovlev 2001; Noll et al. 2005). The communities isolated from *Sphagnum* sp. represent both the R and K strategy of life. Thus, they can be used in different environmental conditions, especially SA3, which comprises a mixture of type I and II methanotrophs.

An important element of this work was the culture of the methanotrophic communities isolated from *Sphagnum* sp. plants and recognition of their properties, because essentially, a consortium achieves the same role as does a genetically manipulated (or engineered) strain, but with superior stability. Genetically engineered strains are invariably both physiologically unbalanced and fastidious. (Hamer 2010). As a matter of fact, methanotrophs are not able to exist in the form of pure cultures, because these are their unnatural conditions of life. For

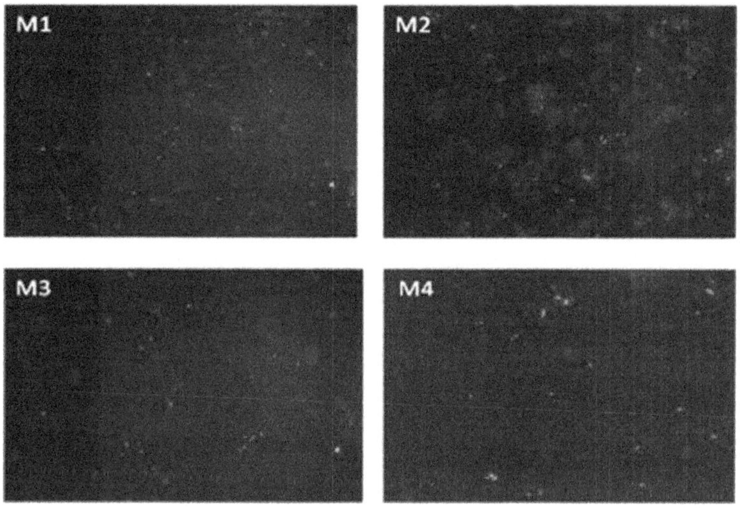

Figure 5 Whole-cell specific hybridization of the endophytic metanotrophic population with probes: Mg84 (type I, pink), Ma 450 + Mg705 (type II, green) and DAPI staining (bacteria, dark blue) by FISH. M1 – *S. flexuosum* , M2– *S. magellanicum* , M3– *S. fallax* , M4 – *S. magellanicum*.

growth, methanotrophs require associated microorganisms necessary for their normal functioning (Hoefman et al. 2010). This hypothesis was confirmed by FISH analysis of the cultured methanotrophic endophytes (Figure 5). Very numerous associated non-methanotrophic bacteria (dark blue cells) were observed, which probable support the growth of methanotrophic bacteria in the tested conditions (Eller et al. 2001). In the literature, more effective neutralization of harmful factors was confirmed, when strains formed a mutualistic type of correlations in biofilm or aggregates (Tay et al. 2004). The use of a consortium of micro-organisms, which exhibit different genomic and phenotypic diversity and the ability to aggregate, shall ensure much more efficient purification of the environment for several reasons: 1) bacterial cells are more resistant to toxic factors, 2) physiological diversity of microorganisms leads to increased degradation of contaminants in a single bioreactor and 3) separating of aggregated cells from a mixture is easy to perform (Tay et al. 2004).

Negative contrast imaging of the cultured endophyte community by transmission microscopy showed the existence of structures facilitating communication between cells. These crystalline-like structures have been described as connecting lines referred to as nanopods (Shetty et al. 2011, Figure 4). So far, these organelles, which are used to transport the virulence factors over long distances, have been described only for pathogens such as *Pseudomonas aeruginosa* and *Legionella pneumophila* (Bomberger et al. 2009; Ellis and Kuehn 2010). The structure of nanopods of non-pathogenic bacteria has been identified for CS1 *Delftia* sp. 4 isolated from Wisconsin soil contaminated with polycyclic hydrocarbons (Shetty et al. 2011). Nanopods have a crystalline-like outer surface and exhibit inner structures with varied (from spherical to spiral) morphology.

Table 3 Identification of cultured endophytic methanotrophs isolated from different *Sphagnum* species based on molecular genetic analysis

Isolate	Identity	Plant species	Max score	Max iden	Accession number (GenBank)
SF1	*Methylocystis. echinoides*	*S. flexuosum*	850	99%	AJ4558502.1
	Methylocystis sp. DWT		850	99%	AJ868423.1
	Methylomonas sp. LC1		588	85%	DQ119049.1
SM2	*Methylomonas sp.* R-45373	*S. magellanicum* 2	1153	100%	FR798959.1
	Methylocystis sp. B3		863	99%	DQ119049.1
	Methylomonas sp. LW15		1144	99%	AF150794.1
SA3	*Methylocystis echinoides*	*S. fallax*	787	96%	AJ458502
	Methylosinus sp. B3R		603	98%	AB636301
SM4	*Methylomonas sp.* R-45372	*S. magellanicum* 4	941	94%	FR798959.1
	Methylomonas sp. LC 1		1079	99%	DQ119049.1

Furthermore, using the three-dimensional electron cryotomography technique, (Shetty et al. 2011) demonstrated that nanopods have tubular architecture, unlike the linear, filamentous construction characteristic of flagella or pili. Nanopods of *Delftia* sp. 4 projecting from cell surfaces were abundant and often spanned the space between neighboring bacteria, as for the cell surfaces of the methanotrophic endophytes (Figure 3). (Shetty et al. 2011) suggested that formation of nanopods and the S-layer by bacterial cells are partially dependent on each other. The presence of the S-layer has been confirmed in many species of methanotrophic, for example halotolerant or thermotolerant, as well as non-halophilic strains (Khmelenina et al. 1999; Khmelenina et al. 2010). Our studies also indicate that methanotrophic endophytes have the S-layer, which, in contrast to carbohydrates, was observed in the presence of ruthenium red (Khmelenina et al. 1999).

There are data indicating that bacteria able to produce nanopods have diverse lifestyles, for instance *Delftia* sp. Cs1-4, *D. acidovorans* SPH1, and *A. delafieldii*, which are free-living soil bacteria. In contrast, *A. avenae* subsp. *avenae* ATCC 19860, *A. avenae* subsp. *Citrulli*, and *V. eiseniae* are biotrophs and they live associated with eukaryotic organisms. Nanopods are produced by plant pathogens and earthworm symbionts (*V. eiseniae* EF01-2), which inhabit environments of decomposing organic material (Pinel et al. 2008). The existence of these structures in methanotrophic endophytes is probable. Notably, considering the lifestyles of these microorganisms as well as the ecological niche (high decomposition of organic matter) from which they were isolated, it will be interesting to determine how, or if, bacteria tailor nanopods for unique functions in methanotrophs and recognize their structure.

In summary, contemporary physical/chemical treatment methods for decreasing the CH_4 concentration such as activated carbon adsorption or incineration are either inefficient or costly at the low concentrations typically found in emissions from waste treatment and animal farming (López et al. 2013). Therefore, properly tailored bioengineering can become a low-cost and environmentally friendly alternative to the physical/chemical methods for the abatement of CH_4 (López et al. 2013). We have described a unique methanotrophic community, which can be a good example for future microbial biotechnology, which has not been quite recognized until recently as far as natural environmental processes were concerned. The proposal presented can indicate the scope of ideas and concepts in process-oriented environmental biotechnology.

Competing interests

The authors declare that they have no competing interests.

Authors' contributions

ZS have made substantial contributions to conception and design, or acquisition of data, or analysis and interpretation of data. AK carried out the molecular genetic studies, participated in the sequence alignment, drafted the manuscript, in the design of the study and performed the statistical analysis. Both authors read and approved the final manuscript.

Acknowledgements

This work was supported by the National Science Centre grant in Poland (No 2011/01/N/NZ9/06811and N 305 29 94 40).

References

Andrews JH, Harris RF (1986) R-selection and K-selection and microbial ecology. Adv Microb Ecol 9:99–147

Bomberger JM, MacEachran DP, Coutermarsh BA, Ye SY, O'Toole GA, Stanton BA (2009) Long-distance delivery of bacterial virulence factors by *Pseudomonas aeruginosa* outer membrane vesicles. PLoS Pathog 5(4):e1000382. doi: 10.1371/journal.ppat.1000382

Eller G, Stubner S, Frenzel P (2001) Group-specific 16S rRNA targeted probes for the detection of type I and type II methanotrophs by fluorescence in situ hybridisation. FEMS Microbiol Lett. 198(2):91–7

Ellis TN, Kuehn MJ (2010) Virulence and immunomodulatory roles of bacterial outer membrane vesicles. Microbiol Mol Biol R 74:81–94. doi:10.1128/MMBR.00031-09

Golovlev EL (2001) Ecological strategy of bacteria: specific nature of the problem. Microbiology 70:379–383

Hamer G (2010) Methanotrophy: from the environment to industry and back. Chem. Eng. J. 160:391–397. doi:10.1016/j.cej.2010.04.008

Hanson RS, Hanson TE (1996) Methanotrophic bacteria. Microbiol Rev 60:439–471

Hoefman S, Boon N, de Vos P, Heylen K (2010) Portecting the fragile: Presevation of methanotrophic bacteria. Cryobiology 61:362–408. doi: 10.1016/j.cryobiol.2010.10.023. 0 0

Jiang H, Chen Y, Jiang P, Zhang C, Smith TJ, Murrell JC, Xing XH (2010) Methanotrophs: multifunctional bacteria with promising applications in environmental bioengineering. Biochem. Eng. J. 49:277–288. doi:10.1016/j.bej.2010.01.003

Khmelenina VN, Shchukin VN, Reshetnikov AS, Mustakhimov II, Suzina NE, Eshinimaev BT, Trotsenko YA (2010) Structural and functional features of methanotrophs from hypersaline and alkaline lakes. Microbiology (Moscow) 79:472–482. doi: 10.1134/S0026261710040090

Khmelenina VN, Kalyuzhnaya MG, Sakharovsky VG, Suzina NE, Trotsenko YA, Gottschalk G (1999) Osmoadaptation in halophilic and alkaliphilic methanotrophs. Arch Microbiol 172:321–329

López JC, Quijano G, Souza TSO, Estrada JM, Lebrero R, Muñoz R (2013) Biotechnologies for greenhouse gases (CH4, N2O, and CO2) abatement: state of the art and challenges. Appl Microbiol Biotechnol 97:2277–2303. doi:10.1007/s00253-013-4734-z

Minamisawa K (2006) A milestone for endophyte biotechnology. Nat Biotechnol 24:1357–358. doi:10.1038/nbt1106-1357

Noll M, Matthies D, Frenzel P, Derakshani M, Liesack W (2005) Succession of bacterial community structure and diversity in a paddy soil oxygen gradient. Environ. Microbiol. 7(3):382–395. doi:10.1111/j.1462-2920.2005.00700.x

Omel'chenko MV, Vasiljeva LV, Zawarzin GA (1993) Psychrophilic methanotroph from Tundra soil. Curr Microbiol 27:255–259

Pinel N, Davidson SK, Stahl DA (2008) *Verminephrobacter eiseniae* gen. nov., sp nov., a nephridial symbiont of the earthworm Eisenia foetida (Savigny). Int J Syst Evol Micr 58:2147–2157. doi:10.1099/ijs.0.65174-0

Sambrook J, Russell DW (2001) Molecular Cloning: A Laboratory Manual, 3. Cold Spring Harbor Laboratory Press, New York

Santl-Temkiv T, Finster K, Hansen BM, Pasic L, Karlson UG (2013) Viable methanotrophic bacteria enriched from air and rain can oxidize methane at cloud-like conditions. Aerobiologia 29:373–384. doi: 10.1007/s10453-013-9287-1

Shetty A, Chen S, Tocheva EI, Jensen GJ, Hickey WJ (2011) Nanopods: a new bacterial structure and mechanism for deployment of outer membrane vesicles. PLoS ONE 6(6):1–7. Doi: 10.1371/journal.pone.0020725

Stahl DA, Amann R (1991) Development and Application of Nucleic Acid Probes. In: Stackebrandt E, Goodfellow M (ed) Nucleic Acid Techniques in Bacterial Systematics. Wiley and Sons, New York, p 248

Steenbergh AK, Meima MM, Kamst M, Bodelier PLE (2010) Biphasic kinetics of a methanotrophic community is a combination of growth and increased

activity per cell. FEMS Microbiol. Ecol. 71:12–22. doi:10.1111/j.1574-6941.2009.00782.x
Stępniewska Z, Szmagara A, Niewiarowska M (2006) The Environmental Requirements of Methanotrophic Bacteria Inhabiting Coal Mine Dump Rock. International Workshop "Pathways of Pollutant from Landfills to air and Water-Soil Systems and Mitigation Strategies of Their Impact on the Ecosystems". The Conference Proceedings Kazimierz Dolny, pp 17–20
Tay JH, Pan S, He Y, Tay STL (2004) Effect of organic loading rate on aerobic granulation. I: Reactor performance. J. Environ. Eng. 130:1094–1101. doi: 10.1061/(ASCE)0733-9372(2004)130:10(1094)
Trotsenko YA, Khmelenina VN (2005) Aerobic methanotrophic bacteria of cold ecosystems. FEMS Microbiol. Ecol. 53:15–26. doi:10.1016/j.femsec.2005.02.010
Trotsenko YA, Medvedkova KA, Khmelenina VN, Eshinimaev BT (2009) Thermophilic and thermotolerant aerobic merthanotrophs. Microbiology (Moscow) 78:387–401
Trotsenko YA, Murrell JC (2008) Metabolic aspects of aerobic obligate methanotrophy. Adv. Appl. Microbiol. 63:183–229. Doi: 10.1016/S0065-2164(07)00005-6

14

Unraveling the kinetic diversity of microbial 3-dehydroquinate dehydratases of shikimate pathway

Chang Liu[1,2], Yi-Ming Liu[1,2], Qing-Lan Sun[3], Cheng-Ying Jiang[1,2] and Shuang-Jiang Liu[1,2*]

Abstract

3-Dehydroquinate dehydratase (DHQase) catalyzes the conversion of 3-dehydroquinic acid to 3-dehydroshikimic acid of the shikimate pathway. In this study, 3180 prokaryotic genomes were examined and 459 DHQase sequences were retrieved. Based on sequence analysis and their original hosts, 38 DHQase genes were selected for chemical synthesis. The selected DHQases were translated into new DNA sequences according to the genetic codon usage bias by both *Escherichia coli* and *Corynebacterium glutamicum*. The new DNA sequences were customized for synthetic biological applications by adding Biobrick adapters at both ends and by removal of any related restriction endonuclease sites. The customized DHQase genes were successfully expressed in *E. coli*, and functional DHQases were obtained. Kinetic parameters of K_m, k_{cat}, and V_{max} of DHQases were determined with a newly established high-throughput method for DHQase activity assay. Results showed that DHQases possessed broad strength of substrate affinities and catalytic capacities. In addition to the DHQase kinetic diversities, this study generated a DHQase library with known catalytic constants that could be applied to design artificial modules of shikimate pathway for metabolic engineering and synthetic biology.

Keywords: 3-Dehydroquinate dehydratase (DHQase); Kinetic constants; Shikimate pathway; Biobrick; Synthetic biology

Introduction

Shikimate pathway widely exists in microbes and plants, but not animals. This pathway is involved in the synthesis of aromatic amino acids, vitamins, as well as lignin (Herrmann and Weaver 1999; Vanholme et al. 2012). The pathway consists of seven catalytic steps, condensing erythrose-4-phosphate (E4P) and phosphoenolpyruvate (PEP) finally into chorismate. 3-Dehydroquinate dehydratase (DHQase, EC 4.2.1.10) catalyzes the third step, i.e., reversible transformation of DHQ to form 3-dehydroshikimic acid. DHQases belong to the family of lyases, and cleave carbon-oxygen bonds. So far as known, DHQases involve in not only shikimate pathway but also other metabolic processes such as the quinate pathway for synthesis of 4-hydroxybenzoic acid (Giles et al. 1985; Giles et al. 1991). According to their origins and catalytic features, DHQases are classified into either type I or type II. Type I DHQases are heat-liable dimeric (Roszak et al. 2002), and mainly occur in plants and fungi. Type II DHQases are heat-stable dodecameric (Roszak et al. 2002) and widely occur in bacteria for shikimate pathway or in fungi for quinate catabolism (Giles et al. 1991).

Many investigations of DHQases have been focused on their structures and catalytic mechanisms (Blomberg et al. 2009; Bottomley et al. 1996; Devi et al. 2013; Lee et al. 2002; Pan et al. 2012; Roszak et al. 2002), or on structure-based design of inhibitors to DHQase activity (Blanco et al. 2012; 2014; Dias et al. 2011; Peon et al. 2010). These investigations have generated increasing numbers of DHQase structures with high resolution and have significantly advanced the understanding of DHQase catalytic mechanisms (Chaudhuri et al. 1986; Deka et al. 1994; Euverink et al. 1992; Hawkins et al. 1993; Lee et al. 2003; Moore et al. 1993; Roszak et al.

* Correspondence: liusj@im.ac.cn
[1]State Key Laboratory of Microbial Resources, Chinese Academy of Sciences, Beijing 100101, China
[2]Environmental Microbiology Research Center, Chinese Academy of Sciences, Beijing 100101, China
Full list of author information is available at the end of the article

2002; Singh and Christendat 2006). Type I DHQases catalyze the dehydrogenation of DHQ through a *cis*-elimination (Chaudhuri et al. 1991; Gourley et al. 1999; Leech et al. 1995), while type II DHQases take a *trans*-elimination mechanism (Blomberg et al. 2009; Bottomley et al. 1996).

Recent study has revealed that overexpression of DHQase enhanced transformation of quinic acid into shikimic acid in *Gluconobacter oxydans* (Nishikura-Imamura et al. 2014). The kinetic properties of DHQases such as K_m, V_{max}, and k_{cat} are important particularly to design new biocatalysts and to predict the validity and efficiency of newly constructed metabolic networks. So far, only a small number of DHQases from prokaryotes such as *Escherichia coli*, *Mycobacterium tuberculosis*, and *Streptomyces coelicolor* were characterized for their catalytic properties (Harris et al. 1996; Kleanthous et al. 1992; Moore et al. 1993; Richards et al. 2006; White et al. 1990), and the catalytic and kinetic properties of the majority of microbial DHQases remain still unknown.

In this study, we aimed to investigate the kinetic diversities of DHQases and to expand the toolbox of catalytic parts for synthetic biology. By data-mining of 3180 prokaryotic genomes from NCBI genome database, 459 putative DHQases were targeted. Thirty-eight DHQases were further selected and standardized according to "Biobricks" requirements (Knight 2003; Shetty et al. 2011; Sleight et al. 2010). DHQase kinetic constants were determined with a newly established high-throughput method. Our results showed that DHQases are highly diverse in catalytic kinetics.

Materials and methods

Genome data-mining for DHQase genes

The amino acid sequences of the type II DHQase (NP_599670) from *Corynebacterium glutamicum* ATCC13032 and of the type I DHQase (NP_416208) from *E. coli* K-12 were used as seed sequences to retrieve putative DHQase sequences from NCBI genome database with a filter condition of threshold E value $\leq 10^{-10}$. The retrieved sequences were then screened and redundant copies were removed. To increase the credibility of functional DHQase prediction, the retrieved sequences were further filtered by removal of putative DHQase sequences from which host organisms have incomplete shikimate pathway in their genomes.

Design and chemical synthesis of DHQase genes

The selected DHQase amino acid sequences were reverse-translated into DNA sequences, and were recoded with referring codon usage bias of *E. coli* and *C. glutamicum*. The obtained DNA sequences were optimized for expression in *E. coli* and *C. glutamicum* by check for RNA secondary structure with software UNAFold (Markham and Zuker 2008). Any predicted secondary structures were eliminated by codon replacements. The new DHQase genes were further customized, by linking to Biobrick adapters (Knight 2003) at both ends. The customized DNA sequences of the DHQases are accessible at http://www.genoportal.org/bbdb/ under the accession numbers of SBB_00477 ~ SBB_004481, and were chemically synthesized (Sangon Biotech, China), and were cloned in *E. coli*. The DNA sequences of the chemically synthesized DHQase genes were confirmed by DNA sequencing.

Bacterial strain, plasmids, and growth condition

For cloning and expression of DHQase genes, *E. coli* BL21 (DE3) (TransGen Biotech, China) and plasmid pET28a$^+$ (Novagen, Germany) were used. Expression of DHQase genes was induced with 0.1 mmol/L of IPTG when culture reached OD_{600nm} of 0.6-0.8. After addition of IPTG, culture was further incubated at 16°C, overnight. Routine cultivation of *E. coli* proceeded in Luria-Bertani (LB) medium at 37°C and at 200 rpm rotatory shaking. To maintain the stability of pET28a$^+$ and its derivatives, kanamycin at final concentration of 100 μg/ml were added into LB medium.

Preparation of cellular lysates and purification of DHQase proteins

E. coli cells were harvested by centrifugation at 10,000 rpm, and suspended in 50 mmol/L Tris–HCl buffer, pH 8.0. After addition of 0.01% (w/v) protease inhibitor cocktail (Amresco, the United States), cell suspension was treated with ultrasonication (work for 3 sec, stop for 5 sec, 100 repeats). The cellular debris was removed by centrifugation at 10,000 rpm for 10 min, and the supernatant was filtered with 0.22 μm filters (Millipore, the United States). The filtered supernatant was collected and DHQase proteins were purified using HisPur™ Ni-NTA Spin Columns Kit (Thermo Scientific, the United States). The purified DHQase proteins were stored at −80°C in 50 mmol/L Tris–HCl (pH 8.0) buffer containing 25% glycerol. All procedures were operated at 4°C unless indicated.

Protein concentrations were determined with Bio-Rad Protein Assay (BIO-RAD, the UK).

Tricine sodium dodecyl sulfate polyacrylamide gel electrophoresis was used to evaluate the DHQase expression and the purity during protein purification. Preparation of gels (4% sample gel and 10% separating gel) and operation of electrophoresis were conducted according to Schägger (2006). Gels were visualized with Coomassie brilliant blue G-250 staining, and were scanned with PC scanner (T68, Founder, China) for imagine analysis.

Table 1 Distribution of type I and type II DHQases in prokaryotic genomes[a]

Sources	DHQase in a genome	Number of genomes
Bacteria	Type I only	60
	Type II only	306
	Both Type I and II	25
Archaea	Type I	43

[a]Notes: Redundant genomes representing the same species were removed.

DHQase activity assays

A high-throughput method of DHQase activity assay was established in this study. The method has the same principal for measurement as White et al. (1990), but with new systems. Synergy H4 Hybrid Multi-Mode Microplate Reader (BioTek, the UK) and 96 Well UV-Plate (Corning Costar, the United States) were used to monitor the changes of OD_{234nm} of multi-samples. DHQase catalysis was optimized in volume of 100 μL and with various DHQ concentrations (0.08 to 1.0 mmol/L). Kinetic constants K_m and V_{max} were calculated according to Lineweaver-Burk plot.

Sequence alignments and construction of phylogenetic tree

The amino acid sequences of DHQases were retrieved from NCBI genome database. Amino acid sequence alignment of DHQases were performed with MUSCLE (Edgar 2004), and the graphic display of alignments were made by ESPript3.0 (Gouet et al. 1999). The phylogenetic tree of amino acid sequences from all putative DHQases were firstly constructed by Phylip Package (Abdennadher and Boesch 2007) under Linux system with method of maximum likelihood and bootstrap replications of 1000, and further annotated using iTOL v2.2.2 (Letunic and Bork 2011).

Results

DHQases are phylogenetically diverse

3180 genomes were explored for putative DHQases, and 459 putative DHQase were targeted, including 128 type I and 331 type II DHQases. Noticeably, all archaeal DHQases were type I. Bacterial genomes carried both type I and type II DHQases, which 79.6% of the bacterial DHQases were type II and 20.4% were type I. Of the 391

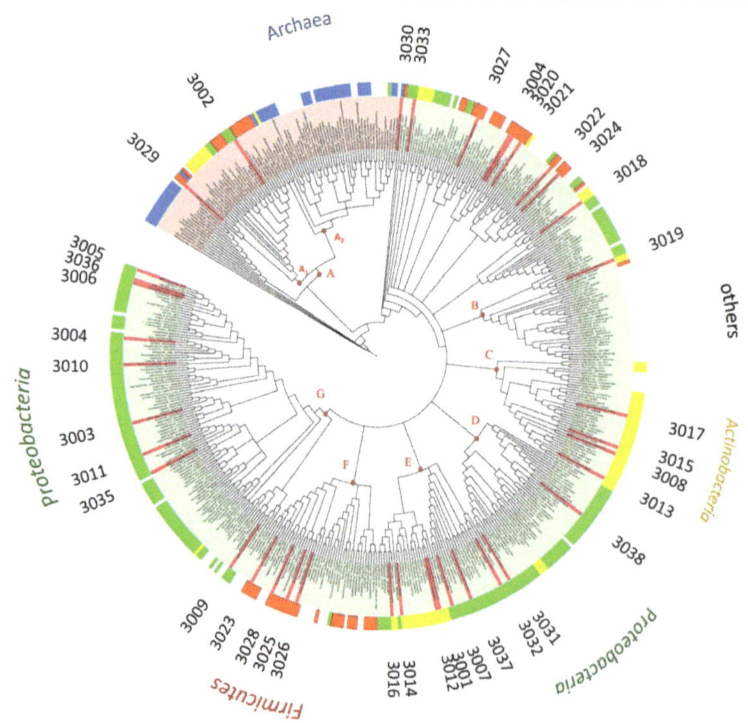

Figure 1 Circle phylogenetic tree of 459 DHQases, constructed with Maximum Likelihood method with 1000 bootstrap replications. The salmon color range covers all species with type I DHQases, the light green range covers all species with type II DHQases. The codes for the 38 selected DHQases were assigned outside circles. For the outmost color ring, colored strips indicate the origins (at phylum levels) of the DHQases: *Firmicutes* is marked with red strips, *Proteobacteria* is green, *Actinobacteria* is yellow. All archaea including *Thaumarchaeota,Crenarchaeota*, and *Euryarchaeota* are showed in dark blue. Species from other phylum is left with no color strips. Capital letters A-G indicated7 big clades with bootstrap support values >70%.

Table 2 Selected DHQase, their origins and theoretical molecular masses of translational products

Codes for DHQases[a]	Origins	NCBI accession numbers of	Accession IDs at www.genoportal.org	Theoretical molecular mass (kDa)	Sequence identity
Type I					
3002	*Escherichia coli*	YP_489955.1	SBB_00449	27.5	100%
3029	*Staphylococcus aureus*	YP_005744190.1	SBB_00476	26.9	30.0%
3030	*Metallosphaera cuprina*	YP_004409005.1	SBB_00477	23.7	23.5%
Type II					
3001	*Corynebacteriumglutamicum*	YP_224725.1	SBB_00448	15.4	100.0%
3003	*Alcanivorax borkumensis*	YP_693728.1	SBB_00450	17.5	50.7%
3004	*Bacillus pseudofirmus*	YP_003425282.1	SBB_00451	16.5	50.0%
3005	*Citrobacter koseri*	YP_001456151.1	SBB_00452	16.5	53.4%
3006	*Enterobacter aerogenes*	YP_004591144.1	SBB_00453	16.5	53.0%
3007	*Gluconobacter oxydans*	YP_190876.1	SBB_00454	17.4	51.1%
3008	*Mycobacterium tuberculosis*	YP_003031384.1	SBB_00455	15.8	52.6%
3009	*Psychrobacter*sp. PRwf-1	YP_001279670.1	SBB_00456	18.4	52.8%
3010	*Psychromonas ingrahamii*	YP_944417.1	SBB_00457	16.5	53.0%
3011	*Xylella fastidiosa*	NP_297340.2	SBB_00458	16.7	52.2%
3012	*Arthrobacter crystallopoietes*	EMY34512.1	SBB_00459	16.3	71.9%
3013	*Micromonas poralupini*	CCH17589.1	SBB_00460	15.4	47.8%
3014	*Nocardia brasiliensis*	YP_006813099.1	SBB_00461	15.6	64.0%
3015	*Propionibacterium acnes*	YP_056375.1	SBB_00462	15.8	43.7%
3016	*Streptomyces acidiscabies*	WP_010352491.1	SBB_00463	16.8	61.4%
3017	*Thermomonas poracurvata*	YP_003300530.1	SBB_00464	16.6	45.3%
3018	*Atopobium rimae*	EEE18006.1	SBB_00465	15.5	41.7%
3019	*Acidimicrobium ferrooxidans*	YP_003110305.1	SBB_00466	16.4	45.8%
3020	*Bacillus subtilis*	NP_390327.1	SBB_00467	16.4	49.3%
3021	*Geobacillus* sp. Y4.1MC1	YP_003988579.1	SBB_00468	16.0	48.6%
3022	*Alicyclobacillus acidocaldarius*	EED07485.1	SBB_00469	16.3	49.6%
3023	*Butyrivibrio crossotus*	EFF68279.1	SBB_00470	22.8	47.1%
3024	*Halanaerobium hydrogeniformans*	YP_003995185.1	SBB_00471	16.6	51.4%
3025	*Clostridium clariflavum*	YP_005045936.1	SBB_00472	15.5	53.1%
3026	*Natranaerobius thermophilus*	YP_001917881.1	SBB_00473	15.8	43.7%
3027	*Ruminococcus champanellensis*	YP_007829441.1	SBB_00474	16.1	49.6%
3028	*Thermodesulfobium narugense*	YP_004438087.1	SBB_00475	16.8	46.3%
3031	*Comamonas testosteroni*	YP_003279351.1	SBB_00478	15.9	52.1%
3032	*Pseudomonas putida*	NP_745147.1	SBB_00479	16.2	59.4%
3033	*Halomonas elongata*	YP_003899044.1	SBB_00480	15.5	55.8%
3034	*Vibrio vulnificus*	NP_935927.1	SBB_00481	18.7	55.6%
3035	*Acidithiobacillus caldus*	YP_004750222.1	SBB_00482	16.9	50.7%

Table 2 Selected DHQase, their origins and theoretical molecular masses of translational products *(Continued)*

3036	*Klebsiella pneumoniae*	EMH96211.1	SBB_00483	16.5	51.5%
3037	*Magnetospirillum magneticum*	YP_422069.1	SBB_00484	14.5	50.4%
3038	*Zymomonas mobilis*	YP_162472.1	SBB_00485	15.8	52.9%

The amino acid sequence identities of DHQases to the one from *C. glutamicum* (type II) or from *E. coli* (type I) were calculated based on alignment with BLAST searches. The NCBI numbers refer to the wildtype DHQases, and the customized DNA sequences of these DHQases are accessible at http://www.genoportal.org/bbdb.
[a]The first digit of the code represents DHQase catalyzes the 3rd step of shikimate pathway, and the following digits are the order when that DHQases were selected.

genomes explored, 306 genomes (78.3%) harbored only type II DHQases, 60 genomes (15.3%) harbored only type I, and 25 genomes (6.4%) harbored both type I and type II (Table 1).

To display the phylogenetic relationship of all 459 DHQase sequences and taxonomic distribution, their amino acid sequences were used to construct a phylogenetic tree (Figure 1). As seen from Figure 1, seven clades (from cluster A to G) of DHQases were recognized. The majority of type I DHQases fell into the clade A, which contained two sub-clades A_1 and A_2. Sub-clade A_1 covered bacterial type I DHQases from *Proteobacteria, Actinobacteria,* and *Firmicutes.* Sub-clade A_2 was mainly archaeal DHQases from *Thaumarchaeota*, *Crenarchaeota*, and *Euryarchaeota.* Clades B through G was type II DHQases of different bacterial phyla. DHQases of the clades C and F were mainly from *Actinobacteria* and *Firmicutes*, respectively. DHQases of clades D and G were mainly from *Proteobacteria.* DHQases of clades B and E were from *Proteobacteria, Actinobacteria* and other bacterial phyla.

Cloning and expression of customized DHQase genes in *E. coli* cells

From all retrieved 459 DHQases, 38 were further selected (Table 2).The selection was carried out according to the following criterions: 1) Both type I and type II DHQases were included, and the majority was bacterial type II DHQases; 2) Host genomes harbored a complete shikimate pathway, or taxonomically representative (*Archaea, Actinobacteria, Firmicutes,* and *Proteobacteria*), or from unique habitats. From the 38 DHQases, 35 were of type II, and 3 were of type I. The theoretical molecular masses of type II DHQases were ranged from 15 to 18 kDa, and their amino acid sequence identities to DHQase from *C. glutamicum* were 41.7-71.9%. The selected DHQases were recorded by referring the codon usages of *E. coli* and of *C. glutamicum* (Nakamura 2007). The new DNA sequences were edited by curing restriction endonucleases (*Nde*I, *Xho*I, *Eco*RI, *Not*I, *Xba*I, *Spe*I and *Pst*I) sites and any predicted secondary mRNA structures. Further, the sequences were customized for "Biobricks" (Knight 2003) by attaching the prefix5′-GCAGAATTCGCGGCCGCTTCTAGA-3′ and the suffix 5′-ACTAGTAGCGGCCGCCTGCAG-3′.

All 38 DHQase genes were cloned with *E. coli.* Except those DHQases of *A. ferrooxidans*, *B. pseudofirmus*, *S. aureus* and *Z. mobilis* that occurred as either inclusion bodies or no synthesis in *E. coli* cells, all other DHQase were synthesized and purified. The purified DHQases actively catalyzed the conversion of DHQ into 3-dehydroshikimic acid.

DHQases have broad ranges of kinetic parameters

Catalytic properties such as V_{max}, K_m, k_{cat}, and k_{cat}/K_m of the selected DHQases were determined and are listed in Table 3. Results showed that their kinetic parameters are distributed in broad ranges, which suggested a kinetic diversity of DHQases regarding catalytic capacity and efficiency, and substrate affinity. The V_{max} values of DHQases at the given enzyme concentrations in this study were determined to be 0.57-66.7 μmol/L/s. The K_m values were determined to be 37.2-2226.5 μmol/L. These results indicated that natural DHQases had evolved broad catalytic properties, probably for reasons to serve their hosts at different growing conditions. The DHQase (Code 3012) from *A. crystallopoietes* showed the highest catalytic capacity ($V_{max} = 66.7$ μmol/L/s) and moderate affinity to 3-dehydroquinate ($K_m =$ 448.5 μmol/L). The DHQase (Code 3031) from *C. testosteroni* showed highest affinity to 3-dehydroquinate ($K_m = 37.2$ μmol/L) among the tested DHQases. The kinetic parameter k_{cat} signifying the catalytic efficiency of an enzyme, and it was found that the DHQase (Code 3016) from *S. acidiscabies* was the most efficient one among the tested DHQases and its k_{cat} value was determined to be 211.83 s^{-1}. When all the tested DHQases were evaluated by k_{cat}/K_m values, the DHQase (Code 3016) had the highest k_{cat}/K_mvalue (1.50 L/μmol/s) that represented the best catalytic specificity and efficiency among all the tested DHQases. In addition, the specific activities of DHQases at 3-dehydroquniate concentration of 0.5 mM were determined and DHQase (Code 3016) from *S. acdiscabies* showed the highest activity (501.87 units/mg). Figure 2 depicts a whole image of kinetic parameter distribution of DHQases. Apparently, the kinetic parameters of all tested DHQases varied significantly, reflecting a long-span distribution of their catalytic properties and kinetic diversity.

Table 3 The kinetic parameters of 3-dehydroquinate dehydratases

Codes for DHQases*	V_{max} (μmol/L/s)	K_m (μmol/L)	k_{cat} (s^{-1})	k_{cat}/K_m (L/μmol/s)	Specific Activity** (U/mg)
3001	2.56	281.43	52.26	0.19	119.14
3002	4.57	187.68	29.53	0.16	57.34
3003	1.34	210.71	13.53	0.06	34.72
3004	——	——	——	——	——
3005	15.50	870.37	208.96	0.24	93.91
3006	6.39	478.67	102.15	0.21	187.69
3007	2.43	777.75	77.06	0.10	84.95
3008	2.46	624.52	23.96	0.04	46.37
3009	4.24	203.24	31.67	0.16	75.06
3010	0.80	202.55	10.59	0.05	9.39
3011	0.57	158.79	7.15	0.05	15.06
3012	66.67	448.50	185.43	0.41	369.99
3013	7.04	159.89	4.63	0.03	86.50
3014	3.60	151.51	3.83	0.03	91.93
3015	4.18	46.14	10.00	0.22	67.71
3016	12.42	141.53	211.83	1.50	501.87
3017	8.25	1218.56	11.92	0.01	30.19
3018	0.68	400.82	9.06	0.02	15.64
3019	——	——	——	——	——
3020	0.96	107.07	10.46	0.10	62.27
3021	2.16	133.97	5.33	0.04	15.90
3022	8.26	601.64	119.98	0.20	195.81
3023	2.86	67.88	42.20	0.62	68.99
3024	6.23	526.07	75.11	0.14	131.08
3025	5.89	368.46	59.61	0.16	117.89
3026	2.95	304.54	3.80	0.01	5.16
3027	6.00	255.54	50.20	0.20	116.07
3028	1.16	180.72	6.88	0.04	17.66
3029	——	——	——	——	——
3030	1.14	92.76	3.86	0.04	8.58
3031	1.45	37.16	5.95	0.16	19.82
3032	12.35	617.28	78.87	0.13	105.18
3033	2.64	150.25	57.84	0.38	155.93
3034	3.78	345.90	37.97	0.11	71.87
3035	11.14	2226.50	192.27	0.09	136.57
3036	9.64	131.63	102.70	0.78	192.10
3037	3.87	119.97	73.18	0.61	365.73
3038	——	——	——	——	——

*The first digit of the code represents DHQase catalyzes the 3rd step of shikimate pathway, and the following digits are the order when that DHQase was selected.
**Specific activities were determined at substrate concentration of 0.5 mM. "-" indicated no activity was detected.

Discussion

Kinetic diversity of enzymes is the results of natural revolution in life, similar to their phylogenetic diversity (Zhi et al. 2014). But the kinetic diversity of enzymes had not yet been explored before. This study took DHQase as an example to explore the kinetic diversity of enzymes involved in shikimate pathway. The results from this study demonstrated that DHQases from various hosts had very broad ranges of affinity (K_m) to substrate, catalytic efficiency and catalytic capacity (k_{cat}

		Ratio of k_{cat}/K_m (%)	Ratio of Specific Activity (0.5 mM substrate) (%)	Ratio of k_{cat} (%)	Ratio of K_m (%)	Ratio of V_{max} (%)
Bacillus pseudofirmus	3004	0	0	0	0	0
Acidimicrobium ferrooxidans	3019	0	0	0	0	0
Staphylococcus Aureus	3029	0	0	0	0	0
Zymomonas mobilis	3038	0	0	0	0	0
Thermomonospora curvata	3017	5	25	23	433	323
Natranaerobius thermophilus	3026	7	4	7	108	115
Atopobium rimae	3018	12	13	17	142	27
Nocardia brasiliensis	3014	14	77	7	54	141
Micromonospora lupini	3013	16	73	9	57	275
Thermodesulfobium narugense	3028	21	15	13	64	45
Mycobacterium tuberculosis	3008	21	39	46	222	96
Geobacillus sp. Y4.1MC1	3021	21	13	10	48	84
Metallosphaera cuprina	3030	22	7	7	33	45
Xylella fastidiosa	3011	24	13	14	56	22
Psychromonas ingrahamii	3010	28	8	20	72	31
Alcanivorax borkumensis	3003	35	29	26	75	52
Acidithiobacillus caldus	3035	47	115	368	791	435
Bacillus subtilis	3020	53	52	20	38	37
Gluconobacter oxydans	3007	53	71	147	276	95
Vibrio vulnificus	3034	59	60	73	123	148
Pseudomonas putida	3032	69	88	151	219	483
Halanaerobium hydrogeniformans	3024	77	110	144	187	244
Psychrobacter sp. PRwf-1	3009	84	63	61	72	166
Escherichia coli	3002	85	48	57	67	179
Comamonas testosteroni	3031	86	17	11	13	57
Clostridium clariflavum	3025	87	99	114	131	230
Corynebacterium glutamicum	3001	100	100	100	100	100
Ruminococcus champanellensis	3027	106	97	96	91	235
Alicyclobacillus acidocaldarius	3022	107	164	230	214	323
Enterobacter aerogenes	3006	115	158	195	170	250
Propionibacterium acnes	3015	117	57	19	16	164
Citrobacter koseri	3005	129	79	400	309	606
Halomonas elongata	3033	207	131	111	53	103
Arthrobacter crystallopoietes	3012	223	311	355	159	2607
Magnetospirillum magneticum	3037	329	307	140	43	151
Butyrivibrio crossotus	3023	335	58	81	24	112
Klebsiella pneumoniae	3036	420	161	197	47	377
Streptomyces acidiscabies	3016	806	421	405	50	486

Figure 2 Demonstration of kinetic diversity of DHQases. Kinetic parameters were shown as ratio (in %) of value of any DHQase to that of code 3001. Four sets of kinetic parameters were displayed with color gradients.

and V_{max}). It was considered that other enzymes of the shikimate pathway would also have broad ranges of kinetic constants. Such broad ranges of kinetic constants are reflections of natural evolution and adaptation of enzymes, and founded the cornerstones of diverse metabolic fluxes in different lives. Therefore, it would be theoretically feasible to engineer metabolic processes quantitatively by mining enzymatic kinetic database. Such strategy is basically different from the current used strategy, ie., manipulation at levels of genetic transcription and translation (Rytter et al. 2014; Salis et al. 2009; Sohoni et al. 2014).

Recently, synthetic biology has emerged as a new discipline to manipulate biological systems for application and to understand the law of life (Andrianantoandro et al. 2006; Cheng and Lu 2012). The basic units for synthetic biology are standardized biological modules and parts, and standardized enzymes serve as catalytic parts for assembling novel and artificially designed biological systems (Cooling et al. 2010). Such catalytic part library is represented by the chemically synthesized genes of methyl halide transferases (Bayer et al. 2009). In this study, 38 selected DHQase genes were chemically synthesized and customized as Biobricks for future applications. All DHQases have Biobrick adapters, so they are compatible to other Biobrick parts/modules. Our study is part of an on-going project of design and construction of artificial modules for shikimate pathway using synthetic biological tools, aiming to create modules for shikimate pathway that would stimulate industrial applications for bioproduction of aromatic-related primary and secondary compounds. The pipeline for DHQase sequence design, chemical synthesis and purification established from this study has been generalized for other enzymes of the shikimate pathway, and construction of a catalytic part library of shikimate pathway is in progress.

Previous reports described that type I DHQases are represented by DHQases from plants and fungi (Hawkins et al. 1993; Weaver and Herrmann 1997), we have found that all 43 archaeal DHQases in this study were belonging to type I, although this study was focused on type II DHQases. Type II DHQases had been classified into two groups in terms of their kinetic constants k_{cat} (Pan et al. 2012). The first group included DHQase from *S. coelicolor* and had relative high k_{cat} values larger than 100 s^{-1}, whereas the sencond group included DHQase from *M. tuberculosis* with a k_{cat} lower than 10 s^{-1}. As demonstrated in this study, the k_{cat} values of DHQases distributed contineouslyfrom 4 to over 200 s^{-1}, which rendered it unrealistic to devide them into two groups based on k_{cat} values.

Alignments of type II DHQases revealed a number of conserved amino acid residues that are potentially important for catalytic efficiency and capacity. Specifically, the Tyr24 residue facilitated the proton abstraction from substrate, and then the His101 residues promoted the dehydrogenation by donating proton to 1-hydroxyl on C1 of substrate as a general acid (Blomberg et al. 2009; Pan et al. 2012). The residue His101was found in all the type II DHQases. But Tyr24 was replaced by Phe residue in DHQase from *B. subtilis* (Code 3020) (Data not shown). It was found that this DHQase (Code 3020) were functional and catalyzed the conversion of DHQ to 2-dehydroshikimate. The Phe residue was found in eighteen *B. subtilis* genomes as well as other bacilli genomes such as *B. mojavensis*, *B. tequilensis*, *B. vallismortis*, *B. atrophaeus*, demonstrating a natural evolution of Tyr24 into a Phe24 residue in members of the genus *Bacillus*. This natural evolution resulted in a low efficient but still active DHQase, which might be a result of adaption to the *in vivo* metabolic fluxes of bacilli cells. The catalytic mechanism of the Phe-DHQase has not been explored. Since Phe residue is rather stable and hard to be deprotonated, the catalytic mechanism of the Phe-DHQases might be different from that of the previous characterized DHQases.

Competing interests
The authors declare that they have no competing interests.

Authors' contributions
CL carried out the studies of gene cloning and expression, enzyme purification and activity determination. YL carried out the sequence analysis, gene recoding and participated gene cloning and expression, enzyme purification and activity determination. QLS carried out genome data mining and participated sequence analysis. Both CYJ and CL helped to draft the manuscript. SJL conceived of the study and finalized the manuscript. All authors read and approved the final manuscript.

Acknowledgements
This work was supported by 973 Project from Ministry of Science and Technology (No. 2012CB7211-04).

Author details
[1]State Key Laboratory of Microbial Resources, Chinese Academy of Sciences, Beijing 100101, China. [2]Environmental Microbiology Research Center, Chinese Academy of Sciences, Beijing 100101, China. [3]Bioinformatic Center at Institute of Microbiology, Chinese Academy of Sciences, Beijing 100101, China.

References
Abdennadher N, Boesch R (2007) Deploying PHYLIP phylogenetic package on a large scale distributed system. Paper presented at the CCGrid 2007: Seventh IEEE International Symposium on Cluster Computing and the Grid

Andrianantoandro E, Basu S, Karig DK, Weiss R (2006) Synthetic biology: new engineering rules for an emerging discipline. Mol Syst Biol doi:10.1038/msb4100073

Bayer TS, Widmaier DM, Temme K, Mirsky EA, Santi DV, Voigt CA (2009) Synthesis of methyl halides from biomass using engineered microbes. J Am Chem Soc 131:6508–6515

Blanco B, Sedes A, Peon A, Lamb H, Hawkins AR, Castedo L, Gonzalez-Bello C (2012) Synthesis of 3-alkyl enol mimics inhibitors of type II dehydroquinase: factors influencing their inhibition potency. Org Biomol Chem 10:3662–3676

Blanco B, Sedes A, Peon A, Otero JM, van Raaij MJ, Thompson P, Hawkins AR, Gonzalez-Bello C (2014) Exploring the water-binding pocket of the type II dehydroquinase enzyme in the structure-based design of inhibitors. J Med Chem 57:3494–3510

Blomberg LM, Mangold M, Mitchell JBO, Blumberger J (2009) Theoretical study of the reaction mechanism of *Streptomyces coelicolor* Type II dehydroquinase. J Chem Theory Comput 5:1284–1294

Bottomley JR, Hawkins AR, Kleanthous C (1996) Conformational changes and the role of metals in the mechanism of type II dehydroquinase from *Aspergillus nidulans*. Biochem J J319:269–278

Chaudhuri S, Lambert JM, McColl LA, Coggins JR (1986) Purification and characterization of 3-dehydroquinase from *Escherichia coli*. Biogeosciences J239:699–704

Chaudhuri S, Duncan K, Graham LD, Coggins JR (1991) Identification of the active-site lysine residues of two biosynthetic 3-dehydroquinases. Biogeosciences J275:1–6

Cheng AA, Lu TK (2012) Synthetic biology: an emerging engineering discipline. Annu Rev Biomed Eng 14:155–178

Cooling MT, Rouilly V, Misirli G, Lawson J, Yu T, Hallinan J, Wipat A (2010) Standard virtual biological parts: a repository of modular modeling components for synthetic biology. Bioinformatics 26:925–931

Deka RK, Anton IA, Dunbar B, Coggins JR (1994) The Characterization of the shikimate pathway enzyme dehydroquinase from *Pisum sativum*. FEBS Lett 349:397–402

Devi AS, Ebihara A, Kuramitsu S, Yokoyama S, Kumarevel T, Ponnuraj K (2013) Crystal structure of type I 3-dehydroquinate dehydratase of *Aquifex aeolicus* suggests closing of active site flap is not essential for enzyme action. Biochem Biophys Res Commun 432:350–354

Dias MVB, Snee WC, Bromfield KM, Payne RJ, Palaninathan SK, Ciulli A, Howard NI, Abell C, Sacchettini JC, Blundell TL (2011) Structural investigation of inhibitor designs targeting 3-dehydroquinate dehydratase from the shikimate pathway of *Mycobacterium tuberculosis*. Biochem J 436:729–739

Edgar RC (2004) MUSCLE: a multiple sequence alignment method with reduced time and space complexity. BMC Bioinforma 5:1–19

Euverink GJW, Hessels GI, Vrijbloed JW, Coggins JR, Dijkhuizen L (1992) Purification and characterization of a dual function 3-dehydroquinate dehydratase from *Amycolatopsis methanolica*. J Gen Microbiol 138:2449–2457

Giles NH, Case ME, Baum J, Geever R, Huiet L, Patel V, Tyler B (1985) Gene organization and regulation in the Qa (Quinic acid) gene-cluster of *Neurospora crassa*. Microbiol Rev 49:338–358

Giles NH, Geever RF, Asch DK, Avalos J, Case ME (1991) Organization and regulation of the Qa (Quinic acid) genes in *Neurospora crassa* and other fungi. J Hered 82:1–7

Gouet P, Courcelle E, Stuart DI, Metoz F (1999) ESPript: analysis of multiple sequence alignments in PostScript. Bioinformatics 15:305–308

Gourley DG, Shrive AK, Polikarpov I, Krell T, Coggins JR, Hawkins AR, Isaacs NW, Sawyer L (1999) The two types of 3-dehydroquinase have distinct structures but catalyze the same overall reaction. Nat Struct Biol 6:521–525

Harris JM, GonzalezBello C, Kleanthous C, Hawkins AR, Coggins JR, Abell C (1996) Evidence from kinetic isotope studies for an enolate intermediate in the mechanism of type II dehydroquinases. Biogeosciences J319:333–336

Hawkins AR, Lamb HK, Moore JD, Charles IG, Roberts CF (1993) The pre-chorismate (shikimate) and quinate pathways in filamentous fungi - theoretical and practical aspects. J Gen Microbiol 139:2891–2899

Herrmann KM, Weaver LM (1999) The shikimate pathway. Annu Rev Plant Phys 50:473–503

Kleanthous C, Deka R, Davis K, Kelly SM, Cooper A, Harding SE, Price NC, Hawkins AR, Coggins JR (1992) A comparison of the enzymological and biophysical properties of two distinct classes of dehydroquinase enzymes. Biochem J J282:687–695

Knight TF (2003) Idempotent vector design for standard assembly of Biobricks. http://hdl.handle.net/1721.1/21168

Lee WH, Perles LA, Nagem RA, Shrive AK, Hawkins A, Sawyer L, Polikarpov I (2002) Comparison of different crystal forms of 3-dehydroquinase from

Salmonella typhi and its implication for the enzyme activity. Acta Crystallogr D Biol Crystallogr 58:798–804

Lee BI, Kwak JE, Suh SW (2003) Crystal structure of the type II 3-dehydroquinase from *Helicobacter pylori*. Proteins 51:616–617

Leech AP, James R, Coggins JR, Kleanthous C (1995) Mutagenesis of active-site residues in Type-I dehydroquinase from *Escherichia coli* - stalled catalysis in a histidine to alanine mutant. J Biol Chem 270:25827–25836

Letunic I, Bork P (2011) Interactive tree of life v2: online annotation and display of phylogenetic trees made easy. Nucleic Acids Res 39:475–478

Markham NR, Zuker M (2008) UNAFold: software for nucleic acid folding and hybridization. Methods Mol Biol 453:3–31

Moore JD, Hawkins AR, Charles IG, Deka R, Coggins JR, Cooper A, Kelly SM, Price NC (1993) Characterization of the Type-I dehydroquinase from *Salmonella typhi*. Biogeosciences J295:277–285

Nakamura Y (2007) Condon usage database. http://www.kazusa.or.jp/codon/

Nishikura-Imamura S, Matsutani M, Insomphun C, Vangnai AS, Toyama H, Yakushi T, Abe T, Adachi O, Matsushita K (2014) Overexpression of a type II 3-dehydroquinate dehydratase enhances the biotransformation of quinate to 3-dehydroshikimate in Gluconobacter oxydans. Appl Microbiol Biotechnol 98:2955–2963

Pan Q, Yao Y, Li ZS (2012) Theoretical study of the reaction mechanism of Mycobacterium tuberculosis type II dehydroquinate dehydratase. Comput Theor Chem 1001:60–66

Peon A, Otero JM, Tizon L, Prazeres VFV, Llamas-Saiz AL, Fox GC, van Raaij MJ, Lamb H, Hawkins AR, Gago F, Castedo L, Gonzalez-Bello C (2010) Understanding the key factors that control the inhibition of Type II dehydroquinase by (2R)-2-benzyl-3-dehydroquinic acids. ChemMedChem 5:1726–1733

Richards TA, Dacks JB, Campbell SA, Blanchard JL, Foster PG, McLeod R, Roberts CW (2006) Evolutionary origins of the eukaryotic shikimate pathway: gene fusions, horizontal gene transfer, and endosymbiotic replacements. Eukaryot Cell 5:1517–1531

Roszak AW, Robinson DA, Krell T, Hunter IS, Fredrickson M, Abell C, Coggins JR, Lapthorn AJ (2002) The structure and mechanism of the type II dehydroquinase from *Streptomyces coelicolor*. Structure 10:493–503

Rytter J, Helmark S, Chen J, Lezyk M, Solem C, Jensen P (2014) Synthetic promoter libraries for *Corynebacterium glutamicum*. Appl Microbiol Biotechnol 98:2617–2623

Salis HM, Mirsky EA, Voigt CA (2009) Automated design of synthetic ribosome binding sites to control protein expression. Nat Biotechnol 27:946–U112

Schägger H (2006) Tricine-SDS-PAGE. Nat Protoc 1:16–22

Shetty R, Lizarazo M, Rettberg R, Knight TF (2011) Assembly of biobrick standard biological parts using three antibiotic assembly. Methods Enzymol 498:311–326

Singh SA, Christendat D (2006) Structure of arabidopsis dehydroquinate dehydratase-shikimate dehydrogenase and implications for metabolic channeling in the shikimate pathway. Biochemistry 45:7787–7796

Sleight SC, Bartley BA, Lieviant JA, Sauro HM (2010) In-fusion BioBrick assembly and re-engineering. Nucleic Acids Res 38:2624–2636

Sohoni SV, Fazio A, Workman CT, Mijakovic I, Lantz E (2014) Synthetic promoter library for modulation of actinorhodin production in *Streptomyces coelicolor* A3(2). PLos ONE 9:e99701

Vanholme R, Storme V, Vanholme B, Sundin L, Christensen JH, Goeminne G, Halpin C, Rohde A, Morreel K, Boerjan W (2012) A systems biology view of responses to lignin biosynthesis perturbations in *Arabidopsis*. Plant Cell 24:3506–3529

Weaver LM, Herrmann KM (1997) Dynamics of the shikimate pathway in plants. Trends Plant Sci 2:346–351

White PJ, Young J, Hunter IS, Nimmo HG, Coggins JR (1990) The purification and characterization of 3-dehydroquinase from *Streptomyces coelicolor*. Biochem J 265:735–738

Zhi XY, Yao JC, Li HW, Huang Y, Li WJ (2014) Genome-wide identification, domain architectures and phylogenetic analysis provide new insights into the early evolution of shikimate pathway in prokaryotes. Mol Phylogenet Evol 75:154–164

Physiological and growth response of rice plants (*Oryza sativa* L.) to *Trichoderma spp.* inoculants

Febri Doni[1*], Anizan Isahak[2], Che Radziah Che Mohd Zain[1] and Wan Mohtar Wan Yusoff[1]

Abstract

Trichoderma spp., a known beneficial fungus is reported to have several mechanisms to enhance plant growth. In this study, the effectiveness of seven isolates of *Trichoderma spp.* to promote growth and increase physiological performance in rice was evaluated experimentally using completely randomized design under greenhouse condition. This study indicated that all the *Trichoderma spp.* isolates tested were able to increase several rice physiological processes which include net photosynthetic rate, stomatal conductance, transpiration, internal CO_2 concentration and water use efficiency. These *Trichoderma spp.* isolates were also able to enhance rice growth components including plant height, leaf number, tiller number, root length and root fresh weight. Among the *Trichoderma spp.* isolates, *Trichoderma sp.* SL2 inoculated rice plants exhibited greater net photosynthetic rate (8.66 $\mu molCO_2\ m^{-2}\ s^{-1}$), internal CO_2 concentration (336.97 ppm), water use efficiency (1.15 $\mu moCO_2/mmoH_2O$), plant height (70.47 cm), tiller number (12), root length (22.5 cm) and root fresh weight (15.21 g) compared to the plants treated with other *Trichoderma* isolates tested. We conclude that beneficial fungi can be used as a potential growth promoting agent in rice cultivation.

Keywords: *Trichoderma spp*; Rice; Physiological response; Growth response

Introduction

Rice growth is influenced by a combination of genotype, environment and management factors. Balancing and improving soil fertility is one of the main factors in enhancing rice growth and yield. The intensive cultivation of rice that depends on chemical fertilizers and pesticides have led to the decrease in soil fertility and deteriorating soil health. The excessive use of chemical fertilizers in the current decades has led to soil toxicity through the presence of toxic heavy metals and adversely affecting the health of rice plants (Habibah *et al.* 2011). To overcome the decrease in soil fertility and deterioration in soil health caused by the use of chemical fertilizers, it is necessary to look for alternative ways to improve soil fertility and stimulate the growth of rice plants.

Microbes have been reported to be a key factor in maintaining soil quality and increasing rice yield and growth. The use of microbes to enhance rice growth while making the plant resistant to pathogens has been reported as an eco-friendly way to maintain the ecosystem. For decades, the application of microbes in a sustainable agroecological manner has increased rapidly due to their ability to act as plant growth promoters (Sakthivel and Gnanamanickam 1987; Carreres *et al.*, 1996; Malik *et al.*, 1997; Anhar *et al.*, 2011; Pedraza *et al.*, 2009; Amprayn *et al.*, 2012; Anizan *et al.*, 2012).

Trichoderma spp. recently was suggested as a Plant Growth Promoting Fungi (PGPF) due to their ability to produce siderophores, phosphate-solubilizing enzymes, and phytohormones (Doni *et al.*, 2013). *Trichoderma* species play important roles in decomposition, mycoparasitism, and even in cellulose degradation (Jiang *et al.*, 2011; Druzhinina *et al.*, 2006; Samuels, 1996). *Trichoderma spp.* was reported to be able to increase growth in plants such as strawberries, tomatoes, soya beans, apples, cotton and gray mangroves (Saravanakumar *et al.*, 2013; Raman, 2012; John *et al.*, 2010; Shanmugaiah *et al.*, 2009; Morsy et al. 2009; Porras *et al.*, 2007). However, very little research has been conducted on the potential of *Trichoderma spp.* for improving rice growth.

Recently, we reported the successful isolation of several *Trichoderma spp.* and which have been proven to

* Correspondence: doni@siswa.ukm.edu.my
[1]School of Biosciences and Biotechnology, Faculty of Science and Technology, Universiti Kebangsaan Malaysia, 43600 Bangi, Selangor, DE, Malaysia
Full list of author information is available at the end of the article

possess positive plant growth promotion potential in enhancing rice seed germination and vigour (Doni *et al.*, 2014). This research was conducted to examine the effect of *Trichoderma spp.* on rice growth and physiological response.

Materials and methods

This experiment was conducted at the Fermentation Technology Laboratory, Genetics Molecule Laboratory and Greenhouse, School of Biosciences and Biotechnology, Faculty of Science and Technology, Universiti Kebangsaan Malaysia. A completely randomized design (with nine treatments and three replications) was used for this experiment. Rice plants were placed in soil with different *Trichoderma spp.* inoculants and soil with NPK fertilizer 100 kg ha^{-1}. The NPK fertilizer used is in accordance to the rate employed in Micheal *et al.*, (2013) for rice variety MRQ74. Sterilized soil without any application was used for control.

Trichoderma spp. inoculants preparation

Local isolates of *Trichoderma spp.* which was isolated from System of Rice Intensification Paddy Field, Sik, Kedah, Malaysia namely *Trichoderma sp.* SL1, *Trichoderma sp.* SL2 (Public Accession number: UPMC 1021), *Trichoderma sp.* SL3, *Trichoderma sp.* SL4, *Trichoderma sp.* SL5, *Trichoderma sp.* SL6 and *Trichoderma sp.* SL7 were obtained from the Fermentation Technology Laboratory, School of Biosciences and Biotechnology, Faculty of Science and Technology, Universiti Kebangsaan Malaysia. Each isolate was grown separately in potato dextrose broth and incubated for seven days in a thermo shaker at a speed of 200 rpm and 30°C. After incubation, *Trichoderma spp.* inoculants were filtered using filter paper and stored in sterile polyethylene plastic bags.

Soil preparation and *Trichoderma spp.* inoculation

Autoclaved homogenous sandy clay loam soils were used for this experiment; the dosage of *Trichoderma spp.* inoculants was set at 5 g per 1 kg soil. The inoculated soil was immediately placed in 15 × 15 cm plastic containers.

Rice plant preparation

Rice seeds of the variety MRQ74 were surface sterilized with 70% ethanol, followed by 5% sodium hypochlorite and washed by sterilized distilled water. The rice seeds were grown in autoclaved sandy clay loam soil under green house condition with 30 ± 4°C temperature, 320 ± 3 µmol light intensity, 80 ± 3% humidity and 11 h 11 m 17 s ± 9 s photoperiod, and placed in a seedling tray. Seven day-old rice seedlings of MRQ74 were transplanted singly in 15 × 15 cm plastic containers containing different *Trichoderma spp.* treatments, NPK treatment and control. Water was maintained at 2 cm level from the soil surface and actively aerated by physically disturbing and breaking-up the soil surface once every ten days.

Measurement of rice growth and physiological components

Rice physiological and growth components were measured 30 days after transplanting. Measurements for the physiological characteristics were made on flag leaves. Measurements of net photosynthesis (µmol CO_2 m^{-2} s^{-1}), leaf stomatal conductance (mol H_2O m^{-2} s^{-1}), and transpiration rate (mmol H_2O m^{-2} s^{-1}) were monitored using a LICOR 6400 portable photosynthesis system (Lincoln, Nebraska, USA) and infrared gas analyser (IRGA). To assess the trade-off between CO_2 uptake and water loss, instantaneous water-use efficiency (WUE) was calculated as the ratio between photosynthetic rate and transpiration rate (µmol CO_2/µmol H_2O). These measurements were taken on a clear sunny day between 09:45 am until 11:30 am under a saturated light condition (solar radiation > 1200 µmol.m^{-2}.s^{-1}). Plant height (cm) measured from ground level to the tip of the longest leaf, tiller number and leaf number were counted for each treatment and control. For rice root length (cm) and root fresh weight (g) measurements, the rice plants were separated carefully from the soil. Root length was measured from the base of the stem to the longest root using a ruler and root fresh weight was measured using digital scales. Rice root dry weight (g) measurement was done after rice roots were dried in the oven at a temperature of 65°C for seven days.

Statistical analyses

All data were statistically analyzed using one-way analysis of variance (ANOVA). The significance of the effect of the treatment was determined using F-test and to determine the significance of the difference between the means of the treatments, least significant difference (LSD) was calculated at 5% probability level. Regression relationship was determined using the data analysis SPSS software version 20.

Results

Rice plant growth performance

The results of the experiments showed that the treatment of rice plants with *Trichoderma spp.* has a significant effect on rice plant growth performance (Table 1). Significant increase in plant height was observed for the *Trichoderma spp.* treated rice plants, registering plant heights in the range of 65.77 – 70.47 cm whilst the means for NPK treated rice plants and control were recorded at 28.9 cm and 63.1 cm respectively. Significantly greater values for leaf number and tiller number were also observed for *Trichoderma spp.* treated rice plants. The highest value for leaf number was 38 whilst for tiller number was 13 for the *Trichoderma spp.* treated plants.

Table 1 Comparison of plant height, leaf number, tiller number, root length, root dry weight and root wet weight in different treatments

Treatment	Plant height (cm)	Leaf number	Tiller number	Root length (cm)	Root dry weight (g)	Root fresh weight (g)
Trichoderma sp. SL1	69.23(3.29)	36(3.46)	12(1.53)	21.43(1.440)	2.68(1.39)	9.83(1.73)
Trichoderma sp. SL2	70.47(5.66)	37(4.93)	12(2.65)	22.5(1.76)	4.27(3.33)	15.21(3.20)
Trichoderma sp. SL3	65.77(1.75)	24(2.65)	9(1.15)	15.43(1.10)	1.94(1.13)	8.18(1.68)
Trichoderma sp. SL4	66.47(1.39)	30(5.20)	10(1.73)	17.17(0.65)	2.74(1.28)	11.57(0.49)
Trichoderma sp. SL5	66.50(4.30)	32(2.52)	13(3)	17.37(4.58)	2.74(0.91)	14.8(0.77)
Trichoderma sp. SL6	68.97(3.54)	37(2.65)	11(0)	21.9(2.71)	2.2(0.84)	12.96(3.51)
Trichoderma sp. SL7	67(1.15)	38(3.79)	12(1.73)	21.52(.30)	2.71(0.46)	15.01(3.38)
NPK	28.9(0.72)	3(0.58)	1(0)	9.33(0.91)	0.62(0.33)	3.07(0.76)
Control	63.1(0.96)	28(1)	8(1.53)	14.57(3.93)	1.94(0.37)	8.51(0.37)
$LSD_{0.05}$	55.46	33.08	13.02	9.18	ns	9.57

ns: Not significant.
Standard deviations are given in parentheses ($n = 27$).

On the other hand, the NPK treated rice plant registered 3 leaf number, whilst the control recorded 28. No tiller was observed for the NPK treated rice, whilst the control registered a tiller number of 8. Root length was found to be significantly increased for the *Trichoderma spp.* treated rice plants with a range between 17.17–22.5 cm whilst the NPK treated rice plants recorded root length of only 9.33 cm. For the control, root length was 14.57 cm. Root fresh weight was significantly higher for the *Trichoderma spp.* treated rice plants as well compared to NPK treated and control. The *Trichoderma spp.* treated rice plants registered the highest value which was 15.21 g whilst the root fresh weight for NPK treated plants and control were 3.07 g and 8.51 g respectively. Among the *Trichoderma sp.* SL2 treated rice plants showed the greatest increase in plant height, tiller number, root length and root fresh weight, while *Trichoderma sp.* SL7 had the greatest effect in enhancing the number of leaves. However, no significant improvement in root dry weight was seen for the *Trichoderma spp.* treated rice plants.

Rice plant physiological characteristic

The general assessment from this experiment is that the application of *Trichoderma spp.* to rice plants significantly increased the physiological properties of the rice plants (Table 2). Net photosynthetic rate was high in rice plants inoculated with *Trichoderma spp.* compared to NPK treatment and control. The range of net photosynthetic rate in rice inoculated with *Trichoderma spp.* was 6.74 – 8.79 μmol CO_2 m^{-2} s^{-1} while NPK treatment and control recorded values of 2.09 μmol CO_2 m^{-2} s^{-1} and 6.21 μmol CO_2 m^{-2} s^{-1} respectively. Significantly high values of stomatal conductance were recorded for rice plants inoculated with *Trichoderma sp.* SL3 and *Trichoderma sp.* SL6 registering 1237.88 and 1084.76 mmol H_2O m^{-2} s^{-1} respectively (Table 2). Stomatal conductance for NPK

Table 2 Comparison of net photosynthetic rate, stomatal conductance, transpiration and internal CO_2 concentration in different treatments

Treatment	Net photosynthetic rate (μmolCO_2 m^{-2} s^{-1})	Stomatal conductance (mmolH_2O m^{-2} s^{-1})	Internal CO_2 concentration (ppm)
Trichoderma sp. SL1	8.79(0.010)	979.08(0.00009)	358.91(0.037)
Trichoderma sp. SL2	8.66(0.007)	412.40(0.00070)	336.97(0.086)
Trichoderma sp. SL3	8.47(0.018)	1237.88(0.0022)	363.79(0.046)
Trichoderma sp. SL4	6.88(0.009)	752.65(0.0024)	358.60(0.023)
Trichoderma sp. SL5	7.38(0.043)	499.54(0.0008)	348.35(0.172)
Trichoderma sp. SL6	8.60(0.012)	1084.76(0.0014)	361.75(0.039)
Trichoderma sp. SL7	6.74(0.024)	712.38(0.0005)	361.23(0.076)
NPK	2.09(0.007)	340.16(0.0004)	376.69(0.045)
Control	6.21(0.016)	818.30(0.0002)	365.74(0.051)
$LSD_{0.05}$	0.10	0.005	3.70

Standard deviations are given in parentheses ($n = 36$).
All means were significantly different between treatments at $p < 0.05$.

treatment and control were 340.16 and 818.30 mmol H_2O m^{-2} s^{-1} respectively. Inoculation of rice plants with *Trichoderma spp.* also significantly influenced the internal CO_2 concentration of the rice plants. Results showed that *Trichoderma sp.* SL2 has the lowest CO_2 concentration (Table 2). Rice plants inoculated with *Trichoderma spp.* significantly enhanced the water use efficiency compared to NPK treatment and control (Figure 1). The highest water use efficiency was observed for *Trichoderma sp.* SL2 treated plants which registered 1.15 μmol CO_2/mmol H_2O.

Discussion

Rice plant growth performance

The benefit of *Trichoderma spp.* in improving plant growth can be realised through several mechanisms which include mycoparasitism, antibiosis, degradation of toxins, inactivation of pathogenic enzymatic pathways, resistance to pathogens, enhanced nutrient uptake, solubilization, sequestration of inorganic nutrients, and enhanced root hair development (Harman, 2006; Lorito *et al.*, 2010). In this research we tested the ability of *Trichoderma spp.* to enhance rice growth. The results showed that rice plants inoculated with *Trichoderma spp.* significantly increased rice growth components. Plant height of *Trichoderma spp.* inoculated rice plants was higher compared to NPK treatment and control. The ability of *Trichoderma spp.* to produce phytohormones is the key factor in the increase in rice plant height as reported by Chowdappa *et al.* (2013). Rice plants treated with *Trichoderma spp.* also have better nutrient uptake as suggested by Saba *et al.* (2012). Better nutrient uptake will enhance the physiological processes within the rice plants treated with *Trichoderma spp.* leading to good growth performance (Figure 2). Leaf number and tiller number were significantly higher in *Trichoderma spp.* treated rice plants compared to NPK treatment and control. The enhancement of leaf and tiller number by *Trichoderma spp.* were made possible because of the ability of the *Trichoderma spp.* to act through several mechanisms such as environmental buffering (against pH, drought, waterlogging, cold and heat), P solubilization and siderophore production (Neumann and Laing, 2006). Furthermore, Shukla *et al.* (2012) reported that *Trichoderma harzianum* significantly increased the ability of rice plants to tolerate drought stress and increase rice water-holding capacity. In this research, these mechanisms are believed to be contributing factors that led to higher leaf number and tiller number.

Figure 2 The use of *Trichoderma sp.* SL2 resulted better growth compared to control.

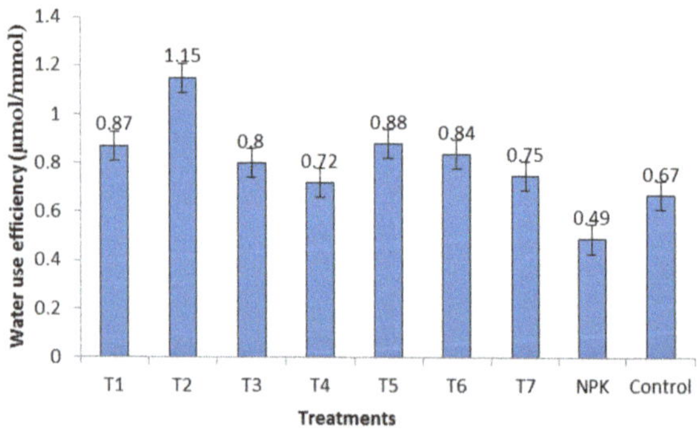

Figure 1 Changes in the water use efficiency by the different treatments.

Trichoderma spp. applied to rice plants reported in this research significantly increased rice root length compared to NPK treatment and control (Figure 3). *Trichoderma sp.* SL2 treated rice plants showed an impressive increase in root length compared to the plants treated with the other strains. Nawrocka and Malolepsza (2013) stated that the ability of *Trichoderma spp.* hyphae to release elicitors may contribute to signals being transmitted within the plant such as salicylic acid (SA), jasmonic acid (JA) and reactive oxygen species (ROS). Elicitors released by *Trichoderma spp.* are also involved in triggering expressions of defense protein within the plant (Thakur and Sohal, 2013). In this way, plant immunity against pathogens is induced and in turn improves plant growth. Tchameni *et al.* (2011) showed that the inoculation of *Trichoderma spp.* in cacao plants may induce cacao plant resistance against *Phytophthora megakarya* and increase cacao root length. In addition, Cai *et al.* (2013) reported that harzianolide

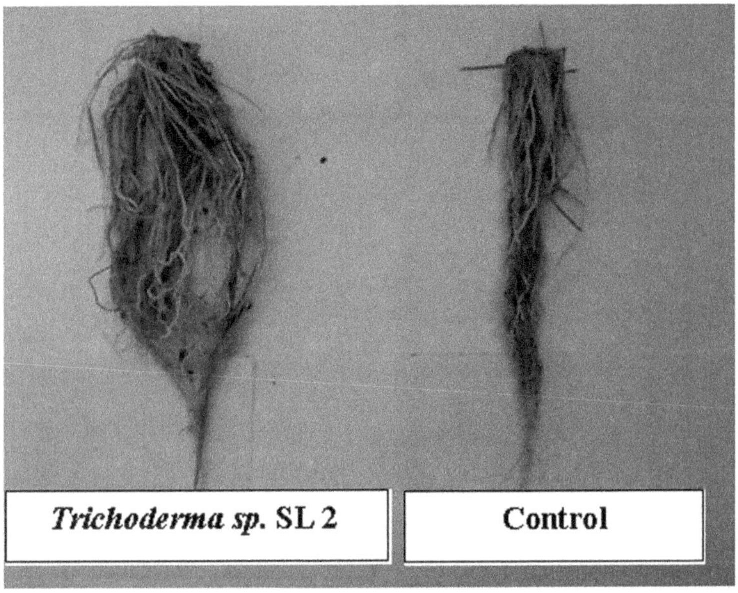

Figure 3 Rice plant root inoculated by *Trichoderma sp.* SL 2 recorded large mass compared to control.

produced by *Trichoderma spp.* can improve the early stage of plant development through the enhancement of root length.

The fresh weight of rice root plants treated with *Trichoderma spp.* was significantly greater than the NPK treated plants and control. However, the weight of dry rice root plants was not significant among treatments. Rice plants inoculated with *Trichoderma sp.* SL2 have the highest increase in root fresh weight. The capacity of *Trichoderma spp.* to produce growth hormones such as auxins and gibberelins were reported as the main factor that contributes to the ability of *Trichoderma spp.* to support root growth and increase water absorption from soil (Arora *et al.*, 1992; Contreras-Cornejo *et al.*, 2009; Martínez-Medina *et al.*, 2011). The present findings are also in agreement with previous research by Viterbo *et al.* (2010) that revealed the role of *Trichoderma asperellum* in promoting canola seedling root elongation via ACC deaminase (ACCD) activity.

Our study also reports a lack of response of the rice variety MRQ74 to NPK fertilizer. Recently, research done by Khairiah *et al.* (2013) revealed that the heavy metal content of the rice variety MRQ 74 plants in chemical fertilizers-used rice field area was high due to excessive usage of NPK fertilizers. Previous research by Yap *et al.* (2009) also suggested that the use of NPK fertilizers in rice cultivation significantly increase heavy metal content in rice plants. Further, Yadav (2010) stated that heavy metals may cause oxidative stress inside the plants leading to cellular damage. Our research is in agreement with previous researches which revealed the lack response of this variety to NPK fertilizer.

Rice plants physiological characteristics

Rice growth performance is subjected to environmental factors which affect the physiological processes inside rice plant cells. Improving rice physiological characteristics is considered to be desirable due to its agronomic importance towards the achievement of high rice yield (Makino, 2011; Li *et al.*, 2012). From Table 2, net photosynthetic rates, stomatal conductance, and internal CO_2 concentration were significantly different between treatments. High photosynthetic rates were observed for *Trichoderma sp.* SL1 and *Trichoderma sp.* SL2. It is known that large amounts of sucrose exudates provided by rice roots to *Trichoderma sp.* SL1 and *Trichoderma sp.* SL2 can facilitate good root colonization by *Trichoderma spp.* as well as promote good coordination of defense mechanism of rice plants (Vargas *et al.*, 2009). The results imply that the root physiological activity of *Trichoderma sp.* SL2 is the highest among the *Trichoderma spp.* studied. The results also revealed that *Trichoderma sp.* SL2 has the lowest internal CO_2 concentration and the highest photosynthetic rate. This indicates that the activity of carboxylation by CO_2 fixation for glucose production as carbohydrate metabolism in the rice plants was very active. The results agree with the findings of Thakur *et al.* (2010).

Research by Thakur *et al.* (2010) found that high photosynthetic rates coupled with low transpiration rates indicate high water use efficiency. This report supports our experiment on *Trichoderma sp.* SL2 which showed high values for these parameters. The activity of *Trichoderma spp.* that contributes to the enhancement of root growth and distribution was also considered as a key factor to the prolonged photosynthetic activity and the delayed senescence in rice plants (Mishra and Salokhe, 2011). Further, *Trichoderma spp.* was recently reported as having the potential to degrade cellulose (Jiang *et al.*, 2011). Cellulose degradation may release a large amount of N in rice plant rhizosphere. High N concentration uptake has positive correlation with photosynthetic rate.

In Figure 2, water use efficiency of *Trichoderma sp.* SL2 was the highest among the *Trichoderma spp.* isolates registering at 1.15 μmolmmol^{-1}. Stomatal conductance was significantly correlated with photosynthesis. Stomatal conductance plays an importance role in generating photosynthesis in rice plants because H_2O and CO_2 which are involved in photosynthetic process must pass through the stomata before they enter mesophyll cells and chloroplast stroma (Fu *et al.*, 2008). Further, Harman *et al.* (2004) stated that mechanisms employed by *Trichoderma spp.* in enhancing nutrient availability by solubilization and chelation of minerals can increase plant metabolism leading to the enhancement of plant physiological activity.

The present study concludes that *Trichoderma spp.* have the potential to enhance rice physiological processes and growth. In this respect, the present experiment proved that *Trichoderma sp.* SL2 was the best strain compared to six others strains.

Competing interest
All the authors declare that they have no competing interest.

Authors' contributions
FD and AI carried out the green house study and conducted the statistical analysis, and CRCMZ and WMWY drafted the manuscript. All authors read and approved the final manuscript.

Acknowledgments
We wish to thank Captain Zakaria Kamantasha and Tuan Hj. Marzuki Md. Zin from SRI Lovely, Sik, Kedah, for providing rice seeds for this research. This study was financially funded by Universiti Kebangsaan Malaysia under grant ETP-2013-070, Komuniti-2012-001, Komuniti-2012-007 and DLP-2013-023.

Author details
[1]School of Biosciences and Biotechnology, Faculty of Science and Technology, Universiti Kebangsaan Malaysia, 43600 Bangi, Selangor, DE, Malaysia. [2]School of Environmental and Natural Resources Sciences, Faculty of Science and Technology, Universiti Kebangsaan Malaysia, 43600 Bangi, Selangor, DE, Malaysia.

References
Amprayn K-O, Michael TR, Kecskés M, Pereg P, Nguyen HT, Kennedy IR (2012) Plant growth promoting characteristics of soil yeast (*Candida tropicalis* HY) and its effectiveness for promoting rice growth. App Soil Ecol 61:295–299

Anhar A, Doni F, Advinda L (2011) Respons pertumbuhan tanaman padi (*Oryza sativa* L.) terhadap introduksi *Pseudomonad flouresen*. Eksakta 1(1):1–11

Anizan I, Ahmad A, Rosenani AB, Habibah J (2012) SRI rice crop establishment. Trans Malaysian Soc Plant Physiol 20:11–15

Arora DK, Elander RP, Mukherji KG (1992) Fungal Biotechnology. In: Handbook of Applied Mycology. Marker Dekker, New York, p 4p

Cai F, Yu G, Wang P, Wei Z, Fu L, Shen Q, Chen W (2013) Harzianolide, a novel plant growth regulator and systemic resistance elicitor from *Trichoderma harzianum*. Plant Physiol Biochem 73:106–113

Carreres RGT, Sendra J, Ballesteros R, Fernández Valiente E, Quesada A, Nieva M, Leganés F (1996) Effect of nitrogen rates on rice growth and biological nitrogen fixation. J Agric Sci 127:295–302

Chowdappa P, Kumar SPM, Lakshmi MJ, Upreti KK (2013) Growth stimulation and induction of systemic resistance in tomato against early and late blight by *Bacillus subtilis* OTPB1 or *Trichoderma harzianum* OTPB3. Biol Control 65:109–117

Contreras-Cornejo HA, Macías-Rodríguez L, Cortés-Penagos C, López-Bucio J (2009) *Trichoderma virens*, a plant beneficial fungus, enhances biomass production and promotes lateral root growth through an auxin-dependent mechanism in arabidopsis. Plant Physiol 149:1579–1592

Doni F, Al-Shorgani NKN, Tibin EMM, Abuelhassan NN, Anizan I, Che Radziah CMZ, Wan Mohtar WY (2013) Microbial involvement in growth of paddy. Curr Res J Biol Sci 5(6):285–290

Doni F, Anizan I, Che Radziah CMZ, Salman AH, Rodzihan MH, Wan Mohtar WY (2014) Enhancement of rice seed germination and vigour by *Trichoderma spp.* Res J App Sci Eng Technol 7(21): (in press)

Druzhinina IS, Kopchinsky AG, Kubicek CP (2006) The first 100 *Trichoderma* species characterized by molecular data. Mycoscience 47:55–64. doi:10.1007/s10267-006-0279-7

Fu Y, Zheng-wei L, Zhi-chun W, Yuan C (2008) Relationship between diurnal changes of net photosynthetic rate and influencing factors in rice under saline sodic stress. Rice Sci 15(2):119–124

Habibah J, Lee PT, Khairiah J, Ahmad MR, Fouzi BA, Ismail BS (2011) Speciation of heavy metals in paddy soils from selected areas in Kedah and Penang, Malaysia. African J Biotechnol 10(62):13505–13513

Harman GE (2006) Overview of mechanisms and uses of *Trichoderma spp.* Phytopathol. doi:10.1094/PHYTO-96-0190

Harman GE, Petzoldt R, Comis A, Chen J (2004) Interactions between *Trichoderma harzianum* strain T22 and maize inbred line Mo17 and effect of this interaction on diseases caused by *Pythium ultmum* and *Colletotricum graminicola*. Phytopathol 94(2):147–153

Jiang X, Geng A, He N, Li Q (2011) New Isolate of *Trichoderma viride* strain for enhanced cellulolytic enzyme complex production. J Biosci Bioeng 111:121–127

John RP, Tyagi RD, Prévost D, Brar SK, Pouleur S, Surampalli RY (2010) Mycoparasitic *Trichoderma viride* as a biocontrol agent against *Fusarium oxysporum* f. sp. adzuki and *Pythium arrhenomanes* and as a growth promoter of soybean. Crop Prot 29:1452–1459

Khairiah J, Azie RR, Habibah J, Zulkifle I, Ismail BS (2013) Heavy metal content of paddy plants in Langkawi, Kedah, Malaysia. Aust J Basic Appl Sci 7(2):123–127

Li X, Bu N, Li Y, Ma L, Xin S, Zhang L (2012) Growth, photosynthesis and antioxidant responses of endophyte infected and non-infected rice under lead stress conditions. J Hazard Mater 213:55–61. doi:10.1016/j.jhamat2012.01.052

Lorito M, Woo SL, Harman GE, Monte E (2010) Translational research on *Trichoderma*: from 'omics to the field. Annu Rev Phytopathol 48:395–517

Makino A (2011) Photosynthetic, grain yield, and nitrogen utilization in rice and wheat. Plant Physiol 155:125–129

Malik KA, Rakhshanda B, Mehnaz B, Rasul G, Mirza MS, Ali S (1997) Association of nitrogen-fixing, Plant-Growth-Promoting Rhizobacteria (PGPR) with Kallar grass and rice. Plant Soil 194:37–44

Martínez-Medina A, Roldán A, Albacete A, Pascual JA (2011) The Interaction with Arbuscular Mycorrhizal Fungi or *Trichoderma harzianum* alters the shoot hormonal profile in melon plants. Phytochemistry 72:223–229. doi:10.1016/j.phytochem.2010.11.008

Micheal JSAS, Juraimi AS, Selamat A, Man A, Anwar MP, Uddin MK (2013) Critical period of weed control in aerobic rice system. AJCS 7(5):665–673

Mishra A, Salokhe VM (2011) Rice growth and physiological responses to SRI water management and implications for crop productivity. Paddy Water Environ 9:41–52. doi:10.1007/s10333-010-0240-4

Morsy EM, Abdel-Kawi KA, Khalil MNA (2009) Efficiency of *Trichoderma viride* and *Bacillus subtilis* as biocontrol agents gainst *Fusarium solani* on tomato plants. Egypt J Phytopathol 37(1):47–57

Nawrocka J, Malolepsza U (2013) Diversity in plant systemic resistance induced by Trichoderma. Biol Control 67:149–156. doi:10.1016/j.biocontrol.2013.07.005

Neumann B, Laing M (2006) *Trichoderma*: An Ally in the Quest for Soil System Sustainability. In: Uphoff N, Fernandes E, Herren H, Husson O, Laing M, Palm

C, Pretty J, Sanchez P, Sanginga N, Thies J (eds) Biological Approaches to Sustainable Soil System. Taylor & Francis, Boca Raton, FL, pp 491–500

Pedraza RO, Bellone CH, de Bellone SC, Sorte PMFB, dos Santos Teixeira KR (2009) *Azospirillum* Inoculation and Nitrogen Fertilization Effect on Grain Yield and on the Diversity of Endophytic Bacteria in the Phyllosphere of Rice Rainfed Crop. European J Soil Biol 45:36–43

Porras M, Barrau C, Romero F (2007) Effects of soil solarization and *Trichoderma* on strawberry production. Crop Prot 26:782–787

Raman J (2012) Response of *Azotobacter, Pseudomonas* and *Trichoderma* on Growth of Apple Seedling. International Conference on Biological and Life Sciences IPCBEE, IACSIT Press, Singapore

Saba HDV, Manisha M, Prashant KS, Farham H, Tauseff A (2012) *Trichoderma* – a promising plant growth stimulator and Biocontrol agent. Mycosphere. doi:10.5943/mycosphere/3/4/14

Sakthivel N, Gnanamanickam SS (1987) Evaluation of *Pseudomonas fluorescens* for suppression of sheath rot disease and for enhancement of grain yields in rice (*Oryza sativa* L.). Appl Environ Microbiol 53:2056–2059

Samuels GJ (1996) *Trichoderma*: a review of biological systemics of the genus. Mycol Res 100(8):923–935

Saravanakumar K, Shanmuga Arasu V, Kathiresan K (2013) Effect of *Trichoderma* on soil phosphate solubilization and growth improvement of *Avicennia marina*. Aquat Bot 104:101–105

Shanmugaiah V, Balasubramanian N, Gomathinayagam S, Manoharan PT, Rajendran A (2009) Effect of single application of *Trichoderma viride* and *Pseudomonas fluorescens* on growth cotton plants. African J Agri Res 4(11):1220–1225

Shukla N, Awasthi RP, Rawat L, Kumar J (2012) Biochemical and physiological responses of rice (*Oryza sativa* L.) as influenced by *Trichoderma harzianum* under drought stress. Plant Physiol Biochem 54:78–88

Tchameni SN, Ngonkeu MEL, Begoude BAD, Wakam Nana L, Fokom R, Owona AD, Mbarga JB, Tchana T, Tondje PR, Etoa FX, Kuaté J (2011) Effect of *Trichoderma asperellum* and Arbuscular Mycorrhizal Fungi on Cacao growth and resistance against black pod disease. Crop Prot 30:1321–1327. doi:10.1016/j.cropro.2011.05.003

Thakur M, Sohal BS (2013) Role of elicitors in inducing resistance in plants against pathogen infection: a review. ISRN Biochem :1–10. doi:10.1155/2013/762412

Thakur AK, Uphoff N, Antony E (2010) An assessment of physiological effects of system of rice intensification (sri) practices compared with recommended rice cultivation practices in india. Cambridge J 46(1):77–98

Vargas WC, Mandawe JC, Kenerley CM (2009) Plant derived sucrose is a key element in the symbiotic association between *Trichoderma virens* and maize plants. Plant Physiol 151:792–808

Viterbo A, Landau U, Kim S, Chernin L, Chet I (2010) Characterization of AAC deaminase from the biocontrol and plant growth-promoting agent *Trichoderma asperellum* T203. FEMS Microbiol Lett 305:42–48

Yadav SK (2010) Heavy metals toxicity in plants: an overview on the role of glutathione and phytochelatins in heavy metal stress tolerance of plants. S Afr J Bot 76:167–179. doi:10.1016/j.sajb.2009.10.007

Yap DW, Adezrian J, Khairiah J, Ismail BS, Ahmad-Mahir R (2009) The uptake of heavy metals by paddy plants (*Oryza sativa*) in Kota Marudu, Sabah, Malaysia. Am Eurasian J Agric Environ Sci 6(1):16–19

16

Identification and characterization of endophytic bacteria from corn (*Zea mays* L.) roots with biotechnological potential in agriculture

Vivian Jaskiw Szilagyi-Zecchin[1*], Angela Cristina Ikeda[1], Mariangela Hungria[2], Douglas Adamoski[1], Vanessa Kava-Cordeiro[1], Chirlei Glienke[1] and Lygia Vitória Galli-Terasawa[1]

Abstract

Six endophytic bacteria of corn roots were identified as *Bacillus* sp. and as *Enterobacter* sp, by sequencing of the 16S rRNA gene. Four of the strains, CNPSo 2476, CNPSo 2477, CNPSo 2478 and CNPSo 2480 were positive for the nitrogen fixation ability evaluated through the acetylene reduction assay and amplification of *nifH* gene. Two *Bacillus* strains (CNPSo 2477 and CNPSo 2478) showed outstanding skills for the production of IAA, siderophores and lytic enzymes, but were not good candidates as growth promoters, because they reduced seed germination. However, the same strains were antagonists against the pathogenic fungi *Fusarium verticillioides*, *Colletotrichum graminicola*, *Bipolaris maydis* and *Cercospora zea-maydis*. As an indication of favorable bacterial action, *Enterobacter* sp. CNPSo 2480 and *Bacillus* sp. CNPSo 2481 increased the root volume by 44% and 39%, respectively, and the seed germination by 47% and 56%, respectively. Therefore, these two strains are good candidates for future testing as biological inoculants for corn.

Keywords: Molecular phylogeny; 16S rRNA; *nifH*; Plant growth promotion; Antagonism

Introduction

Literature shows that some bacteria, which live attached to plants have the ability to promote plant growth (Compant et al. 2010; George et al. 2012; Beneduzi et al. 2013). These microorganisms have attracted attention because of the need to reduce the use of chemicals, especially when considering the context of sustainable agriculture and environmental protection (Vale et al. 2010). One of the strategies is to exploit the benefits that several microorganisms may give to plants when added as inoculants (Lucy et al. 2004).

Bacteria promoting plant growth can act directly, through one or more mechanisms, including biological nitrogen fixation (Ashraf et al. 2011; Verma et al. 2013), phosphate solubilization (Rodriguez et al. 2004; Krey et al. 2013), production of hormones such as auxins, gibberellins and zeatin (Cassan et al. 2009), or act indirectly by means of biological control of pathogens (Wang et al. 2009). In addition, some studies point out to the preferred action with endophytic microorganisms or adapted isolates, due to easier colonization and the lower risk to introduce exogenous organisms (Enebak et al. 1998; Khalid et al. 2004).

Several studies have shown the positive effects of endophytic bacteria inoculation in plants, e.g. sugarcane (*Saccharum* spp.), leading to increased contribution of biological nitrogen fixation, to promotion of root development, increased biomass and productivity (Oliveira et al. 2003); soybean (*Glycine max* (L.) Merr.), with bacteria capable to inhibit growth and sporulation of pathogenic fungi (Assumpção et al. 2009); tomato (*Lycopersicum esculentum* L.), with bacteria increasing plant height, leaf area, leaf number, together with fresh and dry plant weight (Barreti et al. 2008).

Benefits provided by the inoculation of corn (*Zea mays* L.) with selected strains have also been described, including the promotion of growth due to the increased availability of nutrients, provided by inoculation with *B. subtilis* in seeds (Canbolat et al. 2006). In addition,

* Correspondence: vivian.szilagyi@gmail.com
[1]Departamento de Genética, Universidade Federal do Paraná (UFPR), Setor de Ciências Biológicas, Avenida Coronel Francisco Heráclito dos Santos, 210, Curitiba, Paraná 81531-970, Brazil
Full list of author information is available at the end of the article

there are reports of strains of *Azospirillum brasilense* increasing corn yield by 24% to 30%, compared to the non-inoculated control (Hungria et al. 2010).

However, for the selection of isolates that have a greater chance of being efficient in the process of colonization and growth promotion, the isolates have first to be identified with preliminary in vitro biochemical analysis (Pal et al. 2001; Berg et al. 2002; Bernardes et al. 2010).Taxonomy of selected microorganisms with biotechnological potential in agriculture is thus a key step and, for this reason, sequencing of the 16S rRNA has been successfully used to identify genera such as *Bacillus* (Porwal et al. In 2009, Deepa et al. 2010), *Rhizobium, Burkholderia* (Ferreira et al. 2011), *Enterobacter, Pantoea Serratia*, among others (Tian et al. 2009).

The aim of this study was to evaluate and characterize the properties of endophytic bacteria isolated from corn roots, with the potential to have biotechnological interest for their use in agriculture.

Material and methods

Biological material

The seven bacteria studied belong to the Laboratory of Microorganisms Genetics Collection (LABGEM) of the Federal University of Parana, and are deposited at the Diazotrophic and Plant Growth Promoting Bacteria Culture Collection of Embrapa Soja (WDCC Collection # 1054), with the record numbers from CNPSo 2475 to CNPSo 248. GeneBank accession numbers for the 16S rRNA go from JQ821359 to JQ821365. These bacteria were isolated from corn roots samples by Ikeda et al. (2013). For the isolation, corn roots from field trial conducted at Semilia Genetics and Breeding LTDA, located in the southern Brazilian region of Campo Largo-PR, were surface-sterilized (Petrini, 1986) and aseptically cut, then transferred to Petri dish plates containing solid culture media. Several controls confirmed that the sterilization procedure was effective.

The fungi *Cercospora zea-maydis* (MMBF 51/08) and *Bipolaris maydis* (MMBF 46/08) were kindly provided by the Biological Institute of São Paulo, taken from the Micoteca "Mario Barreto Figueiredo". The *Fusarium verticillioides* (CMM-1060) and *Colletotrichum graminicola* (CMM-1061) were deposited in the Culture Collection of Phytopathogenic Fungi "Prof. Maria Menezes", with GenBank accession numbers for the 18S rRNA of KC335155 and KC335154.

Bacterium identification by analyzing the 16S rRNA gene

Extraction of bacterial DNA

Genomic DNA was obtained using the Raeder and Broda protocol (1985), with the following modifications: bacteria were grown overnight in 4 mL of DYGS medium (Rodrigues Neto et al. 1986) and then centrifuged for 2 min at 10.000 *g*. Then, a 2 mL pellet of extraction buffer (125 mMol/L Tris–HCl pH 8.0; 2 Mol/L NaCl; 50 mMol/L EDTA pH 8.0 and 1% SDS pH 7.2) was added, preheated at 60°C. After drying, the DNA was re-suspended in 50 μL of ultrapure water.

The integrity was checked by electrophoresis on agarose gel 0.8% (w/v), stained with GelRedTM (Biotium, USA), diluted in a sample buffer for 100X, observed in a UV transluminator (Ultraviolet Benchtop transilluminators) and photo-documentation application (Program Digidoc it Quantification). Purity was evaluated by spectrophotometer (per wavelength) NanoDrop® 2000 (Thermo Scientific).

Amplification and purification of the coding region of the 16S rRNA gene

The amplification was performed by PCR (Polymerase Chain Reaction) with primers fD1 (5′AGA GTTTGAT CCTGGCTCAG 3′) and rD1 (5′AAGGAGGTGATC CAGCC 3′) (Weisburg et al. 1991). The cycle included an initial denaturation at 95°C for 2 min, 30 cycles of 15 s at 94°C, 45 s at 93°C, 45 s at 55°C, 2 min at 72°C, followed by a final extension of 5 min at 72°C. Purification with 7.5 M ammonium acetate was performed according to Menna et al. (2006). Ultrapure water was used to re-suspend the DNA. The amount and purity were assessed as described above on agarose gel 1.5% (w/v), adjusting the final concentration at 16 ng/μL.

Sequencing and analysis of the 16S rRNA gene

For the second PCR reaction, the following primers, designed by Prof. Leonardo M. Cruz (Dept. Biochemistry, Federal University of Paraná, Curitiba, PR, Brazil), were used: 362f (5′ CTCCTACGGGAGGCAGCAGTG GGG 3′), 786f (5′ CGAAAGCGTGGGGAGCAAAC AGG 3′), in addition to the fD1 primer (Weisburg et al. 1991). Eighty ng of the purified product from the first PCR reaction, 0.25 mM of primer and 2 μL ET mixture were used for sequencing. Ultrapure water was added to reach the final volume of 10 μL. Amplification was performed for 35 cycles of 30 s at 94°C, 15 s at 50°C and 90 s at 60°C.

After amplification, purification was performed with Sephadex™ filtration gel, G-50 medium (GE® Healthcare) and then the product was submitted to electrophoresis with an automated DNA sequencing analyser, model MegaBACE1000 (Amersham Biosciences®).

The quality of the bases and the fragment assembly were analyzed with the PhredPhrapConsed software (Ewing et al.1998). Alignments were carried out with the PRANK software (Löytynoja and Goldman 2010). The percentage of similarity was obtained from the RDP (Ribosomal Database Project).

The phylogenetic analysis was made using Bayesian inference with the MrBayes 3.1.2 software (Ronquist and

Huelsenbeck 2003) using parallel computing (Altekar 2004), with four chains. The evolutionary model was selected through the JModelTest program (Posada and Kandral 1998). The number of generations and "burn-in" were defined by analysis of multiple sequences, with topological stability assessment by the AWTY package (Nylander et al. 2008) and analyzed by the parametric stability software, Tracer 1.5. With appropriate values of "burn-in", the final trees were summarized with the software package SumTrees DendroPy 3.8.1 (Sukumaran and Holder 2010).

Amplification of the nitrogenase gene *nifH*

The PCR was performed with primers *nifH*F (5′ TAC GGNAARGGSGGNATCGGCAA 3′) and *nifH*I (5′ AGC ATGTCYTCSAGYTCNTCCA 3′) (Boldface letters denote degenerate positions N = A + C + G + T; R = A + G; S = G + C; Y = C + T; W = A + T), as follows: 1.5 mM $MgCl_2$; 20 μM (each) dATP, dCTP, dTTP,dGTP; 0.1 μM (each) primers; 0.04 U Taq DNA polymerase/μL. The modified steps for amplification were: an initial denaturation at 94°C for 2 min, 30 cycles of 1 min 94°C, 1 min at 55°C, 3 min at 72°C, followed by a final extension of 5 min at 72°C (Laguerre et al. 2001).

With another pair of primers *nifH*-F (5′ AAAGGYGG WATCGGYAARTCCACCAC 3′) and *nifH*-R (5′ TTGT TSGCSGCRTACATSGCCATCAT 3′), the following steps were performed in the PCR: a 4-min hot start at 97°C before *Taq* DNA polymerase was added; 1 cycle of denaturation at 96°C for 20 s, annealing at 65°C for 30 s, and elongation at 72°C for 30 s; 2 cycles of denaturation at 96°C for 20 s, annealing at 62°C for 30 s, and elongation at 72°C for 35 s; 3 cycles of denaturation at 96°C for 20 s, annealing at 59°C for 30 s, and elongation at 72°C for 40 s; 4 cycles of denaturation at 96°C for 20 s, annealing at 56°C for 30 s, and elongation at 72°C for 45 s; 5 cycles of denaturation at 96°C for 20 s, annealing at 53°C for 30 s, and elongation at 72°C for 50 s; 25 cycles of denaturation at 94°C for 20 s, annealing at 50°C for 45 s, and elongation at 72°C for 60 s; and an extension at 72°C for 10 min (Rösch et al. 2002). Products of the PCR were confirmed in 1.5% (w/v) agarose gels. A Low DNA Mass Ladder (Invitrogen®) and photo-documentation, as described above.

Enzymatic and physiological characterization of bacteria

Biological nitrogen fixation

Biological nitrogen fixation was verified according to the methodology of Döbereiner et al. (1995), by using NFb semi-solid medium (Baldani et al. 1986). The target was the formation of a sub-surface pellicle.

The nitrogen-fixation ability was confirmed by using the ARA (acetylene reduction assay), as described by Boddey (1987). Each bacterial isolate was inoculated into 10 mL vials containing the respective 4 mL semi-solid N-free medium NFb (Baldani et al. 1986), TBNR (Seldin et al. 1984) and LGD (Döbereiner 1980). All isolates were incubated at 30°C in the dark. After 24 h, 10% (v/v) of the air phase was replaced with acetylene (Burris, 1972) and then the vials were incubated again. After the addition of acetylene, three vials for each isolate were sampled over 1 h to determine the amount of ethylene. Ethylene was measured using a Clarus 600 gas chromatograph (PerkinElmer®) with a Col-Elite-Alumina column and a flame ionization detector connected to a chromatography data computer system. *Azospirillum brasilense* Ab-V5 was used as positive control.

Production of indoleacetic acid

Quantification of indoleacetic acid (IAA) was performed according to Kuss et al. (2007). Single bacterial colonies were inoculated in 2 mL of DYGS medium supplemented with 2 mg/mL of tryptophan and incubated at 30°C for 72 h at 180 rpm. IAA was measured by mixing 1 mL broth with 1 mL of Salkowsky's reagent (Loper and Schroth, 1986) and incubated for 30 min at room temperature. Bacterial cells were separated from the supernatant by centrifugation at 10,000 rpm for 5 min and absorbance at 530 nm was determined. A standard IAA curve was plotted with commercial auxin (Sigma Aldrich, USA) and correlated with the absorption of the bacterial strains. Values are expressed in μg/mL.

Siderophore production and solubilization of inorganic calcium phosphate

The confirmation of siderophore production was performed according to Schwyn Neilands (1987) universal method, with Chrome Azurol S-(CAS) in Petri dishes, with DYGS solid medium. The spot was inoculated with test organism and incubated at 30°C for 3–5 days. Development of a yellow–orange halo around the colony was considered as a positive result.

Solubilization of calcium phosphate was done according to Sylvester-Bradley et al. (1982). Bacteria were grown in GY (glucose yeast medium) (10 g glucose, 2 g yeast extract and 15 g agar per liter). Two other solutions were prepared separately, one containing 5 g K_2HPO_4 in 50 mL of distilled water, and the other containing 10 g $CaCl_2$ in 100 mL of distilled water. These solutions were added to one liter of GY medium just before pouring onto Petri dishes, and together they formed an insoluble layer of calcium phosphate that made the medium opaque. Clear halos around the colony were recorded as a positive result (after 12 days).

Production of ammonia and hydrocyanic acid

Confirmation of the production of ammonia was done by colorimetric reaction, by adding 1 mL Nessler's reagent to a 72-h-old culture grown in peptone broth and recording the presence of the yellowish brown color (Dye 1962).

For hydrocyanic acid (HCN), the Bakker and Schippers (1987) method was followed in King B medium (King et al. 1954), supplemented with 4.4 g/L of glycine. HCN production was inferred with change in the color of the filter paper previously dipped in 2% sodium carbonate prepared in 0.05% picric acid, and it was rated visually depending on the intensity of the color change from yellow to dark brown.

Detection of lytic enzymes

Bacteria were seeded in a mineral medium containing 0.5% carboxymethylcellulose as sole carbon source for detection of cellulose activity. Colloidal chitin at 0.08%, obtained according to Moura et al. (2005) for the detection of chitinase and 0.5% of laminarin for the detection of β-1,3-glucanase (Renwick et al. 1991) were used.

The amylolytic and pectinolytic activities were evaluated after Hankin and Anagnostakis (1975), in a minimal M9 medium (Sambrook and Russell 2001) containing 0.5% of yeast extract and 0.2% soluble starch (v/w) to test for amylase, and pectin 0.5% (v/w) to test for pectinase.

The proteolytic activity of strains was verified through the degradation of casein in half-skimmed milk agar (Berg et al. 2002). The appearance of a halo around the colonies was recorded as positive (after seven days).

Antifungal activity

The paired cultures test was performed according to Shiomi et al. (2008), with potato dextrose agar (PDA) medium, using two equidistant grooves and a 5 mm disc of fungi mycelia placed in the center of the plate, grown for five days. The inhibition activity was evaluated by measuring the radius of fungal growth and comparing it with the control that was not treated with bacteria.

Germination of inoculated corn seeds

A bacterial cell suspension was incubated at 30°C for 24 hours at 150 rpm in DYGS medium. Then the suspension was centrifuged at 5.000 *g* and re-suspended in saline solution (0.85% NaCl). The concentration was adjusted to 2×10^9 CFU/mL, in a spectrophotometer, to OD_{540} nm, from an exponential growth curve previously performed. One mL of this solution was inoculated into 100 commercial hybrid corn seeds SX2530 (Semilia LTD), previously surface disinfested (Shiomi et al. 2008); tissues were also internally disinfected with warm water to eliminate fungi (Daniels 1983). To confirm the disinfection, samples of seeds were germinated in Petri dishes with PDA medium. The amount of cells in the seed was measured by doing plate counts (CFU). Bacteria were extracted by using 0.085% saline solution agitated by vortexing (Silva and Reis 2004). Seeds were germinated in an incubator at 28°C, on filter paper moistened with distilled water. Evaluations of the length (cm) and volume (cm^3) of the root and of the hypocotyl were made after 4 and 7 days (Cassán et al. 2009) using the Win-rhizo v.4.0 software (Regent Systems, Quebec, Canada). Statistical analysis were performed on the Assistat 7.6 Beta software, and the data were set to 5% confidence level ($p > 0.05$).

Results

Molecular phylogeny of the bacterial isolates

After aligning the sequences of the gene 16S rRNA, a region of 1512 bp was used for phylogenetic analysis, along with type strains taken from the RDP. Strain CNPSo 2480 grouped with the *Enterobacter* branch with 100% Bayesian posterior probability (BPP). This same strain had a 96% BPP and 98% similarity with type strain *E. asburiae* JCM6051^T. The six other bacterial strains belong to the genus *Bacillus*. Among them, CNPSo 2481 showed 100% BPP and 98% similarity with *B. pumilus* DSMZ27^T (Figure 1).

The sequences of strains CNPSo 2479 (LGMB227), CNPSo 2476 (LGMB163), CNPSo 2477 (LGMB128), CNPSo 2480 (LGMB184), CNPSo 2481 (LGMB186), CNPSo 2475 (LGMB155), CNPSo 2478 (LGMB135) were deposited at the Genbank and received the accession numbers of JQ821359 to JQ821365.

Amplification of nifH gene

Strains *Bacillus* CNPSo 2477 and CNPSo 2476 were positive for *nifH* gene amplification (primers *nifH*F and *nifH*I), producing an amplified fragment of about 780 bp. *Enterobacter* sp. CNPSo 2480 was positive for the two pairs of primers tested (*nifH*F and *nifH*I; *nifH*-F and *nifH*-R). *Bacillus* sp CNPSo 2478, amplified with the second primer, resulted in a fragment of about 400 pb.

Enzymatic and physiological characterization of bacteria

In relation to the nitrogen fixation capacity evaluated in vitro, all strains were able to grow in NFb semi-solid medium and showed the typical pellicle. But only *Enterobacter* sp. CNPSo 2480 showed the ability to reduce acetylene to ethylene in NFb N-free semi-solid medium. *Bacillus* CNPSo 2476, CNPSo 2477and CNPSo 2478 showed the same property in TBNR and LGD.

Only *Enterobacter* sp. CNPSo 2480 did not produce siderophore and *Bacillus* sp. CNPSo2481 produced ammonia.

All strains were positive for IAA production with values between 35.1 and 96.3 μg/mL, especially *Bacillus* CNPSo 2477 and CNPSo 2478, with 105.2 and 105.11 μg/mL respectively.

Lytic enzymes activities were detected in the following strains: cellulase CNPSo 2477, CNPSo 2475 and CNPSo

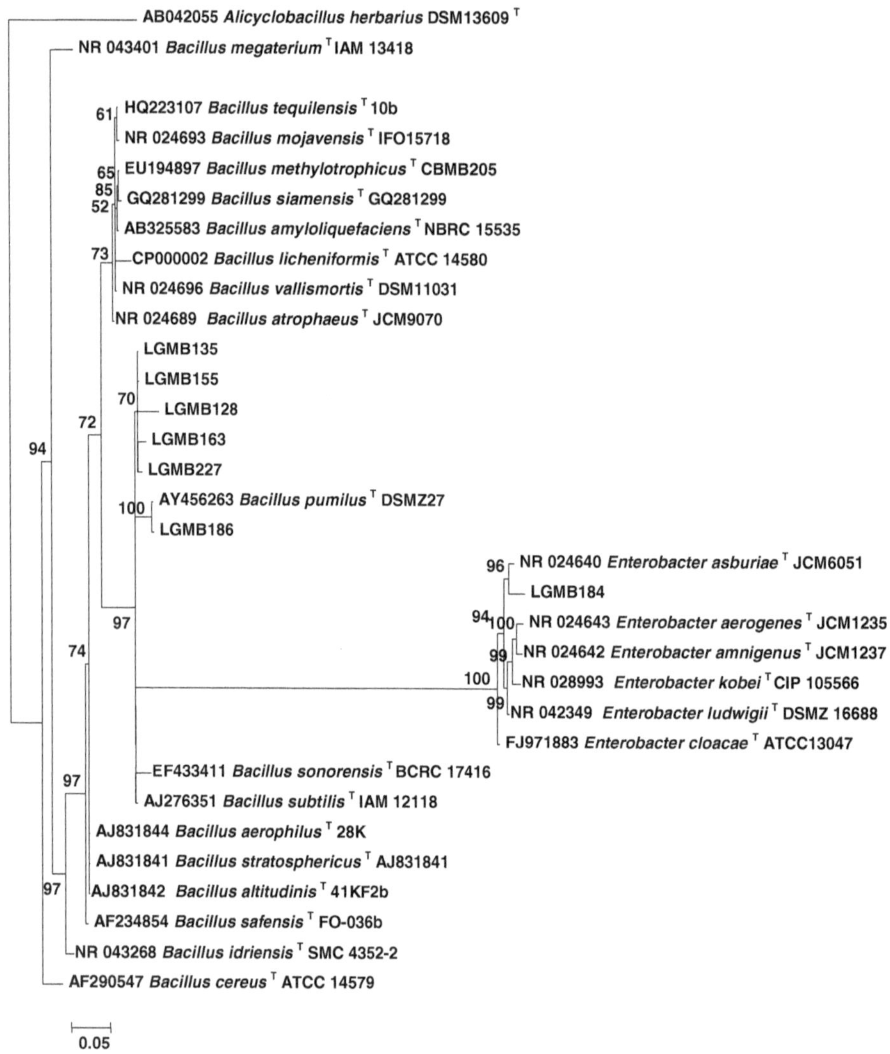

Figure 1 Phylogenetic tree based on the 16S rRNA gene sequence of the isolates in this study and reference strains obtained from the RDP. Used were 1512 bp fragment after alignment. Data generated using Bayesian inference, with outgroup *Alicyclobacillus herbarius*. Suportes branches less than 50% were not shown.

2476; amylase CNPSo 2477, CNPSo 2478, CNPSo 2475 and CNPSo 2476; pectinase CNPSo 2477, CNPSo 2478, CNPSo 2475, CNPSo 2476 and CNPSo 2479; protease, for degradation of casein CNPSo 2478, CNPSo 2475, CNPSo 2476, CNPSo 2481 and CNPSo 2479. All strains were negative for all phosphate solubilization tests, production of HCN, chitinase and β1,3-glucanase.

Antifungal activity

The seven bacterial strains significantly reduced the growth of fungi, with inhibition above 70% of the following pathogens: CNPSo 2477 (*B. maydis, C. graminicola* and *C. zea-maydis*); CNPSo 2478 (*B. maydis* and *C. graminicola*); CNPSo 2475 (*C. zea-maydis*); and CNPSo 2476, CNPSo 2480, CNPSo 2481, CNPSo 2479 (*C. graminicola* and *C. zea-maydis*).

Germination of inoculated corn seeds

The inoculation resulted in 10^7 CFU/seed. The strains *Bacillus* sp. CNPSo 2477 and CNPSo 2478 reduced the germination percentage by about 27%. In contrast, germination increased to approximately 52% with CNPSo 2475, CNPSo 2480 and CNPSo 2481. CNPSo 2476 and CNPSo 2479 increased germination to about 36.5% and 15.8%, respectively. The bacteria CNPSo 2476, CNPSo 2480 and CNPSo 2481 promoted root volume increases (14.6%, 43.9% and 39.02% respectively). CNPSo 2480 and CNPSo 2481 also promoted an increment in root length. No changes were observed in hypocotyl length and volume.

Discussion

Sequencing of 16S rRNA allows the accurate identification of bacterial genera endophytes from various plant species, including: sugar cane (Ratón et al. 2011), corn (Pereira et al. 2011) and rice (*Oryza sativa* L.) (Beneduzi et al. 2008). However, for the precise definition of species and even subspecies such as *Bacillus subtilis* the analysis of other genes, such as *gyrA* (Chun and Bae 2000) is needed.

All strains from this study showed an indication of the nitrogen fixation by the formation of the typical pellicule in semi-solid N-free medium. However, in only four strains (CNPSo 2476, CNPSo 2477, CNPSo 2478 and CNPSo 2480) the *nifH* gene was amplified. The nitrogen fixation genes are found in different phylogenetic groups (Affourtit et al. 2001). Among them, *nifH* is one of the oldest and most functional (Rosado et al. 1998) and its amplification by degenerative primers is a useful tool to confirm fixation potential (Zehr et al. 1996). However, if amplification does not occur with these primers, it does not mean that the strains are not capable of nitrogen fixation, because the gene may have different nucleotide sequences between species and even within the same species Zehr et al. (2003). In our study, the four strains that amplified *nifH* with different primers were also positive to the ARA test. This result indicated that strains have the possibility to fix nitrogen. Montañez et al. (2009) also verified the same situation in bacteria with nitrogen fixation capability by the reduction of acetylene to ethylene and confirmed by PCR the presence of *nifH* gene. The group of Bacilli was reported to be capable of nitrogen fixation, Ratón et al. (2011) and Rana et al. (2011) working with ARA-positive strains, indicated suitable environments for searching new inoculants or candidates for biotechnological purposes.

In our study, all strains produced IAA, although in different amounts. Bacteria that produced lower levels (CNPSo 2480 and CNPSo 2481), had bigger influence on root elongation and root volume, characteristics of great interest that provide greater surface area for the absorption of nutrients. The effects of the auxin depend on its concentration, i.e. when it is low it can stimulate growth and when it is high can be inhibitory, same for seed germination (Arshad and Frankenberger 1991). This may occur because the IAA can modify the endogenous auxin of the plant to an optimal or deleterious level (Patten and Glick 1996). For example, Sarwar and Kremer (1995), by comparing the production of auxin between promoting and inhibiting bacteria growth, noted that the latter produced high levels and inhibited root growth of *Convolvulus arvensis*; Baranzani and Friedman (1999) observed the same effect with lettuce (*Lactuca sativa* L.). Araujo and Guerreiro (2010), working with *Bacillus*, found that most strains of corn identified as growth promoters did not correspond to those producing higher levels of IAA. The species *Pseudomonas fluorescens*, whose production of IAA was only 15.63 µg/mL, increased root length in corn plants, according to Hernández-Rodríguez et al. (2008).

The ability to produce siderophores has been reported in *Bacillus* (Pereira and Castro-Silva, 2010; Kumar and Kumar 2012) and *Enterobacter* (Tian et al. 2009; Arruda et al. 2013). Out of the strains tested in this study, only *Enterobacter* sp. (CNPSo 2480) did not produce siderophores. Siderophores produced by bacteria that promote plant growth may act through two mechanisms: directly promoting growth, as iron availability to plants is generally low, therefore organic chelators produced by bacteria will help absorption (Powell et al. 1980), and indirectly, by inhibiting the availability of iron to pathogens, thereby limiting pathogen growth (Ahmad et al. 2008).

Lytic enzymes such as β-1,3-glucanases, cellulases, proteases, amylases and chitinases are connected to hyperparasitic activities (Kim and Chung 2004) since they have the ability to degrade structural fungal cell walls (Oppenheim and Chet 1992). Pectinases, as well as cellulases, also assist in penetration of bacteria into the plant host. They can also promote the induction of systemic resistance (Hallmann et al. 1997). In our study, these enzymes were detected in strains of the genus *Bacillus*, corroborating Ratón et al. (2011). The inhibitory action on fungal pathogens can also be related to the ammonia released by bacteria, as reported by Fravel (1988) with *Enterobacter cloacae* on *Verticillium dahliae* and *Pythium ultimum*.

In Brazil, among the field phytopathogens that affect corn, *F. verticilliodes* is the most common (Peixoto et al. 1998). The following pathogens are also common: *C. graminicola*, *B. maydis* (Pinto 1998) and *C. zea-maydis*, and have become more important in recent decades in Brazil (Pereira et al. 2005). Among the strains tested in our study, *Bacillus* sp. CNPSo 2477 and CNPSo 2476 were the most efficient in the inhibition against all four fungi tested. They are known to have an antagonist effect against various fungi, for example, *F. moniliforme*, *F. graminearum*, *M. phaseolina* (Pal et al. 2001), *F. oxysporum* and *Colletotrichum truncatum* (Araujo and Guerreiro 2010). One of the advantages of the biological control strategy using endophytic bacteria is that they can act in the same niche, in direct competition with the pathogens (Bacon and Hinton, 2002).

Based on the 16 s rRNA sequence, the endophytic bacteria from our study were identified as *Enterobacter asburie* and *Bacillus pumilus*, while the other strains were confirmed as belonging to the genus *Bacillus*. We found that *Bacillus* CNPSo 2477 and CNPSo 2476 have good antagonist performance against corn phytopathogens. We also showed that *Bacillus* sp. CNPSo 2481 and *Enterobacter* sp. CNPSo 2480, have shown the ability to promote plant growth, leading to increases in the percentage of seed germination and improving the early development of seedlings. Therefore, these strains are very promising for use as inoculants in corn. Future studies are needed to test the biotechnological potential of these strains under field conditions, in the hope that they will contribute as an alternative source of biological fertilizer and biological control. It is also interesting to investigate which mechanisms would be related to fungal inhibition activity detected in our study.

Competing interests

All authors declare have no competing interests.

Acknowledgments

Our thanks go to: Semilia Genetics and Breeding LTDA, specially Dr. Francisco Terasawa Junior, for experiments in obtaining fungal strains. Also, the team of the Biotechnology Laboratory of Embrapa Soybean Soil, namely Jacqueline Delamuta and Renan Ribeiro, for support with the sequencing. Renata Rodrigues Gomes and Maicon Wons Andrew Fernandes, for their help in the antagonism tests. This work had financial support from the National Council for Scientific and Technological Development (CNPq), Project Repensa (562008/2010-2) and Coordination for the Improvement of Higher Education Personnel (CAPES).

Author details

[1]Departamento de Genética, Universidade Federal do Paraná (UFPR), Setor de Ciências Biológicas, Avenida Coronel Francisco Heráclito dos Santos, 210, Curitiba, Paraná 81531-970, Brazil. [2]Embrapa Soja, Cx. Postal 231, Londrina, Paraná 86001-970, Brazil.

References

Affourtit J, Zehr JP, Paerl HW (2001) Distribution of nitrogen-fixing microrganisms along the Neuser river estuary, North Carolina. Microb Ecol 41:114–123

Ahmad F, Ahmad I, Khan MS (2008) Screening of free-living rhizospheric bacteria for their multiple plant growth promoting activities. Microbiol Res 1631:73–181

Altekar G, Dwarkadas S, Huelsenbeck JP, Ronquist F (2004) Parallel metropolis coupled Markov chain Monte Carlo for Bayesian phylogenetic inference. Bioinformatics 20:407–415

Araujo FF, Guerreiro RT (2010) Bioprospecção de isolados de *Bacillus* promotores de crescimento de milho cultivado em solo autoclavado e natural. Ciênc Agrotec 34(4):837–844

Arruda L, Beneduzi A, Martins A, Lisboa B, Lopes C, Bertolo F, Vargas LK (2013) Screening of rhizobacteria isolated from maize (*Zea mays* L.) in Rio Grande do Sul State (South Brazil) and analysis of their potential to improve plant growth. Appl Soil Ecol 63:15–22

Arshad M, Frankenberger WT (1991) Microbial production of plant hormones. Plant Soil 133:1–8

Ashraf MA, Rasool M, Mirza MS (2011) Nitrogen fixation and indole acetic acid production potential of bacteria isolated from rhizosphere of sugarcane (*Saccharum officinarum* L.). Adv Biol Res 5(6):348–355

Assumpção LC, Lacava PT, Dias ACF, Azevedo JL, Menten JOM (2009) Diversidade e potencial biotecnológico da comunidade bacteriana endofítica de semente de soja. Pesq Agropec Bras 44:503–510

Bacon CW, Hinton DM (2002) Endophytic and biological control potential of *Bacillus mojavensis* and related species. Biological Control 23(3):274–284

Bakker AW, Schippers B (1987) Microbial cyanide production in the rhizosphere in relation to potato yield reduction and *Pseudomonas* spp-mediated plant growth-stimulation. Soil Biol Biochem 19:451–457

Baldani JI, Baldani VLD, Seldin L, Döbereiner J (1986) Characterization of *Herbaspirillum seropedicae* gen nov, sp nov, a root-associated nitrogen-fixing bacterium. Int J Syst Bacteriol 30:86–93

Barazani O, Friedman J (1999) Is IAA the major root growth factor secreted from plantgrowth-mediating bacteria. J Chem Ecol 25(10):2397–2406

Barretti PB, Souza RM, Pozza EA (2008) Bactérias endofíticas como agentes promotores do crescimento de plantas de tomateiro e de inibição in vitro de *Ralstonia solanacearum*. Ciênc Agrotec 32:731–739

Beneduzi A, Peres D, Vargas LK, Bodanese-Zanettini MH, Passaglia LMP (2008) Evaluation of genetic diversity and plant growth promoting activities of nitrogen-fixing Bacilli isolated from rice fields in South Brazil. Appl Soil Ecol 39:311–320

Beneduzi A, Moreira F, Costa PB, Vargas LK, Lisboa BB, Favreto R, Baldani JI, Passaglia LMP (2013) Diversity and plant growth promoting evaluation abilities of bacteria isolated from sugarcane cultivated in the South of Brazil. Appl Soil Ecol 63:94–104

Berg G, Roskot N, Steidle A, Eberl L, Zock A, Smalla K (2002) Plant-dependent genotypic and phenotypic diversity of antagonistic rhizobacteria isolated from different *Verticillium* host plants. Appl Environ Microbiol 68 (7):3328–3338

Bernardes FS, Patrício FRA, Santos AS, Freitas SS (2010) Indução de resistência sistêmica por rizobactérias em cultivos hidropônicos. Summa Phytopathol 36(2):115–121

Boddey RM (1987) Methods for quantification of nitrogen fixation associated with gramineae. Crit Rev Plant Sci 6:209–266

Burris RH (1972) Nitrogen fixation assaymethods and techniques. Meth Enzymol 24:415–431

Canbolat M, Bilen S, Çakmakçi R, Sahin F, Aydi A (2006) Effect of plant growth-promoting bacteria and soil compaction on barley seeding growth, nutrient uptake, soil properties and rhizosphere microflora. Biol Fertil Soils 42(3):350–357

Cassán F, Maiale S, Masciarelli O, Vidal A, Luna V, Ruiz O (2009) Cadaverine production by *Azospirillum brasilense* and its possible role in plant growth promotion and osmotic stress mitigation. Eur J Soil Biol 45:12–19

Chun J, Bae KS (2000) Phylogenetic analysis of *Bacillus subtilis* and related taxa based on partial *gyrA* gene sequences. Antonie van Leeuwenhoek 78:123–127

Compant S, Clément C, Sessitsch A (2010) Plant growth-promoting bacteria in the rhizo and endosphere of plants: their role, colonization, mechanisms involved and prospects for utilization. Soil Biol Biochem 42:669–678

Daniels BA (1983) Elimination of *Fusarium moniliforme* from corn seed. Plant Disease 67(6):609–611

Deepa CK, Dastager SG, Pandey A (2010) Plant growth-promoting activity in newly isolated *Bacillus thioparus* (NII-0902) from Western ghat forest, India. World J Microbiol Biotechnol 26(12):2277–2283. doi:10.1007/s11274-010-0418-3

Döbereiner J (1980) Forage grasses and grain crops. In: Bergersen FJ (ed) Methods for evaluating biological nitrogen fixation. Wiley, Chichester, pp 535–555

Döbereiner J, Baldani VLD, Baldani JI (1995) Como isolar e identificar bactérias diazotróficas de plantas não-leguminosas. Embrapa, Brasília

Dye DW (1962) The inadequacy of the usual determinative tests for identification of *Xanthomonas* sp. New Zeal J Sci 5:393–416

Enebak SA, Wei G, Kloepper JW (1998) Effects of plant growth promoting rhizobacteria on loblolly and slash pine seedlings. Forest Sci 44:139144

Ewing B, Hillier L, Wendl MC, Green P (1998) Basecalling of automated sequencer traces using phred I Accuracy assessment. Genome Res 8(3):175–185

Ferreira PAA, Bomfeti CA, Soares BL, Moreira FMS (2011) Efficient nitrogen-fixing *Rhizobium* strains isolated from amazonian soils are highly tolerant to acidity and aluminium. World J Microbiol Biotechnol. doi:10.1007/s11274-011-0997-7

Fravel DR (1988) Role of antibiosis in the biocontrol of plant diseases. Ann Rev Phytopathol 26:75–91

George P, Gupta A, Gopal M, Thomas L, Thomas GV (2012) Multifarious beneficial traits and plant growth promoting potential of *Serratia marcescens* KiSII and *Enterobacter* sp. RNF 267 isolated from the rhizosphere of coconut palms (*Cocos nucifera* L.). World J Microb Biot 29(1):109–117

Hallmann J, Quadt-Hallmann A, Mahaffee WF, Kloepper JW (1997) Bacterial endophytes in agricultural crops. Can J Microbiol 43:895–914

Hankin L, Anagnostakis SL (1975) The use of solid media for detection of enzyme production by fungi. Mycologia 67:597–607

Hernandez-Rodriguez A, Heydrich-Perez M, Acebo-Guerrero Y, Velazquez-Del Valle MG, Hernandez-Lauzardo NA (2008) Antagonistic activity of Cuban native rhizobacteria against *Fusarium verticillioides* (Sacc.) Nirenb. in maize (*Zea mays* L.). Appl Soil Ecol 39(2):180–186

Hungria M, Campo RJ, Souza EM, Pedrosa FO (2010) Inoculation with selected strains of *Azospirillum brasilense* and *A. lipoferum* improves yields of maize and wheat in Brazil. Plant Soil 331:413–425

Ikeda AC, Bassani LL, Adamoski D, Stringari D, Kava-Cordeiro V, Glienke C, Steffens MBR, Hungria M, Galli-Terasawa LV (2013) Morphological and genetic characterization of endophytic bacteria isolated from roots of different maize genotypes. Microb Ecol 65:154–160. doi:10.1007/s00248-012-0104-0

Khalid A, Arshad M, Zahir ZA (2004) Screening plant growth promoting rhizobacteria for improving growth and yield of wheat. J Appl Microbiol 96:473–480

Kim PI, Chung KC (2004) Production of an antifungal protein for control of *Colletotrichum lagenarium* by *Bacillus amyloliquefaciens* MET0908. FEMS Microbiol Lett 234(1):177–183

King EO, Ward MK, Raney DE (1954) Two simple media for the demonstration of pyocyanin and fluorescin. J Lab Clin Med 44(2):301–307

Krey T, Vassilev N, Baum C, Eichler-Löbermann B (2013) Effects of long-term phosphorus application and plant-growth promoting rhizobacteria on maize phosphorus nutrition under field conditions. Eur J Soil Biol 55:124–130

Kumar MV, Kumar A (2012) Plant growth promoting and phytostimulatory potential of *Bacillus subtilis* and *Bacillus amyloliquefaciens*. JABS 7(7):509–519

Kuss AV, Kuss VV, Lovato T, MI F (2007) Fixação de nitrogênio e produção de ácido indol acético in vitro por bactérias diazotróficas endofíticas. PAB 42:1459–1465

Laguerre G, Nour SM, Macheret V, Sanjuan J, Drouin P, Amarger N (2001) Classification of rhizobia based on *nodC* and *nifH* gene analysis reveals a close phylogenetic relationship among *Phaseolus vulgaris* symbionts. Microbiol 147:981–993

Loper JE, Schroth MN (1986) Influence of bacterial sources of indole-3-acetic acid on root elongation of sugar beet. Phytopathology 76:386–389

Löytynoja A, Goldman D (2010) webPRANK: a phylogenyaware multiple sequence aligner with interactive alignment browser. BMC Bioinformatics. doi:10.1186/1471-2105-11-579

Lucy M, Reed E, Glick BR (2004) Application of free living plant growth promoting rhizobacteria. Antonie van Leewenhoek 86:1–25

Menna P, Hungria M, Barcellos FG, Bangel EV, Hess PN, Martínez-Romero E (2006) Molecular phylogeny based on the 16S rRNA gene of elite rhizobial strains used in brazilian commercial inoculants. Syst Appl Microbiol 29:315–332

Montañez A, Abreu C, Gill PR, Hardarson G, Sicardi M (2009) (2009) Biological nitrogen fixation in maize (*Zea mays* L.) by 15 N isotope-dilution and identification of associated culturable diazotrophs. Biol Fertil Soils 45:253–263. doi:10.1007/s00374-008-0322-2

Moura JM, Ferreira AFC, Silva FMM, Rizzi J, Pinto LAA (2005) Obtenção de quitina a partir de carapaças de siri (*Maia squinado*): uso de um planejamento experimental na etapa de desmineralização. Vetor 15(1):717

Nylander JAA, Wilgenbush JC, Warren DL, Swofford DL (2008) AWTY (Are We There Yet?): A system for graphical exploration of MCMC convergence in Bayesian phylogenetics. Bioinformatics 24:581–583

Oliveira ALM, Canuto EL, Reis VM, Baldani JI (2003) Response of micropropagated sugarcane varieties to inoculation with endophytic diazotrophic bacteria. Braz J Microbiol 34:59–61

Oppenheim AB, Chet I (1992) Cloned chitinases in fungal plant-pathogen control strategies. Trends Biotechnol 10:392–394

Pal KK, Tilak KVBR, Saxena AK, Dey R, Singh CS (2001) Suppression of maize root diseases caused by *Macrophomina phaseolina*, *Fusarium moniliforme* and *Fusarium graminearum* by plant growth promoting rhizobacteria. Microbiol Res 156:209–223

Patten CL, Glick BR (1996) Bacterial biosynthesis of indole-3-acetic acid. Can J Microbiol 42:207–220

Peixoto AR, Torres SB, Karasawa N (1998) Qualidade sanitária de sementes de milho produzidas no submédio São Francisco. Rev Bras Sementes 20(1):12–15

Pereira BA, Castro-Silva MA (2010) Rizobactérias formadoras de endósporos associadas a *Tibouchina urvilleana*de áreas impactadas por rejeitos da mineração do carvão. Rev Bras Ciênc Solo 34:563–567

Pereira AOP, Camargo RV, Camargo LEA (2005) Doenças do milho (*Zea mays*). In: Kimati H, Amorim L, Bergamim-Filho A, Camargo LEA (eds) Manual de Fitopatologia - Doenças de Plantas cultivadas, 2º vol. Ceres, Piracicaba, pp 477–488

Pereira P, Ibáñez F, Rosenblueth M, Etcheverry M, Martínez-Romero E (2011) Analysis of the bacterial diversity associated with the roots of maize (*Zea mays* L) through culture-dependent and culture-independent methods. ISRN Ecology. doi:105402/2011/938546

Petrini O (1986) Taxonomy of endophytic fungi of aerial plant tissues. In: Den V (ed) Fokkema NJ, Heuvel J. Cambridge University, Cambridge, Microbiology of the phyllosphere. Cambridge, pp 75–87

Pinto NFJA (1998) Patologia de sementes de milho, Circular Técnica 29. Embrapa, Sete Lagoas

Porwal S, Lal S, Cheema S, Kalia VC (2009) Phylogeny in aid of the present and novel microbial lineages: diversity in *Bacillus*. PLoS ONE 4(2):e4438. doi:10.1371/journal.pone.0004438

Posada D, Krandal KA (1988) MODELTEST: testing the model of DNA substitution. Bioinformatics 14:817–818

Powell KA, Faull JL, Renwick A (1990) The commercial and regulatory challenge. In: Horny D (ed) Biological control of soil borne plant pathogens. CAB International Wallingford, United Kingdom, pp 445–463

Raeder U, Broda P (1985) Rapid preparation of DNA from filamentous fungi. Lett Appl Microbiol 1:17–20

Rana A, Saharan B, Joshi M, Prasanna R, Kumar K, Nain L (2011) Identification of multi-trait PGPR isolates and evaluating their potential as inoculants for wheat. Ann Microbiol 61:893–900, doi:1 0.1007/s13213-011-0211-z

Ratón TMO, Yano R, Gámez OR, Floh EIS, Díaz MJS, Barbosa HR (2011) Isolation and characterisation of aerobic endospore forming Bacilli from sugarcane rhizosphere for the selection of strains with agriculture potentialities. World J Microbiol Biotechnol. doi:101007/s11274-011-0965-2

Renwick A, Campbell R, Coe S (1991) Assessment of in vivo screening systems for potencial biocontrol agents o *Gaeumannomyces graminis*. Plant Pathol 40:524–532

Rodrigues Neto J, Malavolta Júnior VA, Victor O (1986) Meio simples para isolamento e cultivo de *Xanthomonas campestrispv citri* tipo B. Summa Phytopatol 12:16

Rodriguez H, Gonzalez T, Goire I, Bashan Y (2004) Gluconic acid production and phosphate solubilization by the plant growth-promoting bacterium *Azospirillum* spp. Naturwissenschaften 91:552–555

Ronquist F, Huelsenbeck JP (2003) MRBAYES 3: Bayesian phylogenetic inference under mixed models. Bioinformatics 19:1572–1574

Rosado AS, Duarte GF, Seldin L, VanElsas JD (1998) Genetic diversity of nifH gene sequences in Paenibacillus azotofixans strains and soil samples analyzed by denaturing gradient gel electrophoresis of PCR amplified gene fragments. Appl Environ Microbiol 64:2770–2779

Rösch C, Mergel A, Bothe H (2002) Biodiversity of denitrifying and dinitrogen-fixing bacteria in an acid forest soil. Appl Environ Microbiol 68(8):3818–3829. doi:10.1128/AEM.68.8.3818-3829.2002

Sambrook J, Russell D (2001) Molecular cloning: a laboratory manual, 3rd edn. Cold Spring Harbor Laboratory Press, New York

Sarwar M, Kremer RJ (1995) Enhanced suppression of plant growth through production of L-tryptophan-derived compounds by deleterious rhizobacteria. Plant Soil 172:261–269

Schwyn B, Neilands JB (1987) Universal chemical assay for the detection and determination of siderophores. Anal Biochem 160:4756

Seldin L, van Elsas JD, Penido EGC (1984) Bacillus azotofixans sp. nov., a nitrogen-fixing species from Brazilian soils and grass roots. Int J Syst Bacteriol 34:451–456

Shiomi HF, Melo IS, Minhoni MTA (2008) Seleção de bactérias endofíticas com ação antagônica a fitopatógenos. Scientia Agraria 9:535–538

Silva MF, Reis VM (2004) Técnicas usadas para quantificar a população de bactérias diazotróficas inoculadas em sementes de milho (*Zea mays*). Embrapa, Seropédica

Sukumaran J, Holder MT (2010) DendroPy: a Python library for phylogenetic computing. Bioinformatics 26:1569–1571

Sylvester-Bradley R, Asakawa N, La Torraca S, Magalhães FMM, Oliveira L, Pereira RM (1982) Levantamento quantitativo de microrganismos solubilizadores de fosfatos na rizosfera de gramíneas e leguminosas forrageiras na Amazônia. Acta Amaz 12:15–22

Tian F, Ding Y, Zhu H, Yao L, Du B (2009) Genetic diversity of siderophore-producing bacteria of tobacco rhizosphere. Braz J Microbiol 40(2):276–284

Vale M, Seldin L, Araújo FF, Lima R (2010) Plant growth promoting rhizobacteria: fundamentals and applications. In: Maheshwari DK (ed). Plant growth and health promoting bacteria Springer, Berlin, pp 21–43

Verma JP, Yadav J, Tiwari KN, Kumar A (2013) Effect of indigenous *Mesorhizobium* spp. and plant growth promoting rhizobacteria on yields and nutrients uptake of chickpea (*Cicer aritenium* L.) under sustainable agriculture. Ecol Eng 51:282–286

Wang S, Huijun W, Junqing Q, Lingli M, Jun L, Yanfei X, Xuewen G (2009) Molecular mechanism of plant growth promotion and induced systemic resistance to tobacco mosaic virus by *Bacillus* spp. J Microbiol Biotechnol 19(10):1250–1258

Weisburg WG, Barns SM, Pelletier DA, Lane DJ (1991) 16S ribosomal DNA amplification for phylogenetic study. J Bacteriol 173:697–703

Zehr JP, Capone DG (1996) Problems and promises of assaying the genetic potential for nitrogen fixation in the marine environment. Microb Ecol 32:263–281

Zehr JP, Jenkins BD, Short SM, Steward GF (2003) Nitrogenase gene diversity and microbial community structure: A cross-system comparison. Environ Microbiol 5:539–554

Diversity and biochemical features of culturable fungi from the coastal waters of Southern China

Li Li[1,2], Purnima Singh[1], Ying Liu[1], Shenquan Pan[2] and Guangyi Wang[1,3,4*]

Abstract

Fungi play a major role in various biogeochemical cycles of terrestrial and marine ecosystems. However, fungi in marine environments remain to be one of the most under-studied microbial groups. This study investigates the diversity of planktonic fungi from the coastal habitat off Pearl River Delta (China) using culture-dependent approach. A total of 22 fungi and 9 yeast isolates were recovered from 30 seawater and 2 sediment samples. Microscopic and ITS rRNA gene sequence analyses revealed that most of the fungi belonged to the phylum *Ascomycota* and *Basidiomycota* with a very small percentage (3%) of the subphylum *Mucoromycotina* of the Phylum *Zygomycota*. Most of these fungal isolates exhibited considerable production of extracellular enzymes, cellulase, lipase and laccase. Fungal isolates of two genera *Mucor* and *Aspergillus* sp. demonstrated pelletization capability over a wide range of pH, suggesting them as potential agents towards algae harvesting and wastewater treatment.

Keywords: Marine-derived fungi; Diversity; Hydrolytic enzymes; Pelleterization

Introduction

Coastal marine habitats have been characterized as the most variable, highly diverse and rich in primary production (Jickells 1998; Danovaro and Pusceddu 2007). The primary production in coastal habitats sometimes reaches very high levels resulting into availability of a great fraction of organic matter for consumers as detritus even after consumption of herbivores (Newell 1982). Different types of microbes in coastal waters degrade a large proportion of this detritus actively (Manini et al. 2003; Pusceddu et al. 2003). Among these microbes, heterotrophic bacteria and archaea have been described for their degradation abilities towards such detritus to a greater extent (Moran and Miller 2007; Mou et al. 2008). In spite of being a significant component of coastal waters, the diversity and ecology of heterotrophic eukaryotes however has not been received much attention (Giovannoni and Stingl 2005; Hallam et al. 2006; Fenchel 2008; Strom 2008).

Among eukaryotes, fungi have been reported to exhibit as individual filaments or aggregates in coastal waters (Gutiérrez et al. 2010). However, in comparison with terrestrial environments, fungi in the world's oceans remain largely unknown (Gao et al. 2010). Despite of a few reports on diversity of fungi from the oceans, the diversity and ecology of their planktonic forms (mycoplankton) have barely been explored (Richards et al. 2012). Mycoplankton include free-living filamentous fungi, yeasts, fungal-like protists, and those associated with planktonic particles or phytoplankton (Wang and Johnson 2009; Gao et al. 2010).

Fungi are a key component of the biosphere, fulfilling a wide range of biogeochemical and ecological functions in natural environments (Christensen 1989; Pang and Mitchell 2005). They are best known as decomposers of organic matter and play major role in nutrient regeneration in the detrital ecosystems. The filamentous mycelia may greatly enhance the efficient mineralization of particulate organic matter (Tisdall and Oades 1982; Damare and Raghukumar 2008) and thus benefit the growth of planktonic microbial communities (Kiørboe and Jackson 2001; Gutiérrez et al. 2010). The biomass of planktonic fungi has been reported to be comparable with prokaryotes including both Bacteria and Archaea (Gutiérrez et al. 2011).

Fungi occupy distinct ecological niches from that of bacterioplankton in detritus ecosystems with the ability to utilize large lignocellulose-predominated substrates

* Correspondence: gywang@tju.edu.cn
[1]School of Environment and Energy, Peking University Shenzhen Graduate School, Shenzhen, 518055, China
[3]Tianjin University Center for Marine Environmental Ecology, School of Environmental Science and Engineering, Tianjin University, Tianjin 300072, China
Full list of author information is available at the end of the article

with high C: N ratio (Newell 1994; Raghukumar 2004). They possess the ability to penetrate relatively persistent particulate detritus much more efficiently than bacterioplankton (Raghukumar 2004). The endophytic fungi have been demonstrated to reside within marine plants intra or intercellularly, and produce a variety of bioactive and chemically active metabolites (Kaul et al. 2013). The bioactive metabolites produced by endophytic fungi originate from different biosynthetic pathways and belong to different groups of terpenoids, steroids, quinones, phenols and coumarins (Kaul et al. 2013). Therefore, the endophytes represent a potential chemical reservoir for anticancer, antioxidant, antiviral and insecticidal compounds for pharmaceutical and agrochemical industries. Two new benzopyranones, diaportheone A and B, were obtained via bioassay-guided isolation of the secondary metabolites from the endophytic fungus *Diaporthe* sp. P133 isolated from *Pandanus amaryllifolius* leaves (Bungihan et al. 2011). These benzopyranones have been successfully used as antimicrobial compounds towards several microorganisms. One of the fungal isolate belonging to *Aureobasidium pullulans* has been exhibited as a reservoir of biotechnologically active products in previous reports (Chi et al. 2009).

Considering the crucial role of planktonic fungi in versatile oceanic biogeochemical cycles, their diversity needs to be addressed from different ecosystems. The coastal habitats of Pearl River Delta, China have been highly productive ecosystems of China, being the major source for fishing industries. However, recently there has been an increased pollution level detected in these ecosystems which may further enhance detritus levels available for degradation. Therefore, the mycoplankton diversity studies from these still unexplored habitats may provide greater insight on potential fungal isolates playing significant role in ecological cycles of Pearl River Delta. This study is the first report on diversity of planktonic fungi based on culture-dependent approach from the coastal habitats of Pearl River Delta.. Potential fungal isolates were further investigated for the production of different extracellular enzymes such as laccase, lipase and cellulase in order to understand their active role in ecological cycles of coastal ecosystems.

Materials and methods

Sample collection and isolation of fungi

Seawater and sediment samples were collected from coastal marine habitats of Pearl River Delta region of China during March, 2012 (Table 1). These samples were carried in sterile, screw capped plastic bottles and bags immediately back to the laboratory for isolation. The isolation of fungi from the seawater samples was done within one hour of collection using the membrane filtration technique. Briefly, 15 ml triplicate water samples were filtered through sterile 0.45 μm cellulose ester membranes (Millipore, USA). These membranes were then placed on solid media plates, Malt Extract Agar (MEA), Sabouraud Dextrose Agar (SDA), Potato Dextrose Agar (PDA), Czapek Dox Agar (CDA) and Corn Meal Agar (CMA), supplemented with antibiotics (0.075% streptomycin and 0.05% ampicillin) to suppress bacterial growth. Sediment samples (0.1 g) were suspended in 10 ml sterile seawater and 100 μl of the resulting suspension was plated directly on the above media plates containing antibiotics. The plates were incubated at room temperature (28°C) and examined daily for the growth of fungi. Fungal colonies that developed were subcultured onto fresh MEA plate for pure, single colony isolation and identification. The identification of filamentous fungi was done by macroscopic and microscopic morphology (Additional file 1: Figure S1). Three promising strains (*Rhodosporidium* sp. PKU Y5, *Rhodotorula* sp. PKU Y7 and *Cladosporium* sp. PKU F16) have been deposited in China General Microbiological Culture Collection Center (CGMCC No. 2.5198, CGMCC No. 2.5199 and CGMCC No.3.17121).

Isolation and sequencing of ITS rRNA gene from the fungal isolates

All the isolated fungi were grown in MEB for 4–5 days for DNA isolation. Yeasts were grown in YPD (yeast extract peptone and dextrose) medium and shaken at 170 rpm for 3–4 days. Mycelia and cells were harvested, lyophilized and crushed in a mortar and pestle to fine powder. Isolation of DNA was carried out using the Ezup Soil DNA extraction kit (Sangon Biotech, China), following the manufacturer's guideline. The small subunit ITS rRNA gene was amplified by polymerase chain reaction (PCR) in the DNA T100™ Thermal cycler (Bio-Rad, USA) using the ITS rRNA gene specific primers ITS1 (5'-TCCGTAGGTGAACCTGCGG-3') and ITS4 (5'-TCCTCCGCTTATTGATATGC-3') (White et al. 1990). One microliter of DNA (~25 ng) was added to 50 μl reaction volume containing 25 μl of Taq PCR mix (Generay, China) , 23 μl dd water and 5 pmoles of each primer. The PCR program was run for initial denaturation step at 95°C for 3 min, followed by 35 cycles of 1 min at 94°C, 0.5 min at 50°C and 1 min at 72°C, and a final extension at 72°C for 5 min. The PCR products were purified using Gel DNA extraction kit (NewTopBio, China). Amplified products were transformed into *Escherichia coli* DH5α cells (Invitrogen, Carlsbad, CA, USA), following the manufacturer's instructions. Transformants were grown overnight at 37°C in Luria-Bertani broth containing 100 mg of ampicillin. The presence of insert was confirmed by PCR with M13 forward and reverse primers. One ml of the broth containing the clone was added to 25 ml of PCR reaction mixture. PCR protocol included an initial hot start incubation (5 min at 94°C) followed by 34 cycles of

Table 1 Details regarding colony forming units (CFUs) of fungal colonies on the media plates

Sampling date	Location	Lat(°N)	Long (°E)	Habitat	Depth (m)	Temperature (°C)	Salinity (ppt)	Media	No. of colonies	CFU/L
2012-03-05	Shenzhen Bay	22°31'19.776"	113°57'4.284"	Seawater	0	16.25	31.02	MEA	3	120
		22°31'19.776"	113°57'4.284"	Seawater	0	16.25	31.02	SDA	2	80
		22°31'19.776"	113°57'4.284"	Sediment	0	16.25	31.02	MEA	12	480
		22°31'19.776"	113°57'4.284"	Sediment	0	16.25	31.02	SDA	55	2200
	Daya Bay	22°35'34.224"	114°30'28.800"	Seawater	0	16.25	31.02	MEA	4	160
		22°35'34.224"	114°30'28.800"	Seawater	0	16.25	31.02	SDA	5	200
		22°35'34.224"	114°30'28.800"	Sediment	0	16.25	31.02	MEA	6	240
		22°35'34.224"	114°30'28.800"	Sediment	0	16.25	31.02	SDA	8	320
2012-03-21	Mirs Bay	22°29'44.02"	114°27'34.38"	Seawater	0	17.37	31.37	MEA	1	40
		22°29'44.02"	114°27'34.38"	Seawater	0	17.37	31.37	SDA	3	120
		22°29'44.02"	114°27'34.38"	Seawater	5	17.37	31.37	MEA	2	80
		22°29'44.02"	114°27'34.38"	Seawater	5	17.37	31.37	SDA	4	160
		22°29'44.02"	114°27'34.38"	Seawater	10	17.37	31.37	MEA	3	120
		22°29'44.02"	114°27'34.38"	Seawater	10	17.37	31.37	SDA	5	200
	Mirs Bay	22°31'33.14"	114°27'51.59"	Seawater	0	17.74	31.37	MEA	3	120
		22°31'33.14"	114°27'51.59"	Seawater	0	17.74	31.37	SDA	2	80
		22°31'33.14"	114°27'51.59"	Seawater	5	17.74	31.37	MEA	4	160
		22°31'33.14"	114°27'51.59"	Seawater	5	17.74	31.37	SDA	13	520
		22°31'33.14"	114°27'51.59"	Seawater	10	17.74	31.37	MEA	3	120
		22°31'33.14"	114°27'51.59"	Seawater	10	17.74	31.37	SDA	10	400
	Mirs Bay	22°31'32.02"	114°28'26"	Seawater	0	17.69	31.38	MEA	2	80
		22°31'32.02"	114°28'26"	Seawater	0	17.69	31.38	SDA	8	320
		22°31'32.02"	114°28'26"	Seawater	5	17.69	31.38	MEA	5	200
		22°31'32.02"	114°28'26"	Seawater	5	17.69	31.38	SDA	7	280
		22°31'32.02"	114°28'26"	Seawater	10	17.69	31.38	MEA	5	200
		22°31'32.02"	114°28'26"	Seawater	10	17.69	31.38	SDA	17	680
	Daya Bay	22°34'19.99"	114°31'30"	Seawater	0	19.24	31.00	MEA	2	80
		22°34'19.99"	114°31'30"	Seawater	0	19.24	31.00	SDA	3	120
		22°34'19.99"	114°31'30"	Seawater	5	19.24	31.00	MEA	2	80
		22°34'19.99"	114°31'30"	Seawater	5	19.24	31.00	SDA	7	280
		22°34'19.99"	114°31'30"	Seawater	10	19.24	31.00	MEA	3	120
		22°34'19.99"	114°31'30"	Seawater	10	19.24	31.00	SDA	12	480
	Daya Bay	22°34'34.37"	114°30'30.46"	Seawater	0	19.69	30.86	MEA	3	120
		22°34'34.37"	114°30'30.46"	Seawater	0	19.69	30.86	SDA	9	360
		22°34'34.37"	114°30'30.46"	Seawater	10	19.69	30.86	MEA	10	400
		22°34'34.37"	114°30'30.46"	Seawater	10	19.69	30.86	SDA	8	320

denaturation at 94°C for 30 s, annealing at 55°C for 30 s, and extension at 72°C for 1 min followed by a final extension at 72°C for 5 min. Clones containing positive insert were further processed for plasmid isolation and purification using Millipore plasmid preparation kit (Millipore, USA). Clones containing positive insert were sent to BGI (Shenzhen, China) for sequencing analysis using M13 primers.

Phylogenetic analyses

Forward and reverse sequences were edited and assembled using Chromas Pro version 1.34 (Technelysium Pty Ltd, Tewantia, Queensland, Australia). The final sequences were compared to the nucleotide sequences of reference organisms available in the GenBank database using Blastn (Altschul et al. 1990). The ITS1-5.8S-ITS4 gene sequences obtained for the organisms were aligned with their closest match using the program, ClustalW (Thompson et al. 1994). Gaps and ambiguously aligned sequences were removed manually from further analyses. Phylogenetic analyses were carried out using distance setting (Maximum parsimony) in MEGA 4 software (Tamura et al. 2007) with 1,000 bootstrap replicates. The resulting ITS1-5.8S-ITS4 gene sequences were submitted to GenBank under the accession number of KC113282-KC113312.

Qualitative assay for extracellular enzymes

The enzymatic activity of all the fungal isolates was analyzed in this study. The strains were screened initially using qualitative plate assay for three different enzymes (laccase, cellulase and lipase) by streaking them on the media plates, supplemented with specific substrates. Laccase activity was detected using MEA plates amended with ABTS (2, 2'-azino-bis-3-ethylbenzothiazoline-6-sulfonic acid) (Srinivasan et al. 1995). The fungal isolates were grown on these media plates at room temperature for 4 days. Green color produced around the fungal colonies on media plates indicated laccase activity.

Cellulase (CMCase) activity was detected using carboxymethylcellulose (CMC)-MEA plates (Carder et al. 1986). The CMC-MEA plates comprised 0.5%

Table 2 Phylogenetic affiliations of culturable fungi based on of ITS rDNA gene sequences

Isolate ID	GenBank accession No.	Closest identified relative		Phylum of closest relative	Source of isolation of the closest relative	% Identity
		Species	(GenBank accession No.)			
PKU F1	KC113282	*Mucor* sp. FJ09	(HQ019160)	*Zygomyceta*	Water hyacinth leaf	99
PKU F2	KC113283	*Hypocrea koningii* strain JZ-25	(HQ637343)	*Ascomycota*	Soil from Sichuan	99
PKU F3	KC113284	*Arthrinium phaeospermum* isolate T57	(FJ462766)	*Ascomycota*	Not known	99
PKU F4	KC113285	*Diaporthe* sp. H4236	(GU595056)	*Ascomycota*	Mangrove in China	98
PKU F5	KC113286	*Phoma* sp. ZH2.1	(FJ450059)	*Ascomycota*	*Argyrosomus argentatus*	99
PKU F6	KC113287	*Trichoderma piluliferum* strain wxm37	(HM037939)	*Ascomycota*	River water	99
PKU F7	KC113288	*Trichoderma asperellum* isolate T29	(JN108927)	*Ascomycota*	Rhizospheric soil	99
PKU F8	KC113289	*Aspergillus nomius* strain NRRL 26885	(JF824686)	*Ascomycota*	*Philanthus triangulum*	99
PKU F9	KC113290	*Aspergillus flavipes* isolate BCT2-3	(JQ082507)	*Ascomycota*	Shark gills	100
PKU F10	KC113291	*Dothideomycetes sp. OY307*	(FJ571450)	*Ascomycota*	Not known	99
PKU F11	KC113292	*Arthrinium phaeospermum* isolate DFFSCS004	(JX156350)	*Ascomycota*	Deep-sea sediments	99
PKU F12	KC113293	*Hypocrea lixii* isolate SZMC 20858	(JX173851)	*Ascomycota*	Hungarian vegetables	100
PKU F13	KC113294	*Trichoderma hamatum* strain LXM1	(GQ220703)	*Ascomycota*	Saline-alkali soil	100
PKU F14	KC113295	*Cladosporium* sp. JS1043	(AM176680)	*Ascomycota*	Deep sea	100
PKU F15	KC113296	*Curvularia* sp. B34	(HQ696021)	*Ascomycota*	Moso Bamboo Seeds	99
PKU F16	KC113297	*Cladosporium sphaerospermum* isolate IJL07	(EU823317)	*Ascomycota*	Soybean	99
PKU F17	KC113298	*Arthrinium* sp. LH11	(HQ832842)	*Ascomycota*	Tea plants	99
PKU F18	KC113299	*Ascomycota* sp. AR-2010 isolate TR063	(HQ608095)	*Ascomycota*	Soil	92
PKU F19	KC113300	*Pleosporales* sp. LH241	(HQ832825)	*Ascomycota*	Tea plants	100
PKU F20	KC113301	*Cladosporium cladosporioides* isolate 2728	(EU272532)	*Ascomycota*	Plant	99
PKU F21	KC113302	*Pyrenochaeta* sp. CF-2008	(EU885415)	*Ascomycota*	Corneal ulcer	99
PKU F22	KC113303	*Aspergillus terreus* strain 1B61	(KF572455)	*Ascomycota*	Geothermal soil	99
PKU Y1	KC113304	*Rhodotorula mucilaginosa* strain KDLYC24-1	(HQ909092)	*Basidiomycota*	Not known	99
PKU Y2	KC113305	*Rhodotorula glutinis var. salinaria* strain ZH7-(7)	(FJ487944)	*Basidiomycota*	Mangrove in China	99
PKU Y3	KC113306	*Aureobasidium pullulans* strain KDLYC4-9	(HQ909088)	*Ascomycota*	Not known	99
PKU Y4	KC113307	*Rhodosporidium sphaerocarpum* isolate 2223	(HQ670691)	*Basidiomycota*	Shrimp Culture	99
PKU Y5	KC113308	*Rhodosporidium diobovatum* strain IWBT-Y840	(JQ993385)	*Basidiomycota*	grapevines	99
PKU Y6	KC113309	*Trichosporon brassicae*	(NR_073251)	*Basidiomycota*	Not known	99
PKU Y7	KC113310	*Rhodotorula mucilaginosa* strain WC53-2	(EF190221)	*Basidiomycota*	Not known	99
PKU Y8	KC113311	*Candida parapsilosis* IFM52618	(AB109284)	*Ascomycota*	Braziland Japan	99
PKU Y9	KC113312	*Rhodotorula mucilaginosa* strain R-4685	(KF726105)	*Basidiomycota*	Aortic valve	100

carboxymethylcellulose-sodium salt (CMC-Na), 1.0% glucose, 0.15% peptone, 0.01% yeast extract, 100% seawater, and 2.0% agar. After growing the fungal isolates for 3 days at room temperature, these plates were stained with 0.1% Congo red solution for 20 min at room temperature. The resulting plates were washed twice with 1.0 M NaCl, and were kept overnight at 4°C. Clear zones around the colonies indicated the CMCase activities (Nagano and Fraser 2011; Carder et al. 1986). Lipase activity was detected using MEA plates supplemented with 0.01% phenol red, 1% olive oil and 10 mM $CaCl_2$. The pH was adjusted to 7.3-7.4 with 1.0 M NaOH (Singh et al. 2006). After incubation for 3 days, a change in color from pink to yellow indicated the lipase activity.

Quantitative analysis of enzymatic activities

On the basis of qualitative screening, four fungal strains (PKU F16, PKU F18, PKU Y5 and PKU Y8) were selected for quantitative studies. Fungal inoculums were prepared by growing isolates in MEB medium and yeast in YPD medium, respectively, for four days at 28°C. These inoculums were further subcultured to fresh MEB medium containing individual enzyme specific substrates and incubated on shaker at 30°C, 150 rpm for 4 days. All these experiments were performed in triplicates. Individual enzymes were quantified in the supernatants of the isolates.

The laccase activity was assayed using glycine-HCl (pH 3.0) buffer and ABTS as substrate (Niku-Paavola et al. 1988). Five hundred µl of crude culture filtrate was incubated with equal volumes of buffer containing ABTS and laccase activity was measured at 405 nm. The enzyme units were expressed as µM of substrate transformed per minute per liter of culture filtrate i.e. as enzyme units per liter of culture filtrate (UL^{-1}). In the absence of the enzyme activity, no increase in the rate of absorbance was observed.

Cellulase activity was assayed following the method described by Raghukumar et al. (1994). A reaction mixture containing 200 µl culture filtrate and 200 µl of 0.5% CMC in 0.05 M sodium phosphate buffer, pH 7 was incubated at 37°C for 30 min. To terminate the reaction, 1 ml of DNS (dinitrosalicylic acid) reagent was added to above reaction mixture and then was boiled for 5 min. CMCase activity was measured at 575 nm. One unit of CMCase activity was defined as the amount of enzyme liberating 1 µM of reducing sugar per minute under the above assay conditions.

For lipase assay, 6 mg *para*-nitrophenyl palmitate (pNPP), dissolved in 2 ml isopropanol and 18 ml 50 mM sodium phosphate buffer (pH 8) was used as the substrate. One ml each of culture filtrate and sodium phosphate buffer was added to 1 ml of the pNPP substrate solution. After incubating at 37°C for 30 min, the released product, pNP (*para*-nitrophenol) was measured at 410 nm. One unit of lipase activity was defined as the amount of enzyme liberating 1 µM of pNP per minute under these assay conditions.

Fungal pelletization experiment

Three of the fungal isolates (*Mucor* sp., PKU F1), (*Aspergillus* sp., PKU F8) and (*Cladosporium* sp., PKU F14) were investigated for pellet formation in this study. The spore suspension from these species was obtained by rinsing the mycelia on media plate with distilled water containing 10% Tween 20. The number of spores in the suspension was counted using an optical microscope (Olympus BX53 Manual fluorescence microscope). The spore suspensions were added to the 250-mL Erlenmeyer flasks containing 50 ml of nutrient media (Malt Extract Broth). The pellet formation experiments were performed by placing these Erlenmeyer flasks on a horizontal shaker (100 rpm) at room temperature for 4 days. The pellet formation was analyzed at different spore concentrations and pH ranges (2–8).

Results

Culturable diversity

Fungi were isolated from all the seawater depths in the present study (Table 1). However, maximum numbers of fungal colonies were recovered from surface sediments (0 m depth) and seawater samples at 10 m depth. Fungi were also recovered from sediment samples using particle plating technique (Table 1). SDA was found to be better media than MEA for the isolation of fungi (Table 1). However, there was no statistical significance observed between different depths, media and the number of fungal colonies. A total of 22 fungi and 9 yeasts were isolated in this study. The fungal isolates mostly belonged to *Ascomycota*, *Basidiomycota* and *Zygomycota* based on ITS rRNA gene analysis (Table 2, Figure 1). Fungal isolates showing similarity with the phylum

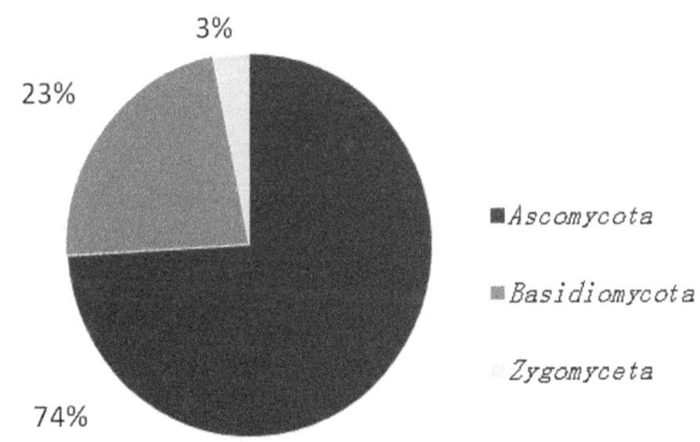

Figure 1 Phylum affiliation of culturable fungi with the existing sequences of NCBI database.

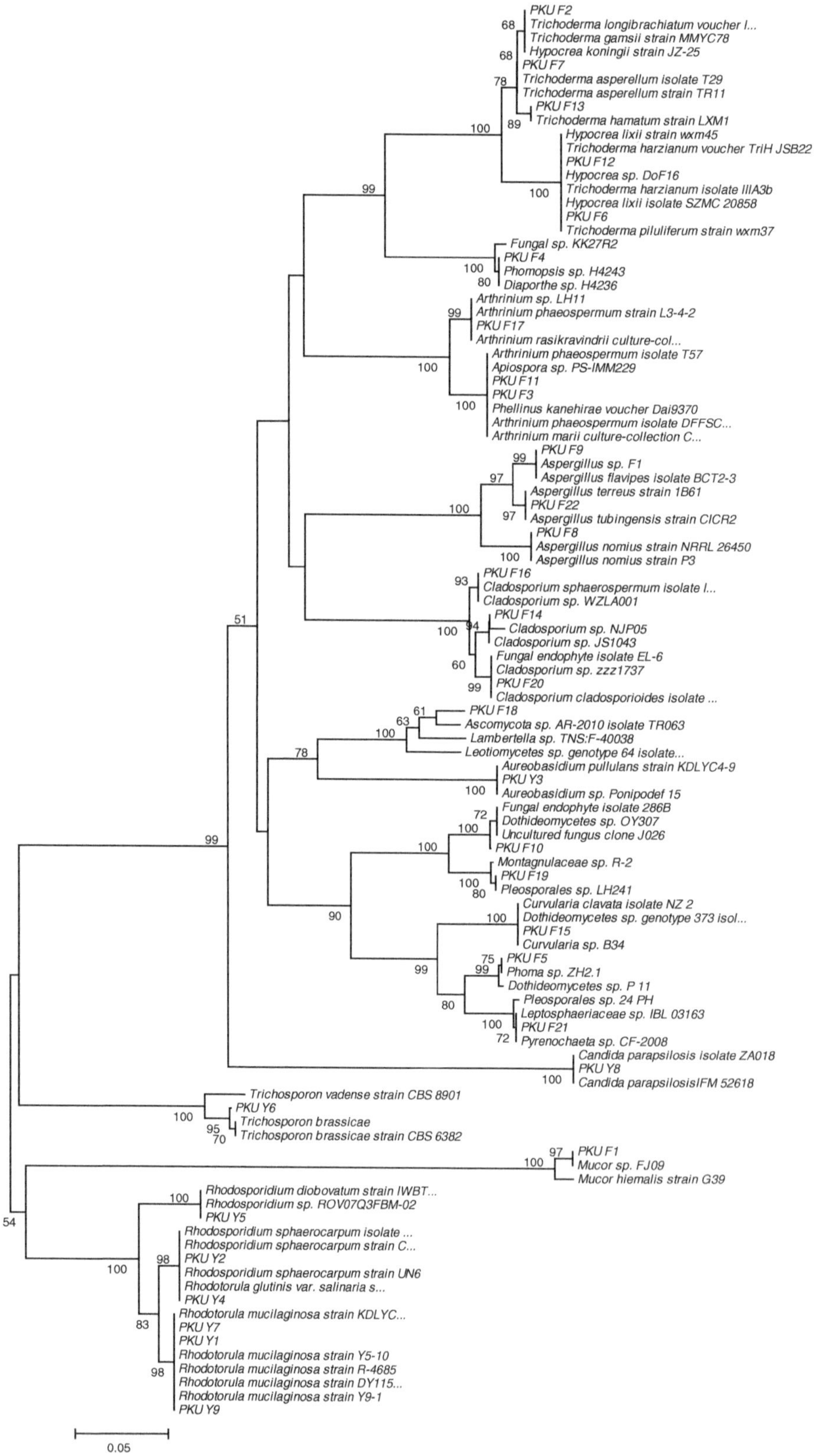

Figure 2 NJ phylogenetic tree based on ITS rRNA genes from 22 fungi and 9 yeast derived from coastal ecosystems of Pearl River Delta. Numbers at nodes indicate bootstrap values of neighborjoining analysis for 1,000 replicates (values below 50% not shown).

Ascomycota were dominating among above three, accounting for 74%. Members of *Basidiomycota* and *Zygomycota* made up for 23% and 3%, respectively (Figure 1). Of the Ascomycetes, isolates belonged to 13 genera i.e. *Aspergillus, Hypocraea, Arthrinium, Diaporthe, Phoma, Trichoderma, Dothideomycetes, Cladosporium, Curvularia, Pleosporales, Pyrenochaeta, Aureobasidium* and *Candida*. Isolates of Basidiomycota were affiliated with *Rhodotorula, Rhodosporidium* and *Trichosporon* sp. Only *Mucor* sp. of *Zygomycota* was identifed in this study (Table 2, Figure 2). All the fungal ITS rRNA gene sequences showed 100 or 99% identity with the existing sequences of NCBI database except PKU F18, showing 92% identity with *Ascomycota* sp. AR-2010 (Table 2).

Enzymatic activities

Qualitative analyses of different enzymes demonstrated positive result for most of the fungal isolates (Table 3, Figure 3). The majority of fungal isolates (~84%) exhibited cellulase and lipase activities. In comparison, only a few of the isolates (~38%) showed positive laccase activity (Table 3, Figure 3). The quantitative analyses for production of these three enzymes were done for four of these fungal isolates (Table 4). *Cladosporium* sp. (PKU F16), *Rhodosporidium* sp. (PKU Y5) and *Candida* sp. (PKU Y8) showed considerable production of lipase enzyme, reaching up to 21.94 U ml^{-1} for *Candida* sp. In contrast, cellulase production was not very high for any of these isolates, being maximum of 1.3 U ml^{-1}

Table 3 Qualitative assay of enzyme activities on media plates amended with specific substrates

Isolate ID	Laccase (4th day)	Lipase (3rd day)	Cellulase (3rd day)	Source of isolation	Depth (m)
PKU F1	-	+	+	Seawater	0
PKU F2	+	-	+	Seawater	0
PKU F3	+	++	++	Seawater	0
PKU F4	++	+	+	Seawater	0
PKU F5	-	++	-	Sediment	0
PKU F6	-	-	+	Seawater	0
PKU F7	++	+	++	Sediment	0
PKU F8	-	+	++	Sediment	0
PKU F9	-	++	++	Sediment	0
PKU F10	+	-	-	Sediment	0
PKU F11	+	++	+++	Seawater	10
PKU F12	-	++	++	Seawater	10
PKU F13	-	++	+++	Seawater	0
PKU F14	++	+++	++	Seawater	0
PKU F15	+	-	-	Seawater	10
PKU F16	+++	+++	++	Seawater	5
PKU F17	-	++	+	Seawater	10
PKU F18	+++	+	+++	Seawater	0
PKU F19	-	+	-	Seawater	0
PKU F20	++	++	++	Seawater	0
PKU F21	+	++	-	Seawater	0
PKU F22	-	+++	++	Seawater	0
PKU Y1	-	++	+	Seawater	0
PKU Y2	-	+	++	Seawater	0
PKU Y3	-	++	+	Seawater	10
PKU Y4	-	+	++	Seawater	0
PKU Y5	-	++	++	Seawater	10
PKU Y6	-	++	++	Seawater	5
PKU Y7	-	-	++	Seawater	0
PKU Y8	-	+++	+++	Seawater	0
PKU Y9	-	++	++	Seawater	10

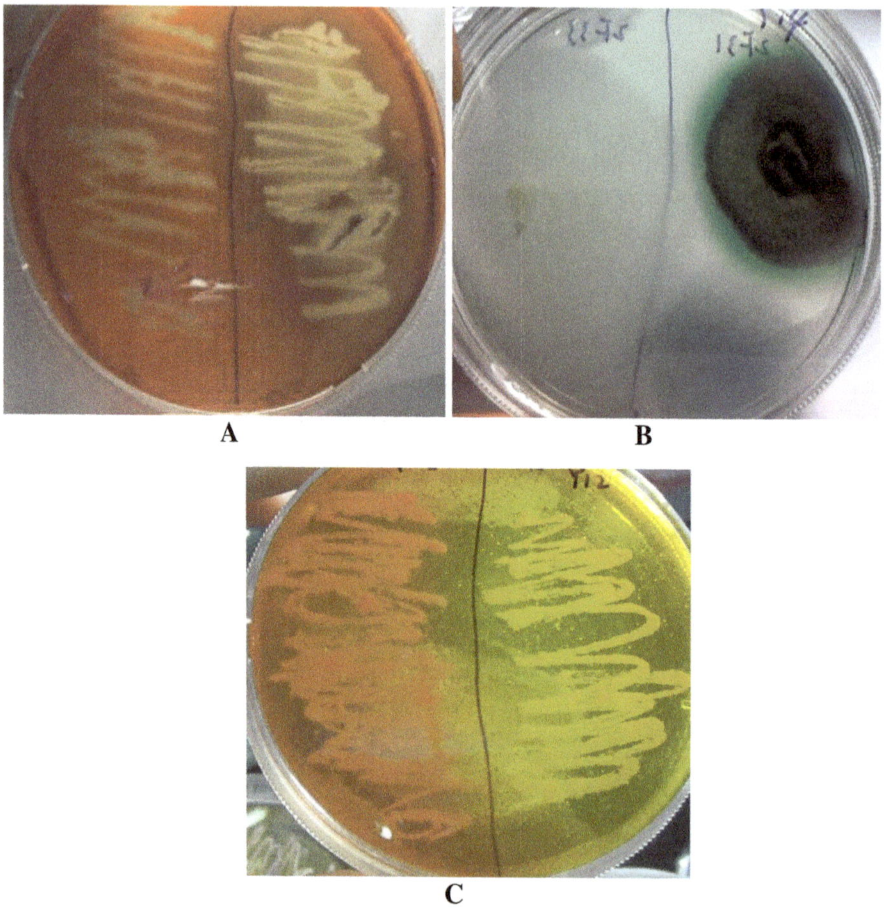

Figure 3 Qualitative assay of enzyme activities of fungi on substrate specific media plates. A. Plate assay for cellulose activities (PKU Y7 *Rhodotorula mucilaginosa* sp. (left) and PKU Y9 *Rhodotorula mucilaginosa* sp. showing good cellulose activity (rght)); **B**. Plate assay for laccase activities (PKU F11 *Arthrinium phaeospermum* sp. showing no laccase activity (left); PKU F16 *Cladosporium sphaerospermum* showing considerable laccase activities (right)); **C**. Plate assay for lipase activities on plate (PKU Y7 *Rhodotorula mucilaginosa* sp. showing little lipase activity (left); PKU Y9 *Rhodotorula mucilaginosa* sp. showing good lipase activity (right)).

demonstrated by *Candida* sp. PKU F16 (*Cladosporium* sp.) and PKU F18 (*Ascomycota* sp.) were the best producers of laccase enzyme showing activities of 14.6 U ml^{-1} and 10.5 U ml^{-1} respectively (Table 4). All the isolates showed maximum production of these extracellular enzymes on 6th day of their growth (Table 4).

Fungal pellet formation

Among three fungal isolates i.e. *Aspergillus* sp. (PKU F8), *Mucor* (PKU F1) and *Cladosporium* sp. (PKU F14), PKU F1 was found to be the best agent for pellet formation (Table 5, Additional file 1: Figure S2). The pH range of 6 and 8 were most suitable for all of these four isolates for pellet formation. Pellet sizes were comparative bigger with higher concentration of spore inoculums for *Mucor* sp. (Table 5). However, *Aspergillus* sp. did not show much variation with inoculum spore concentrations (Table 5). Maximum size of pellet of 7 mm was demonstrated by *Aspergillus* sp. (PKU F8) at pH 8.

Discussion

Diversity of fungi has been reported from various environments such as freshwater (Gulis et al. 2006), marine environments such as coastal waters (Gao et al. 2010), deep-sea sediments (Damare et al. 2006; Singh et al. 2010), hypersaline waters (Buchalo et al. 2000), methane hydrates (Lai et al. 2007), oxygen deficient ecosystems (Cathrine and Raghukumar 2009), mangroves and salt marshes (Hyde et al. 1998; Raghukumar 2004), and hydrothermal vents (Le Calvez et al. 2009). This study is the first report on culturable diversity of fungi from coastal ecosystems of Pearl River Delta, China.

Despite of being isolated from coastal environment, the diversity of fungi was observed to be comparatively less, resulting into a total of 22 filamentous fungi and 9 yeast isolates. The isolated fungi belonged mostly to *Ascomycota*, *Basidiomycota* and at a very less percentage of *Zygomycota* (Table 2). Previous studies have also reported abundance of above fungal phylotypes in coastal

Table 4 Quantitative assay of enzyme activities for four fungal isolates

Isolate ID	Fungal genera	Growthdate	Lipase ($Umin^{-1} ml^{-1}$)	Cellulase ($Umin^{-1} ml^{-1}$)	Laccase ($Umin^{-1} ml^{-1}$)
PKU F16	*Cladosporium sp.*	6th day	14.384 ± 0.994	0.843 ± 0.084	10.495 ± 1.300
		9th day	12.285 ± 0.772	0.443 ± 0.032	5.105 ± 0.943
PKU F18	*Ascomycota* sp.	6th day	6.605 ± 0.959	1.133 ± 0.096	14.609 ± 1.039
		9th day	4.630 ± 0.812	1.001 ± 0.049	7.581 ± 1.642
PKU Y5	*Rhodosporidium sp.*	6th day	17.174 ± 0.742	0.333 ± 0.040	-
		9th day	12.186 ± 1.705	0.371 ± 0.028	-
PKU Y8	*Candida sp.*	6th day	21.940 ± 1.582	1.311 ± 0.013	-
		9th day	18.767 ± 1.554	1.132 ± 0.029	-

seawater ecosystems (Gao et al. 2010). The results obtained in the present study are also in concordance with earlier reports where surface water of coastal ecosystems has been reported to contain higher mycoplankton diversity compared with open-ocean ecosystems (Gao et al. 2010). However, use of only culture-dependent approach in this study for estimation of diversity may have limited the isolation of other genera of fungi which are present as uncultivable forms in the oceanic habitats. Previous studies have reported a greater percentage of fungal diversity when assayed using combined approach of culture-dependent and culture-independent methods (Singh et al. 2012). Therefore, a detailed study on fungal diversity from Pearl River Delta using both the approaches may provide a greater insight on hidden, still unexplored mycoplankton communities in future. The diversity of fungi has also been reported to vary with the nutrient contents of any particular habitat, which can be available as detritus for the fungal population (Newell 1982).

Isolation of fungi was attempted using five different media (MEA, SDA, CMA, PDA and CMA). However, fungal isolates were recovered only on two media i.e., MEA and SDA (Table 3). In contrast, the fungal isolates were obtained with all the above five media during isolation from deep-sea sediments by Singh et al. (2012). Most of the isolates affiliated with the existing sequences of NCBI data base at the percentage identity of 100 or 99. However, the 92% identity of the fungal sequence PKU F18 with *Ascomycota* sp., suggests its probability of being novel. On the contrary, the insufficient database for ITS sequences may also be one of the reasons for such low similarity values (Zachow et al. 2009; Anderson et al. 2003).

The production of extracellular enzymes i.e., laccase. cellulase and lipase was demonstrated by most of the fungal isolates in the present study, suggesting their active role in various ecological cycles of coastal ecosystems off China. Laccase in one of the important lignin degrading enzymes, demonstrated to decolorize a range of dyes and toxic industrial effluent in earlier reports (Nyanhongo et al. 2002). The fungal isolates PKU F16 and PKU F18, showing considerable quantitative production of laccase in the present study render them possible candidates for industrial application (Table 4). Cellulose is the most common substrate present in the seawater column in the form of plant biomass. It is found in nature exclusively in plant cell walls, although it is produced by some animals also e.g., tunicates and few bacteria (Lynd et al. 2002). Fungi are well known agents of decomposition of organic matter composed of cellulosic substrate in particular (Lynd et al. 2002). Therefore, cellulase production by the fungal isolates in the present study suggest their active involvement in the mineralization and leaf litter degradation in the coastal sea water habitats. Lipases are ubiquitous enzymes of considerable

Table 5 Details regarding pellet formation by the fungal isolates

Isolate ID	Species	No. of spores (per L)	Pellet color	Different pH			
				PH = 2	PH = 4	PH = 6	PH = 8
				Pellet size(mm)	Pellet size(mm)	Pellet size(mm)	Pellet size(mm)
PKU F1	*Mucor* sp. FJ09	$4.00*10^6$	yellow	-	cotton- shaped	5	4
		$1.60*10^7$	yellow	-	cotton- shaped	2 pellets (20*10)/(5*10)	1 pellet (20*10)
PKU F8	*Aspergillus nomius* strain	$1.59*10^7$	white	<0.5	4-6	3	3-7
		$6.34*10^7$	white	<0.5	1-2	3	4
PKU F14	*Cladosporium* sp.	$3.85*10^8$	black	<0.1	0.5	0.5-1	1-2
		$1.54*10^9$	black	-	-	-	1

industrial and physiological significance. They catalyze the hydrolysis of triacylglycerols to glycerol and free fatty acids. Lipases are widely used in the processing of fats and oils, detergents and degreasing formulations, food processing, the synthesis of fine chemicals and pharmaceuticals, paper manufacture, and production of cosmetics, and pharmaceuticals (Rubin and Dennis 1997a, b; Kazlauskas and Bornscheuer 1998). Lipase production up to 21.9 U ml^{-1} by *Candida* sp. in the present study reveals the potential of this fungus towards lipolytic degradation abilities. Additionally, the optimization of media and nutrient condition for enhanced lipase production from above isolate may be applied in future to maximize their industrial application and commercialization.

Fungi have been characterized widely for their efficient role in pellet formation towards harvesting of algae and wastewater treatment (Zhou et al. 2012). The microalgae cells can be processed into a broad spectrum of biofuels by the transesterification process. These biofuels include biodiesel, green diesel and gasoline, being produced by transformation of algal biomass using various technologies (Chisti 2007). However, many challenges have restricted the development of algal biofuel technology to commercial practicality that could allow for sustainable production and utilization (Brennan and Owende 2010). Among these, one is the harvesting process of algae, which can be improved by application of agents, causing aggregation of algal cells. The pellet formation capabilities exhibited by three of the fungal isolates in the present study (Table 5, Additional file 1: Figure S2) opens new avenues for their efficient utilization towards algae harvesting and wastewater treatment. In addition, the pellet formation capabilities of fungi have been shown to be affected widely by various factors such as pH, inoculum concentration and trace metals (Zhou et al. 2000). A co-cultivation method using fungi of the present study with algal strains under different pH, trace metals and inoculum concentration for efficient optimization of the harvesting process is suggested for future studies.

In conclusion, the fungal diversity obtained from this study was low. The fungal isolates belonged to three major phyla i.e., *Ascomycota*, *Basidiomycota* and *Zygomycota*, with *Ascomycota* being the dominant forms. The different qualitative as well as quantitative levels of extracellular enzymes produced by these isolates suggests them as significant component of the ecological cycles of coastal ecosystems off Pearl River Delta. Finally, the production of enzymes and pellet formation abilities of a few fungal isolates indicate their possible utilization in biotechnological industries.

Additional file

Additional file 1: Figure S1. Microscopic photograph of fungal isolates. **Figure S2.** Pellet formation by different fungal isolates using spores as inoculums at PH 6.

Competing interests
The authors declare that they have no competing interests.

Author's contributions
GW & PS conceived and designed the experiments; LL & YL performed the experiments; PS & LL analyzed the data; GW & PS contributed reagents/materials/analysis tools; PS, LL, GW & SQP wrote the paper. All authors read and approved the final manuscript.

Acknowledgements
This work was partially funded by National Science Foundation of China grant 31170109 (GYW) and Shenzhen Development and Reform Commission grant 835 (GYW).

Author details
[1]School of Environment and Energy, Peking University Shenzhen Graduate School, Shenzhen, 518055, China. [2]Department of Biological Science, National University of Singapore, Singapore 117543, Singapore. [3]Tianjin University Center for Marine Environmental Ecology, School of Environmental Science and Engineering, Tianjin University, Tianjin 300072, China. [4]Department of Microbiology, University of Hawaii at Manoa, Honolulu, HI 96822, USA.

References
Altschul SF, Gish W, Miller W, Myers EW, Lipman DJ (1990) Basic local alignment search tool. J Mol Biol 215:403–410
Anderson IC, Campbell CD, Prosser JI (2003) Potential bias of fungal 18S rDNA and internal transcribed spacer polymerase chain reaction primers for estimating fungal biodiversity in soil. Environ Microbiol 5:36–47
Brennan L, Owende P (2010) Biofuels from microalgae-a review of technologies for production, processing, and extractions of biofuels and co-products. Renew Sustain Energy Rev 14:557–577
Buchalo AS, Nevo E, Wasser SP, Volz PA (2000) Newly discovered halophilic fungi in the Dead Sea (Israel). In: Seckbach J (ed) Journey to diverse microbial worlds. Kluwer, Dordrecht, The Netherlands, pp 239–252
Bungihan ME, Tan MA, Kitajima M, Kogure N, Franzblau SG, Cruz TEE, Takayama H, Nonato MG (2011) Bioactive metabolites of *Diaporthe* sp. P133, an endophytic fungus isolated from *Pandanus amaryllifolius*. J Nat Med 65:606–609
Carder KL, Steward RG, Paul JH, Vargo GA (1986) Relationships between chlorophyll and ocean color constituents as they affect remote-sensing reflectance models. Limnol Oceanogr 31:403–413
Cathrine SJ, Raghukumar C (2009) Anaerobic denitrification in fungi from the coastal marine sediments off Goa, India. Mycol Res 11:100–109
Chi Z, Wang F, Chi Z, Yue L, Liu G, Zhang T (2009) Bioproducts from *Aureobasidium pullulans*, a biotechnologically important yeast. Appl Microbiol Biotechnol 82:793–804
Chisti Y (2007) Biodiesel from microalgae. Biotechnol Adv 25:294–306
Christensen M (1989) A view of fungal ecology. Mycol 81:1–19
Damare V, Raghukumar S (2008) Abundance of thraustochytrids and bacteria in the equatorial Indian Ocean, in relation to transparent exopolymeric particles (TEPs). FEMS Microbiol Ecol 65:40–49
Damare S, Raghukumar C, Raghukumar S (2006) Fungi in deep-sea sediments of the Central Indian Basin. Deep-Sea Res Part I 53:14–27
Danovaro R, Pusceddu A (2007) Biodiversity and ecosystem functioning in coastal lagoons: does microbial diversity play any role? Estuar Coast Shelf Sci 75:4–12
Fenchel T (2008) The microbial loop-25 years later. J Exp Mar Biol Ecol 366:99–103
Gao Z, Johnson ZI, Wang G (2010) Molecular characterization of the spatial diversity and novel lineages of mycoplankton in Hawaiian coastal waters. ISME J 4:111–120

Giovannoni SJ, Stingl U (2005) Molecular diversity and ecology of microbial plankton. Nature 437:343–348
Gulis V, Kuehn K, Suberkropp K (2006) The role of fungi in carbon and nitrogen cycles in freshwater ecosystems. In: Gadd GM (ed) Fungi in biogeochemical cycles. Cambridge University Press, New York, pp 405–435
Gutiérrez MH, Pantoja S, Quiñones RA, González RR (2010) First record of filamentous fungi in the coastal upwelling ecosystem off central Chile. Gayana 74:66–73
Gutiérrez MH, Pantoja S, Tejos E, Quiñones RA (2011) Role of fungi in processing marine organic matter in the upwelling ecosystem off Chile. Mar Biol 158:205–219
Hallam SJ, Mincer TJ, Schleper C, Preston CM, Roberts K, Richardson PM, DeLong EF (2006) Pathways of carbon assimilation and ammonia oxidation suggested by environmental genomic analyses of marine Crenarchaeota. PLoS Biol 4:520–536
Hyde KD, Jones EBG, Leano E, Pointing SB, Poonyth AD, Vrijmoed LLP (1998) Role of fungi in marine ecosystems. Biodivers Conserv 7:1147–1161
Jickells TD (1998) Nutrient biogeochemistry of the coastal zone. Science 281:217–222
Kaul S, Gupta S, Ahmed M, Dhar MK (2013) Endophytic fungi from medicinal plants: a treasure hunt for bioactive metabolites. Phytochem Rev 11:487–505
Kazlauskas RJ, Bornscheuer UT (1998) Biotransformations with lipases. In: Rehm HJ, Pihler G, Stadler A, Kelly PJW (eds) Biotechnology, vol 8. VCH, New York, pp 37–192
Kiørboe T, Jackson GA (2001) Marine snow, organic solute plumes, and optimal sensory behaviour of bacteria. Limnol Oceanogr 46:1309–1318
Lai X, Cao L, Tan H, Fang S, Huang Y, Zhou S (2007) Fungal communities from methane hydrate-bearing deep-sea marine sediments in South China Sea. ISME J 1:756–762
Le Calvez T, Burgaud G, Mahe S, Barbier G, Vandenkoornhuyse P (2009) Fungal diversity in deep-sea hydrothermal ecosystems. Appl Environ Microbiol 75:6415–6421
Lynd LR, Weimer PJ, Van Zyl WH, Pretorius IS (2002) Microbial cellulose utilization: fundamentals and biotechnology. Microbiol Mol Biol Rev 66:506–577
Manini E, Fiordelmondo C, Gambi C, Pusceddu A, Danovaro R (2003) Benthic microbial loop functioning in coastal lagoons: a comparative approach. Oceanolog Acta 26:27–38
Moran MA, Miller WL (2007) Resourceful heterotrophs make the most of light in the coastal ocean. Na Rev Microbiol 5:792–800
Mou XZ, Sun SL, Edwards RA, Hodson RE, Moran MA (2008) Bacterial carbon processing by generalist species in the coastal ocean. Nature 451:708–711
Nagano T, Fraser P (2011) No-nonsense functions for long noncoding RNAs. Cell 145:178–181
Newell RC (1982) The energetics of detritus utilisation in coastal lagoons and nearshore waters. In: Laserre P, Postma H (eds) Proceedings of International Symposium on Coastal Lagoos; 8–14 September 1981; Oceanologica Acta (Special issue). SCOR/IABO/UNESCO, Bordeaux, pp 347–355
Newell SY (1994) Ecomethodology for organo-osmotrophs, prokaryotic unicellular versus eukaryotic mycelial. Microb Ecol 28:151–157
Niku-Paavola ML, Karhunen E, Salola P, Raunio V (1988) Lignolytic enzymes of the white-rot fungus Phlebia radiata. Biochem J 254:877–884
Nyanhongo GS, Gomes J, Gübitz GM, Zvauya R, Read J, Steiner W (2002) Decolorization of textile dyes by laccases from a newly isolated strain of Trametes modesta. Wat Res 36:1449–1456
Pang KL, Mitchell JI (2005) Molecular approaches for assessing fungal diversity in marine substrata. Bot Mar 48:332–347
Pusceddu A, Dell'Anno A, Danovaro R, Manini E, Sara G, Fabiano M (2003) Enzymatically hydrolysable protein and carbohydrate sedimentary pools as indicators of the trophic state of detritus sink systems: a case study in a Mediterranean coastal lagoon. Estuar 26:641–650
Raghukumar S (2004) The role of fungi in marine detrital processes. In: Ramaiah N (ed) Marine Microbiology: Facets & Opportunities. National Institute of Oceanography, Goa, India, pp 91–101
Raghukumar S, Sharma S, Raghukumar C, Sathe-Pathak V, Chandramohan D (1994) Thraustochytrid and fungal component of marine detritus. IV. Laboratory studies on decomposition of leaves of the mangrove Rhizophora apiculata Blume. J Exp Mar Biol Ecol 183(1):113–131
Richards TA, Jones MD, Leonard G, Bass D (2012) Marine fungi: their ecology and molecular diversity. Ann Rev Mar Sci 4:495–522
Rubin B, Dennis EA (1997a) Lipases: Part A. Biotechnology Methods in enzymology, vol 284. Academic Press, New York, pp 1–408
Rubin B, Dennis EA (1997b) Lipases: Part B. Enzyme characterization and utilizationMethods in enzymology, vol 286. Academic Press, New York, pp 1–563
Singh R, Gupta N, Goswami VK, Gupta R (2006) A simple activity staining protocol for lipases and esterases. Appl Microbiol Biot 70:679–682
Singh P, Raghukumar C, Verma P, Shouche Y (2010) Phylogenetic diversity of culturable fungi from the deep-sea sediments of the Central Indian Basin and their growth characteristics. Fungal Divers 40:89–102
Singh P, Raghukumar C, Verma P, Shouche Y (2012) Fungal diversity in deep-sea sediments revealed by culture-dependent and culture-independent approaches. Fungal Ecol 5:543–553
Srinivasan C, Dsouza TM, Boominathan K, Reddy CA (1995) Demonstration of laccase in the white rot basidiomycete phanerochaete chrysosporium BKM-F1767. Appl Environ Microbiol 61:4274–4277
Strom SL (2008) Microbial ecology of ocean biogeochemistry: A community perspective. Science 320:1043–1045
Tamura K, Dudley J, Nei M, Kumar S (2007) MEGA4: molecular evolutionary genetics analysis (MEGA) software version 4.0. Mol Biol Evol 24:1596–1599
Thompson JD, Higgins DG, Gibson TJ (1994) CLUSTAL W: improving the sensitivity of progressive multiple sequence alignment through sequence weighting, position-specific gap penalties and weight matrix choice. Nucleic Acids Res 22:4673–4680
Tisdall JM, Oades JM (1982) Organic matter and water-stable aggregates in grassland soils. J Soil Sci 33:141–163
Wang GY, Johnson ZI (2009) Impact of parasitic fungi on the diversity and functional ecocology of marine phytoplankton. In: Kersey TW, Munger SP (eds) Marine Phytoplankton. Nova Science Publishers, Hauppauge, NY, pp 211–228
White TJ, Bruns T, Lee S (1990) Amplification and direct sequencing of fungal ribosomal RNA genes for phylogenetics. PCR Protocols: a Guide to Methods and Applications 18:315–322
Zachow C, Berg C, Muller H, Meincke R, Komon-Zelazowska M, Druzhinina IS, Kubicek CP, Berg G (2009) Fungal diversity in the rhizosphere of endemic plant species of Tenerife (Canary Islands): relationship to vegetation zones and environmental factors. ISME J 3:79–92
Zhou Y, Du J, Tsao GT (2000) Mycelial Pellet Formation by *Rhizopus oryzae* ATCC 20344. Appl Biochem Biotechnol 84:779–789
Zhou WG, Cheng YL, Li Y, Wan YQ, Liu YH, Lin XY, Ruan R (2012) Novel fungal pelletization assisted technology for algae harvesting and wastewater treatment. Appl Biochem Biotechnol 167:214–228

Genetic transformation of *Knufia petricola* A95 - a model organism for biofilm-material interactions

Steffi Noack-Schönmann[1], Tanja Bus[2], Ronald Banasiak[1], Nicole Knabe[1], William J Broughton[1], H Den Dulk-Ras[3], Paul JJ Hooykaas[3] and Anna A Gorbushina[1,4*]

Abstract

We established a protoplast-based system to transfer DNA to *Knufia petricola* strain A95, a melanised rock-inhabiting microcolonial fungus that is also a component of a model sub-aerial biofilm (SAB) system. To test whether the desiccation resistant, highly melanised cell walls would hinder protoplast formation, we treated a melanin-minus mutant of A95 as well as the type-strain with a variety of cell-degrading enzymes. Of the different enzymes tested, lysing enzymes from *Trichoderma harzianum* were most effective in producing protoplasts. This mixture was equally effective on the melanin-minus mutant and the type-strain. Protoplasts produced using lysing enzymes were mixed with polyethyleneglycol (PEG) and plasmid pCB1004 which contains the hygromycin B (HmB) phosphotransferase (*hph*) gene under the control of the *Aspergillus nidulans trpC*. Integration and expression of *hph* into the A95 genome conferred hygromycin resistance upon the transformants. Two weeks after plating out on selective agar containing HmB, the protoplasts developed cell-walls and formed colonies. Transformation frequencies were in the range 36 to 87 transformants per 10 µg of vector DNA and 10^6 protoplasts. Stability of transformation was confirmed by sub-culturing the putative transformants on selective agar containing HmB as well as by PCR-detection of the *hph* gene in the colonies. The *hph* gene was stably integrated as shown by five subsequent passages with and without selection pressure.

Keywords: DNA transfer; Fungal cell-walls; Protoplasts; Hygromycin resistance; Black yeast; Sub-aerial biofilms; Stress-protective morphology; Ancestor of opportunistic pathogens & lichens

Introduction

Melanised microcolonial fungi (MCF) are a taxonomically diverse group of Ascomycetes that colonise sub-aerial rock surfaces and other varied materials in all regions of the world, from the Antarctic to hot deserts (Staley et al. 1982; Gorbushina et al. 1993; Urzi et al. 1995; Wollenzien et al. 1995; Sterflinger and Prillinger 2001; Ruibal 2004; Selbmann et al. 2005; Gorbushina 2007; Ruibal et al. 2008). MCF are characterised by simple stress-protective morphologies including a peculiar compact colonial structure, protective cell walls and multiple secondary metabolic products that support stress tolerance – carotenoids, melanins, mycosporines and compatible solutes (e.g. Gorbushina et al. 2008). For all these reasons MCF are unequalled amongst eukaryotes in their ability to withstand extreme heat, desiccation and UV radiation. MCF interact with accompanying phototrophic algae, cyanobacteria and heterotrophic bacteria to form a stable sub-aerial (SABs) microbial community (Gorbushina et al. 2005; Gorbushina 2007). At the extreme, MCF survival can depend on the small amounts of nutrients available in atmospheric aerosols, gases and air-borne particles – as well as on other SAB inhabitants (Gorbushina 2007). As a consequence of MCF stress tolerance, SABs are pioneer settlers on diverse surfaces exposed to air including artificial materials. Eventually SABs cause discoloration and deterioration of facades and roofs of buildings and are thus an important aspect of material science.

Knufia petricola A95 is a well suited model fungus to study organism-organism and organism-material interactions in SABs (Gorbushina and Broughton, 2009;

* Correspondence: anna.gorbushina@bam.de
[1]Department 4 (Materials & Environment), Federal Institute for Materials Research and Testing (Bundesanstalt für Material-forschung und -prüfung, BAM), Unter den Eichen 87, 12205 Berlin, Germany
[4]Department of Biology, Chemistry and Pharmacy & Department of Earth Sciences, Freie Universität Berlin, Malteserstrasse 74-100, 12249 Berlin, Germany
Full list of author information is available at the end of the article

Nai et al. 2013). As the rock-inhabiting *K. petricola* A95 belongs to an ancestral lineage of the Chaetothyriales (Ascomycetes) and is a predecessor of opportunistic pathogens and lichens (Gueidan et al. 2008), it is also of potential importance for understanding the evolution of Ascomycete life styles. Nevertheless, to be a tractable model organism, a reliable and efficient transformation system is essential (Perez-Nadales et al. 2014).

SABs are principal determinants of weathering of rocks (Gorbushina and Krumbein, 2000) and other materials exposed to air (e.g. Noack-Schönmann et al. 2014). SABs are important in the material sciences not only because they are omni-present on all exposed materials, but also because they degrade the surfaces to which they adhere. Weathering results and transparent surfaces become opaque restricting the passage of radiation (Hallbauer and Jahns, 1977; Jones and Wilson, 1985; Chen et al. 2000; Avakian et al. 1981; Jongmans et al. 1997; Gadd et al. 1999; Burford et al. 2003; Gorbushina et al. 1993; Noack-Schönmann et al. 2014; Shirakawa, Zilles, Mocelin, Gaylarde, Gorbushina, Heidrich, Giudice, Del Negro, John. Microbial colonisation affects the efficiency of photovoltaic panels in a tropical environment, submitted to Science of the Total Environment). Diagnostic and prognostic tools for biofilm development on different materials and under various environmental conditions require easily accessible and genetically modifiable SABs. The establishment of a DNA-transfer- and integration-system into *K. petricola* A95 will set the stage for a variety of investigations into the role of MCFs in SAB formation and biofilm-induced material deterioration.

An efficient transformation system is particularly important in dual-component systems like sub-aerial biofilms (SAB) where rock-inhabiting fungi form close inter- and intra-cellular contacts with phototrophic cells (Gorbushina et al. 2005). Gorbushina and Broughton (2009) proposed a model biofilm comprising the cyanobacterium *Nostoc punctiforme* and the MCF *Knufia petricola* (A95) as a new test system for the simulation of environment-biofilm interactions and their impact on materials. As an experimental model it is however essential that both partners be accessible to genetic manipulation. A transformation system exists for *N. punctiforme* (e.g. Risser and Meeks, 2013) and here we report a method for stably transforming *K. petricola*.

Materials and methods

Strains and media

K. petricola A95 (= CBS 123872) was isolated from the surface of a marble rock near the Philopappos monument on Musaios Hill, Athens (Greece) (Gorbushina et al. 2008). It is maintained in the culture collection of the Centraalbureau for Schimmelcultures (CBS 123872) in Utrecht and the Federal Institute for Materials Research and Testing in Berlin (A95). A spontaneous pink mutant of A95 deficient in melanin synthesis was isolated (A95p). The characteristic orange/pink colour of the mutant (Figure 1B) is caused by carotenoids (Gorbushina et al. 2008) that are normally masked by the melanised cell-wall. The type-strain (A95) and A95p were cultivated in liquid malt-extract medium (MEB)[2% (w/v) malt extract, 0.1% (w/v) peptone, 2% (w/v) glucose] at 25°C and 100 rpm. The cultures were refreshed each week as described in Nai et al. (2013). A one week old culture was homogenised along with 10 steel balls (Ø 5 mm) in a ball mill (Retsch™, Retsch GmbH, 42781 Haan, Germany) for 30 s^{-1} and 10 min then diluted 1:100 with fresh malt extract medium.

Formation of protoplasts

300 ml Erlenmeyer flasks containing 100 ml of MEB were inoculated with 1 ml of one-week-old freshly dispersed A95 and A95p cultures grown in MEB (experiments were duplicated for both strains). The cultures were grown for three days at 25°C and 100 rpm, spun down at 3,000 × g for 10 min and the resulting biomass washed three times with protoplast buffer (0.01 M *Tris–HCl*, 0.01 M $MgSO_4$, 1M KCl, pH 7.0 (Walz and Kück, 1995)). Different enzyme combinations were added to 2 × 10^7 cfu A95 in 20 ml protoplast buffer. The enzymes used were: (i) a mixture of Vinoflow (Novozymes, Bagsvaerd, Denmark), Yatalase (Takara, Shiga Japan) and ß-glucanase (*Aspergillus niger*) (Sigma-Aldrich, Steinheim, Germany)(each 10 $mg.ml^{-1}$) (ii) Lysing enzymes from *Trichoderma harzianum* (Sigma-Aldrich)(10$mg.ml^{-1}$) and (iii) a mixture of Driselase (Sigma-Aldrich), Lyticase (Sigma-Aldrich) and ß-glucanase (*A.niger*) (Sigma-Aldrich) also at 10 $mg.ml^{-1}$. The A95-enzyme mixes were incubated at 27°C and 100 rpm for 24 h. The efficiency of protoplast formation (ratio of protoplasts to the number of A95 cells still surrounded by cell walls) was determined in 10 μl aliquots by counting in a Neubauer chamber. For transformation, the protoplasts were centrifuged for 5 min at 3,000 × g, washed several times with transformation buffer [1M sorbitol, 80 mM $CaCl_2$, pH 7.0 (Walz and Kück, 1993)] and re-suspended in the same buffer at a final concentration of 5 × $10^7.ml^{-1}$.

Recombinant plasmids

The hygromycin selection vector for PEG mediated transformation pCB1004 (Carroll et al. 1994) was used. pCB1004 contains the hygromycin B phosphotransferase gene (*hph*) under control of the *A. nidulans trpC* promoter (Mullaney et al. 1985) but without the *trpC* terminator. pCB1004 also contains a chloramphenicol resistance cassette, a functional *lacZ* gene and a multiple cloning site (Figure 2).

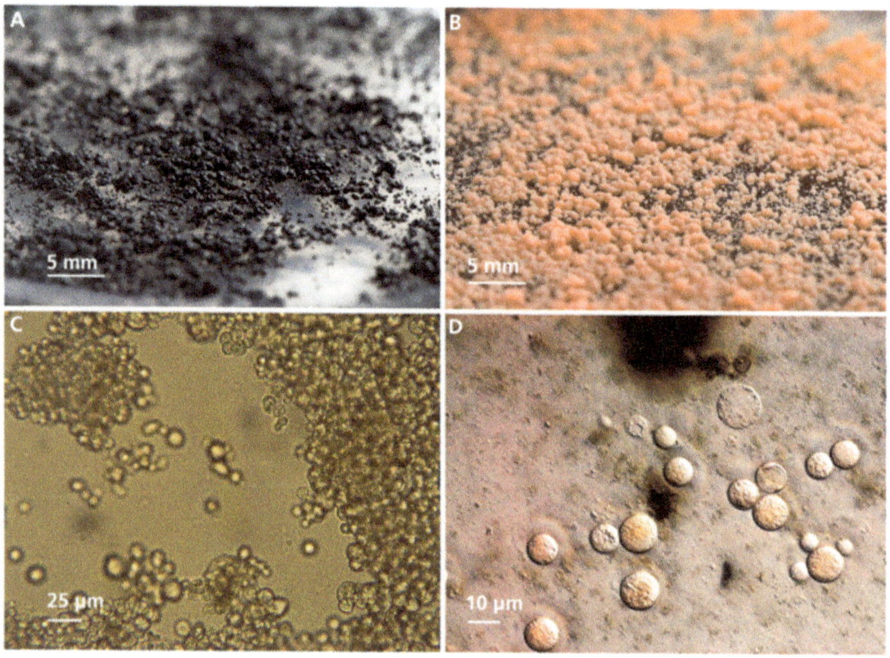

Figure 1 ***Knufia petricola*** **A95 morphology. A**- Being a typical MCF, *K. petricola* A95 retains protective pigmentation and restricted colony growth even under favourable growth conditions on a Petri dish; **B** - As a result of a spontaneous mutation, the A95 pink mutant (A95p) is deficient in melanin synthesis. The orange/red colour is caused by carotenoids that are normally masked by melanins. **C**- Micrograph of A95 cells with melanin. The ball-shaped, compact A95 cells are surrounded by a thick cell-wall and are embedded in extracellular polymeric substances that are arranged in compact clusters of cells. **D**- A95 protoplasts prepared from similar cells to those shown in Figure 1C. A95 protoplasts were most effectively isolated by digestion of the cell-wall using lysing enzymes from *Trichoderma harzianum* and prevented from bursting in a 1M KCl buffer.

PEG mediated transformation

50 µl (≈ 10^6) protoplasts, re-suspended in transformation buffer, were incubated with 10–20 µg of plasmid DNA for 20 min at 4°C. 100 µl of 24% (w/v) PEG 6000 was added, gently mixed and incubated for 30 min at room temperature. Five mL MEBS [malt extract broth with 10.8% (w/v) sucrose] containing 150 µl transformation mix was distributed over five MEAS-plates [malt-extract agar with 10.8% (w/v) sucrose]. After incubation at 25°C for 24 h, the plates were overlain with 5 ml of selective agar [0.8 M NaCl, 0.4% w/v) agar, final concentration of 0.25 mg of HygB.ml^{-1}, pH 7.0] and incubated for 3 to 4 weeks at 25°C.

Isolation of fungal DNA

A95 cells were frozen in liquid nitrogen and ground with a mortar and pestle. 100 mg of the ground cells was mixed with 500 µl Genomic Lysis Buffer (OmniPrep™ for Fungus Kit, G-Biosciences, St. Louis, MO 63132–1429, U.S.A.) and total DNA isolated using the same kit according to the manufacturer's instructions. Five randomly selected A95 transformants were used for total DNA isolation and amplification of the *hph* gene.

Amplification of the hygromycin B phosphotransferase gene (*hph*)

On the assumption that the hygromycin B phosphotransferase gene (*hph*) was integrated randomly into the genome of A95, total genomic DNA of the transformants was amplified with the primer pair hphRforw (5'-ATGCCTGAACTCACCGCGAC-3') and hphRrev (5'-CTATTCCTTTGCCCTCGGACGAG-3'). PCR reactions were carried out with *ReproFast*-DNA Polymerase (Genaxxon Bioscience, 89007 Ulm, Germany) under the following conditions: initial denaturation at 96°C for 30s, followed by 30 cycles of denaturation at 96°C for 30s, annealing at 54°C for 30s, extension at 72°C for 2 min, terminated by 10 min extension at 72°C. The PCR products were sequenced.

Mitotic stability of the A95 transformants

Five randomly selected colonies were transferred to 5 ml MEA, mixed with 10 steel balls (Ø 5 mm), vigorously shaken in a Retsch™ mill for 10 min at 30 Hz and subcultured on MEA with and without hygromycin B. The putative transformants were subcultured five times on selective or non-selective agar. After five rounds of

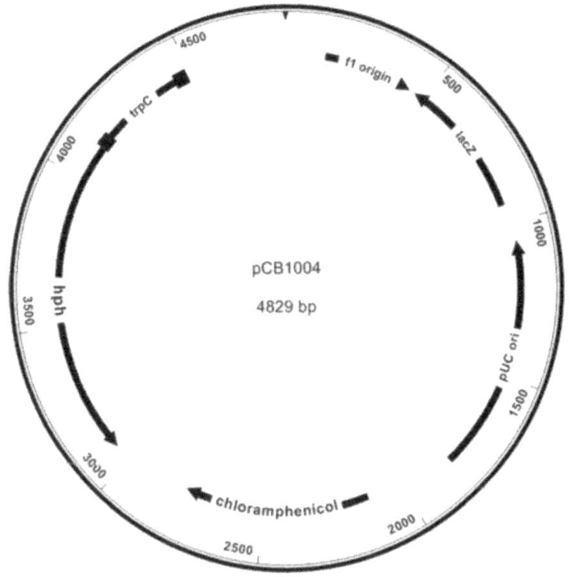

Figure 2 Plasmid map of pCB1004. Plasmid pCB1004 was constructed by Carroll et al. (1994). The vector has a chloramphenicol resistance- and a functional *lacZ*-gene for blue-white screening. *hph* = Hygromycin B resistance; *trpC* = *A. nidulands* promoter.

non-selective subculture, the transformants were transferred to MEA plates containing 50 μg.ml^{-1} hygromycin B. This experiment was repeated twice.

Results

Different ways of transforming A95 were tested (S. Noack-Schönmann et al. unpublished results). These included the commonly used binary Ti-vector system of *A. tumefaciens* (De Groot et al. 1998, Michielse 2005a; 2005b; 2008) which failed to yield transformants in our hands. As A95 is strongly melanised, problems in penetrating the melanised and as a consequence thick cell wall were expected. Chemical weakening of the cell wall with DMSO before addition of *A. tumefaciens* was also unsuccessful. Bombardment of A95 with microprojectiles (Lorito et al. 1993; Te'o et al. 2002) coated with pTAS5 and pCB1004 plasmid DNA did not yield transformants. Finally different enzyme preparations containing various chitinases and β-glucanases were tested to establish an efficient protocol for the isolation of protoplasts. Armed with an effective protocol for the production of protoplasts, we used the vector pCB1004 carrying the hygromycin resistance gene and polyethylene glycol to transfer hygromycin resistance to A95.

Protoplast isolation

All enzymes and enzyme mixtures tested yielded protoplasts (Figure 1D and Table 1). At best, the protoplast yields were ten percent of the intact cells subjected to the enzyme treatment (Table 1). Clearly, the mixture Driselase, ß-Glucanase and Lyticase was only about half as efficient as the other two treatments. On the other hand, there were no significant differences in protoplast yields obtained with the lysing enzymes of *T. harzianum* and the mixture ß-Glucanase, Vinoflow and Yatalase. As preparation of the *T. harzianum* mix is simpler and more cost effective, we used this procedure in all further experiments. Black A95 cells (A95) and cells of a pink melanin-deficient A95 mutant (A95p) (Figure 1) were subjected to treatment with lysing enzymes of *T. harzianum* as described above. Although higher than in preliminary experiments, protoplast yields were the same for both cell types (Table 2).

Table 1 Efficiency of protoplast isolation from A95 cells using different cell-wall lysing enzymes

Enzymes used for protoplast isolation	Starting number of A95 cells (cfu.mL^{-1})	Number of protoplasts isolated.mL^{-1}
Lysing enzymes from *Trichoderma harzianum*	2×10^7	2.0×10^6
β-Glucanase, Vinoflow, Yatalase	2×10^7	1.9×10^6
Driselase, β-Glucanase, Lyticase	2×10^7	1.1×10^6

Transformation of A95 protoplasts

Colonies were visible on hygromycin B containing MEAS agar two to three weeks after transformation of A95 protoplasts with pCB1004 (Figure 3A). Five randomly selected, transformed colonies were able to grow when plated out again on hygromycinB-containing agar. Total DNA from some of these re-picked colonies was amplified by PCR using the primer pair hphRforw/hphRrev resulting in an amplicon of ≈ 1,000 bp, which is consistent with the size (1,020 bp) of the *hph*-gene (Figure 3B). The sequences of all PCR products matched completely those of *hph*. About 55 transformants were obtained from 10^6 protoplasts in a total volume of 150 μl containing 10 to 20 μg plasmid DNA.

Mitotic stability

All randomly selected A95 transformants maintained hygromycin resistance after multiple passages on MEA

Table 2 Efficiency of protoplast isolation from A95 and a pink A95 derivative

	Starting number of cells (cfu.mL^{-1})	Number of protoplasts isolated.mL^{-1}
A95	5×10^6	1.1×10^6
Pink derivative of A95	5×10^6	1.1×10^6

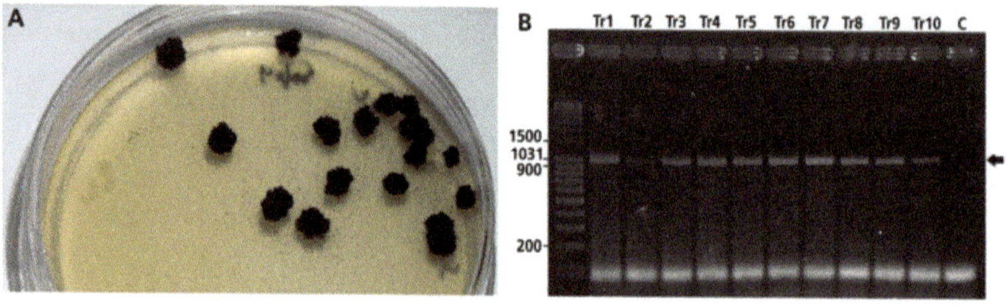

Figure 3 A95 transformants. A- three weeks after PEG mediated transformation of A95 protoplasts with pCB1004, single A95 colonies were observed on malt-extract agar plates containing hygromycin. **B**- PCR products of the *hph*-gene (1,019 bp) seen on agarose gels. Polymerase chain reaction detection of the *hph*-gene from DNA of five independent putative transformants confirmed integration of *hph* into the genome of A95 transformants. Arrow indicates location of *hph* DNA bands.

with or without hygromycin. After subculturing five times without selection pressure (MEA without hygromycin), five formerly hygromycin-resistant colonies were still able to grow when plated out again on hygromycin containing MEA.

Discussion

An assembly of methods that are used for transformation of filamentous fungi were tested in preliminary experiments, but only PEG-mediated transformation of protoplasts showed promise. We performed PEG-mediated transformation of A95 protoplasts with pTAS5 (de Groot et al. 1998) and pCB1004 (Carroll et al. 1994), but only obtained transformants using pCB1004. Perhaps the TAS5 vector (11.8 kb) was too large to escape restriction on entering the A95 protoplasts or the *gpd*A promoter was less efficient in A95 than the *trp*C promoter.

Incubation of protoplasts with pCB1004 and PEG allowed integration of the hygromycin B resistance cassette (under control of the *trp*C promoter) into the genome of A95. All hygromycin B resistant colonies selected maintained resistance to hygromycin during five rounds of subculturing on selective or non-selective agar. Most probably transformation efficiencies can be improved by for example optimising the ratio between applied plasmid-DNA and the number of cells. Various preparations containing a complex mixture of hydrolytic enzymes were tested (Table 1) for preparation of transformable protoplasts. Lysing enzymes of *Trichoderma harzianum* were most effective in isolating protoplasts from A95 cells (Table 1). Further improvement of the protoplast isolation protocol by testing different enzyme concentrations and various incubation temperatures could also be tried. Interestingly, melanin from the pigmented walls did not interfere with protoplast isolation since the efficiency of protoplast formation was the same using type A95 cells and a melanin-deficient mutant (Table 2).

Stable integration of the hygromycin-resistance gene into the A95 genome showed that illegimate recombination is possible in *Knufia petricola* A95. Whether this means that homologous recombination also occurs at a frequency that would permit gene-replacement remains to be determined. Nevertheless, the methods described here permit exploration of A95 using reverse genetic tools. Armed with these tools, it should be possible to discover the mechanisms which A95 and other rock-inhabiting melanised fungi use to colonise inhospitable sub-aerial surfaces so successfully.

Competing interests

The authors declare that they have no competing interests.

Authors' contributions

SNS performed many types of experiments to produce protoplasts and transform them, actively participated in the design of the study and drafted a preliminary manuscript containing Figures 1 and 3. TB prepared many batches of protoplasts. NK conducted the transformation experiments and prepared Figures 2 and 3. RB assisted in all experiments conducted at the BAM and attempted transformation using the "GeneGun". HDDR, PJJH and WJB initiated protoplast isolation and *Agrobacterium*-mediated transformation experiments in the Leiden laboratories. They also participated in the design of later experiments. WJB shaped the presentation of the material written. AAG conceived the study, participated in its design and coordination, helped to interpret the data and write the manuscript. All authors read and approved the final manuscript.

Acknowledgements

We thank Dr. Zheng Wang for the plasmid pCB1004, Prof. Thomas Schmülling and Prof. Wolfgang Schuster (Freie Universität Berlin) for use of their Gene Gun™ equipment in transformation experiments and Thorsten Doil (BAM) for collating the figures. This work was supported financially by internal funds of the BAM.

Author details

[1]Department 4 (Materials & Environment), Federal Institute for Materials Research and Testing (Bundesanstalt für Material-forschung und -prüfung, BAM), Unter den Eichen 87, 12205 Berlin, Germany. [2]Department of Biotechnology, University of Applied Sciences Jena, Carl-Zeiß-Promenade 2, 07745 Jena, Germany. [3]Institute of Biology Leiden, Universiteit Leiden, Sylviusweg 72, 2333 BE Leiden, The Netherlands. [4]Department of Biology, Chemistry and Pharmacy & Department of Earth Sciences, Freie Universität Berlin, Malteserstrasse 74-100, 12249 Berlin, Germany.

References

Avakian ZA, Karavaiko GI, Mel'nikova EO, Krutsko VS, Ostroushko II (1981) Role of microscopic fungi in the process of weathering of pegmatite deposit rocks and minerals. Mikrobiologiya 50:156–162

Burford EP, Fomina M, Gadd GM (2003) Fungal involvement in bioweathering and biotransformation of rocks and minerals. Miner Mag 67:1127–1155

Carroll AM, Sweigard JA, Valent B (1994) Improved vectors for selecting resistance to hygromycin. Fungal Genet Newsl 41:22

Chen J, Blume HP, Beyer L (2000) Weathering of rocks induced by lichen colonization-a review. Catena 39:121–146

De Groot MJA, Bundock P, Hooykaas PJJ, Beijersbergen GM (1998) *Agrobacterium tumefaciens*-mediated transformation of filamentous fungi. Nat Biotechnol 16:839–842

Gadd GM (1999) Fungal production of citric and oxalic acid: importance in metal speciation, physiology and biogeochemical processes. Adv Microb Physiol 41:47–92

Gorbushina AA (2007) Life on the rocks. Environ Microbiol 9:1613–1631

Gorbushina AA, Broughton WJ (2009) Microbiology of the atmosphere-rock interface: How biological interactions and physical stresses modulate a sophisticated microbial ecosystem. Ann Rev Microbiol 63:431–450

Gorbushina AA, Krumbein WE (2000) Subaerial microbial mats and their effects on soil and rock. P. 161–170. In: Riding RE, Awramik SM (eds) Microbial Sediments. Springer, Berlin, p 331

Gorbushina AA, Krumbein WE, Panina L, Soukharjevski S, Wollenzien U (1993) On the role of black fungi in colour change and biodetioration of antique marbles. Geomicrobiol J 11:205–221

Gorbushina AA, Beck A, Schulte A (2005) Microcolonial rock inhabiting fungi and lichen photobionts: evidence for mutualistic interactions. Mycol Res 109:1288–1296

Gorbushina AA, Kotlova ER, Sherstneva OA (2008) Cellular responses of microcolonial rock fungi to long-term desiccation and subsequent rehydration. Stud Mycol 61:91–97

Gueidan C, Villasenor CR, de Hoog GS, Gorbushina AA, Untereiner WA, Lutzoni F (2008) A rock-inhabiting ancestor for mutualistic and pathogen-rich fungal lineages. Stud Mycol 61:111–119

Hallbauer DK, Jahns HM (1977) Attack of lichens on quartzitic rock surfaces. Lichenologist 9:119–122

Jones D, Wilson MJ (1985) Chemical activity of lichens on mineral surfaces – a review. Int Biodeterior Biodegradation 21:99–104

Jongmans AG, van Breemen N, Lundstrom U, van Hees PAW, Finlay RD, Srinivasan M, Unestam T, Giesler R, Melkerud PA, Olsson M (1997) Rock-eating fungi. Nature 389:682–683

Lorito M, Hayes CK, Di Pietro A, Harman GE (1993) Biolistic transformation of *Trichoderma harzianum* and *Gliocladium virens* using plasmid and genomic DNA. Curr Genet 24:349–356

Michielse CB, Arentshorst M, Ram AF, van den Hondel CA (2005a) *Agrobacterium*-mediated transformation leads to improved gene replacement efficiency in *Aspergillus awamori*. Fungal Genet Biol 42:9–19

Michielse CB, Hooykaas PJ, van den Hondel CA, Ram AF (2005b) *Agrobacterium*-mediated transformation as a tool for functional genomics in fungi. Curr Genet 48:1–17

Michielse CB, Hooykaas PJJ, van den Hondel CAMJJ, Ram AFJ (2008) *Agrobacterium*-mediated transformation of the filamentous fungus *Aspergillus awamori*. Nat Protoc 3:1671–1678

Mullaney EJ, Hamer JE, Yelton MM, Timberlake WE (1985) Primary structure of the *trpC* gene from *Aspergillus nidulans*. Mol Gen Genet 199:37–45

Nai C, Wong HY, Pannenbecker A, Broughton WJ, Benoit I, de Vries RP, Gueidan C, Gorbushina AA (2013) Nutritional physiology of a rock-inhabiting, model micro-colonial fungus from an ancestral lineage of the Chaetothyriales (Ascomycetes). Fungal Genet Biol 56:54–66

Noack-Schönmann S, Spagin O, Gründer K-P, Breithaupt M, Günter A, Muschik B, Gorbushina AA (2014) Sub-aerial biofilms as blockers of solar radiation: spectral properties as tools to characterise material-relevant microbial growth. Intern Biodeterior Biodegradation 86:286–293

Perez-Nadales E, Almeida Nogueira MF, Baldin C, Castanheira S, El Ghalid M, Grund E, Lengeler K, Marchegiani E, Vinod Mehrotra P, Moretti M, Naik V, Oses-Ruiz M, Oskarsson T, Schäfer K, Wasserstrom L, Brakhage AA, Gow NAR, Kahmann R, Lebrun M-H, Perez-Martin J, Di Pietro A, Talbot NJ, Toquin V, Walther A, Wendland J (2014) Fungal model systems and the elucidation of pathogenicity determinants. Fungal Genet Biol 70:42–67

Risser DD, Meeks JC (2013) Comparative transcriptomics with a motility-deficient mutant leads to identification of a novel polysaccharide secretion system in *Nostoc punctiforme*. Mol Microbiol 87:884–893

Ruibal CV (2004) Isolation and Characterization of Melanized, Slow-growing Fungi from Semiarid Rock Surfaces of Central Spain and Mallorca. Ph.D. Dissertation. Universidad Autónoma de Madrid; Merck, Sharp & Dohme de Espana, Madrid

Ruibal CV, Platas G, Bills GF (2008) High diversity and morphological convergence among melanised fungi from rock formations in the Central Mountain System of Spain. Persoonia 21:93–110

Selbmann L, de Hoog GS, Mazzaglia A, Friedmann E, Onofri S (2005) Fungi at the edge of life: cryptoendolithic black fungi from Antarctic desert. Stud Mycol 51:1–32

Staley JT, Palmer F, Adams JB (1982) Microcolonial fungi: common inhabitants on desert rocks? Science 215:1093–1095

Sterflinger K, Prillinger H (2001) Molecular taxonomy and biodiversity of rock fungal communities in an urban environment (Vienna, Austria). Antonie Van Leeuwenhoek 80:275–286

Te'o VSJ, Bergquist PL, Nevalainen KMH (2002) Biolostic transformtaion of Trichoderma reesei using the Bio-Rad seven barrels Hepta Adaptor system. J Microbiol Meth 51:393–399

Urzi C, Wollenzien U, Criseo G, Krumbein WE (1995) Biodiversity of the rock inhabiting microbiota with special reference to black fungi and black yeasts. In: Allsopp D, Colwell RR, Hawksworth DL (eds) Microbial Diversity and Ecosystem Function. CAB International, Wallingford, pp 289–302

Walz M, Kück U (1993) Targeted integration into the *Acremonium chrysogenum* genome: disruption of the *pcbC* gene. Curr Genet 24:421–427

Walz M, Kück U (1995) Transformation of *Sordaria macrospora* to hygromycin B resistance: characterization of transformants by electrophoretic karyotyping and tetrad analysis. Curr Genet 29:88–95

Wollenzien U, de Hoog GS, Krumbein WE, Urzi C (1995) On the isolation of microcolonial fungi occurring on and in marble and other calcareous rocks. Sci Total Environ 167:287–294

19

Purification and characterization of a novel lipopeptide from *Streptomyces amritsarensis* sp. nov. active against methicillin-resistant *Staphylococcus aureus*

Deepika Sharma[1], Santi M Mandal[2] and Rajesh Kumari Manhas[1*]

Abstract

Nowadays antimicrobial lipopeptides are being widely exploited for developing potential therapeutic agents for treating bacterial infections. In the present study, we have purified and characterized an antimicrobial lipopeptide produced by *Streptomyces amritsarensis* sp. nov. (= MTCC 11845^T = JCM 19660^T). The lipopeptide was purified using silica gel chromatography, size exclusion chromatography and reverse phase- HPLC. The MS/MS analysis of the lipopeptide revealed that it has amino acid sequence as Ala-Thr-Gly-Ser-His-Gln and a long chain fatty acid tail with six times repeated the molecular mass of 161 Da which is corresponding to -$C_{12}H_{19}$. Based on the molecular mass (878.5 Da) and amino acid composition, the lipopeptide was identified as a novel lipopeptide. The MIC values of purified lipopeptide against *Bacillus subtilis* (MTCC 619), *Staphylococcus epidermidis* (MTCC 435), *Mycobacterium smegmatis* (MTCC 6) and clinical strain, Methicillin Resistant *Staphylococcus aureus* (MRSA) were found to be 10, 15, 25 and 45 μg/ml, respectively. It was completely stable at 70°C for 1 h and retained 81.8% activity after autoclaving (121°C for 15 min). It did not show any change in its activity profile between pH 5.0 - 9.0 and is stable to trypsin, proteinase K and lipase enzymes. It was found to be non-mutagenic against *Salmonella typhimurium* (TA98) and did not show cytotoxicity when checked against Chinese hamster ovary (CHO) cell line. In addition to antibacterial activity it also exhibits biosurfactant activity.

Keywords: Antibacterial; Novel; Lipopeptide; Thermostable; Non-toxic

Introduction

In the last two decades, exploring the possibility of developing new classes of antimicrobial compounds has emerged prominently due to pathogens acquired resistance to emerging antibiotics. Presently available antimicrobial compounds are getting old and less efficient, only few truly original replacements are available. This is due to a number of different reasons including daunting R&D costs for putting a new molecule on a highly competitive market and the inherent difficulty of identifying innovative antibiotic targets (Nathan and Goldberg 2005; Payne et al. 2007). Therefore, screening and characterization of the novel antimicrobial compounds especially, peptides from microorganisms, have drawn attention (Pirri et al. 2009; Laverty et al. 2011). Among antimicrobial peptides, lipopeptides are small molecules and have been considered as potential source of future antibiotics because of their different mechanisms of action as compared to conventional antibiotics (Baindara et al. 2013).

Antimicrobial lipopeptides represent an old class of antibiotics that were discovered over 50 years ago and consist of a hydrophilic cyclic peptide portion attached to a fatty acid chain. They are biosynthesized by large multi-enzymes called non-ribosomal peptide synthetases *via* non-ribosomal pathways (Pirri et al. 2009). All antibiotics belonging to this novel class contain multiple nonproteinogenic amino acids as well as different lipid tails; this yields remarkable structural diversity (Strieker and Marahiel 2009). Lipopeptides easily bind to the bacterial surface bilayer and alter the local lipid organizational

* Correspondence: rkmanhas@rediffmail.com
[1]Department of Microbiology, Guru Nanak Dev University, Amritsar, Punjab 143005, India
Full list of author information is available at the end of the article

linking on negatively charged fatty acids resulting in restructuring of the lipid bilayer and prevent cellular processes (Horn et al. 2012; Mandal et al. 2013).

Actinobacteria especially, species of the genus *Streptomyces* are reported to produce diverse antimicrobial lipopeptides with their applications in pharmaceutical industries. Amphomycin was the first lipopeptide to be discovered from *Actinoplanes friuliensis*, followed by a number of related antibiotics, including crystallomycin, aspartocin, glumamycin, laspartomycin, tsushimycin, and the best studied so far, daptomycin (Schneider et al. 2009). As apparently 99% of the microbial species are still unexplored (Davies 1999; Watve et al. 2001) therefore, possibility of discovering actinobacteria producing potent and novel lipopeptides still remains. In the light of this, present study was aimed at purification and characterization of a novel antimicrobial lipopeptide produced by *Streptomyces amritsarensis*.

Materials and methods

Microorganism and maintenance

Streptomyces amritsarensis (= MTCC 11845^T = JCM 19660^T) was isolated from soil and identified using polyphasic taxonomic approach (Sharma et al. 2014). The culture was maintained on starch casein nitrate agar slopes at 4°C and as mycelial fragments and spores in 20% v/v glycerol at -80°C. All the test organisms except clinical isolates were procured from Microbial Type Culture Collection and Gene Bank (MTCC), Institute of Microbial Technology (IMTECH), Chandigarh, India. Clinical isolates *viz.* multi-drug resistant *E. coli*, MRSA and VRE were procured from local hospitals. They were maintained on nutrient agar slopes at 4°C.

Growth and lipopeptide production profile

S. amritsarensis was grown in Starch Casein Nitrate (SCN) broth at 28°C with continuous shaking at 180 rpm for 10 days. Cell free supernatants were collected at one day intervals by centrifuging culture broths at 10,000 rpm for 30 min at 4°C and used for detection of lipopeptide antimicrobial activity using Kirby-Bauer disk diffusion susceptibility test (Bauer et al. 1966). Filter paper discs (6 mm) impregnated with 10 µl of supernatants (filtered through 0.22 µm filter, Pall Lifesciences) were placed on Mueller-Hinton agar plate seeded with *Bacillus subtilis* (MTCC 619) and incubated at 37°C for 24 h. For the determination of growth, absorbance of culture broths were read at 600 nm and growth curve was prepared. Protein contents of culture supernatants were determined using Lowry method (Lowry et al. 1951).

For large scale production of lipopeptide batch fermentation of *Streptomyces* strain was carried out in SCN broth at 28°C on a rotary shaker at 180 rpm. The flasks were inoculated with 2% by volume of seed culture, grown at 28°C for two days. Fermentation was terminated on 4th day, at which the maximum production was obtained. The active compound from cell free supernatant was adsorbed on resin XAD-4 (5%) at 15°C under shaking conditions for 2 days and recovered from resin by eluting with methanol. The methanol fraction was dried using a rotavapour (BUCHI Rota vapor R-200) and dried residue was re-dissolved in small volume of methanol.

Purification of lipopeptide

For the purification of the lipopeptide silica gel column chromatography (60-120 mesh) was carried out. Column (35 × 1.0 cm) was packed with silica gel using chloroform as solvent. The methanol extract was loaded at the top of the column and eluted step-wise with 100% $CHCl_3$, 95:5, 90:10, 75:25, 70:30, 50:50 (v/v) of $CHCl_3$: CH_3OH solvent and 100% CH_3OH (200 ml each) at a flow rate of 2 ml/min. Fractions were concentrated and redissolved in the same solvent ratio from which they were recovered. Fractions showing antimicrobial activity were pooled and further purified by subjecting to size exclusion chromatography using Toyopearl resin HW-40 and methanol as an eluent. The methanol fractions as such were subjected to antimicrobial activity, active fractions were combined and solvent was evaporated using a rota vapour, the residue was re-dissolved in Milli-Q water. Further purification was achieved through reverse phase HPLC (1260 Infinity, Agilent Technologies, USA) using a semi-preparative C18 column (Pursuit 10C18 250 × 21.2 mm) and acetonitrile: water (5:5) as mobile phase. Collected fractions were concentrated by speed vacuum and screened for antimicrobial activity. Lipopeptide was purified to homogeneity using reverse phase- HPLC (Agilent 1100 series) with a ZORBAX 300-SB18 column (4.6 mm × 250 mm, particle size 5 µm), at a flow rate of 1 ml/min. The solvent system used was 0.1% aqueous TFA (A) and acetonitrile containing 0.1% TFA (B). The gradient of solvent B used to run the column was as follows: 0-60% for 0–45 min, 60-80% for 45–55 min and 80-100% for 55–60 min. The elution from the column was monitored at 215 nm in a diode array detector and all the peaks of HPLC chromatogram were collected using a fraction collector (GILSON, France) coupled with the system. Collected fractions were concentrated by speed vacuum and screened for antimicrobial activity.

MALDI-TOF-MS and sequencing of lipopeptide

Lipopeptide was characterized using Matrix-assisted laser desorption ionization (MALDI). The purified lipopeptide was lyophilized and re-suspended in methanol. Solution (4 µl) was mixed with 4 µl of matrix (CHCA, 10 mg/ml), 1.0 µl of this mixture solution was spotted onto the MALDI 100 well stainless steel sample plate and allowed to air dry prior to the MALDI analysis

(Mandal et al. 2009). MALDI mass spectra was obtained using a Voyager time-of-flight mass spectrometer (Applied Biosystem, USA), equipped with 337 nm N2 laser and operated in accelerating voltage 20 kV. The spectra were recorded in positive ion linear mode. To check the reproducibility of the spectrum, sample was separately spotted several times.

For peptide MS/MS sequencing, lipopeptide was incubated with 10% NaOH in methanol at room temperature for 16 h to cleave the lactone ring. The cleaved lipopeptide was lyophilized, again extracted with methanol and allowed for mass spectrometry analysis. The spectra were recorded in the post-source decay (PSD) ion mode as an average of 100 laser shots with a grid voltage of 75%. The reflector voltage was reduced in 25% steps and guide wire was reduced 0.02–0.01% with an extraction delay time of 100 ns.

Fatty acid analysis of lipopeptide by GC-MS

Lipopeptide was subjected to acid hydrolysis by incubating the lipopeptide (5 mg) with 0.5 ml of 6 M HCl at 90°C for 18 h in sealed tubes. The fatty acids were extracted with ether and esterified with 0.95 ml methanol and 0.05 ml of 98% H_2SO_4 at 65°C for 6 h. Fatty acid methyl esters were obtained after extraction with n-hexane and analyzed by GC-MS with a Clarus 500 GC (PerkinElmer, USA). Helium was used as carrier gas at a flow rate of 1.0 ml/min. The column temperature was maintained at 120°C for 3 min and thereafter gradually increased (8°C/min) to 260°C.

Determination of antimicrobial activity

Sensitivity of test organisms to purified lipopeptide was measured in terms of zone of inhibition using Kirby-Bauer disk diffusion susceptibility test (Bauer et al. 1966). The plates containing Mueller-Hinton agar, yeast malt agar and potato dextrose agar were seeded with test bacteria, yeasts and fungi, respectively. Filter paper discs (6 mm) impregnated with 10 μg of lipopeptide were placed on media plates. The diameter of the resultant zone of inhibition was measured in mm after 24- 48 hours of incubation. Each experiment was performed in duplicates and repeated thrice. Various test organisms used in the study included *Bacillus subtilis* (MTCC 619), *Mycobacterium smegmatis* (MTCC 6), *Staphylococcus epidermidis* (MTCC 435), *Escherichia coli* (MTCC 1885), *Klebsiella pneumoniae* sub sp. *pneumoniae* (MTCC 109), *Enterobacter aerogenes* (MTCC 111), *Salmonella typhi* (MTCC 733), multi-drug resistant *E. coli*, MRSA, VRE, *Candida albicans* (MTCC 3017), *Rhodotorula rubra* (MTCC 248), *Colletotrichum acutatum* (MTCC 1037), *Cercospora beticola* (GenBank acc. no. KJ461435), *Fusarium oxysporum* f. sp. *dianthi* (MTCC 6659), and *Alternaria brassicicola* (MTCC 2102).

Determination of minimum inhibitory concentration

The MIC of purified lipopeptide was evaluated by using a microtiter plate dilution assay. Test bacteria were grown to logarithmic phase under optimal conditions (up to 0.3 OD) and the test was performed in triplicates. To each well of the microtiter plate, 200 μl of fresh nutrient medium and 50 μl of bacterial suspension were added. Subsequently, different dilutions (50 μl) of freshly prepared samples were added to each well. The first column of the microtiter plate was left as a blank, containing fresh medium only. The microtiter plates were incubated at 37°C and OD was measured at 600 nm at 24 and 48 h using ELISA microplate reader (Bio-rad, Model 680XR). The lowest concentration that inhibited growth of the test strain and did not show any increase in absorption after 48 h was considered as MIC of the lipopeptide for that strain.

Effect of pH, temperature and enzymes on lipopeptide activity

Sensitivity to temperature was determined by incubating the purified lipopeptide at different temperatures *viz.* 50°C, 60°C, 70°C, 80°C , 90°C, 100°C and 121°C for different durations. To determine the optimum pH for activity, purified compound was incubated over pH range of 3.0–12.0 for 30 min at 37°C. The sensitivity of the lipopeptide (1.0 mg/ml) to enzymes was tested against proteinase K (~30 U/mg), trypsin (~10,000 U/mg) and lipase (~9 U/mg). All enzymes were purchased from Sigma Aldrich. Enzyme solutions were prepared at 1.0 mg/ml in 50 mM phosphate buffer (pH 7.0). Equal volumes of enzyme solution and antimicrobial compound (20 μl each) were mixed and incubated at 37°C for 12 h. The enzyme reaction was terminated by heating reaction mixture at 80°C and residual activity was determined by disc diffusion method.

Safety evaluation of lipopeptide

Mutagenicity of the lipopeptide was evaluated by Ames test (Maron and Ames 1983). This *Salmonella* reverse mutation test is based on histidine dependence and mutations in *Salmonella typhimurium* (TA98/ MTCC 1251, IMTECH, Chandigarh). Concentrations of the lipopeptide used for checking toxicity were 50 and 100 μg 0.1 ml^{-1} $plate^{-1}$. The overnight grown, 0.1 ml bacterial culture and 0.1 ml of lipopeptide were added to 2.0 ml of top agar. The contents were mixed and poured onto glucose minimal agar plates immediately. The plates were inoculated and incubated at 37°C for 48 h. The experiment was repeated to confirm the results. To determine the spontaneous reversion which is characteristic of the tester strain (TA 98), negative control (0.1 ml bacterial culture + 0.1 ml DMSO $plate^{-1}$) was run while 4-Nitro-o-phenylenediamine (20 μg 0.1 ml^{-1} $plate^{-1}$) was used as a positive control mutagen. The mutagenic potential of the lipopeptide was determined by comparing the number of colonies with control plates where no test compound was added.

In vitro cytotoxicity was evaluated using sulforhodamine B dye assay (Skehan et al. 1990). The Chinese hamster ovary cell line (CHO) was used for the assay. Camptothecin (CPT), an anticancer drug was used as standard. The 96-well tissue culture plate, containing different concentrations of lipopeptide (2.5- 25 μg/ml) and CHO cells, was incubated for 48 h and the cell growth was stopped with trichloroacetic acid (50% TCA, 50 μl). The optical density (OD) was recorded at 540 nm on ELISA reader and percent growth inhibition was calculated.

Biosurfactant property of lipopeptide

Biosurfactant property of lipopeptide was determined using the qualitative drop-collapse test (Youssef et al. 2004). In this method, mineral oil (2 μl) was added to 96-well microtitre plate. The plate was equilibrated for 1 h at 37°C and 5 μl of the lipopeptide (10 μg, dissolved in water) was added to the surface of the oil and drop shape was observed after 1 min. The surface tension of a lipopeptide (0.2%, w/v) was measured using Du-Nouy-Ring method (Du Noüy and Pierre 1925).

Results

Production and purification of lipopeptide

Lipopeptide production and growth profile of *S. amritsarensis* is shown in Figure 1. Antimicrobial activity appeared in culture supernatant during late logarithmic phase. A significant increase in production (as measured by inhibition zone) and growth were observed with further incubation. The maximum growth and production were attained after 4 days which remained constant for 7 days and declined slightly with further incubation.

For purification of compound, production was carried out in SCN broth for 4 days. The active compound from supernatant was adsorbed on resin XAD-4 and recovered by eluting with methanol. Antimicrobial compound was partially purified by silica gel chromatography and size exclusion chromatography using Toyopearl resin HW-40. Reverse phase- HPLC of the partially purified compound using a semi-preparative C18 column revealed the presence of five peaks (Additional file 1: Figure S1). After lyophilization, all collected peaks (fractions) were tested for antimicrobial activity. Peaks 1, 2 and 3, demonstrating antimicrobial activity, were further resolved using ZORBAX 300-SB18 column (Figure 2).

Lipopeptide MS/MS sequencing

The primary structure of lipopeptide (peak 3) was elucidated using a combination of mass spectrometry techniques. The peaks obtained for different fragments at m/z 834, 736, 665, 564, 507, 420, 283 and 155 in MS/MS analysis revealed the lipopeptide sequence as Ala-Thr-Gly-Ser-His-Gln. The C-terminal amino acid in peptide is linked to aliphatic chain of $-(CH_2)_7-CH_2-(CH_3)_2-$ with a total mass value of m/z 878.5 Da (Figure 3). Further, MALDI TOF MS analysis clearly showed an addition of 137 Da mass unit corresponding to $-C_{10}H_{17}-$ shown in figure inset (Figure 4). Interestingly, there is a long tail of fatty acid chain with six times repeated the molecular mass of 161 Da which is corresponding to $-C_{12}H_{19}$ (Figure 3). Based on the molecular mass and amino acid composition, the lipopeptide was identified as a novel lipopeptide.

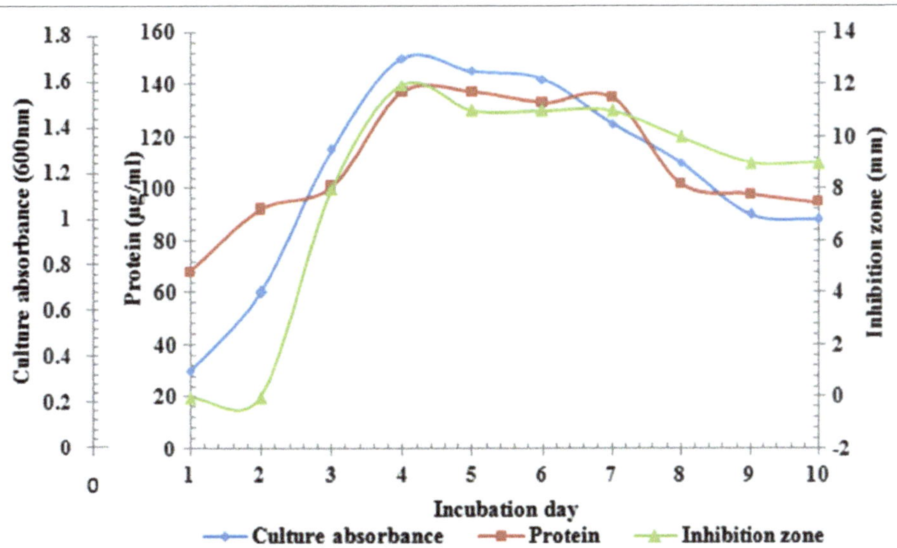

Figure 1 Growth and lipopeptide production profile of *S. amritsarensis*.

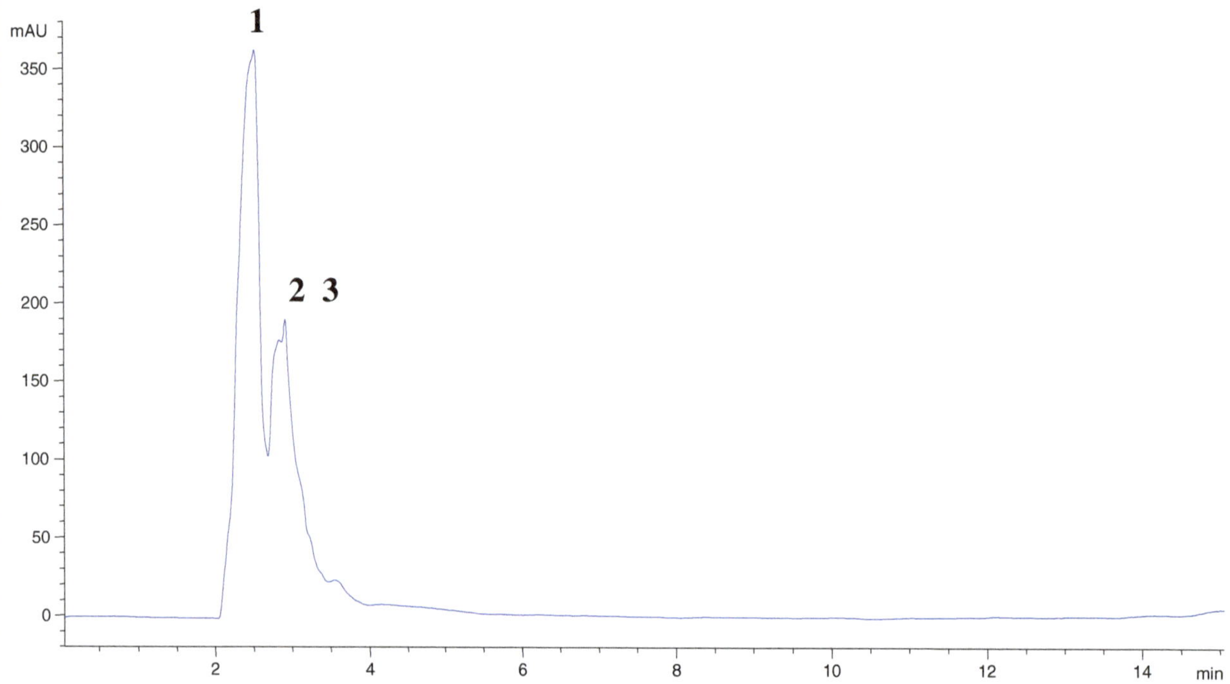

Figure 2 Elution profile of lipopeptide using HPLC reverse phase chromatography on ZORBAX 300-SB18 column monitoring by absorbance at 215 nm. Peak 3 corresponds to lipopeptide elution.

Inhibition spectrum and sensitivity of the lipopeptide

The purified lipopeptide (10 μg) showed activity only against Gram positive bacteria with inhibition zones of 21, 17, 15 and 13 mm against *B. subtilis, S. epidermidis, M. smegmatis* and MRSA, respectively (Additional file 1: Figure S2). It did not show activity against any of the tested Gram-negative bacteria and fungi. The MIC assay for test organisms with purified lipopeptide using micro-titer plates in triplicates revealed lowest MIC value of 10 μg/ml against *B. subtilis* and highest against MRSA i.e 45 μg/ml (Figure 5).

The results of heat stability assay demonstrated that the lipopeptide was completely stable at 70°C for 1 h and a loss of 13.7% was observed at 100°C after 15 minutes incubation. After autoclaving (121°C for 15 min) anti-microbial activity was reduced by 18.2%. It did not show

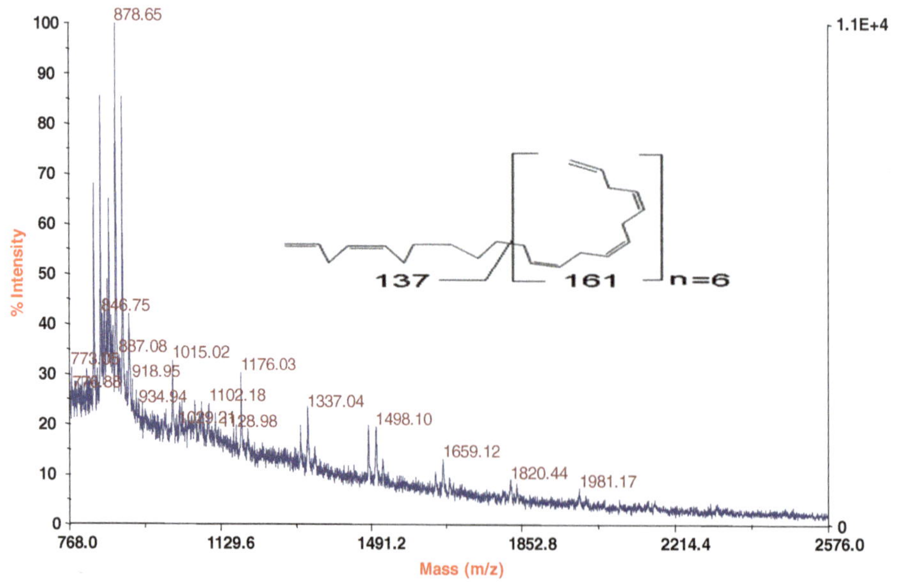

Figure 3 MALD-TOF mass spectrometry analysis of lipopeptide.

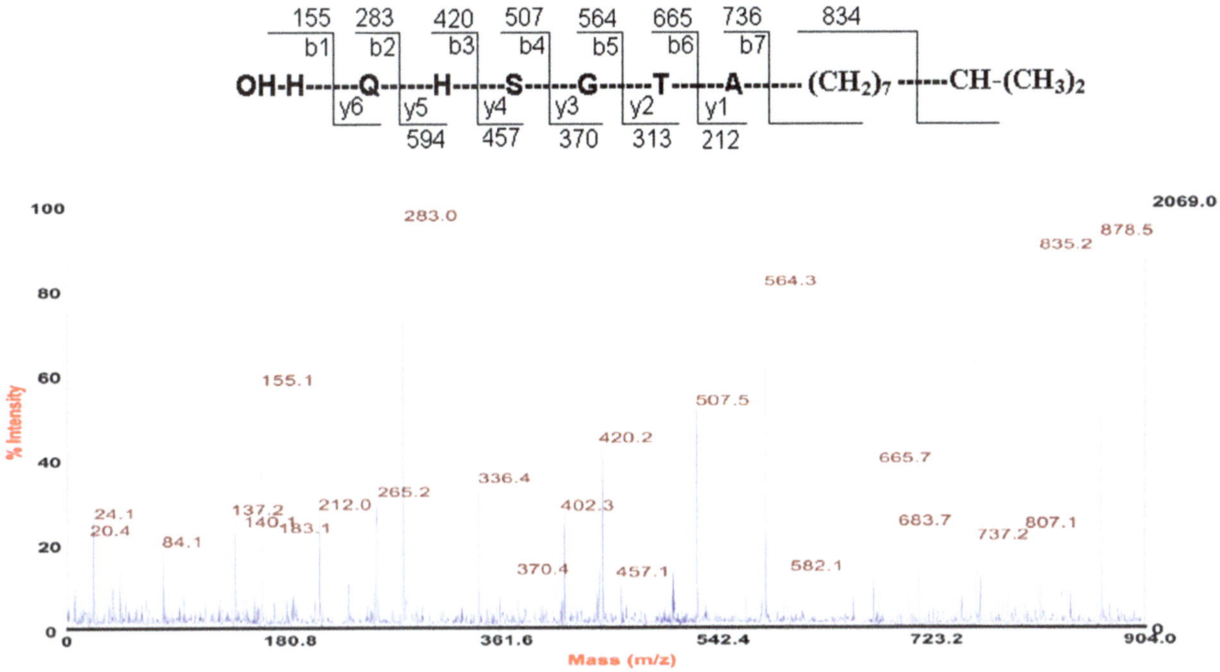

Figure 4 MALDI TOF PSD (MS/MS) spectrum of lipopeptide (Peak 3) and amino acid sequence of the lipopeptide obtained by de novo sequencing.

any change in its activity profile between pH 5.0–9.0 (Table 1). Lipopeptide was found to be resistant to trypsin and lipase, and negligible loss in activity was observed after treatment with proteinase K.

Safety evaluation

Mutagenicity of lipopeptide was checked by Ames test at two concentrations *viz.* 50 and 100 μg/0.1 ml. The number of revertant colonies were counted after 48 hours of incubation at 37°C and compared with the negative control. The number of revertant colonies in the presence of lipopeptide was found to be same as in the negative control (21 ± 2.0) for TA98. However, the number of colonies in presence of positive mutagen (20 μg plate^{-1}) was found to be considerably higher. The results of the present study indicate that the lipiopeptide is non-mutagenic.

The effect of lipopeptide at different concentrations, 2.5–25 μg/ml on chinese hamster ovary (CHO) cell line is shown in Figure 6. It showed 2.9, 16.8, 19.25, 23.28

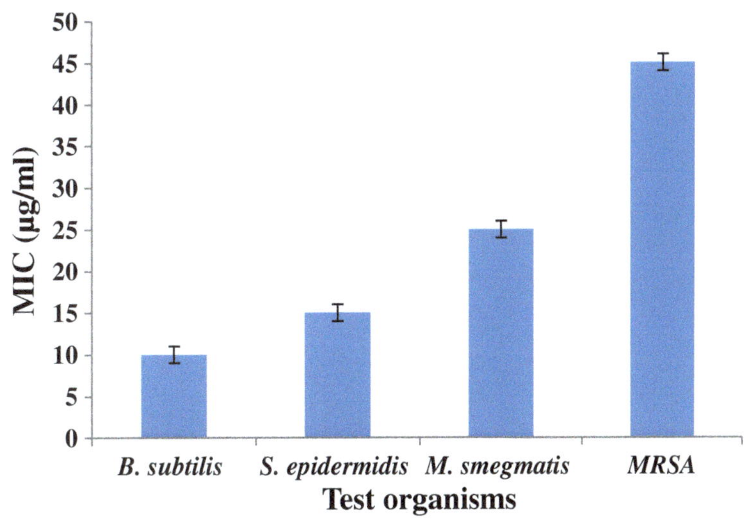

Figure 5 Determination of MIC for lipopeptide.

Table 1 Factors affecting lipopeptide activity

Treatment	Reaction duration/condition	Residual activity (%)
pH		
2	1 h/4°C	0
3	1 h/4°C	72.7
4	1 h/4°C	84.6
5	1 h/4°C	100
6	1 h/4°C	100
7	1 h/4°C	100
8	1 h/4°C	100
9	1 h/4°C	100
10	1 h/4°C	86.3
11	1 h/4°C	68.1
12	1 h/4°C	0
Temperature (°C)		
50	1 h	100
60	1 h	100
70	1 h	100
80	1 h	84.2
90	15 min	95.0
100	15 min	86.3
121	15 min (304 KPa)	81.8

and 28.9% growth inhibition at concentrations of 2.5, 6.25, 12.5, 18.75 and 25 μg/ml, respectively. The IC_{50} value of lipopeptide was found to be 387 μg/ml which is very high as compared to the IC_{50} value of standard drug (0.959 μg/ml). The cytotoxicity % inhibition of standard (camptothecin) is shown in Figure 7. This indicates that it do not exhibit cytotoxicity.

Lipopeptide as biosurfactant

The lipopeptide collapsed the oil drop and lowered surface tension of water from 72 to 37 mN m^{-1} when used at a concentration of 0.2%. These results show that it possesses good surfactant activity.

Discussion

Due to few antibiotics and free availability of effective antibiotics against diverse bacterial species the antimicrobial era is threatened by high levels of antibiotic resistance (Song 2008). Among infections caused by antibiotic resistant Gram-positive bacteria, MRSA and VRE are of particular concern (Rice 2008) and this lead to the discovery of potential antibiotics such as vancomycin, teicoplanin, synercid (quinipristin and dalfopristin), tigecycline and linezolid. Daptomycin is the recent FDA-approved lipopeptide, exhibiting potent activity against a broad range of Gram-positive pathogens especially, MRSA and VRE. However, rare incidences of clinical resistance have also been reported against daptomycin (D'Costa et al. 2012). Since resistance to each new antibiotic ultimately emerges, usually within few years after it is marketed, there is always a need to find new antimicrobial agents to combat antibiotic resistant strains of pathogenic and opportunistic pathogenic microorganisms.

Since the late 1960s, only two novel classes of antibiotics, the oxazolidinones and the cyclic lipopeptides, have entered the market (Rodriguez de Castro et al. 2009). Thus antibiotic market is looking to increase the number of new products with improved effectiveness. Nowadays, a huge amount of resources is being invested in R&D to find novel antimicrobials that can solve the problem of antibiotic resistance (Maria-Neto et al. 2012). The market for antibacterial drugs is highly competitive, and many companies are engaged in the development of anti-MRSA or

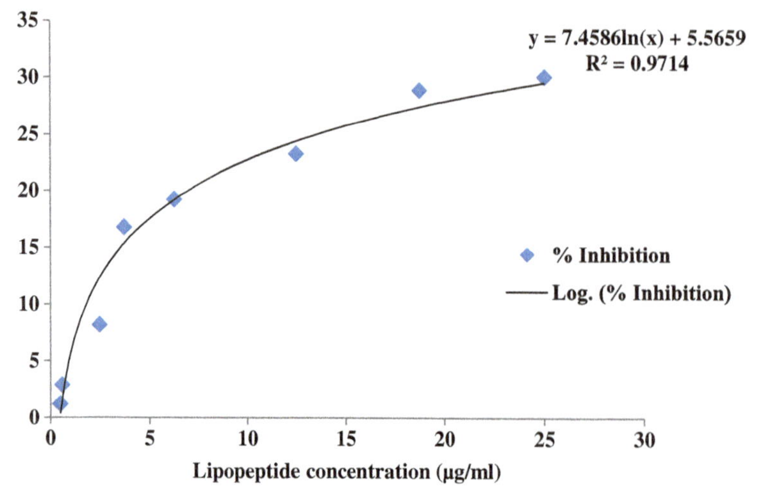

Figure 6 Cytotoxic effect of lipopeptide on CHO cell line.

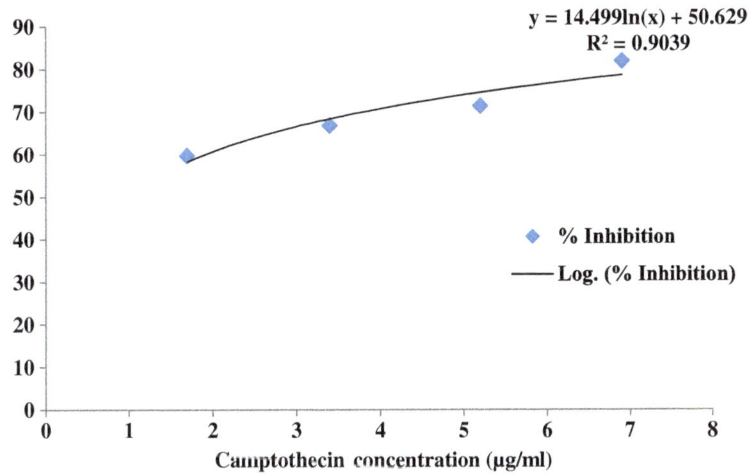

Figure 7 Cytotoxic effect of camptothecin on CHO cell line.

multi-drug-resistant treatments, where lipopeptides are the target compounds of most companies (Mandal et al. 2013). Antimicrobial lipopeptides are nonribosomally synthesized, having macrocyclic peptide cores consisting of eleven to thirteen amino acids, which are rigidified by the formation of a ten-membered ring. They are produced by NRPSs with variation of the fatty acid tail (Strieker and Marahiel 2009). The mechanism of action of lipopeptides is distinct from those of other antibiotics currently on the market. They form pores in membranes of bacteria after oligomerization and these pores may cause transmembrane ion influxes, including Na + and K+, which result in membrane disruption and cell death. Two key properties of lipopeptides are: i) a number of lipopeptides tend to oligomerize and ii) their ability to interact with membranes via their lipid tail (Straus and Hancock 2006). These unique properties of the lipopeptides prevent pathogens to develop resistance against them, thus makes them highly active against multidrug-resistant bacteria (Mangoni and Shai 2011). Generally the bactericidal activity of the lipopeptide increases, with the addition of a lipid tail of appropriate length (typically C10–C12) and lipopeptides containing 14 or 16 carbon atoms in lipid tail length exhibit enhanced antifungal activity in addition to antibacterial activity (Mandal et al. 2013). This can be due to either an increase in the affinity of the lipid tail for the hydrocarbon chains or as a result of the stronger interaction of the cationic peptide with the lipid headgroups (Straus and Hancock 2006).

Antimicrobial lipopeptides have been purified from many bacterial genera including several *Streptomyces* spp. such as *S. violaceus* var. *aspartocinius* (aspartocin), *S. griseoflavus* (tsushimycin), *S. viridochromogens* (laspartomycin), *S. coelicolor* (calcium-dependent antibiotic), *S. roseosporus* (daptomycin), *S. fradiae* (A54145), *S. parvulus var. parvuli* (parvuline) and *S. canus* (amphomycin) (Schneider et al. 2009). The present study reports the production of another novel lipopeptide from a *Streptomyces* sp. The primary structure of lipopeptide was determined using a combination of chemical reactions and mass spectrometry techniques. It consists of six amino acids linked to aliphatic chain of $-(CH_2)_7-CH_2-(CH_3)_2$ and a long tail of fatty acid chain with six times repeated the molecular mass of 161 Da which is corresponding to $-C_{12}H_{19}$. Based on structure and molecular weight (878.5 Da), it is different from already reported lipopeptides *viz.* daptomycin (1620.6 Da), aspartocin (1317 Da), glumamycin (1290.4 Da), tsushimycin (1304.7 Da) and arylomycin A6 (867.4 Da). Lipopeptides vary in their amino acid and/or fatty acid composition and all these variations in length and branching of the fatty acid chains and amino acid substitutions lead to remarkable lipopeptide diversity and activities. Generally, they are reported to be thermostable, resistant to proteolytic enzymes and inhibit the growth of microorganism by altering the membrane integrity (Mandal et al. 2013). Similarly, lipopeptide in the present study is found to be completely stable for 1 h at 70°C, retaining 81.8% activity even after autoclaving (121°C for 15 min). It exhibits absolute activity over a broad pH range of 5.0 –9.0 and found to be resistant to hydrolytic enzymes: trypsin, proteinase K and lipase. Similarly, paenibacterin, an antimicrobial lipopeptide produced by *Paneibacillus* sp. strain is reported to be resistant to trypsin and lipase enzymes (Guo et al. 2012). The strong antibacterial activity of the present lipopeptide can be related to the carbon chain length (C12) of the lipid tail as bactericidal activity of the lipopeptides depends upon the length of the lipid tail. Safety evaluation of the lipopeptide demonstrates its non-cytotoxic and non-mutagenic nature which is a prerequisite for development of a drug.

In conclusion, *Streptomyces amritsarensis* produces a novel antimicrobial lipopeptide that is active against a variety of Gram-positive bacteria especially, MRSA. Its stability and non-toxic nature suggest that it may serve as a new pharmacological drug and an addition to the panoply of lipopeptide group of antibiotics. It may also be used in the cosmetics industry for developing skin-care products and shampoos as it possesses good surface active property and as emulsifier and bio-preservative in the food industry. Further antimicrobial spectrum of the lipopeptide can be enhanced by chemical modifications in the lipid tail length, increasing number of carbon atoms.

Additional file

Additional file 1: Figure S1. Ellution profile of the partially purified compound using HPLC reverse phase chromatography on C18 column monitoring by absorbance at 280 nm. **Figure S2.** Antibacterial activity of lipopeptide against: **(a)** *B. subtilis* **(b)** *S. epidermidis* **(c)** *M. smegmatis* **(d)** MRSA.

Competing interests
The author(s) declare that they have no competing interests.

Authors' contributions
RKM: Conceived the study, designed and performed experiments, critically analyzed data and drafted manuscript. DS: Designed and performed all the experiments, analyzed data and drafted manuscript. SMS: Performed MALDI-TOF-MS and sequencing, elucidated lipopeptide structure and helped to draft manuscript. All authors have read and approved the final manuscript.

Acknowledgement
The authors are thankful to Department of Science and technology (DST), New Delhi, India for providing the financial support (Project no. SR/WOS-A/LS-87/2010). We also acknowledge help of Dr. Saroj Arora, Professor, Department of Botanical & Environmental Sciences, Guru Nanak Dev University, Amritsar help in performing safety evaluation tests.

Author details
[1]Department of Microbiology, Guru Nanak Dev University, Amritsar, Punjab 143005, India. [2]Central Research Facility, Indian Institute of Technology, Kharagpur, West Bengal 721302, India.

References
Baindara P, Mandal SM, Chawla N, Singh PK, Pinnaka AK, Korpole S (2013) Characterization of two antimicrobial peptides produced by a halotolerant *Bacillus subtilis* strain SK. DU. 4 isolated from a rhizosphere soil sample. AMB Express 3(2):1–11
Bauer AW, Kirby WM, Sherris JC, Turck M (1966) Antibiotic susceptibility testing by a standardized single disk method. Am J Clin Path 45:493–496
Davies J (1999) Millennium bugs. Trends Biochem Sci 24:M2–M5
D'Costa VM, Mukhtar TA, Patel T, Koteva K, Waglechner N, Hughes DW, Wright GD, De Pascale G (2012) Inactivation of the lipopeptide antibiotic daptomycin by hydrolytic mechanisms. Antimicrob Agents Chemother 56(2):757–764
Du Noüy, Pierre L (1925) An Interfacial Tensiometer for Universal Use. J Gen Physiol 7(5):625–633
Guo Y, Huang E, Yuan C, Zhang L, Yousef AE (2012) Isolation of a *Paenibacillus* sp. strain and structural elucidation of its broad-spectrum lipopeptide antibiotic. Appl Environ Microbiol 78(9):3156–3165
Horn JN, Sengillo JD, Lin D, Romo TD, Grossfield A (2012) Characterization of a potent antimicrobial lipopeptide via coarse-grained molecular dynamics. Biochim Biophys Acta Biomembr 1818(2):212–218
Laverty G, Gorman SP, Gilmore BF (2011) The potential of antimicrobial peptides as biocides. Int J Mol Sci 12:6566–6596
Lowry OH, Rosebrough NJ, Farr AL, Randall RJ (1951) Protein measurement with the Folin phenol reagent. J Biol Chem 193:265–275
Mandal SM, Dey S, Mandal M, Sarkar S, Maria-Neto S, Franco OL (2009) Identification and structural insights of three novel antimicrobial peptides isolated from green coconut water. Peptides 30:633–637
Mandal SM, Barbosa AE, Franco OL (2013) Lipopeptides in microbial infection control: scope and reality for industry. Biotechnol Adv 31(2):338–345
Mangoni ML, Shai Y (2011) Short native antimicrobial peptides and engineered ultrashort lipopeptides: similarities and differences in cell specificities and modes of action. Cell Mol Life Sci 68:2267–2280
Maria-Neto S, Cândido Ede S, Rodrigues DR, de Sousa DA, da Silva EM, de Moraes LM, de Jesus OA, Magalhães BS, Dias SC, Franco OL (2012) Deciphering the magainin resistance process of Escherichia coli strains in light of the cytosolic proteome. Antimicrob Agents Chemother 56(4):1714–1724
Maron DM, Ames BN (1983) Revised method for Salmonella mutagenicity test. Mutat Res 113(3–4):173–215
Nathan C, Goldberg FM (2005) The profit problem in antibiotic R&D. Nat Rev Drug Discov 4(3):887–891
Payne DJ, Gwynn MN, Holmes DJ, Pompliano DL (2007) Drugs for bad bugs: confronting the challenges of antibacterial discovery. Nat Rev Drug Discov 6:29–40
Pirri G, Giuliani A, Nicoletto S, Pizutto L, Rinaldi A (2009) Lipopeptides as antiinfectives: a practical perspective. Cent Eur J Biol 4(3):258–273
Rice LB (2008) Federal funding for the study of antimicrobial resistance in nosocomial pathogens: no ESKAPE. J Infect Dis 197:1079–1081
Rodriguez de Castro F, Naranjo OR, Marco JA, Violan JS (2009) New antimicrobial molecules and new antibiotic strategies. Semin Respir Crit Care Med 30:161–171
Schneider T, Gries K, Josten M, Wiedemann I, Pelzer S, Labischinski H, Sahl HG (2009) The lipopeptide antibiotic friulimicin B inhibits cell wall biosynthesis through complex formation with bactoprenol phosphate. Antimicrob Agents Chemother 53(4):1610–1618
Sharma D, Mayilraj S, Manhas RK (2014) *Streptomyces amritsarensis* sp. nov., exhibiting broad-spectrum antimicrobial activity. Antonie Van Leeuwenhoek 105(5):943–949
Skehan P, Storeng R, Scudiero D, Monks A, McMohan J, Vistica D, Warren JT, Bokesch H, Kenney S, Boyd MR (1990) New colorimetric cytotoxic assay for anticancer-drug screening. J Natl Cancer Inst 82:1107–1112
Song J (2008) What's new on the antimicrobial horizon? Int J Antimicrob Agents 32:207–213
Straus SK, Hancock REW (2006) Mode of action of the new antibiotic for Gram-positive pathogens daptomycin: Comparison with cationic antimicrobial peptides and lipopeptides. Biochim Biophys Acta 1758:1215–1223
Strieker M, Marahiel MA (2009) The structural diversity of acidic lipopeptide antibiotics. ChemBioChem 10(4):607–616
Watve MG, Tickoo R, Jog MM, Bhole BD (2001) How many antibiotics are produced by the genus *Streptomyces*? Arch Microbiol 176(5):386–390
Youssef NH, Dunacn KE, Nagle DP, Savage KN, Knapp RM, McInerney MJ (2004) Comparison of methods to detect biosurfactant production by diverse microorganism. J Microbiol Methods 56:339–347

20

Systematic analysis of the ability of Nitric Oxide donors to dislodge biofilms formed by *Salmonella enterica* and *Escherichia coli* O157:H7

Massimiliano Marvasi[1], Charles Chen[2], Manuel Carrazana[2], Ian A Durie[1] and Max Teplitski[1,3*]

Abstract

Biofilms in the industrial environment could be problematic. Encased in extracellular polymeric substances, pathogens within biofilms are significantly more resistant to chlorine and other disinfectants. Recent studies suggest that compounds capable of manipulating nitric oxide-mediated signaling in bacteria could induce dispersal of sessile bacteria and provide a foundation for novel approaches to controlling biofilms formed by some microorganisms. In this work, we compared the ability of five nitric oxide donors (molsidomine, MAHMA NONOate, diethylamine NONOate, diethylamine NONOate diethylammonium salt, spermine NONOate) to dislodge biofilms formed by non-typhoidal *Salmonella enterica* and pathogenic *E. coli* on plastic and stainless steel surfaces at different temperatures. All five nitric oxide donors induced significant (35-80%) dispersal of biofilms, however, the degree of dispersal and the optimal dispersal conditions varied. MAHMA NONOate and molsidomine were strong dispersants of the *Salmonella* biofilms formed on polystyrene. Importantly, molsidomine induced dispersal of up to 50% of the pre-formed *Salmonella* biofilm at 4°C, suggesting that it could be effective even under refrigerated conditions. Biofilms formed by *E. coli* O157:H7 were also significantly dispersed. Nitric oxide donor molecules were highly active within 6 hours of application. To better understand mode of action of these compounds, we identified *Salmonella* genomic region *recA-hydN*, deletion of which led to an insensitivity to the nitric oxide donors.

Keywords: Biofilm control; Bacterial signaling; Food-borne pathogens; Nitric oxide

Introduction

Every natural wet surface is a substrate for microbial biofilms. These sessile multicellular microbial consortia are embedded within the self-produced extracellular polymeric substances (EPS) (Costerton et al. 1987; Marvasi et al. 2010; Solano et al. 2002; Wingender et al. 1999). In food handling facilities, biofilms could be particularly problematic. While pathogens do not typically make up the bulk of the biofilms formed in the industrial settings, *Salmonella spp*, *Listeria spp*, pathogenic strains of *E. coli*, *Yersinia spp*, *Pseudomonas spp*, *Shigella spp*, *Staphylococcus spp*, and *Bacillus spp* can be found in biofilms on various contact surfaces (Blanpain-Avet et al. 2011; Shi and Zhu 2009). Because microbes in biofilms are significantly more resistant to chlorine and other disinfectants (Zhang and Mah 2008), there is a pressing need to identify compounds capable of controlling biofilms by other means.

A discovery of the nitric oxide's function in inducing biofilm dispersal led to concerted efforts on the identification of the compounds capable of dislodging biofilms (Landini et al. 2010; McDougald et al. 2012). Several disinfectants, antibiotics and messenger molecules have been studied for their ability to disperse biofilms (Barraud et al. 2006; Barraud et al. 2012; Barraud et al. 2009a; Huynh et al. 2012; McDougald et al. 2012). Nitric oxide (NO) gas and NO donors are currently used clinically (Regev-Shoshani et al. 2010). In bacteria and eukaryotes, nitric oxide is a signaling molecule, active at very low concentrations (Gaupels et al. 2011; Simontacchi et al. 2013). At low concentrations, it is effective as a biofilm dispersant, functioning as a messenger rather than a generic poison (Barraud et al. 2006; Barraud et al. 2009a). Nitric oxide can be delivered to biofilms using NO donor molecules

* Correspondence: maxtep@ufl.edu
[1]Soil and Water Science Department, University of Florida-IFAS, Gainesville, FL, USA
[3]Cancer and Genetics Research Center, Genetics Institute, 2033 Mowry Road, Gainesville, FL 32611, USA
Full list of author information is available at the end of the article

or impregnated onto nanoparticles (Slomberg et al. 2013; Wang et al. 2005). The application of NO donors has the same effect on dispersal of biofilms as the direct addition of nitric oxide and is less toxic than the application of the nitric oxide gas (Barraud et al. 2009a). Over 105 NO donors have been characterized, but only few of them have been tested for controlling industrial biofilms (Wang et al. 2005). Among these, activity of sodium nitroprusside has been recently characterized in detail. It can produce a flux of nitric oxide of 30 pmol cm^{-2} s^{-1}, and this can efficiently reduce the adhesion of *Staphylococcus aureus*, *Staphylococcus epidermidis*, and *E. coli* by 96%, 48%, and 88%, respectively (Charville et al. 2008). Therefore, it appears that nitric oxide could have a universal effect on the dispersal of bacterial biofilm including both Gram-positive and Gram-negative bacteria (Xiong and Liu 2010).

In this work, we tested the effectiveness of NO donors in dispersal of biofilms formed by common foodborne pathogens (non-typhoidal *Salmonella enterica* and enterohaemorrhagic *E. coli*, EHEC) on materials that are common in the food industry. The rationale for this study was based on the reports that the same NO donors have different dispersion potential depending on the bacterial strain, temperature and surface properties (Barraud et al. 2006; Barraud et al. 2009a; Charville et al. 2008; Gilberthorpe and Poole 2008). Because the ultimate goal of these experiments is to identify commercially available compounds for industrial applications, the following criteria were used to select candidate compounds: 1) low/moderate toxicity; 2) have no more than 0.1% of probable, possible or confirmed human carcinogenicity according to the International Agency for Research on Cancer (IARC); 3) low/moderate cost; 4) commercially availability. Based on these criteria, the following compounds were selected: molsidomine (N- (ethoxycarbonyl)-3- (4- morpholinyl)- sydnone imine), MAHMA NONOate (6-(2-Hydroxy-1-methyl-2-nitrosohydrazino)-N-methyl-1-hexanamine), spermine NONOate (N-[4-[1-(3-Aminopropyl)-2-hydroxy-2-nitrosohydrazino]butyl]-1,3-propanediamine), diethylamine NONOate diethylammonium salt, and diethylamine NONOate sodium.

The mechanisms by which NO effects the transition from sessile biofilm organisms to free-swimming bacteria are not entirely clear (Barraud et al. 2006; Barraud et al. 2009a).

Microarray studies have revealed that *P. aeruginosa* genes involved in adherence are down regulated upon exposure to nitric oxide (Firoved et al. 2004), and the involvement of NO in regulating biofilm formation and dispersal in *P. aeruginosa* was also supported by several studies (Barraud et al. 2006; Barraud et al. 2009a; Darling and Evans 2003; Van Alst et al. 2007). Several genes involved in the production and perception of this signaling molecule have been characterized in *Pseudomonas aeruginosa* PAO1. The chemotaxis protein BdlA is involved in biofilm dispersion of *P. aeruginosa*: biofilms formed by the *bdlA* mutant do not detach when exposed to low doses of NO in continuous-flow cultures (Barraud et al. 2009a; Petrova and Sauer 2012). *P. aeruginosa nirS* and *norCB* encode a NO_2^- reductase and NO reductase. A mutation in *nirS* leads to a reduced biofilm dispersion (Barraud et al. 2006), while biofilms formed by a NO reductase-deficient strain Δ*norCB* did not shift to the planktonic state when exposed to endogenous nitric oxide. No homologs of NirS, NorCB are found in *Salmonella*. However, when grown anaerobically with nitrate, *Salmonella* is capable of generating NO after nitrite addition, likely via products of *fnr* and *hmp* genes (Gilberthorpe and Poole 2008). Because the mechanisms of nitric oxide-mediated signaling in *Salmonella* appear to be distinct from those in *P. aeruginosa*, this study also attempted to elucidate genes that are potentially involved in these signaling pathways and contribute to biofilm dispersal in *Salmonella*.

Materials and methods

Bacterial strains and culture media

Escherichia coli EHEC O157:H7 ATCC 43888, *Salmonella enterica* serotovar Typhimurium ATCC14028, sv. Braenderup 04E01347, Braenderup 04E01556, Braenderup 04E00783, sv. Montevideo LJH519, sv. Javiana ATCC BAA-1593 and sv. Newport C6.3 (Noel et al. 2010) were used in this study. When a cocktail of *Salmonella* strains was used, it was a mix of equal volumes of six strains: three strains of the serovar Braenderup (04E01347, 04E01556, 04E00783), sv. Montevideo LJH519, Javiana (ATCC BAA-1593) and Newport (C6.3). pGFP-ON (a strongly fluorescent construct carrying GFP protein expressed from the *Salmonella dppA* promoter (Noel et al. 2010)) was transformed into the strains of interest by electroporation.

S. Typhimurium A9 is derived from *S.* Typhimurium ATCC14028 and lacks the genomic region between 2,974,854 and 2,990,668 nt of NC_003197.1 (*recA* through *hydN*), which was replaced with a kanamycin-resistance cassette. It was constructed by sequential Datsenko and Wanner mutagenesis as in (Santiviago et al. 2009).

All strains were maintained as frozen glycerol stocks, and were sub-cultured into Luria Bertani medium with appropriate antibiotics (50 μg mL^{-1} kanamycin, 100 μg mL^{-1} ampicillin).

Nitric Oxide donors

All were purchased from Sigma Aldrich (St. Luois, MO, USA). For each compound, 1 mmol L^{-1} stock solutions were prepared in phosphate-buffered saline, pH 7.3 (PBS, Fisher, Waltham, MA, USA) and aliquots were stored at −80°C. For the essays, serial dilutions were always prepared fresh in ice-cold PBS just before the experiments and used within 5 minutes of their preparation. The biofilm dispersion potential of the five molecules was tested on polystyrene and polypropylene. The ability of molsidomine to disperse biofilms was also tested on stainless steel.

Biofilm formation and dispersal on plastics

Overnight Luria Bertani cultures of *Salmonella* or *E. coli* strains were diluted 1:100 in CFA medium as described previously (Teplitski et al. 2006), and 100 μL of the diluted cultures were aliquoted into wells of 96-well polypropylene and polystyrene plates (Fisher, Waltham, MA, USA). Plates with bacteria were incubated for 24 hours at 37°C inside a Ziploc bag. Upon completion of the incubation, the medium with planktonic bacteria was removed and serial dilutions of nitric oxide donors in PBS (in 200 μL) were added to the wells with biofilms. Dispersal experiments were conducted at 22°C or 4°C for 6 and 24 hours. Dispersal was measured by staining the remaining biofilms with 1% crystal violet in ethanol, as described previously (Merritt et al. 2005; O'Toole and Kolter 1998).

In parallel, to validate the staining approach, detachment of cells from biofilms was also measured by directly monitoring the increase of fluorescence of planktonic cells of *S.* Typhimurium 14028 pGFP-ON using Victor-2 multimode plate reader with a 485 nm/535 nm excitation/emission filter (Perkin Elmer, Waltham, MA, USA).

Luminescence tests

Effects of selected NO donors on light production by a constitutively luminescent *Salmonella* strain were characterized as indirect assessments of toxicity of the compounds. Two hundred microliters of Luria Bertani broth inoculated with the overnight, 1:50 diluted culture of S. Typhimurium 14028 pTIM2442 (harboring the *luxCDABE* driven by a strong constitutive phage λ promoter, Alagely et al., 2011) were grown in black polystyrene plates (Corning, New York, USA) in presence of serial dilutions of Molsidomine. Molsidomine was diluted in PBS ($9.89gL^{-1}$) (Fisher Scientific, Waltham, MA, USA) to final concentrations of $10\mu molL^{-1}$, $10nmolL^{-1}$, and $10pmolL^{-1}$. PBS was used as a control. Luminescence of *S.* Typhimurium 14028 pTIM2442 was measured over time using Victor-2 multimode plate reader (Perkin Elmer, Waltham, MA, USA). Each experiment included 12 replicas.

Biofilm formation on stainless steel

Biofilms were formed on the stainless steel culture tube closures (Fisher Scientific, Waltham, MA, USA) essentially as described above for plastics with the following modifications: only dilutions of molsidomine were used, and only 24 hours of contact time was tested. Biofilm dispersal was tested by monitoring fluorescence under a multimode microplate reader equipped with a 485 nm/535 nm excitation/emission filter (Perkin Elmer, Waltham, MA, USA).

Additive effect of the disinfectant SaniDate 12.0 with nitric oxide donors

Biofilms of *S.* Typhimurium ATCC14028 were set up as above using overnight cultures of the pathogen diluted 1:100 in the CFA medium in wells of 96-well polypropylene plates (Fisher, Waltham, MA, USA). Plates with bacteria were incubated for 24 hours at 37°C inside a Ziploc bag. Upon completion of the incubation, the medium with planktonic bacteria was removed and $10nmolL^{-1}$ of Molsidomine or MAHMA nonoate were added to the wells with biofilms. As controls, BPS alone was used. Plates were incubated at 22°C for 24 hours. Upon completion of the incubation, planktonic cells were removed, wells were washed twice with PBS and 200 μL of SaniDade 12.0 (BioSafe System, Hartford, CT, USA) diluted as per manufacturer's recommendations and were loaded into the wells. The disinfectant was incubated for 10 minutes, after the incubation time, biofilm dispersal was measured by staining the remaining biofilms with 1% crystal violet in ethanol, as described previously (Merritt et al. 2005; O'Toole and Kolter 1998). 12 replicas for each experiment were done.

Statistical analysis

The statistical software JMP (SAS) package was used to infer the One-way ANOVA analysis ($p < 0.05$). Tukey means separation analysis was inferred in order to group the means.

Results

Biofilm formation on different plastics

More robust biofilms were formed by *Salmonella* and *E. coli* strains on polypropylene than on polystyrene (Additional file 1: Figure S1). On polypropylene and polystyrene, *S. enterica* sv Typhimurium 14028 formed more biofilms under these conditions than *E. coli* O157:H7 (Additional file 1: Figure S1). Biofilm formation is known to vary depending on the surface, the media used to develop biofilms, as well as the protocol adopted (Kroupitski et al. 2009; Teplitski et al. 2006).

Biofilm dispersal by molsidomine

Molsidomine releases nitric oxide and forms polar metabolites rapidly; its half-life is 1 to 2 hours in plasma at pH 7.4 (Rosenkranz et al. 1996). In terms of biofilm dispersal, molsidomine was the most potent molecule. It was effective in dislodging biofilms formed by *S. enterica* sv Typhimurium 14028, the cocktail of the six *Salmonella* outbreak strains and *E. coli* O157:H7. Molsidomine was particularly effective on polypropylene (Figure 1). It

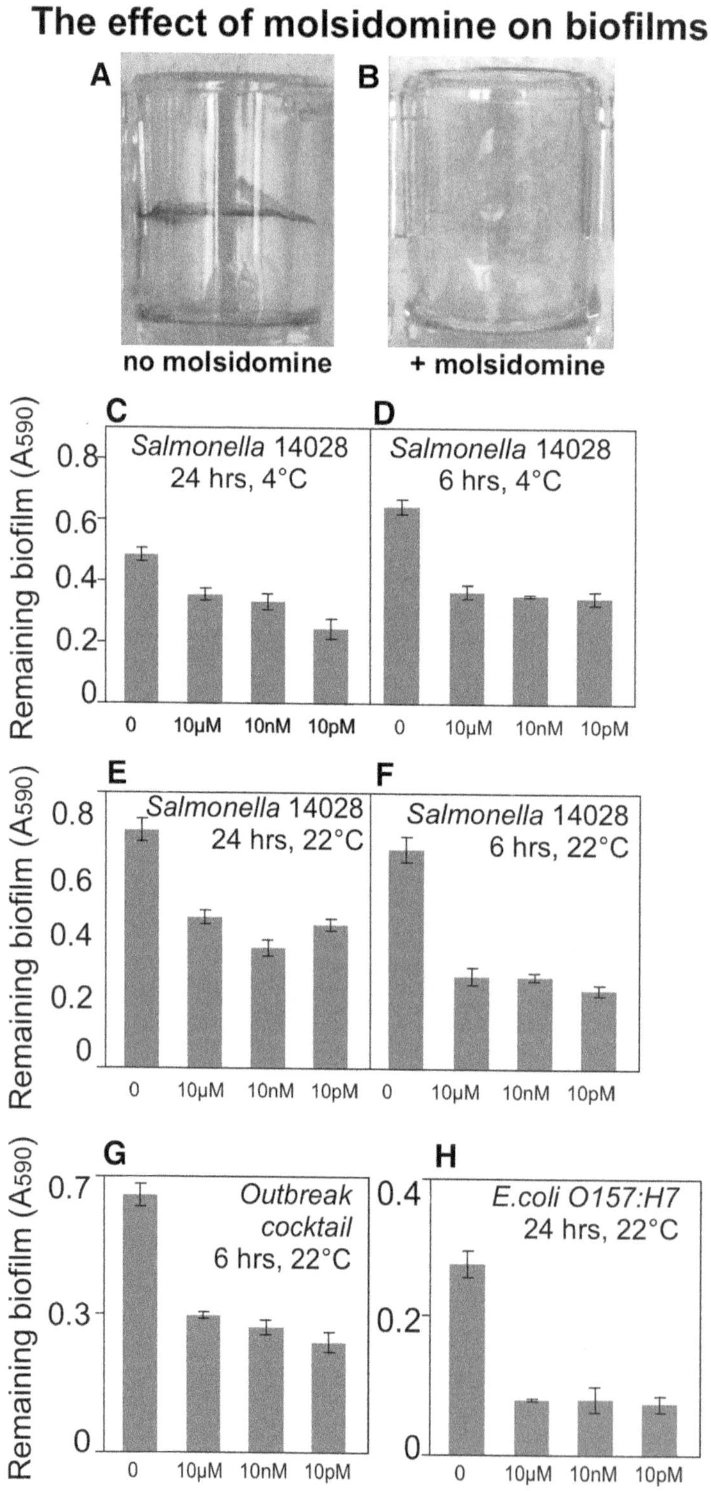

Figure 1 Biofilm dispersal by molsidomine. Exposure to molsidomine doses ranging from 10 μM to 10pM induced biofilm dispersal as measured by staining of the attached cells with crystal violet. "0" indicates control, an existing biofilm not treated with molsidomine. **A.** Biofilm formed by *S. enterica* sv. Typhimurium 14028 in wells of a microtiter plate before treatment with molsidomine. Attached cells are visualized with crystal violet staining. **B.** Residual biofilm after treatment with 10 nM of molsidomine. **C→H.** Induction of biofilm dispersal by molsidomine. Biofilms were pre-formed on polystyrene. Contact time and temperature at which biofilms were exposed to the chemical are listed above each panel. Concentrations of molsidomine are indicated on the x-axis. Error bars are standard errors.

was most effective at 22°C, inducing dispersal of ~50% of biofilms formed by *Salmonella* 14028 and the cocktail of six *Salmonella* outbreak strains after incubation for 6 hours (Figure 1, $p < 0.0001$). Up to 75% of biofilm dispersal was observed after a 24-hrs treatment of *E. coli* O157: H7 biofilms with molsidomine (Figure 1, $p < 0.0001$). Biofilm dispersal by molsidomine was also observed on polystyrene (Additional file 1: Table S1). *Salmonella* biofilms preformed in polystyrene wells were dislodged after 24 hours of contact time at room temperature. Molsidomine was able to induce some dispersal even when biofilms were treated with the compound at 4°C. Intriguingly, the strongest dispersal was observed in response to the treatment with 10 picomolar concentrations of molsidomine (Figure 1). Such potency of the compound at very low concentrations makes it a potentially interesting candidate for commercial applications.

To test whether the decrease in the staining of the attached cells was caused by an increase in the number of planktonic cells, biofilms formed by fluorescent *Salmonella* 14028 pGFP-ON on polypropylene were treated with molsidomine. Fluorescence of the planktonic and attached cells was measured after 0, 3 and 6 hours of exposure to the nitric oxide donor. The increase in total fluorescence of planktonic cells upon treatment of biofilms with molsidomine was statistically significant (Additional file 1: Table S1), reflecting an increase in detachment at the tested concentrations. Similarly, molsidomine treatment of the *Salmonella* biofilms formed on stainless steel resulted in ~0.3 log increase in fluorescence of the planktonic cells (data not shown).

The ability of the constitutively luminescent *Salmonella* construct driven by a phage λ promoter to produce light in the presence of molsidomine was used as an indirect assessment of the toxicity of the compound, and its ability to generally disrupt metabolism or respiration of the bacteria. As shown in Additional file 1: Figure S2, even though modest decrease in luminescence was observed in molsidomine-treated cultures, the reduction of luminescence was not greater than 1-10% of the control.

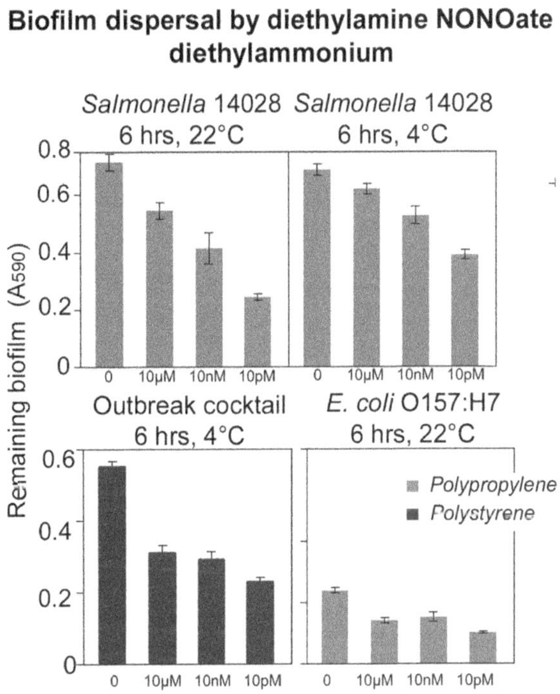

Figure 2 The effect of diethylamine NONOate diethylammonium on biofilms. Remaining biofilms were measured by crystal violet staining, which was solubilized in diluted acetic acid and the absorbance at 590 nm was measured. Concentrations of the nitric oxide donor are on the x-axis. Contact time and temperature at which pre-formed biofilms were treated with diethylamine NONOate diethylammonium are indicated above each panel.

Dose-dependent biofilm dispersal by diethylamine NONOate diethylammonium

Diethylamine NONOate diethylammonium represents a class of molecules which spontaneously dissociate in a pH-dependent manner, with a half-life of 16 minutes at 22-25°C, pH 7.4 to liberate 1.5 moles of NO per mole of the parental compound (Keefer et al. 1996; Maragos et al. 1991). Biofilm dispersal in response to diethylamine NONOate diethylammonium was dose-dependent (Figure 2), however the relationship was inverse: low concentrations of the nitric oxide were associated with higher biofilm dispersion. The strong activity in the picomolar range is similar to what was observed with molsidomine.

Diethylamine NONOate diethylammonium was most potent at dispersing biofilms on polypropylene following a 6-hours incubation. A strong dispersal of biofilms formed by *Salmonella* ATCC 14028 on polypropylene was observed (Figure 2). Biofilms formed by *E. coli* O157:H7 were also dispersed by diethylamine NONOate diethylammonium, but the dispersal was lower when compared with *Salmonella*. The biofilm formed by the cocktail of six *Salmonella* outbreak strains formed on polystyrene was also dispersed (Figure 2 and Additional file 1: Table S2 and Additional file 1: Table S3).

The effect of diethylamine NONOate sodium salt hydrate, MAHMA NONOate and spermine NONOate on biofilms

Of the tested compounds, diethylamine NONOate sodium salt was the least effective biofilm dispersant following 24 and 6 hours incubation. The strongest dispersion was observed in biofilms formed on polypropylene by the cocktail of *Salmonella* strains and *Salmonella* Typhimurium ATCC 14028 (Figure 3). Up to 50% of reduction of biofilm was detected at room temperature for *Salmonella* 14028 and the outbreak strains on polystyrene (Figure 3). On

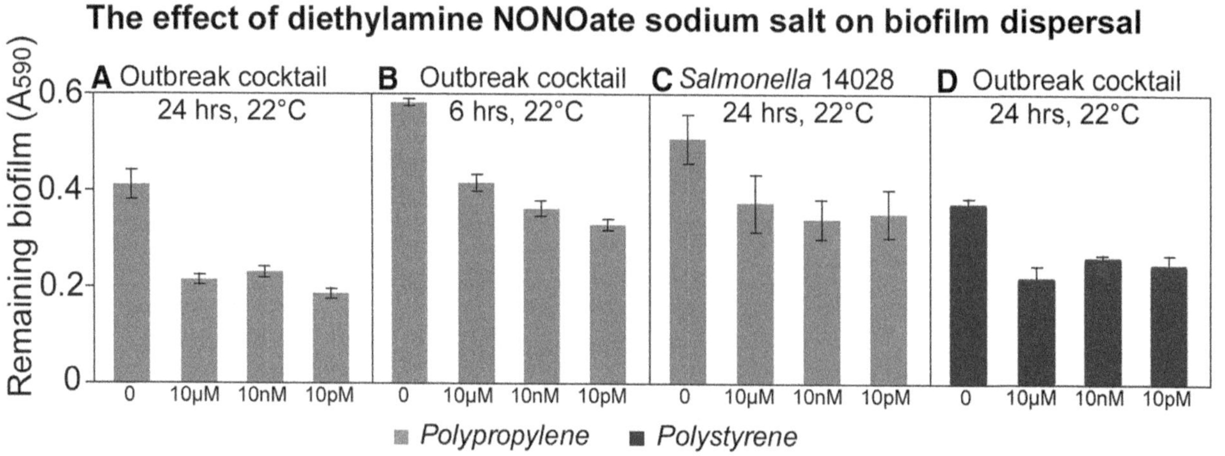

Figure 3 Dispersal of *Salmonella* biofilms by diethylamine NONOate sodium salt. Biofilms were formed overnight by either *S. enterica* sv. Typhimurium ATCC14028 or a cocktail of six *Salmonella enterica* strains (sv. Braenderup 04E01347, Braenderup 04E01556, Braenderup 04E00783, sv. Montevideo LJH519, sv. Javiana ATCC BAA-1593 and sv. Newport C6.3) linked to produce-associated outbreaks of gastroenteritis. Biofilms were pre-formed on either polystyrene (black bars) or polypropylene (grey bars) prior to the treatment with diethylamine NONOate sodium salt. Concentrations of the nitric oxide donor are on the x-axis. Contact time and temperature at which the treatment took place are indicated above each panel. Residual biofilms were quantified by staining with crystal violet. Error bars are standard errors.

polystyrene, biofilm formation by the cocktail of the *Salmonella* strains was reduced by up to 30% when compared with the control. Diethylamine NONOate was also active in dispersing biofilms formed by *E. coli* O157:H7, even though the dispersion was very limited (~20% when compared with the control) (Additional file 1: Table S2 and Additional file 1: Table S3).

The effectiveness of MAHMA NONOate (which has half-life of minutes) was compared to that of sperimne NONOate, which has half-life of several hours. This affects potency and sustainability of the treatment, for MAHMA NONOate showing up to 40 times the potency of spermine NONOate, but the latter showing activaty for hours, compared with few minutes for MAHMA NONOate (Wang et al. 2005). MAHMA NONOate is also an optimal dispersant of preformed biofilm of *Pseudomonas aerouginosa* biofilm (Barnes et al. 2013). Therefore, the two molecules were compared within 6 hours of incubation with *Salmonella* and *E. coli* biofilms. As expected MAHMA NONOate was a more effective dispersant on both the materials (Table 1). However, spermine NONOate was mainly effective at releasing biofilms formed on polypropylene (Table 1).

MAHMA NONOate also appears to be a broadly active molecule in dispersing biofilm formed both on polypropylene and polystyrene (Figure 4): it released up to 70% of *E. coli* biofilms pre-formed on polystyrene at room temperature after a 6-hour treatment (Figure 4). Biofilms formed by *S. enterica* sv Typhimurium ATCC14028 were also dispersed by MAHMA NONOate at room temperature, but compared to spermine NONOate, it was more effective on polystyrene (Figure 4). On polypropylene, up to 50% of *S. enterica* sv Typhimurium ATCC14028 and the cocktail of six *Salmonella* outbreak strains biofilms were dispersed by MAHMA NONOate when incubated for 24 hours at room temperature (Figure 4). Biofilms formed by *E. coli* O157:H7 were also effectively dispersed on polypropylene but less when compared with polystyrene (Figure 4). Effectiveness of MAHMA nonoate in dispersing

Table 1 The effect of Spermine NONOate and MAHMA NONOate on preformed biofilm

Material	Strain	Incubation temperature	Probability > F (p = 0.05)*	
			Spermine NONOate	MAHMA NONOate
Polypropylene	*Salmonella* 14028	4°C	n.s.	n.s.
Polypropylene	*Salmonella* 14028	22°C	n.s.	0.0049
Polypropylene	*E.coli* O157:H7	4°C	0.0273	n.s.
Polypropylene	*E.coli* O157:H7	22°C	0.0486	<.0001
Polypropylene	*Salmonella* cocktail	4°C	0.0036	0.0185
Polypropylene	*Salmonella* cocktail	22°C	0.0003	<.0001
Polystyrene	*Salmonella* 14028	4°C	0.0418	0.0229
Polystyrene	*Salmonella* 14028	22°C	n.s.	<.0001
Polystyrene	*E.coli* O157:H7	4°C	n.s.	n.s.
Polystyrene	*E.coli* O157:H7	22°C	n.s.	n.s.
Polystyrene	*Salmonella* cocktail	4°C	n.s.	0.0184
Polystyrene	*Salmonella* cocktail	22°C	0.0183	<.0001

*Statistically significant effects ($p < 0.05$) of the NO donor treatments when compared with the untreated control; "n.s." not statistically significant biofilm dispersion.

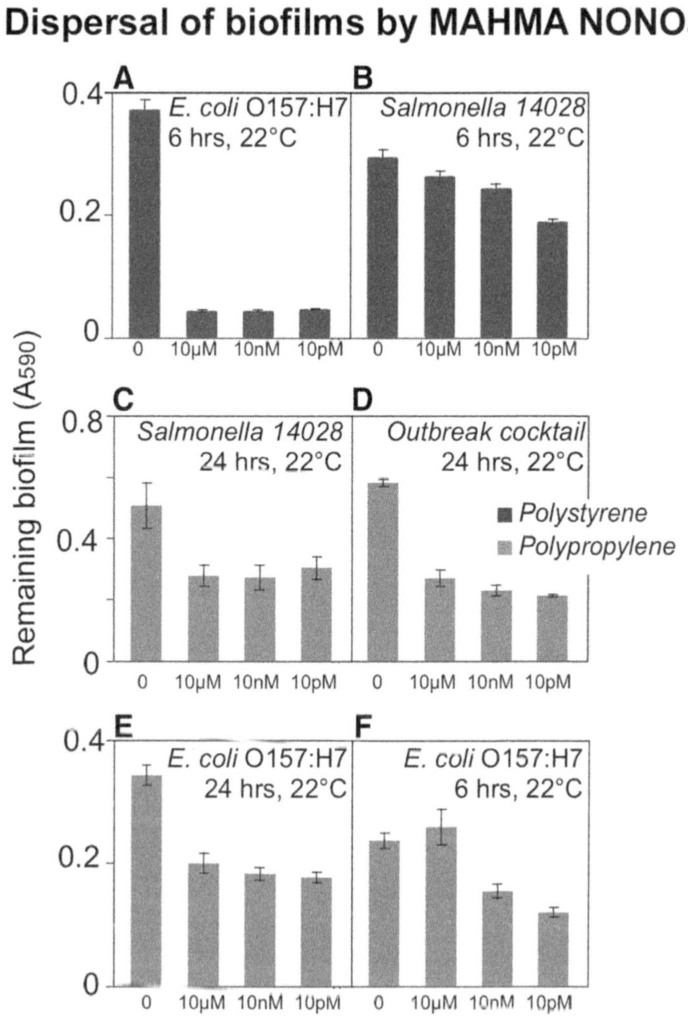

Figure 4 Biofilm dispersal by MAHMA NONOate. Biofilms were formed by *E. coli* O157:H7, *Salmonella* sv. Typhimurium 14028 and a cocktail of the six *Salmonella* strains linked to human produce-related gastroenteritis outbreaks on polystyrene (black bars) or polypropylene (grey bars) prior to the treatment with MAHMA NONOate. Contact times and temperatures at which biofilms were exposed to the nitric oxide donor are listed above each panel. Concentrations of the nitric oxide donor are on the x-axis.

biofilms was further increased by treatment with the microbiocide SaniDate (Figure 5).

The recA-hydN genomic region is involved in Salmonella biofilm dispersal

In order to better understand how nitric oxide donors induce dispersal of *Salmonella* biofilms, the genome of *S.* Typhimurium 14028 was scanned for the homologs of known genes involved in nitric oxide signaling. A mutant lacking the ~15kB *Salmonella enterica* sv Typhimurium ATCC14028 genomic region spanning 15 genes (*recA-hydN*) was studied (Figure 6). This region includes putative NO-reductase machinery: *ygaA*, an anaerobic nitric oxide reductase transcriptional regulator; STM2840, an anaerobic nitric oxide reductase flavorubredoxin; *ygbD*, nitric oxide reductase; *hypF*, hydrogenase maturation protein; *hydN*, an electron transport protein HydN. Proteins encoded region can be involved in transferring electron to NO, its detoxification and generation of nitrous oxide. Consistent with the predicted functions of the deleted genes, the Δ*recA-hydN* mutant formed a more robust biofilm on polystyrene compared to the wild type, and did not respond to the 24-hrs treatment with molsidomine (Figure 6).

Discussion

Controlling biofilms on surfaces of clinical and industrial importance has emerged as an important goal. While an impressive toolbox is potentially available to those aiming to prevent microbes from attaching to surfaces and forming biofilms (Campoccia et al. 2013; Chen et al. 2013), the approaches for controlling existing biofilms are significantly

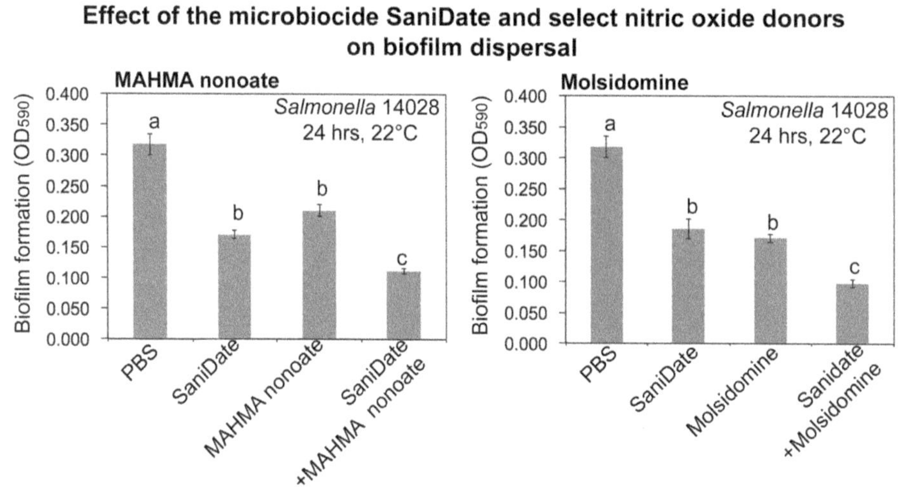

Figure 5 Additive effect of the microbiocide SaniDate and nanomolar concentrations of Molsidomine and MAHMA nonotate. Bars represent the standard error. Different letters represent significant different means (p = 0.05).

more limited. A seminal discovery that self-produced nitric oxide acts as a dispersal cue for *Pseudomonas aeruginosa* biofilms (Barraud et al. 2006) sparked interest in exploring the use of the NO gas, molecules and nanoparticles capable of releasing it (Landini et al. 2010; McDougald et al. 2012). Various nitric oxide generating molecules and nanoparticles can be used effectively to dislodge biofilms formed by Gram-negative *P. aeruginosa* and *E. coli*, Gram-Positive *Staphylococcus aureus* and *S. epidermidis* (Barraud et al. 2009b; Cherayil and Antos 2001; Slomberg et al. 2013) and prevent attachment of zoospores of the green algae *Ulva* (Thompson et al. 2010). However, in some bacteria (*Shewanella oneidensis, Vibrio harveyi*), perception of NO leads to an increased biofilm formation (Landini et al. 2010; Plate and Marletta 2012). Therefore, even though NO is a signal employed by a diversity of organisms, responses to it are not universally conserved. Because of these biological differences in the consequences of NO detection, it is important to establish how and to what extent commercially available nitric oxide donors

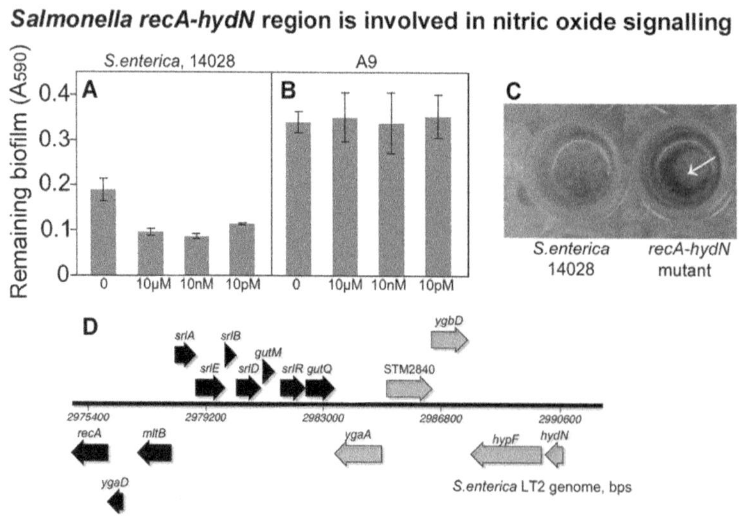

Figure 6 Involvement of the *Salmonella recA-hydN* genomic region in the nitric oxide-mediated signaling. *Salmonella* sv. Typhimurium A9 (mutant lacking a 15kB fragment spanning *recA-hydN* genes) formed more abundant biofilms, which were not responsive to treatment with nitric oxide donor molsidomine. Detachment of the wild type biofilms in response to molsidomine **(A)**. Lack of response to molsidomine by the A9 mutant **(B)**. Appearance of the biofilms formed by the A9 mutant and the wild type **(C)**. Genomic context of the A9 deletion **(D)**. Thick arrows are annotated ORF. Genes with putative functions in nitric oxide signaling are shaded grey.

can be used for controlling biofilms formed by important foodborne bacterial pathogens, known to persist as biofilms in the industrial facilities.

With this study, we focused on the effects of off-the-shelf NO donors on the biofilms formed by seven strains of *S. enterica* and pathogenic *E. coli* on surfaces that mimic those found in food processing facilities. The five compounds selected for these tests based on their low potential toxicity, commercial availability and predicted potency demonstrated varying levels of specificity and efficacy. Of the compounds tested, molsidomine and diethylamine NONOate diethylammonium are the most promising as they are capable of dislodging biofilms formed by *S.* Typhimurium 14028, a cocktail of six *Salmonella* strains isolated from outbreaks and *E. coli* O157:H7 under most of the tested conditions. For most of the compounds tested, their highest activity appears to be in the picomolar range, suggesting that their applications could be further optimized for the economical and effective industrial applications. The activity of the compounds in the nano- and picomolar ranges further suggests that NO acts as a potent cue, consistent with previous reports. Low biofilm-dispersing activity of the nitric oxide donors at higher concentrations is, perhaps, not surprising considering that NO can be bactericidal at a concentration of mg L^{-1} (McDougald et al. 2012; Miller et al. 2009), and the methods employed in this study would not necessarily distinguish between live and dead bacteria within an existing biofilm.

Interestingly, some compounds (molsidomine or diethylamine NONOate diethylammonium) were effective at dispersing biofilms under refrigerated conditions (4°C). Temperature is an important factor affecting the dissociation constant. NONOate (s) are very sensitive to temperature: a 1°C change from 37°C can results in an approximate 13% change in NO release (Ramamurthi and Lewis 1997). Therefore, it is reasonable to hypothesize that when biofilms are exposed to the nitric oxide donors at lower temperatures, they experienced the treatment for a longer period of time, and this increased their potency. The implications of this observation for industrial applications are potentially exciting: the ability of the nitric oxide donors to disperse biofilms at 4°C makes them good candidates for cleaning refrigerated surfaces, common in the food industry and even removing pathogen biofilms from refrigerated foods. While *Salmonella* and *E. coli* are thought to be metabolically inactive under these conditions, their populations in biofilms formed on refrigerated foods remain relatively steady over an extended period of time (Kroupitski et al. 2009). Furthermore, a complement of *Salmonella* genes that were differentially expressed during biofilm formation on cut lettuce during cold storage has been identified (Kroupitski et al. 2013), suggesting that even though the cells of these human pathogens are not dividing, physiological processes and gene expression still take place under refrigerated conditions. It would be of great interest to determine whether these specific genes induced in biofilms on foods at 4-8°C could be subject to manipulation by NO. More broadly, it remains to be determined which of the *Salmonella* genes involved in biofilm formation (Hamilton et al. 2009; Teplitski et al. 2006) are subject to regulation by nitric oxide, and – conversely – which of the known *Salmonella* genes responsive to nitric oxide contribute to the NO-mediated biofilm dispersal (Henard and Vazquez-Torres 2011; Karlinsey et al. 2012; Richardson et al. 2011).

Additional file

Additional file 1: Figure S1. Biofilm formation on different plastics by *Salmonella* and *E. coli* O157:H7. Biofilms were established on polypropylene and polystyrene surfaces in CFA medium for 24 hrs and stained with 1% crystal violet and then washed. The absorbed crystal violet and the biofilm were dissolved in 33% acetic acid and A590 was measured with a spectrophotometer. Error bars represent standard error. Significant different means are displayed with different letters. **Figure S2.** Luminescence of *Salmonella* 14028 pTIM2442 upon exposure to molsidomine. General metabolic state of the cells was assessed using the redox-coupled FMNH2/Luciferase produced by a *S.* Typhimurium ATCC14028 strain harboring high copy number plasmid in which the *luxCDABE* operon is under the phage λ promoter (pTIM2442). Concentrations of molsidomine to which cultures of *Salmonella* 14028 pTIM2442 were exposed are listed on the figure. Error bars represent the standard error of 12 replicas. **Table S1.** Fluorescence of the *Salmonella* cells detached by molsidomine treatment Measurement of *Salmonella* 14028 planktonic cells detached from preformed biofilm during treatment with molsidomine. polypropylene, *Statistically significant effects ($p<0.05$) of molsidomine treatment. **Table S2.** The complete set of experiments on the effect of the NO donors on preformed biofilm after 24 hours of contact time. *Statistically significant effects ($p<0.05$) of the NO donor treatments. n.s., not statistically significant biofilm dispersion. MOL, Molsidomine; M1555, MAHMA NONOate; S150, SPERMINE NONOate; D5431, diethylamine NONOate diethylammonium salt; D184, diethylamine NONOate sodium; S8432 sulfo NONOate disodium salt. **Table S3.** The complete set of experiments on the effect of the NO donors on preformed biofilm after 6 hours of contact time. *Statistically significant effects ($p<0.05$) of the NO donor treatments. n.s., not statistically significant biofilm dispersion. MOL, Molsidomine; M1555, MAHMA NONOate; S150, SPERMINE NONOate; D5431, diethylamine NONOate diethylammonium salt; D184, diethylamine NONOate sodium; S8432 sulfo NONOate disodium salt.

Competing interests
This research was supported by funding provided by the Florida Tomato Committee, Grant #106486 and by the UC-Davis Center for Produce Safety, Grant #2014-308.

Authors' contributions
MM, MT conceived experiments; MM, CC, MC, IAD conducted experiments; MM analyzed data; MM, MT wrote the manuscript. All authors read and approved the final manuscript.

Acknowledgements
We are grateful to M. McClelland for providing A9 mutant.

Author details
[1]Soil and Water Science Department, University of Florida-IFAS, Gainesville, FL, USA. [2]Microbiology and Cell Science Department, University of Florida-IFAS, Gainesville, FL, USA. [3]Cancer and Genetics Research Center, Genetics Institute, 2033 Mowry Road, Gainesville, FL 32611, USA.

References

Alagely A, Rajamani S, Teplitski M (2011) Luminescent reporters and their applications for the characterization of signals and signal-mimics that alter LasR-mediated quorum sensing. Methods Mol Biol 692:113–130. doi:10.1007/978-1-60761-971-0_9

Barnes RJ, Bandi RR, Wong WS, Barraud N, McDougald D, Fane A, Kjelleberg S, Rice SA (2013) Optimal dosing regimen of nitric oxide donor compounds for the reduction of *Pseudomonas aeruginosa* biofilm and isolates from wastewater membranes. Biofouling 29(2):203–212. doi:10.1080/08927014.2012.760069

Barraud N, Hassett DJ, Hwang SH, Rice SA, Kjelleberg S, Webb JS (2006) Involvement of nitric oxide in biofilm dispersal of *Pseudomonas aeruginosa*. J Bacteriol 188(21):7344–7353. doi:10.1128/JB.00779-06

Barraud N, Schleheck D, Klebensberger J, Webb JS, Hassett DJ, Rice SA, Kjelleberg S (2009a) Nitric oxide signaling in *Pseudomonas aeruginosa* biofilms mediates phosphodiesterase activity, decreased cyclic di-GMP levels, and enhanced dispersal. J Bacteriol 191(23):7333–7342. doi:10.1128/JB.00975-09

Barraud N, Storey MV, Moore ZP, Webb JS, Rice SA, Kjelleberg S (2009b) Nitric oxide-mediated dispersal in single- and multi-species biofilms of clinically and industrially relevant microorganisms. Microb Biotechnol 2(3):370–378. doi:10.1111/j.1751-7915.2009.00098.x

Barraud N, Kardak BG, Yepuri NR, Howlin RP, Webb JS, Faust SN, Kjelleberg S, Rice SA, Kelso MJ (2012) Cephalosporin-3′-diazeniumdiolates: targeted NO-donor prodrugs for dispersing bacterial biofilms. Angew Chem Int Ed Engl 51 (36):9057–9060. doi:10.1002/anie.201202414

Blanpain-Avet P, Faille C, Delaplace G, Bénézech T (2011) Cell adhesion and related fouling mechanism on a tubular ceramic microfiltration membrane using *Bacillus cereus* spores. J Membrane Sci 385–386(0):200–216, doi:http://dx.doi.org/10.1016/j.memsci.2011.09.041

Campoccia D, Montanaro L, Arciola CR (2013) A review of the biomaterials technologies for infection-resistant surfaces. Biomaterials 34(34):8533–8554. doi:10.1016/j.biomaterials.2013.07.089

Charville GW, Hetrick EM, Geer CB, Schoenfisch MH (2008) Reduced bacterial adhesion to fibrinogen-coated substrates via nitric oxide release. Biomaterials 29(30):4039–4044. doi:10.1016/j.biomaterials.2008.07.005

Chen M, Yu Q, Sun H (2013) Novel strategies for the prevention and treatment of biofilm related infections. Int J Mol Sci 14(9):18488–18501. doi:10.3390/ijms140918488

Cherayil BJ, Antos D (2001) Inducible nitric oxide synthase and *Salmonella* infection. Microbes Infect 3(9):771–776

Costerton JW, Cheng KJ, Geesey GG, Ladd TI, Nickel JC, Dasgupta M, Marrie TJ (1987) Bacterial biofilms in nature and disease. Annu Rev Microbiol 41:435–464. doi:10.1146/annurev.mi.41.100187.002251

Darling KE, Evans TJ (2003) Effects of nitric oxide on *Pseudomonas aeruginosa* infection of epithelial cells from a human respiratory cell line derived from a patient with cystic fibrosis. Infect Immun 71(5):2341–2349

Firoved AM, Wood SR, Ornatowski W, Deretic V, Timmins GS (2004) Microarray analysis and functional characterization of the nitrosative stress response in nonmucoid and mucoid *Pseudomonas aeruginosa*. J Bacteriol 186(12):4046–4050. doi:10.1128/JB.186.12.4046-4050.2004

Gaupels F, Kuruthukulangarakoola GT, Durner J (2011) Upstream and downstream signals of nitric oxide in pathogen defence. Curr Opin Plant Biol 14(6):707–714. doi:10.1016/j.pbi.2011.07.005

Gilberthorpe NJ, Poole RK (2008) Nitric oxide homeostasis in *Salmonella typhimurium*: roles of respiratory nitrate reductase and flavohemoglobin. J Biol Chem 283(17):11146–11154. doi:10.1074/jbc.M708019200

Hamilton S, Bongaerts RJ, Mulholland F, Cochrane B, Porter J, Lucchini S, Lappin-Scott HM, Hinton JC (2009) The transcriptional programme of *Salmonella enterica* serovar Typhimurium reveals a key role for tryptophan metabolism in biofilms. BMC Genomics 10:599. doi:10.1186/1471-2164-10-599

Henard CA, Vazquez-Torres A (2011) Nitric oxide and *Salmonella* pathogenesis. Front Microbiol 2:84. doi:10.3389/fmicb.2011.00084

Huynh TT, McDougald D, Klebensberger J, Al Qarni B, Barraud N, Rice SA, Kjelleberg S, Schleheck D (2012) Glucose starvation-induced dispersal of *Pseudomonas aeruginosa* biofilms is cAMP and energy dependent. PLoS One 7(8):e42874. doi:10.1371/journal.pone.0042874

Karlinsey JE, Bang IS, Becker LA, Frawley ER, Porwollik S, Robbins HF, Thomas VC, Urbano R, McClelland M, Fang FC (2012) The NsrR regulon in nitrosative stress resistance of *Salmonella enterica* serovar Typhimurium. Mol Microbiol 85(6):1179–1193. doi:10.1111/j.1365-2958.2012.08167.x

Keefer LK, Nims RW, Davies KM, Wink DA (1996) "NONOates" (1-substituted diazen-1-ium-1,2-diolates) as nitric oxide donors: Convenient nitric oxide dosage forms. In: Lester P (ed) Methods in Enzymology. vol Volume 268. Academic Press, New York, pp 281–293

Kroupitski Y, Pinto R, Brandl MT, Belausov E, Sela S (2009) Interactions of *Salmonella enterica* with lettuce leaves. J Appl Microbiol 106(6):1876–1885. doi:10.1111/j.1365-2672.2009.04152.x

Kroupitski Y, Brandl MT, Pinto R, Belausov E, Tamir-Ariel D, Burdman S, Sela Saldinger S (2013) Identification of *Salmonella enterica* genes with a role in persistence on lettuce leaves during cold storage by Recombinase-based in Vivo Expression Technology. Phytopathology 103(4):362–372. doi:10.1094/PHYTO-10-12-0254-FI

Landini P, Antoniani D, Burgess JG, Nijland R (2010) Molecular mechanisms of compounds affecting bacterial biofilm formation and dispersal. Appl Microbiol Biotechnol 86(3):813–823. doi:10.1007/s00253-010-2468-8

Maragos CM, Morley D, Wink DA, Dunams TM, Saavedra JE, Hoffman A, Bove AA, Isaac L, Hrabie JA, Keefer LK (1991) Complexes of NO with nucleophiles as agents for the controlled biological release of nitric oxide: Vasorelaxant effects. J Med Chem 34(11):3242–3247. doi:10.1021/jm00115a013

Marvasi M, Visscher PT, Casillas Martinez L (2010) Exopolymeric substances (EPS) from *Bacillus subtilis*: polymers and genes encoding their synthesis. FEMS Microbiol Lett 313(1):1–9. doi:10.1111/j.1574-6968.2010.02085.x

McDougald D, Rice SA, Barraud N, Steinberg PD, Kjelleberg S (2012) Should we stay or should we go: mechanisms and ecological consequences for biofilm dispersal. Nat Rev Micro 10(1):39–50

Merritt JH, Kadouri DE, O'Toole GA (2005) Growing and analyzing static biofilms. Curr Protoc Microbiol Chapter 1:Unit 1B 1, doi:10.1002/9780471729259.mc01b01s00

Miller C, McMullin B, Ghaffari A, Stenzler A, Pick N, Roscoe D, Ghahary A, Road J, Av-Gay Y (2009) Gaseous nitric oxide bactericidal activity retained during intermittent high-dose short duration exposure. Nitric Oxide 20(1):16–23. doi:10.1016/j.niox.2008.08.002

Noel JT, Arrach N, Alagely A, McClelland M, Teplitski M (2010) Specific responses of *Salmonella enterica* to tomato varieties and fruit ripeness identified by in vivo expression technology. PLoS One 5(8):e12406. doi:10.1371/journal.pone.0012406

O'Toole GA, Kolter R (1998) Initiation of biofilm formation in *Pseudomonas fluorescens* WCS365 proceeds via multiple, convergent signalling pathways: a genetic analysis. Mol Microbiol 28(3):449–461

Petrova OE, Sauer K (2012) Dispersion by *Pseudomonas aeruginosa* requires an unusual posttranslational modification of BdlA. Proc Natl Acad Sci U S A 109 (41):16690–16695. doi:10.1073/pnas.1207832109

Plate L, Marletta MA (2012) Nitric oxide modulates bacterial biofilm formation through a multicomponent cyclic-di-GMP signaling network. Mol Cell 46 (4):449–460. doi:10.1016/j.molcel.2012.03.023

Ramamurthi A, Lewis RS (1997) Measurement and modeling of nitric oxide release rates for nitric oxide donors. Chem Res Toxicol 10(4):408–413. doi:10.1021/tx960183w

Regev-Shoshani G, Ko M, Miller C, Av-Gay Y (2010) Slow release of nitric oxide from charged catheters and its effect on biofilm formation by *Escherichia coli*. Antimicrob Agents Chemother 54(1):273–279. doi:10.1128/AAC.00511-09

Richardson AR, Payne EC, Younger N, Karlinsey JE, Thomas VC, Becker LA, Navarre WW, Castor ME, Libby SJ, Fang FC (2011) Multiple targets of nitric oxide in the tricarboxylic acid cycle of *Salmonella enterica* serovar typhimurium. Cell Host Microbe 10(1):33–43. doi:10.1016/j.chom.2011.06.004

Rosenkranz B, Winkelmann BR, Parnham MJ (1996) Clinical pharmacokinetics of molsidomine. Clin Pharmacokinet 30(5):372–384

Santiviago CA, Reynolds MM, Porwollik S, Choi S-H, Long F, Andrews-Polymenis HL, McClelland M (2009) Analysis of pools of targeted *Salmonella* deletion mutants identifies novel genes affecting fitness during competitive infection in mice. PLoS Pathog 5(7):e1000477. doi:10.1371/journal.ppat.1000477

Shi X, Zhu X (2009) Biofilm formation and food safety in food industries. Trend Food Sci Technol 20(9):407–413, doi:http://dx.doi.org/10.1016/j.tifs.2009.01.054

Simontacchi M, García-Mata C, Bartoli CG, Santa-María GE, Lamattina L (2013) Nitric oxide as a key component in hormone-regulated processes. Plant Cell Rep 32(6):853–866. doi:10.1007/s00299-013-1434-1

Slomberg DL, Lu Y, Broadnax AD, Hunter RA, Carpenter AW, Schoenfisch MH (2013) Role of size and shape on biofilm eradication for nitric oxide-releasing silica nanoparticles. ACS Appl Mater Interfaces 5:19. doi:10.1021/am402618w

Solano C, Garcia B, Valle J, Berasain C, Ghigo JM, Gamazo C, Lasa I (2002) Genetic analysis of *Salmonella enteritidis* biofilm formation: critical role of cellulose. Mol Microbiol 43(3):793–808

Teplitski M, Al-Agely A, Ahmer BMM (2006) Contribution of the SirA regulon to biofilm formation in *Salmonella enterica* serovar Typhimurium. Microbiology 152(11):3411–3424. doi:10.1099/mic.0.29118-0

Thompson SE, Callow ME, Callow JA (2010) The effects of nitric oxide in settlement and adhesion of zoospores of the green alga *Ulva*. Biofouling 26(2):167–178. doi:10.1080/08927010903402420

Van Alst NE, Picardo KF, Iglewski BH, Haidaris CG (2007) Nitrate sensing and metabolism modulate motility, biofilm formation, and virulence in *Pseudomonas aeruginosa*. Infect Immun 75(8):3780–3790. doi:10.1128/IAI.00201-07

Wang PG, Cai TB, Taniguchi N (2005) Nitric Oxide Donors. Wiley-VCH Verlag GmbH & Co, KGaA

Wingender J, Neu TR, Flemming HC (1999) Microbial Extracellular Polymeric Substances: Characterization, Structure, and Function. Springer, New York

Xiong Y, Liu Y (2010) Biological control of microbial attachment: a promising alternative for mitigating membrane biofouling. Appl Microbiol Biotechnol 86(3):825–837. doi:10.1007/s00253-010-2463-0

Zhang L, Mah TF (2008) Involvement of a novel efflux system in biofilm-specific resistance to antibiotics. J Bacteriol 190(13):4447–4452. doi:10.1128/JB.01655-07

Sulfobacillus thermosulfidooxidans strain Cutipay enhances chalcopyrite bioleaching under moderate thermophilic conditions in the presence of chloride ion

Roberto A Bobadilla-Fazzini[1*], Maria Paz Cortés[2,3], Alejandro Maass[3,4] and Pilar Parada[1]

Abstract

Currently more than 90% of the world's copper is obtained through sulfide mineral processing. Among the copper sulfides, chalcopyrite is the most abundant and therefore economically relevant. However, primary copper sulfide bioleaching is restricted due to high ionic strength raffinate solutions and particularly chloride coming from the dissolution of ores. In this work we describe the chalcopyrite bioleaching capacity of *Sulfobacillus thermosulfidooxidans* strain Cutipay (DSM 27601) previously described at the genomic level (Travisany et al. (2012) Draft genome sequence of the Sulfobacillus thermosulfidooxidans Cutipay strain, an indigenous bacterium isolated from a naturally extreme mining environment in Northern Chile. J Bacteriol 194:6327–6328). Bioleaching assays with the mixotrophic strain Cutipay showed a strong increase in copper recovery from chalcopyrite concentrate at 50°C in the presence of chloride ion, a relevant inhibitory element present in copper bioleaching processes. Compared to the abiotic control and a test with *Sulfobacillus acidophilus* DSM 10332, strain Cutipay showed an increase of 42 and 69% in copper recovery, respectively, demonstrating its high potential for chalcopyrite bioleaching. Moreover, a genomic comparison highlights the presence of the 2-Haloacid dehalogenase predicted-protein related to a potential new mechanism of chloride resistance in acidophiles. This novel and industrially applicable strain is under patent application CL 2013–03335.

Keywords: Copper; Chalcopyrite; Bioleaching; Chloride ion; Chloride resistance; *Sulfobacillus*

Introduction

According to the Chilean Copper Commission (COCHILCO, 2014), Chile is the largest copper producer where hydrometallurgy represented 33% of the 2013 local copper production. Copper sulfides processing accounts for more than 90% of world's production. Among the copper sulfides, chalcopyrite constitutes the vast majority of copper resources and reserves, and therefore represents a mineralogical species of great economical importance (Doebrich 2009).

Within the available hydrometallurgical technologies heap leaching is one of the most used, particularly for copper oxides and secondary copper sulfides. The detection and identification of microorganisms able to live and reproduce under the acidic conditions of the heap leaching process, and moreover the biotechnological innovations related to their potential use as catalysts for primary copper sulfide bioleaching, are topics of major interest. Different studies have shown the enhancing effect of mesophilic (Bobadilla Fazzini et al. 2011), moderate thermophilic (Stott et al. 2003) and thermophilic (Konishi et al. 1999) microorganisms for chalcopyrite dissolution compared to abiotic assays. However, most of these studies were performed under lab-scale ideal conditions, not taking into account industrial-scale process variables such as high ionic strength solutions.

The application of heap bioleaching for copper recovery is severely limited due to fresh-water scarcity as in northern Chile, or the saline drains in Western Australia, due to solutions of high ionic strength and concentrations of elements coming from the ore minerals dissolution such as atacamite ($Cu_2Cl(OH)_3$) that releases chloride ions, which are inhibitory for many acidophilic iron-oxidizing microbes.

* Correspondence: rboba001@biosigma.cl
[1]BioSigma S.A., Parque Industrial Los Libertadores, Lote 106, Colina, Santiago de Chile, Chile
Full list of author information is available at the end of the article

In our estimation, during 2013 about 700 million tons of ore with an average copper grade of 0.2% could not be bioleached worldwide due to the inhibitory effect of high ionic strength raffinate solutions over acidophiles. The most studied acidophilic species is *Acidithiobacillus ferrooxidans*, and its sensitivity to low chloride concentration has been described long before, attributed to the loss of the cell outer layer integrity (Lawson et al. 1995). In the case of the widely distributed acidophilic species *Leptospirillum ferriphilum*, ferrous ion biooxidation is completely inhibited by 12 g/L chloride (Gahan et al. 2010). Moreover, prolonged exposure to sodium chloride demonstrated no significant adaptation of mixed chemolithotrophic iron-oxidizing cultures, indicating the lack of an acquired resistance through adaptation (Shiers et al. 2005). Nevertheless, resistant species exist such as *Thiobacillus prosperus*, a halotolerant acidophile that requires chloride for growth (Nicolle et al. 2009) or the marine *Sulfobacillus* sp. TPY able to tolerate 2% (w/v) NaCl (Wang et al. 2012).

With respect to the biochemical mechanism of bacterial chloride resistance, few studies are available particularly on acidophiles. A recent study showed that NaCl stress in the moderate thermophiles *Acidimicrobium ferrooxidans* and *Acidithiobacillus caldus* is tackled by cell membrane adaptation and potential osmoprotectant aminoacid accumulation (Zammit et al. 2012), giving an insight on specific chloride resistance genes.

Based on these facts, a search for chloride resistant iron-oxidizing acidophilic strains is a relevant research topic. In this study we describe the chalcopyrite bioleaching capacity of *Sulfobacillus thermosulfidooxidans* strain Cutipay (DSM 27601) in the presence of chloride ion. Strain Cutipay was previously described with novel copper and arsenic resistance capacities at the genomic level (Travisany et al. 2012).

Material and methods

Strains and culture conditions

Sulfobacillus thermosulfidooxidans DSM 27601 strain Cutipay was isolated from an ore deposit in northern Chile. *Sulfobacillus acidophilus* DSM 10332 was obtained from the DSMZ collection. Both strains were cultivated in shake flasks at 50°C in 9 K medium at pH 1.6 (Bobadilla Fazzini et al. 2011), including the addition of 0.25 g/L yeast extract for mixotrophic growth on ferrous. Microorganisms were inoculated at 1% v/v. The cell number was determined by chamber counting under phase-contrast microscope (Thoma Chamber, depth 0.010 mm) with at least four replicate counts.

Minimum inhibitory concentration assays

Minimum inhibitory concentration assays (MIC) were performed in 6 well plates with 5 mL of 9 K medium at pH 1.6 for mixototrophic growth on ferrous, inoculated with 1,00E + 07 cells/mL for each strain separately. Assays were incubated with agitation at 50°C for 5 d with a top film to prevent evaporation. Different concentrations of chloride (tested as NaCl or KCl), Cu(II) (added as $CuSO_4$) and As(III) (added as As_2O_3) were tested. The lowest concentration at which no growth or iron oxidation was observed corresponds to the MIC that were assayed at least in triplicate.

In silico determination of orthologous protein clusters

The genome sequences and annotation of the predicted proteomes for *Sulfobacillus acidophilus* DSM10332 (GenBank IDs:NC_016884.1 and NC_016888.1) and *Sulfobacillus* sp. TPY (GenBank ID:NC_015757.1) were obtained from the GenBank database. For *Sulfobacillus thermosulfidooxidans* strain Cutipay, predicted proteome sequence was obtained from an in-house annotation of its genome (GenBank ID: ALWJ01000000) as described previously (Travisany et al. 2012). Using these three predicted proteomes, orthologous protein clusters were determined using ORTHOMCL v1.4 (Li et al. 2003) with default parameters.

Bioleaching tests

Bioleaching tests were done in shake flasks, as a minimum in duplicate. Each flask contained 100 mL 9 K medium (pH 1.6), as described above, and supplemented with 1.5 g/L Fe(III) and 2.5 g/L Fe(II) with 1% w/v 300 Tyler mesh sieved chalcopyrite concentrate (85.5% chalcopyrite representing more than 99% of total Cu in the assay (Additional file 1: Table S1)) and 3 g/L chloride (as sodium chloride). Each flask was inoculated with *Sulfobacillus thermosulfidooxidans* DSM 27601 or *Sulfobacillus acidophilus* DSM 10332 (1×10^7 cell/mL each). A non-inoculated control was included. Shake flasks were incubated for 27 d at 50°C with 120 rpm agitation. Each week cell number, Fe(II) by *o*-phenantroline method, total iron and Cu(II) by atomic absorption spectrometry were assayed. Sampling was performed allowing the suspended concentrate to settle, and filtering the supernatant to avoid any solids (Grade 5C filters, Advantec, Japan). Evaporation was compensated with pure water before each sampling time.

Results

In order to determine the phenotypic properties of *Sulfobacillus thermosulfidooxidans* strain Cutipay (DSM 27601) related to heavy metals and chloride resistance, MIC assays were performed and compared to a collection *Sulfobacillus* strain. The results are summarized in Table 1.

Significant differences in chloride MIC using sodium or potassium salts were observed, requiring two to three

Table 1 Minimum Inhibitory Concentrations (MIC) for inhibitory elements chloride, copper and arsenic

Moderate thermophilic species	Chloride (KCl)	Chloride (NaCl)	Cu (II)	As (III)
	[ppm]			
Sulfobacillus thermosulfidooxidans Cutipay (DSM 27601)	10,000	5,000	3,000	<100
Sulfobacillus acidophilus (DSM 10332)	3,000	1,000	1,000	1,000

times of potassium chloride to reach total inhibition compared to the sodium salt. Cutipay showed a higher resistance to chloride (three to five times depending the counter-ion used) and copper (three times) and a lower resistance to arsenic (ten times) compared to *Sulfobacillus acidophilus* DSM 10332. To understand these phenotypes, a whole genome comparison of selected *Sulfobacillus* species was performed looking for specific chloride resistance genes. The genomes of *Sulfobacillus thermosulfidooxidans* strain Cutipay and *Sulfobacillus* sp. TPY, as representatives of chloride-resistant species, were compared together and against *Sulfobacillus acidophilus* DSM10332, a chloride sensitive strain (Figure 1).

The three strains have 2061 of coded proteins in common. An exclusive cluster of 22 proteins was shared between both chloride resistant strains. An interesting protein annotated as Haloacid dehalogenase (sequence identifier gi|339628570|ref|YP_004720213.1|) was recognized. This protein belongs to the haloacid dehalogenase-like (HAD) superfamily that includes the 2-Haloacid dehalogenase that catalyzes the dechlorination of small organic chloro-acids.

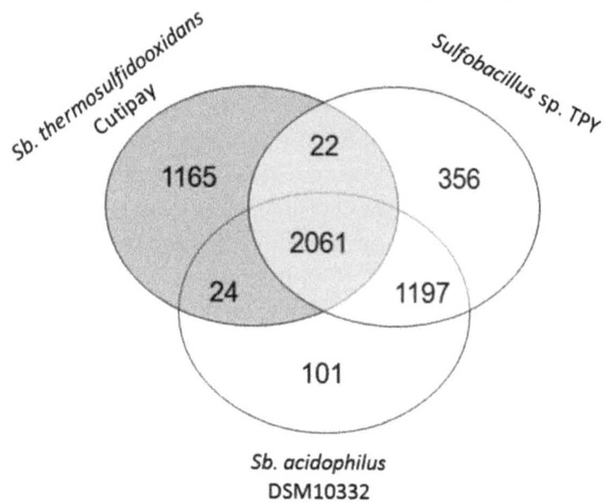

Figure 1 Venn diagram of orthologous genome coded protein clusters for chloride-resistant strains *Sulfobacillus* sp. TPY and *Sulfobacillus thermosulfidooxidans* Cutipay, and chloride-sensitive strain *Sulfobacillus acidophilus* DSM 10332. The diagram shows the number of shared and unique clusters between the three strain genomes.

In order to evaluate the relevance of strain Cutipay to enhance chalcopyrite dissolution under process conditions and compare it to the collection strain DSM 10332, chalcopyrite bioleaching tests were done with chloride ion. Copper recovery was periodically measured, showing 31% copper recovery in the abiotic control at the end of the test. The same effect was observed with *Sulfobacillus acidophilus*, being lower and reaching 26% recovery, probably inhibited by chloride ion. However, a strong increase in copper recovery was observed with strain Cutipay despite chloride present. As shown in Figure 2, chalcopyrite bioleaching was enhanced by strain Cutipay from the beginning, finishing at 44% on day 27. In the comparison with the collection strain *Sulfobacillus acidophilus* DSM 10332 and the abiotic control assays, strain Cutipay increased copper recovery in 18 and 13 percentage points which represented an increment of 69 and 42%, respectively, demonstrating its high potential for chalcopyrite bioleaching in the presence of chloride ion. Figure 3 shows that *Sulfobacillus thermosulfidooxidans* Cutipay was able to oxidize ferrous to ferric ions in the presence of chloride ion maintaining a high oxidation reduction potential, while *Sulfobacillus acidophilus* DSM10332 did not oxidize Fe(II) which is comparable to the control without inoculation. Previous results showed that both species have comparable iron oxidations rates, with same optimum pH of 1.7 and temperature (51°C) (Watling et al. 2008), so the observed effect can be attributed only to chloride ion inhibition.

Discussion

In this study a novel mixotrophic bacterium *Sulfobacillus thermosulfidooxidans* strain Cutipay (DSM 27601) was characterized for heavy metals and chloride resistance. Sodium and potassium chloride salts were separately used for chloride MIC testing, since the counter-ion

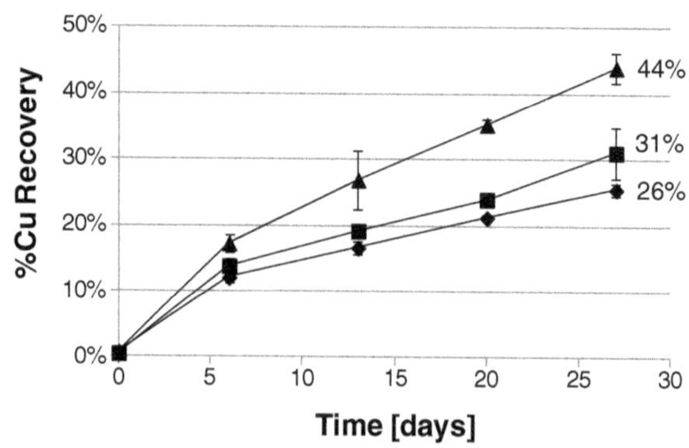

Figure 2 Copper recovery in chalcopyrite bioleaching tests in the presence of chloride ion (3 g/L), incubation at 50°C and addition of *Sulfobacillus thermosulfidooxidans* DSM 27601 strain Cutipay (▲); *Sulfobacillus acidophilus* DSM 10332 (♦). A control without inoculation was included (■).

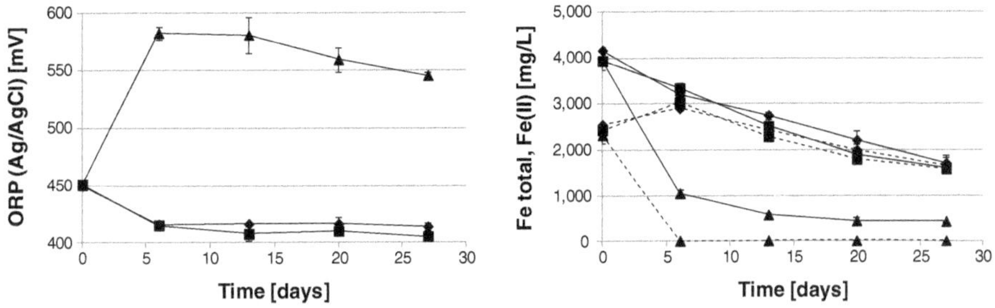

Figure 3 Oxidation Reduction Potential (ORP) (Left panel) and total (continuous lines) and ferrous iron (dotted lines) (Right panel) in chalcopyrite bioleaching assays in the presence of chloride ion (3 g/L) at 50°C. *Sulfobacillus thermosulfidooxidans* DSM 27601 strain Cutipay (▲); *Sulfobacillus acidophilus* DSM 10332 (♦); Control without inoculation (■).

(potassium or sodium) affects the inhibitory effect of anions (Suzuki et al. 1999) and therefore the MIC. These results are only partially in agreement with our previous genomic analyses (Travisany et al. 2012), supporting copper but not arsenic resistance. *Sulfobacillus thermosulfidooxidans* has been reported to be highly resistant to copper, nickel and zinc, while *Sulfobacillus acidophilus* to cobalt (Watling et al. 2008). *Sulfobacillus thermosulfidooxidans* DSM9293 isolated from an ore deposit in Eastern Kazakhstan has been tested for sodium chloride resistance, starting to show inhibition at 7 g/L (4 g chloride/L) (Zammit et al. 2012), while *Sulfobacillus* sp. TPY found in an hydrothermal vent in the Pacific Ocean grows in the presence of 20 g/L NaCl (12 g chloride/L) (Wang et al. 2012). In this study *Sulfobacillus thermosulfidooxidans* Cutipay from an ore deposit in Chile was completely inhibited at 21 and 8 g/L potassium and sodium chloride, respectively, having chloride resistance comparable to previous reports for the same species, but below the marine strain TPY.

Based on the genome comparison, the chloride-resistant representative species have the Haloacid dehalogenase coding gene. Therefore, a possible mechanism of chloride resistance in sulfobacilli may be related to the biodegradation of toxic small chloro-organic acids formed in the presence of chloride, which are produced from the dissolution of gangue material in hydrometallurgical processes (Dopson et al. 2008) that reacts with organic matter such as cell products or debris (Keppler et al. 2000). To our knowledge no chloride resistance mechanism nor related gene has been identified up to date in acidophiles, and therefore this finding may provide new insights into extremophilic microorganisms behavior. On this matter, previous studies have shown the role of membrane and/or amino acid biosynthesis, efflux transporters and CO_2 fixation as multifunctional responses to chloride exposure in acidophilic species (Zammit et al. 2012). However these are general osmo-protection mechanisms observed in different microbial species (Zhou et al. 2013; Paul 2013).

Finally, the bioleaching tests showed that strain Cutipay resistance enhances chalcopyrite bioleaching at 50°C in the presence of chloride, based on its fast iron-oxidizing activity compared to the chloride sensitive strain *Sulfobacillus acidophilus* DSM 10332. Here, only in the case of strain Cutipay all the ferrous iron was completely oxidized. However, a major drop in total iron concentration was observed lowering from 4 to 1 g/L, due to the formation of iron precipitates such as jarosite, as visually observed only on strain Cutipay's inoculated flasks (data not shown). No matter this precipitation and potential surface passivation, chalcopyrite bioleaching enhancement is clearly shown.

Summarizing, in this study a novel mixotrophic bacterium with resistance to chloride was presented, showing a genomic feature related to a potential new mechanism of chloride resistance and proven to enhance chalcopyrite bioleaching at 50°C in the presence of chloride, a relevant inhibitory element present in important industrial bioleaching processes.

Additional file

Additional file 1: Table S1. Chalcopyrite Concentrate mineralogy.

Competing interests

The authors declare that this work was supported by BioSigma 'S.A.', a profit Research, Development and Innovation Company. Currently, BioSigma 'S.A.' is applying to patent *Sulfobacillus thermosulfidooxidans* strain Cutipay (DSM 27601) use for bioleaching purposes under the patent application CL 2013-03335.

Authors' contributions

RABF designed and undertook the microbiological, characterization and bioleaching studies, and conceived and drafted the manuscript. MPC carried out the *In silico* determinations. AM design and coordinated the bioinformatics approach. PP participated in the coordination and helped to draft the manuscript. All authors read and approved the final manuscript.

Acknowledgements

This work was supported by BioSigma 'S.A.'. The authors thank the company for authorizing the submission of the manuscript for publication. We are also grateful to Fondap Grant 15090007, Basal Grant Center for Mathematical Modeling UMI2807 UCHILE-CNRS and the CIRIC-INRIA Chile project.

Author details
[1]BioSigma S.A., Parque Industrial Los Libertadores, Lote 106, Colina, Santiago de Chile, Chile. [2]Laboratory of Bioinformatics and Mathematics of the Genome, Center for Mathematical Modeling (UMI2807-CNRS) and FONDAP Center for Genome Regulation, Santiago de Chile, Chile. [3]Faculty of Mathematical and Physical Sciences, University of Chile, Santiago de Chile, Chile. [4]Department of Mathematical Engineering, Center for Mathematical Modeling (UMI2807-CNRS) and FONDAP Center for Genome Regulation, Santiago de Chile, Chile.

References

Bobadilla Fazzini RA, Levican G, Parada P (2011) Acidithiobacillus thiooxidans secretome containing a newly described lipoprotein Licanantase enhances chalcopyrite bioleaching rate. Appl Microbiol Biotechnol 89:771–780, doi:10.1007/s00253-010-3063-8

Chilean Copper Commission (COCHILCO) (2014) Mining production database., http://www.cochilco.cl/estadisticas/produccion.asp. Accessed 18 November 2014

Doebrich J (2009) Copper–a metal for the ages: U.S. geological survey fact sheet 2009–3031., http://pubs.usgs.gov/fs/2009/3031/. Accessed 18 November 2014

Dopson M, Halinen A, Rahunen N, Bostro D, Sundkvist J, Riekkola-vanhanen M, Kaksonen AH, Puhakka JA (2008) Silicate mineral dissolution during heap bioleaching. 99:811–820. doi:10.1002/bit

Gahan CS, Sundkvist J-E, Dopson M, Sandström A (2010) Effect of chloride on ferrous iron oxidation by a Leptospirillum ferriphilum-dominated chemostat culture. Biotechnol Bioeng 106:422–431, doi:10.1002/bit.22709

Keppler F, Eiden R, Niedan V, Pracht J, Schöler HF (2000) Halocarbons produced by natural oxidation processes during degradation of organic matter. Nature 403:298–301

Konishi Y, Asai S, Tokushige M, Suzuki T (1999) Kinetics of the bioleaching of chalcopyrite concentrate by acidophilic thermophile acidianus brierleyi. Biotechnol Prog 15:681–688, doi:10.1021/bp9900662

Lawson EN, Nicholas CJ, Pellat H (1995) The toxic effects of chloride ions on Thiobacillus ferrooxidans. In: Vargas T, Jerez CA, Wiertz JV, Toledo H (eds) Biohydrometall. Process. vol I, Univ. Chile., pp 165–174

Li L, Stoeckert CJ Jr, Roos DS (2003) OrthoMCL : identification of ortholog groups for eukaryotic genomes., pp 2178–2189, doi:10.1101/gr.1224503.candidates

Nicolle JLC, Simmons S, Bathe S, Norris PR (2009) Ferrous iron oxidation and rusticyanin in halotolerant, acidophilic "Thiobacillus prosperus". Microbiology 155:1302–1309, doi:10.1099/mic. 0.023192-0

Paul D (2013) Osmotic stress adaptations in rhizobacteria. J Basic Microbiol 53:101–110, doi:10.1002/jobm.201100288

Shiers DW, Blight KR, Ralph DE (2005) Sodium sulphate and sodium chloride effects on batch culture of iron oxidising bacteria. Hydrometallurgy 80:75–82, doi:10.1016/j.hydromet.2005.07.001

Stott MB, Sutton DC, Watling HR, Franzmann PD (2003) Comparative leaching of chalcopyrite by selected acidophilic bacteria and archaea. Geomicrobiol J 20:215–230, doi:10.1080/01490450390206692

Suzuki I, Lee D, Mackay B, Harahuc L, Oh JK (1999) Effect of various ions, pH, and osmotic pressure on oxidation of elemental sulfur by Thiobacillus thiooxidans. Appl Environ Microbiol 65:5163–5168

Travisany D, Di Genova A, Sepúlveda A, Bobadilla-Fazzini RA, Parada P, Maass A (2012) Draft genome sequence of the Sulfobacillus thermosulfidooxidans Cutipay strain, an indigenous bacterium isolated from a naturally extreme mining environment in Northern Chile. J Bacteriol 194:6327–6328, doi:10.1128/JB.01622-12

Wang Y, Su L, Zhang L, Zeng W, Wu J, Wan L, Qiu G, Chen XZH (2012) Bioleaching of chalcopyrite by defined mixed moderately thermophilic consortium including a marine acidophilic halotolerant bacterium. Bioresour Technol 121:348–354, doi:10.1016/j.biortech.2012.06.114

Watling HR, Perrot FA, Shiers DW (2008) Comparison of selected characteristics of Sulfobacillus species and review of their occurrence in acidic and bioleaching environments. Hydrometallurgy 93:57–65

Zammit CM, Mangold S, Jonna VR, Mutch LA, Watling HR, Dopson M, Watkin ELJ (2012) Bioleaching in brackish waters–effect of chloride ions on the acidophile population and proteomes of model species. Appl Microbiol Biotechnol 93:319–329, doi:10.1007/s00253-011-3731-3

Zhou A, Baidoo E, He Z, Mukhopadhyay A, Baumohl JK, Benke P, Joachimiak MP, Xie M, Song R, Arkin AP, Hazen TC, Keasling JD, Wall JD, Stahl DA, Zhou J (2013) Characterization of NaCl tolerance in Desulfovibrio vulgaris Hildenborough through experimental evolution. ISME J 7:1790–1802, doi:10.1038/ismej.2013.60

22

Inhibitory effects of H-Ras/Raf-1–binding affibody molecules on synovial cell function

Seiji Shibasaki[1,2,3*], Miki Karasaki[3], Torbjörn Gräslund[4], Per-Åke Nygren[4], Hajime Sano[3] and Tsuyoshi Iwasaki[2,3,5]

Abstract

Affibody molecules specific for H-Ras and Raf-1 were evaluated for their ability to inhibit synovial cell function. Affibody molecules targeting H-Ras (Z_{ras122}, Z_{ras220}, and Z_{ras521}) or Raf-1 (Z_{raf322}) were introduced into the MH7A synovial cell line using two delivery methods: transfection with plasmids encoding the affibody molecules or direct introduction of affibody protein using a cell-penetrating peptide reagent. Interleukin-6 (IL-6) and prostaglandin E2 (PGE2) production by MH7A cells were analyzed by enzyme-linked immunosorbent assay after stimulation with tumor necrosis factor-alpha (TNF-α). Cell proliferation was also analyzed. Phosphorylation of extracellular signal-regulated kinase (ERK) was analyzed by western blot. All affibody molecules could inhibit IL-6 and PGE2 production in TNF-α-stimulated MH7A cells. The inhibitory effect was stronger when affibody molecules were delivered as proteins via a cell-penetrating peptide reagent than when plasmid-DNA encoding the affibody moelcules was transfected into the cells. Plasmid-expressed Z_{ras220} inhibited phosphorylation of ERK in TNF-α-stimulated MH7A cells. Protein-introduced Z_{raf322} inhibited the production of IL-6 and PGE2 and inhibited cell proliferation in MH7A cells. These findings suggest that affibody molecules specific for H-Ras and Raf-1 can affect intracellular signal transduction through the MAP kinase pathway to inhibit cell proliferation and production of inflammatory mediators by synovial cells.

Keywords: Affibody molecules; Inflammation; Interleukin-6; Intracellular signal transduction; Prostaglandin E2; Rheumatoid arthritis

Introduction

Rheumatoid arthritis (RA) is an inflammatory disease characterized by the progressive destruction of joints. In RA, hyperplasia of synovial lining cells is observed along with angiogenesis and inflammatory mononuclear cell infiltration (Kitano et al. 2006). A wide variety of molecules produced by both synovial cells and infiltrating inflammatory mononuclear cells is thought to be involved in RA pathology. Specifically, proinflammatory cytokines such as interleukin-6 (IL-6), IL-17 (Schlegel et al. 2013) and tumor necrosis factor alpha (TNF-α) play a key role in RA pathogenesis, including the activation of synovial fibroblasts, osteoclasts, and inflammatory mononuclear cells, which results in irreversible damage to the soft tissues and bones (Olsen and Stein 2004).

One of the conventional disease-modifying antirheumatic drugs (DMARDs), methotrexate, is the standard for RA pharmacotherapy (Upchurch and Kay 2012). However, its treatment outcomes are not satisfactory and often cause high levels of toxicity and adverse effects for patients (Smolen and Steiner 2003). In recent years, various biological agents have been developed and used clinically as important options for RA treatment (Saag et al. 2008). These biological agents include an anti-CD20 antibody (rituximab) (Mélet et al. 2013), TNF antagonists (e.g., adalimumab, etanercept, golimumab, and infliximab), an IL-6 receptor antagonist (tocilizumab) (Smolen and Aletaha 2011), and a recombinant fusion protein consisting of the extracellular domain of human CTLA-4 (abatacept) (Gómez-Reino 2012). Although these biological agents provide favorable treatment outcomes for RA patients, in view of medical economics, biological therapies are more costly than traditional DMARDs. Thus,

* Correspondence: seiji@huhs.ac.jp
[1]General Education Center, Hyogo University of Health Sciences, 1-3-6 Minatojima, Chuo-ku, Kobe 650-8530, Japan
[2]Graduate School of Pharmacy, Hyogo University of Health Sciences, 1-3-6 Minatojima, Chuo-ku, Kobe 650-8530, Japan
Full list of author information is available at the end of the article

a more cost-effective treatment option is desirable because healthcare payers and providers are conscious of cost in light of the current period of budget constraints (Furneri et al. 2012).

Several biologically active, non-immunoglobulin-based proteins have been created as candidates of novel biological agents. For instance, a monobody is a target-binding protein that is generated from the fibronectin type III domain (Koide et al. 2012). Synthetic ankyrin repeat proteins have a binding scaffold and are potential diagnostic and therapeutic tools (Baumann et al. 2007). In this study, we investigated another class of non-immunoglobulin ligands, referred to as affibody molecules. They are engineered binding proteins based on a scaffold of the three-α-helix bundle Z-domain (Friedman and Ståhl 2009; Nilsson and Tolmachev 2007; Nygren 2008). The 6-kDa Z-domain is derived from the B-domain, one of the five domains in *Staphylococcus aureus* protein A (Gunneriusson et al. 1996).

Affibody molecules associated with signal transduction or transcription in tumor development or immune responses have been screened and applied. For example, affibody molecules against c-Jun, an oncogenic transcription factor, have been selected and are able to detect c-Jun in a melanoma cell line (Lundberg et al. 2009). Furthermore, other affibody molecules against human epidermal growth factor receptor (HER), specifically HER2 (Lindberg et al. 2012) and HER3 (Malm et al. 2013), have been investigated.

We have previously reported the selection of H-Ras-binding or Raf-1-binding affibody molecules (Grimm et al. 2010). Affibody clones Z_{ras122}, Z_{ras220}, and Z_{raf322} were isolated by phage display selection from a library, and the specificity and affinity for H-Ras or Raf-1 proteins were determined in that study. Each molecule was demonstrated to selectively bind Ras or Raf-1 proteins, but not other human control proteins. Z_{ras122}, Z_{ras220}, and Z_{raf322} had binding affinities in the high nanomolar to low micromolar range, as demonstrated by a real-time biospecific interaction analysis. As an alternative or complement to current RA treatment regimens, these affibody molecules might have the potential to inhibit signal transduction mediated by Ras and Raf, since this signaling is thought to be important for efficient synthesis of several inflammatory mediators (Yamamoto et al. 2003).

In this study, we investigated the effect of introducing H-Ras-binding and Raf-1-binding affibody molecules in the MH7A synovial cell line (Tsunemi et al. 2010). First, we subcloned DNA encoding affibody molecules against H-Ras or Raf-1 into a plasmid DNA vector for mammalian expression, and the vectors were introduced into MH7A cells. Then, affibody proteins were produced using an *Escherichia coli* expression system and were introduced into MH7A cells. We investigated inhibition caused by H-Ras- or Raf-1-targeting affibody molecules on the production of the inflammatory mediators IL-6 and prostaglandin E2 (PGE2). Furthermore, we examined proliferation of MH7A cells.

Materials and methods

Cell line

The human MH7A synovial cell line (Riken, Saitama, Japan), which originated from intra-articular soft tissue of the knee joints of an RA patient, was established by transfection with the SV40 T antigen (Miyazawa et al. 1998). MH7A cells were cultured in RPMI 1640 (Sigma, St. Louis, MO, USA) containing 10% heat-inactivated fetal bovine serum (Whittaker, Walkersville, MD, USA), 100 units/mL of penicillin, and 100 μg/mL of streptomycin (Invitrogen, CA, USA) at 37°C in an atmosphere of 5% CO_2 in air.

Plasmid construction for affibodies and introduction into MH7A cells

The plasmids that were used to produce affibodies in mammalian cells were constructed by first amplifying DNA encoding anti-H-Ras and anti-Raf-1 affibodies (Z_{ras122}, Z_{ras220}, Z_{ras521}, and Z_{raf322}) by PCR using primers 5′-AAGGGGATCCACCATGGGCAGCAGCCATCATCA TCA-3′ and 5′-AGGGGTTATGCTAGTTATTGCTCAG CGCGGAATTCTTA-3′ (Grimm et al. 2010). DNA sequences of affibody molecules are listed the additional file (Additional file 1). Each fragment was inserted into the multiple cloning site of pcDNA3.1(+) (Invitrogen). The nucleotide sequences of these constructs were confirmed by sequencing with an ABI PRISM 3100 Genetic Analyzer (Applied Biosystems, Foster City, CA, USA). Constructs were introduced into MH7A cells for transient expression using Lipofectamine 2000 (Invitrogen) following the manufacturer's instructions. Cells used for transfection were seeded at 1.7×10^5 cells/well in a 24-well plate (BD, Franklin Lakes, NJ, USA) and cultured for 1 d before transfection.

Western blot analysis of affibody expression

MH7A cells transfected with affibody-encoding plasmids (1×10^7) were plated in a 24-well plate. After 48 or 72 h of incubation, cells were lysed in Mammalian Lysis Buffer from the Qproteome Mammalian Protein Prep Kit (Qiagen, Hilden, Germany), and protein content was determined using the Bio-Rad Protein Assay Reagent (Bio-Rad, Hercules, CA, USA) with bovine serum albumin as the standard. Each sample was resolved by SDS-PAGE on 20% polyacrylamide gels and then transferred to 0.45-μm nitrocellulose membranes. After blocking overnight at 4°C with Blocking One Solution (Nacalai Tesque, Kyoto, Japan), membranes were incubated with goat polyclonal antibody specific for the affibody (Abcam, Cambridge, UK) as the primary antibody (1:400 dilution in phosphate-buffered saline; PBS) overnight at 4°C. After washing the membranes with Tris-buffered saline–0.05% Tween 20 (washing buffer), target proteins were detected with alkaline phosphatase (AP)-labeled ReserveAP Rabbit anti-goat immunoglobulin G (IgG) (KPL, Gaithersburg, MD, USA; 1:1,000 dilution in 1% PBS containing 0.05% Tween 20). NBT/BCIP (Roche, Mannheim, Germany) was used as the AP colorimetric substrate for visualization.

Measurement of IL-6 or PGE2 levels

MH7A cells (2×10^4) that were transfected with an affibody-encoding plasmid were plated in a 24-well plate and cultured in the presence or absence of 100 ng/mL TNF-α. After 24 or 48 h of incubation, IL-6 and PGE2 levels were determined by assaying supernatants with an Enzyme-Linked Immunosorbent Assay (ELISA) Kit for IL-6 (R&D Systems, Minneapolis, MN, USA) or PGE2 (ENZO Life Sciences, Farmingdale, NY, USA), respectively, according to the instructions.

Cell proliferation studies

Cell viability was determined using a Cell Counting Kit (CCK)-8 (Dojindo, Kumamoto, Japan), in which 2-(2-methoxy-4-nitrophenyl)-3-(4-nitrophenyl)-5-(2,4-disulfophenyl)-2H-tetrazolium monosodium salt was used as a substrate (Ohuchida et al. 2004). Absorbance was measured using a SpectraMax M2 Microplate Reader (Molecular Devices, Sunnyvale, CA, USA) at 450 nm (Shibasaki et al. 2012).

Western blot analysis for phosphorylated extracellular signal-regulated kinase (p-ERK)

Western blot analysis was performed as described above. Briefly, MH7A cells transfected with affibody-encoding plasmids (1×10^7) were plated in a 24-well plate, incubated for 24 h, and then stimulated with 100 ng/mL TNF-α (Sigma). After an additional incubation of 16 or 24 h, the soluble fraction of the cells was analyzed by western blot analysis using IgG against p-ERK (Santa Cruz Biotechnology, Dallas, TX, USA) as a primary antibody. Goat polyclonal IgG against actin (Santa Cruz) was used as a control.

Affibody protein production in *E. coli*

Expression of the pAffi-Z_{ras122}, $-Z_{ras220}$, $-Z_{ras521}$, and -Z_{raf322} plasmids in *E. coli* BL21(DE3) produced H-Ras and Raf-1 affibody molecules with an N-terminal 6 × His tag. The *E. coli* cells were inoculated in 2 mL of Luria-Bertani (LB) broth containing 100 μg/L ampicillin and were incubated in a shaking incubator overnight at 37°C. The 2-mL overnight cultures were used to inoculate fresh LB medium containing 100 μg/L ampicillin (125 mL), and cultures were grown at 37°C to an OD_{600} of 0.6–0.7. Gene expression was then induced by the addition of isopropyl β-D-thiogalactoside (Wako Pure Chemical, Osaka, Japan) to a final concentration of 1 mM. After 3–4 h of incubation at 37°C, the cells were harvested by centrifugation (4,000× *g*, 15 min, 4°C). The cell pellets were subsequently resuspended in 10 mL of B-PER Bacterial Protein Extraction Reagent (Thermo Fisher Scientific, Waltham, MA, USA) and shaken gently for 10 min at room temperature (25°C). Cell debris was pelleted by centrifugation at 27,000 × *g* for 20 min. Supernatants containing soluble proteins were purified by recovering the 6 × His-Z_{ras122}, $-Z_{ras220}$, $-Z_{ras521}$, and -Z_{raf322} fusion proteins after passage through nickel-chelated agarose columns (Thermo Fisher Scientific). The columns were equilibrated with 10 mL of B-PER Bacterial Protein Extraction Reagent before application of the supernatant. After washing the column with B-PER wash buffer, the bound proteins were released with elution buffer (50 mM Tris, 300 mM NaCl, 200 mM imidazole, 10% [v/v] glycerol), according to a previous protocol (Shibasaki et al. 2013). The concentration of each protein was determined by the BCA assay (REF), and the molecular size was confirmed by SDS-PAGE.

Affibody protein introduction into MH7A cells

MH7A cells were seeded at 1.7×10^5 cells/well in a 24-well plate and cultured for 1 d before the introduction of the purified proteins. Each affibody protein (5 μg) was mixed with Xfect Protein Buffer (Takara Bio, Shiga, Japan) to a total volume to 20 μL and then mixed with Xfect Protein Transfection Reagent. As a negative control, PBS was used in the place of the affibody molecule. After substitution of culture medium with serum-free medium for the cells, a diluted protein mixture was added to cultured MH7A cells and incubated at 37°C, according to a previous protocol (Kato et al. 2013).

Statistical analysis

Results are expressed as the mean ± SD. The significance of the difference between the experimental results and control values was determined by Student's *t*-test. *P* values less than 0.05 were considered significant.

Results

Expression of affibody molecules in MH7A cells

Each pcDNA3.1-based plasmid encoding Z_{ras122}, Z_{ras220}, or Z_{ras521} was designed to yield head-to-tail homodimers of the molecules. The plasmid encoding Z_{raf322} was designed to yield a monomeric molecule. Each plasmid was introduced into MH7A cells, and protein expression was confirmed by western blot analysis. The dimeric affibody molecules, Z_{ras122}, Z_{ras220}, and Z_{ras521}, appeared to be approximately 12 kDa, and the monomeric affibody molecule, Z_{raf322}, appeared to be approximately 6 kDa. The Z_{ras122} affibody protein was expressed in a greater quantity than the other affibody proteins at 48 and 72 h after transfection (Figure 1).

Inhibition of IL-6 and PGE2 production by plasmid-expressed affibody proteins

Each plasmid expressing the affibody molecules was introduced into MH7A cells to observe any inhibition of IL-6 production. At 24 h after transfection, TNF-α was added to induce synthesis of IL-6 and PGE2 (Figure 1A). After an additional 24–48 h of culture, the supernatant was collected, and IL-6 production was evaluated by ELISA. Z_{ras220} reduced IL-6 production by approximately 20% relative to the empty plasmid (pcDNA3.1) at 24 and 48 h after TNF-α stimulation; in contrast, expression of Z_{raf322} reduced IL-6 production by approximately 10% (Figure 2A). The inhibition of IL-6 production by each plasmid-expressed affibody was greater at 24 h than at 48 h after TNF-α stimulation (Figure 2A). Possible inhibition of PGE2 production in MH7A cells was also analyzed upon expression of the affibody molecules. Expression of Z_{raf322} reduced PGE2 production by approximately 10% relative to the empty plasmid at 24 h after TNF-α stimulation, while expression of Z_{raf122}. Z_{ras220}, and Z_{ras521} did not significantly affect PGE2 expression at 24 h (Figure 2B). Z_{ras122} and Z_{raf322} reduced PGE2 production by approximately 20% relative to the control plasmid at 48 h after TNF-α stimulation; in contrast to IL-6 production, the inhibition of PGE2 production by each plasmid-expressed affibody was greater at 48 h than at 24 h after TNF-α stimulation (Figure 2B).

Phosphorylation of ERK

ERK1/2 is situated downstream of Ras/Raf in the MAP kinase signal transduction cascade. To determine if affibody molecules affected signaling through this pathway, phosphorylation of ERK1/2 was evaluated in MH7A cells at 16 and 24 h after TNF-α stimulation. Western blot analysis of each cell lysate was performed, using an anti-p-ERK antibody with an anti-actin antibody serving as a control. The expression of Z_{ras220} inhibited phosphorylation of ERK1/2 at 16 h after TNF-α stimulation (Figure 3). This result indicated that Z_{ras220}, but not the other affibody proteins, was able to block ERK1/2 phosphorylation through binding to Ras.

Inhibition of IL-6 and PGE2 production by affibody molecules as proteins in MH7A cells

The affibody molecules were expressed in *Escherichia coli* and purified to >95% purity. SDS-PAGE analysis of the fractions from the last chromatographic step

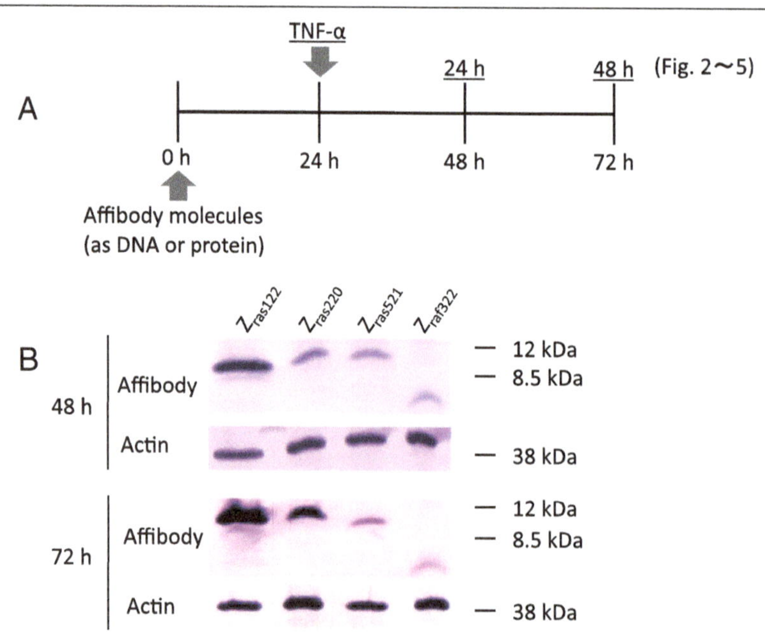

Figure 1 Experimental protocol (A) and western blot analysis for confirmation of cytosolic synthesis of affibody molecules (B). (A) Affibody molecules were introduced into MH7A cells encoded by DNA or as proteins. After 24-h culture, the cells were stimulated with TNF-α (100 ng/mL) and cultured for an additional 24–48 h. **(B)** Western blot analysis of each soluble fraction from MH7A cells harboring plasmid-encoded affibody protein molecules was analyzed at 48 and 72 h after the introduction of the plasmid into MH7A cells. Actin was analyzed as a control.

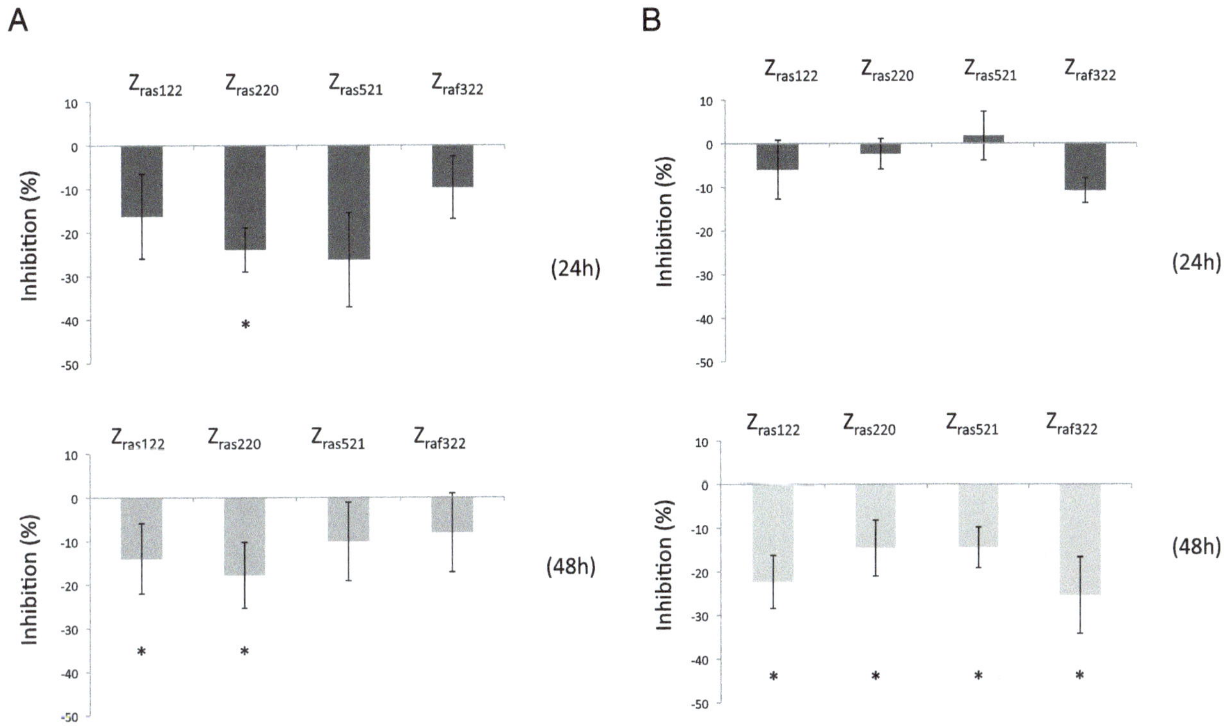

Figure 2 IL-6 (A) and PGE2 (B) production by MH7A cells with plasmid-expressed affibodies. MH7A cells were transfected with plasmids encoding affibodies. After 24 h of culture, the cells were stimulated with TNF-α (100 ng/mL) and cultured for an additional 24–48 h. Data represent the percent inhibition of IL-6 or PGE2 production by MH7A cells. Percent inhibition was calculated as follows: ([(IL-6 or PGE2 levels in MH7A cells transfected with affibody DNA) − (IL-6 or PGE2 levels in MH7A cells transfected with pcDNA3.1)]/[IL-6 or PGE2 levels in MH7A cells transfected with pcDNA3.1]) × 100. Data represent the means ± SD of three independent experiments. *$p < 0.05$.

revealed essentially pure proteins of the expected molecular weight (data not shown). Individually, Z_{ras122}, Z_{ras220}, Z_{ras521}, and Z_{raf322} proteins were introduced into MH7A cells by incubation with the cell-penetrating peptide reagent Xfect. Although the inhibition of IL-6 production by Z_{ras122} was stronger than the inhibition by the other proteins at 24 h, Z_{ras220} protein strongly inhibited IL-6 production by approximately 46% relative to the control penetrating peptide alone at 48 h after TNF-α stimulation (Figure 4A). The Z_{ras220} protein also strongly inhibited PGE2 production by 32% relative to the control penetrating peptide alone at 48 h after TNF-α stimulation (Figure 4B).

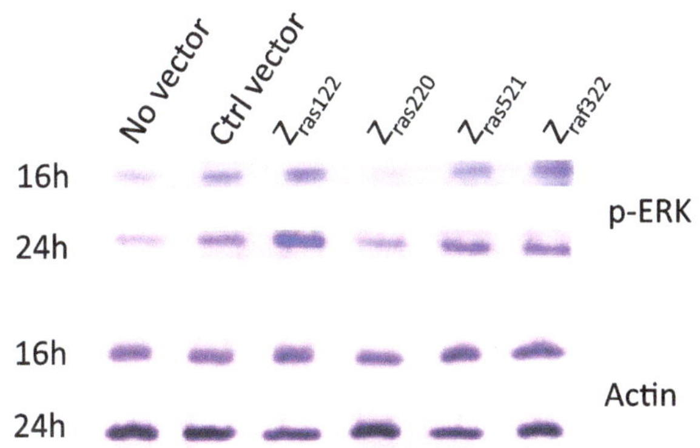

Figure 3 Western blot analysis for the confirmation of ERK1/2 phosphorylation. MH7A cells were transfected with plasmids encoding affibodies. After 24 h of culture, the cells were stimulated with TNF-α (100 ng/mL) and cultured for an additional 24–48 h. Each soluble fraction of MH7A cells harboring plasmid-encoded affibody molecules was analyzed via western blot for the expression of p-ERK and actin at 16 and 24 h after TNF-α stimulation. Representative data are shown for p-ERK and actin expression by MH7A cells.

Effect of affibody molecules on cell proliferation

The proliferation of MH7A cells was also investigated upon expression of the affibody molecules by plasmid transfection. No change in proliferation could be observed at 24 or 48 h after TNF-α stimulation for any of the constructs, compared to the cells harboring the empty plasmid (Figure 5A). We next investigated a possible effect on MH7A cell proliferation after introduction of affibody proteins. MH7A cells transfected with a control peptide proliferated 151% from 0–24 h, while those transfected with Z_{raf322} proteins proliferated almost 0% from 0–24 h (Figure 5B). Furthermore, during 0–48 h, control cells and Z_{raf322} proteins-transfected cells proliferated 327% and 291%, respectively. These results indicate an inhibitory effect of Z_{raf322} protein introduction on proliferation.

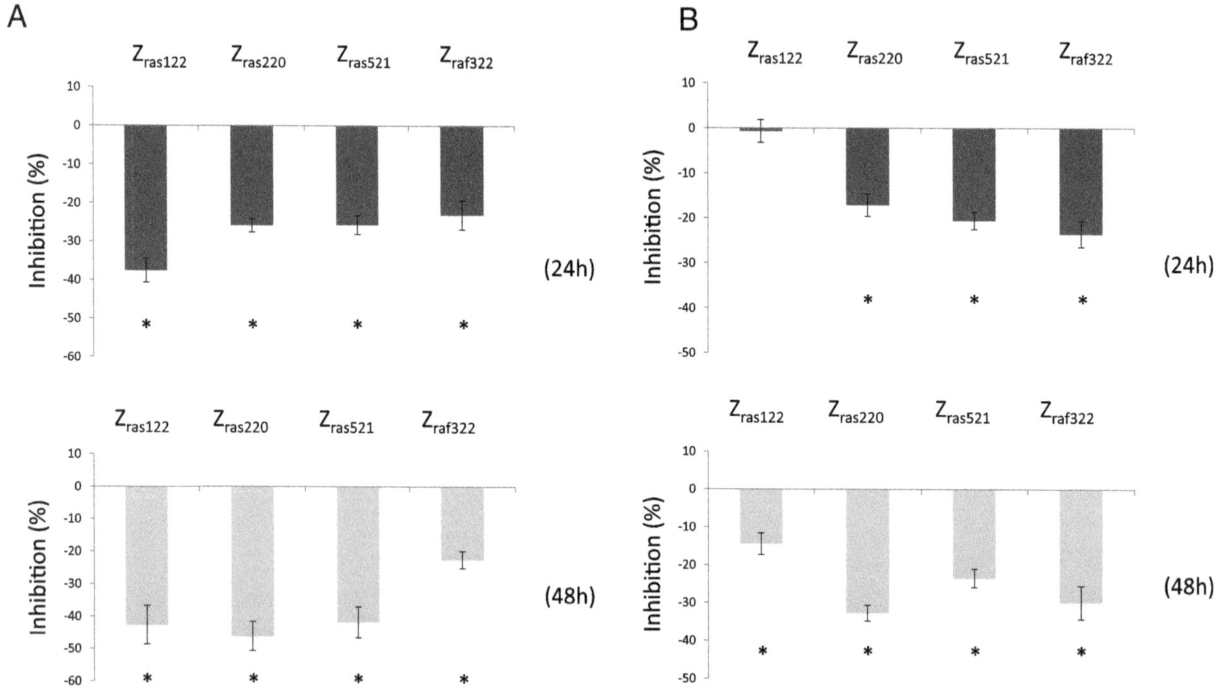

Figure 4 IL-6 (A) and PGE2 (B) production by affibody protein-introduced MH7A cells. MH7A cells were transfected with affibody proteins. After 24 h of culture, the cells were stimulated with TNF-α (100 ng/mL) and cultured for an additional 24–48 h. IL-6 and PGE2 levels were analyzed, and percent inhibition of IL-6 or PGE2 production was determined. Percent inhibition was calculated as follows: ([(IL-6 or PGE2 levels in affibody protein-introduced MH7A cells) – (IL-6 or PGE2 levels in MH7A cells introduced with standard peptides)]/([IL-6 or PGE2 levels in MH7A cells introduced with standard peptides]) × 100. Data represent the means ± SD of three independent experiments. *$p < 0.05$.

Discussion

In this study, we investigated whether affibody molecules targeting H-Ras or Raf-1 can inhibit cell proliferation and the production of inflammatory mediators *in vitro*. To date, it has been demonstrated that intracellularly expressed affibody molecules can bind to target molecules and inhibit their functions. For example, an affibody molecule that has an affinity for epidermal growth factor receptor (EGFR) retains EGFR in the secretory compartment. This constraint led to decreased phosphorylation of EGFR and a reduced proliferation rate (Vernet et al. 2009). An affibody molecule targeting HER-3 inhibits proliferation of a cancer cell line (Göstring et al. 2012). However, we demonstrated for the first time that an affibody targeting intracellular signal transduction molecules could inhibit cell proliferation and the production of inflammatory mediators. We demonstrated that H-Ras- (Z_{ras122}, Z_{ras220}, Z_{ras521}) or Raf-1- (Z_{raf322}) targeting affibody molecules inhibited the production of IL-6 and PGE2 in a synovial cell line. These affibody molecules were introduced into the cells via DNA-encoding plasmids or as purified proteins. In both cases, inhibition of IL-6 and PGE2 production was observed for almost all affibody molecules.

Several previous reports indicated suppression of synovial cell function by Ras inhibition. For instance, gene transfer of a dominant negative mutant of the *Ras* gene (Yamamoto et al. 2003) or gene silencing using a locked nucleic acid against Ras (de Launay et al. 2010) in synovial cells inhibited the production of inflammatory mediators, such as IL-6. Consistent with these reports, we observed that affibody molecules binding to H-Ras or Raf-1 could inhibit the production of IL-6. PGE2 production by RA synovial cells is regulated by Ras (Na et al. 2004). Therefore, we analyzed PGE2 levels and demonstrated that these affibody molecules also inhibited PGE2 production. Matrix metalloproteinase (MMP)-3 is a protease that is produced by synovial cells upon TNF-α stimulation, and its production is regulated by Ras (Abreu et al. 2009). We analyzed MMP-3 levels and demonstrated that these affibody molecules also inhibited MMP-3 production (data not shown). These results indicate that affibody molecules binding to H-Ras or Raf-1 can inhibit inflammatory mediators or proteases, such as IL-6, PGE2, and MMP-3, which are crucial to RA pathogenesis.

Regarding the expression of affibody molecules in MH7A cells, we observed stronger expression of Z_{ras122} than other affibody proteins. In addition, Z_{ras122} protein levels increased, while Z_{raf322} and Z_{ras521} protein levels decreased between 24 and 48 h (Figure 1). These results might be related to the stability of the Z_{ras122} affibody protein afforded by the plasmid construct. Supporting this notion, we

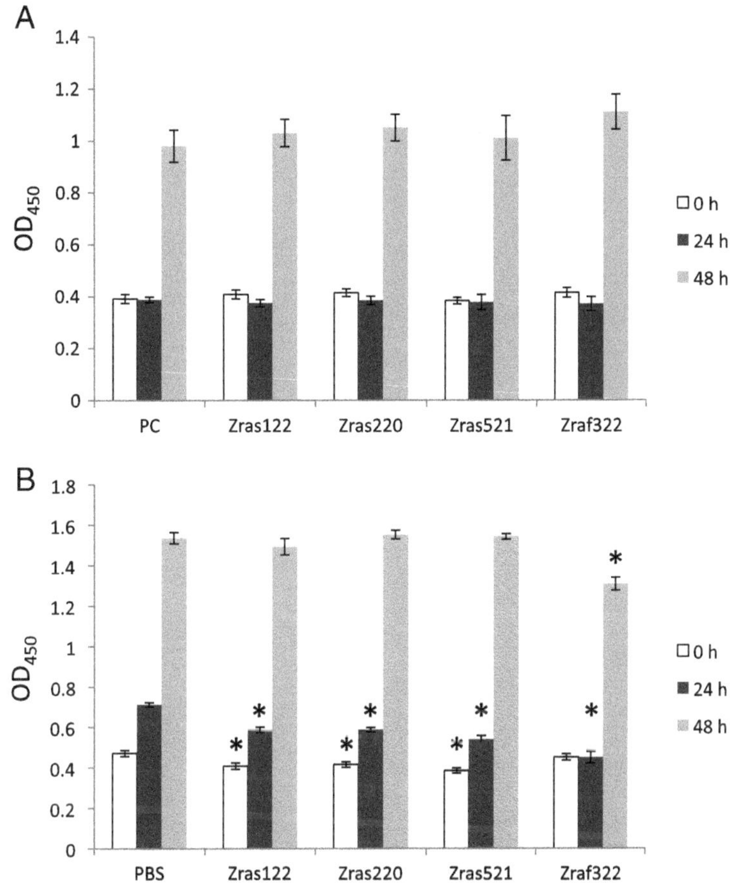

Figure 5 Cell proliferation assay for MH7A cells transfected with plasmid-expressed affibody molecules (A) and MH7A cells introduced affibody molecules as protein (B). After 24 h of culture, the cells were stimulated with TNF-α (100 ng/mL) and cultured for an additional 24–48 h. Cell proliferation was analyzed with a CCK assay. Data represent the means ± SD of three independent experiments. *p <0.05.

observed that Z_{ras220} and Z_{ras521} inhibited IL-6 production by 23.9% and 26.2% relative to the control vector treatment at 24 h, while these inhibitory effects were decreased to 17.8% and 10.1% at 48 h after TNF-α stimulation, respectively. In contrast, inhibition by Z_{ras122} was maintained up to 48 h after TNF-α stimulation (Figure 2A), suggesting increased stability of the affibody. With regard to PGE2 production, inhibition by all affibodies at 48 h increased compared with that at 24 h after TNF-α stimulation. Of interest, the timing of maximal inhibition was different for IL-6 and PGE2, but the underlying mechanism is still unknown. The difference could be attributed to the difference in the signal transduction cascade of these inflammatory molecules.

To determine whether affibody molecules inhibit signal transduction via ERK, we examined the phosphorylation of ERK1/2, a downstream molecule of Ras/Raf signaling. Z_{ras220} inhibited ERK1/2 phosphorylation in the early stage (16 h) after stimulation by TNF-α (Figure 3), although we did not observe distinct inhibition by other affibodies. This transient inhibition of ERK1/2 phosphorylation might be caused relatively by Z_{ras220} via interruption of the interaction between Raf-1 and MEK1/2 (Chen et al. 2001). However, this inhibition on ERK1/2 phosphorylation could not be kept after 16–24 h, so that differences of phosphorylation among all affibody molecules here should be investigate more precisely in further study.

We utilized two different methods to deliver affibody molecules into synovial cells: 1) transfection of plasmids encoding the affibody molecules and 2) introduction of affibody proteins using a cell-penetrating peptide reagent. Protein-introduced affibody molecules exhibited stronger inhibition of IL-6 and PGE2 production than plasmid-expressed affibody molecules (Figures 2 and 4). Moreover, the inhibition resulting from the introduced proteins was less variable than that resulting from plasmid transfection. When cell proliferation was examined using plasmid-encoded affibody molecules, distinctive differences were not observed (Figure 5A). On the contrary, introduction of Z_{raf322} proteins partially inhibited cell proliferation (Figure 5B), suggesting that a plasmid-based expression system did not generate a sufficient concentration of affibody in MH7A cells to inhibit signal transduction. In the experiments using plasmids, cell proliferation was not observed at 24 (Figure 5A), while in the experiments using proteins, cell proliferation was observed at the same period (Figure 5B). This might be based on higher cytotoxicity of liposome compared with peptides used as a carrier (Suhorutsenko et al. 2011). These results indicate that protein-introduced affibody molecules more effectively inhibited cytokine production and cell proliferation in synovial cells than plasmid-expressed affibody molecules. Although the effect on synovial cell proliferation was not strong, our results demonstrate that it is possible to inhibit synovial cell proliferation by selecting useful affibody clones and by improving transduction efficiency of affibody protein molecules.

Inhibition of IL-6 and PGE2 production caused by the protein-introduced affibodies varied. For example, at 48 h after TNF-α stimulation, inhibition of IL-6 production was greater than 40% for three affibodies (Z_{ras122}, Z_{ras220}, and Z_{ras521}), whereas that of PGE2 was approximately 30% for two affibodies (Z_{ras220} and Z_{raf322}). These results could be attributed to the fact that the PGE2 receptor, which couples with G proteins, is integrated into a positive feedback loop for PGE2 production (Nishimura et al. 2013); thus, the effect of an affibody molecule on PGE2 production might be weaker than on IL-6. The differential effectiveness of each affibody molecule on synovial cell function might also be explained by differences in the amino acid sequences of the affibodies themselves. Each clone was selected using a phage display library of Z variants, and the amino acid sequences encoded by each clone are different.

In conclusion, we demonstrated that affibody molecules could inhibit cell proliferation and the production of inflammatory mediators by blocking the Ras/Raf signal transduction cascade. Although inhibition of cell function could be enhanced, and delivery methods should be examined further, we propose that these molecules are candidates to be investigated as a molecular-targeting therapy for RA. Studies are in progress to select useful affibody clones, to improve delivery methods of affibody protein molecules, and to verify the efficiency of affibodies in an RA model.

Additional file

Additional file 1: DNA sequence of affibody molecules in this study.

Competing interests
The authors declare that they have no competing interest.

Authors' contributions
SS, TG and PN carried out the genetic construction of affibody molecules. MK carried out Western blot analysis and ELISA. SS and MK carried out cell proliferation assays and productions of affibody molecules using *E. coli*. HS and TI participated in the design, coordination of the study. SS, TG, PN and TI contributed in the drafting of the manuscript. All authors read and approved the final manuscript.

Acknowledgements
This work was partially supported by the Ministry of Education, Culture, Sports, Science and Technology, Japan, with a Grant-in-Aid for Young Scientists (B).

Author details
[1]General Education Center, Hyogo University of Health Sciences, 1-3-6 Minatojima, Chuo-ku, Kobe 650-8530, Japan. [2]Graduate School of Pharmacy, Hyogo University of Health Sciences, 1-3-6 Minatojima, Chuo-ku, Kobe 650-8530, Japan. [3]Division of Rheumatology, Department of Internal Medicine, Hyogo College of Medicine, 1-1Mukogawa-cho, Nishinomiya 663-8501, Japan. [4]Department of Molecular Biotechnology, School of Biotechnology, Royal Institute of Technology (KTH), AlbaNova University Center, SE-106 91 Stockholm, Sweden. [5]Division of Pharmacotherapy, Department of Pharmacy, School of Pharmacy, Hyogo University of Health Sciences, 1-3-6 Minatojima, Chuo-ku, Kobe 650-8530, Japan.

References
Abreu JR, de Launay D, Sanders ME, Grabiec AM, van de Sande MG, Tak PP, Reedquist KA (2009) The Ras guanine nucleotide exchange factor RasGRF1 promotes matrix metalloproteinase-3 production in rheumatoid arthritis synovial tissue. Arthritis Res Ther 11:R121
Baumann MJ, Stadler BM, Vogel M (2007) Potential applications of designed ankyrin repeat proteins in diagnostics and therapeutics. Exp Opin Med Diagn 1:409–421
Chen J, Fujii K, Zhang L, Roberts T, Fu H (2001) Raf-1 promotes cell survival by antagonizing apoptosis signal-regulating kinase 1 through a MEK-ERK independent mechanism. Proc Natl Acad Sci U S A 98:7783–7788
de Launay D, Vreijling J, Hartkamp LM, Karpus ON, Abreu JR, van Maanen MA, Sanders ME, Grabiec AM, Hamann J, Ørum H, Vervoordeldonk MJ, Fluiter K, Tak PP, Reedquist KA (2010) Silencing the expression of Ras family GTPase homologues decreases inflammation and joint destruction in experimental arthritis. Am J Pathol 177:3010–3024
Friedman M, Ståhl S (2009) Engineered affinity proteins for tumour-targeting applications. Biotechnol Appl Biochem 53:1–29
Furneri G, Mantovani LG, Belisari A, Mosca M, Cristiani M, Bellelli S, Cortesi PA, Turchetti G (2012) Systematic literature review on economic implications and pharmacoeconomic issues of rheumatoid arthritis. Clin Exp Rheumatol 30:S72–S84
Gómez-Reino J (2012) Biologic monotherapy as initial treatment in patients with early rheumatoid arthritis. Rheumatology 51(Suppl 5):v31–v37
Göstring L, Malm M, Höidén-Guthenberg I, Frejd FY, Ståhl S, Löfblom J, Gedda L (2012) Cellular effects of HER3-specific affibody molecules. PLoS One 7:e40023
Grimm S, Lundberg E, Yu F, Shibasaki S, Vernet E, Skogs M, Nygren PÅ, Gräslund T (2010) Selection and characterisation of affibody molecules inhibiting the interaction between Ras and Raf in vitro. N Biotechnol 27:766–773
Gunneriusson E, Samuelson P, Uhlen M, Nygren PA, Ståhl S (1996) Surface display of a functional single-chain Fv antibody on staphylococci. J Bacteriol 178:1341–1346
Kato T, Tanaka M, Oba M (2013) Protein transfection study using multicellular tumor spheroids of human hepatoma Huh-7 cells. PLoS One 8:e82876
Kitano M, Hla T, Sekiguchi M, Kawahito Y, Yoshimura R, Miyazawa K, Iwasaki T, Sano H, Saba JD, Tam YY (2006) Sphingosine 1-phosphate/sphingosine 1-phosphate receptor 1 signaling in rheumatoid synovium: regulation of synovial proliferation and inflammatory gene expression. Arthritis Rheum 54:742–753
Koide A, Wojcik J, Gilbreth RN, Hoey RJ, Koide S (2012) Teaching an old scaffold new tricks: monobodies constructed using alternative surfaces of the FN3 scaffold. J Mol Biol 415:393–405

Lindberg H, Hofström C, Altai M, Honorvar H, Wållberg H, Orlova A, Ståhl S, Gräslund T, Tolmachev V (2012) Evaluation of a HER2-targeting affibody molecule combining an N-terminal HEHEHE-tag with a GGGC chelator for 99mTc-labelling at the C terminus. Tumour Biol 33:641–651

Lundberg E, Brismar H, Gräslund T (2009) Selection and characterization of Affibody ligands to the transcription factor c-Jun. Biotechnol Appl Biochem 52:17–27

Malm M, Kronqvist N, Lindberg H, Gudmundsdotter L, Bass T, Frejd FY, Höidén-Guthenberg I, Varasteh Z, Orlova A, Tolmachev V, Ståhl S, Löfblom J (2013) Inhibiting HER3-mediated tumor cell growth with affibody molecules engineered to low picomolar affinity by position-directed error-prone PCR-like diversification. PLoS One 8:e62791

Mélet J, Mulleman D, Goupille P, Ribourtout B, Watier H, Thibault G (2013) Rituximab-induced T cell depletion in patients with rheumatoid arthritis: association with clinical response. Arthritis Rheum 65:2783–2790

Miyazawa K, Mori A, Okudaira H (1998) Establishment and characterization of a novel human rheumatoid fibroblast-like synoviocyte line, MH7A, immortalized with SV40 T antigen. J Biochem 124:1153–1162

Na HJ, Lee SJ, Kang YC, Cho YL, Nam WD, Kim PK, Ha KS, Chung HT, Lee H, Kwon YG, Koh JS, Kim YM (2004) Inhibition of farnesyltransferase prevents collagen-induced arthritis by down-regulation of inflammatory gene expression through suppression of p21(ras)-dependent NF-kappaB activation. J Immunol 173:1276–1283

Nilsson FY, Tolmachev V (2007) Affibody molecules: new protein domains for molecular imaging and targeted tumor therapy. Curr Opin Drug Discov Devel 10:167–175

Nishimura T, Zhao X, Gan H, Koyasu S, Remold HG (2013) The prostaglandin E2 receptor EP4 is integral to a positive feedback loop for prostaglandin E2 production in human macrophages infected with *Mycobacterium tuberculosis*. FASEB J 27:3827–3836

Nygren PA (2008) Alternative binding proteins: affibody binding proteins developed from a small three-helix bundle scaffold. FEBS J 275:2668–2676

Ohuchida T, Okamoto K, Akahane K, Higure A, Todoroki H, Abe Y, Kikuchi M, Ikematsu S, Muramatsu T, Itoh H (2004) Midkine protects hepatocellular carcinoma cells against TRAIL-mediated apoptosis through down-regulation of caspase-3 activity. Cancer 100:2430–2436

Olsen NJ, Stein CM (2004) New drugs for rheumatoid arthritis. N Engl J Med 350:2167–2179

Saag KG, Teng GG, Patkar NM, Anuntiyo J, Finney C, Curtis JR, Paulus HE, Mudano A, Pisu M, Elkins-Melton M, Outman R, Allison JJ, Suarez Almazor M, Bridges SL, Jr, Chatham WW, Hochberg M, MacLean C, Mikuls T, Moreland LW, O'Dell J, Turkiewicz AM, Furst DE, American College of Rheumatology (2008) American College of Rheumatology 2008 recommendations for the use of nonbiologic and biologic disease-modifying antirheumatic drugs in rheumatoid arthritis. Arthritis Rheum 59:762–784

Schlegel PM, Steiert I, Kötter I, Müller CA (2013) B cells contribute to heterogeneity of IL-17 producing cells in rheumatoid arthritis and healthy controls. PLoS One 8:e82580

Shibasaki S, Fujita A, Usui C, Watanabe S, Kitano S, Sano H, Iwasaki T (2012) Effect of transient expression of fluorescent protein probes in synovial and myoblast cell lines. Springer Plus 1:36

Shibasaki S, Aoki W, Nomura T, Miyoshi A, Tafuku S, Sewaki T, Ueda M (2013) An oral vaccine against candidiasis generated by a yeast molecular display system. Pathog Dis 69:262–268

Smolen JS, Aletaha D (2011) Interleukin-6 receptor inhibition with tocilizumab and attainment of disease remission in rheumatoid arthritis: the role of acute-phase reactants. Arthritis Rheum 63:43–52

Smolen JS, Steiner G (2003) Therapeutic strategies for rheumatoid arthritis. Nat Rev Drug Discov 2:473–488

Suhorutsenko J, Oskolkov N, Arukuusk P, Kurrikoff K, Eriste E, Copolovici DM, Langel U (2011) PepFects, show no evidence of toxicity and immunogenicity in vitro and in vivo. Bioconjug Chem 22:2255–2262

Tsunemi S, Iwasaki T, Kitano S, Imado T, Miyazawa K, Sano H (2010) Effects of the novel immunosuppressant FTY720 in a murine rheumatoid arthritis model. Clin Immunol 136:197–204

Upchurch KS, Kay J (2012) Evolution of treatment for rheumatoid arthritis. Rheumatology 51(Suppl 6):vi28–vi36

Vernet E, Lundberg E, Friedman M, Rigamonti N, Klausing S, Nygren PA, Gräslund T (2009) Affibody-mediated retention of the epidermal growth factor receptor in the secretory compartments leads to inhibition of phosphorylation in the kinase domain. N Biotechnol 25:417–423

Yamamoto A, Fukuda A, Seto H, Miyazaki T, Kadono Y, Sawada Y, Nakamura I, Katagiri H, Asano T, Tanaka Y, Oda H, Nakamura K, Tanaka S (2003) Suppression of arthritic bone destruction by adenovirus-mediated dominant-negative Ras gene transfer to synoviocytes and osteoclasts. Arthritis Rheum 48:2682–2692

23

Identification of a gene, *FMP21*, whose expression levels are involved in thermotolerance in *Saccharomyces cerevisiae*

Toshihide Nakamura[1*†], Mami Yamamoto[1†], Katsuichi Saito[1], Akira Ando[2] and Jun Shima[3]

Abstract

Elucidation of the mechanism of high temperature tolerance in yeasts is important for the molecular breeding of high temperature-tolerant yeasts that can be used in bioethanol production. We identified genes whose expression is correlated with the degree of thermotolerance in *Saccharomyces cerevisiae* by DNA microarray analysis. Gene expression profiles of three *S. cerevisiae* strains showing different levels of thermotolerance were compared, and we chose three of them as candidate genes. Among these genes, *FMP21* was investigated as a thermotolerance-related gene in *S. cerevisiae* by comparing the growth at high temperature with the gene expression in eight strains. The expression ratio of *FMP21* at 37°C was correlated with the doubling time ratio at a coefficient of determination of 0.787. The potential involvement of the Fmp21 in the thermotolerance of yeasts was evaluated. The *FMP21* deletion variant showed a decreased respiratory growth rate and increased thermosensitivity. Furthermore, the overexpression of *FMP21* improved thermotolerance in yeasts. In conclusion, the function of Fmp21 is important for thermotolerance in yeasts.

Keywords: Thermotolerance; Yeast; Gene expression; Growth rate; *FMP21*

Introduction

The yeast *Saccharomyces cerevisiae* has been used in the production of traditional fermented foods and beverages as well as in the industrial production of ethanol. In the process of ethanol production, the temperature of the fermentor is raised by the heat of the fermentation, and the elevated temperature results in decreases in the cell growth, viability, and ethanol productivity of the yeast. Thermotolerant yeast is thus required for efficient ethanol production. The growth properties of *S. cerevisiae* at high temperatures differ according to the strain used, but the optimum temperature of proliferation is around 30°C. Some high-temperature-resistant strains have been isolated from tropical areas and obtained by mutational improvement, breeding, or the cell fusion of existing strains (Pasha et al. 2007; Marullo et al. 2009; Shi et al. 2009). These thermotolerant yeasts were selected based on the cell growth and ethanol production at high temperature.

* Correspondence: nakamrto@affrc.go.jp
†Equal contributors
[1]National Food Research Institute, National Agriculture and Food Research Organization (NARO), 2-1-12 Kannondai, Tsukuba, Ibaraki 305-8642, Japan
Full list of author information is available at the end of the article

Analyses of genes whose expression is induced by high temperature are important for understanding the mechanisms of thermotolerance, and for the use of the genes in breeding thermotolerant yeasts. Genes related to the thermotolerance of *S. cerevisiae* have been identified by DNA microarray analyses (Eisen et al. 1998; Sakaki et al. 2003) and a genome-wide screening of an *S. cerevisiae* deletion mutant collection (Auesukaree et al. 2009). Many genes were found to be important for the tolerance to heat stress (Auesukaree et al. 2009).

Genomic regions related to the high-temperature growth of *S. cerevisiae* were identified by quantitative trait locus (QTL) mapping (Steinmetz et al. 2000; Sinha et al. 2006, 2008). The QTL map contained *MKT1*, *END3*, and *RHO2* as high-temperature growth quantitative trait genes. Other factors such as heat shock proteins and trehalose also function in yeast as protectants contributing to survival under heat stress conditions (Watson 1990; Singer and Lindquist 1998). However, the mechanisms of thermotolerance in yeasts are still unclear.

In the present study, we attempted to identify genes whose expression is correlated with the degree of thermotolerance in *S. cerevisiae*. We compared the gene expression profiles of three *S. cerevisiae* strains that showed different levels of thermotolerance. We investigated *FMP21* as a thermotolerance-related gene in *S. cerevisiae* by comparing the growth at high temperature with the gene expression in eight strains.

Materials and methods

Yeast strains and growth conditions

The yeasts used in this study (*Saccharomyces cerevisiae* NFRI 3122, NFRI 3145, NFRI 3155, NFRI 3205, NFRI 3225, NFRI 3236, and NFRI 3314) were obtained from the Microbiological Bank at our institute (NFRI). *Saccharomyces cerevisiae* S288C was used as a control strain. A wild-type strain, *S. cerevisiae* BY4743 and the deletion strain *fmp21Δ* derived from BY4743 were obtained from EUROSCARF (EUROpean Saccharomyces Cerevisiae ARchive for Functional analysis). The strains were grown in YPD medium (1% yeast extract [Difco Laboratory, Detroit, MI, USA], 2% peptone [Difco Laboratory], and 2% glucose) with shaking at 140 rpm.

Determination of growth rates

The yeast strains were inoculated into YPD medium and incubated at 30°C for 24 h (pre-culture). Cell growth was monitored by measuring the optical density of the cultures at 600 nm (OD_{600}) using a spectrophotometer (Ultrospec 2100 pro; GE Healthcare, Freiburg, Germany). The cells were transferred to YPD medium at a starting OD_{600} of 0.1 and incubated at 30°C with shaking. The experiments were performed in triplicate. Growth rates, expressed as the time required for the cultures to double in optical density (doubling time, T_d), were calculated by comparing the optical density at multiple time points during the linear growth phase and using the following formula:

$$T_d = (t_2 - t_1)/[\log_2(OD_2/OD_1)].$$

Heat treatment for total RNA extraction

To confirm the heat tolerance of the strains, the cells were cultured in YPD medium under high-temperature conditions. Yeast cells were inoculated at an initial cell density of 0.1 at OD_{600} and grown to 1.0 OD_{600}. Aliquots (10 mL) of exponentially growing cells were transferred to preheated conical flasks of replicates of each strain. The flasks were placed in a 37°C or 39°C water bath for 30 min.

DNA microarray analysis

The experimental procedures used for the RNA extraction and DNA microarray analysis have been described previously (Nakamura et al. 2008). The total RNA was quantified using a Bioanalyzer 2100 with an RNA 6000 Nano LabChip kit (Agilent Technologies, Palo Alto, CA, USA). Polyadenylated RNA (mRNA) from total RNA was purified using Oligotex-dT30 < Super > (Takara, Kyoto, Japan) according to the manufacturer's instructions. We performed complementary DNA synthesis and labeling and array hybridization by following the Affymetrix user's manual with a one-cycle target labeling and control reagent kit (Affymetrix, Santa Clara, CA, USA) with 0.4 μg of mRNA as the template material. Labeled cDNA was hybridized to Affymetrix GeneChip Yeast 2.0 arrays (Affymetrix) for 16 h at 45°C with constant rotation at 60 rpm. Washing and staining were performed using a hybridization, wash, and stain kit (Affymetrix) on a GeneChip FS-450 fluidics station (Affymetrix). Fluorescence was detected using the Affymetrix 3000 GeneArray Scanner, and the image analysis of each GeneChip was performed using the GeneChip Operating System 1.4.0 (GCOS) software from Affymetrix with the standard default settings. Probe level data based on the gene expression data from two independent experiments were imported into GeneSpring GX 11.0.2 (Agilent Technologies). Following 75th percentile shift normalization, low-intensity probes subject to high degrees of noise were removed by filtering the lowest 20% of each sample and then by filtering with flags (present and marginal). For the identification of differentially expressed genes, the Filter on Confidence function was used with a *t*-test *P*-value cutoff of 0.05, and a fold change threshold of 1.0 for the gene list was generated using the Filter on Volcano Plot tool. All experiments were done in duplicate with independently grown cells.

Microarray data from the present study have been deposited in the Gene Expression Omnibus (GEO) repository at the National Center for Biotechnology Information (NCBI) under series accession no. GSE33276.

PCR primers and real-time RT-PCR

The oligonucleotide primer pairs listed in Table 1 were designed using Primer3 (http://primer3.sourceforge.net/). Real-time reverse transcription polymerase chain reaction (RT-PCR) was carried out using the LightCycler 2.0 instrument in glass capillaries (Roche, Indianapolis, IN, USA). Two μg of total RNA was reverse-transcribed into cDNA in a 20-μL reaction mixture using a PrimeScript Reverse Transcriptase kit (Takara). The reaction mix consisted of 5 μL standard/diluted cDNA template (1/10 to 1/1000), 0.5 μL 10× primer for the gene [final 0.5 μM primer, and 2.5–4 mM $MgCl_2$], 2 μL water, and 2 μL FAST-START DNA Master SYBR Green I mix (Roche). The reaction conditions were as follows: initial denaturation at 95°C for 10 min, followed by 38 cycles of denaturation at 95°C for 10 s, annealing at 60°C for 10 s, and extension at 72°C for 10 s. The progress of real-time fluorescent PCR was monitored at 530 nm.

The PCR products were verified with both a melting curve analysis and DNA gel electrophoresis. To establish

Table 1 List of primers used in this study

Primer	*Nucleotide sequence (5′-3′)*	*Target ORF*
Real-time PCR experiments		
YER034W-F	CACCCAGTGATAGAAAGTACCG	*YER034W*
YER034W-R	TCCTCATAGTACCTGTCCTCG	*YER034W*
prm5-F	GCAATTTTTGCGGGTTTCCTTA	*PRM5 (YIL117C)*
prm5-R	TCGTCGCTCGATAACGGTA	*PRM5 (YIL117C)*
YBR269C-F	TCGCAAGAGGCAATAGATCAG	*FMP21 (YBR269C)*
YBR269C-R	TCGAATTCAGGTATGGTCTTGG	*FMP21 (YBR269C)*
Act1-F	AGCCTTCTACGTTTCCATCC	*ACT1*
Act1-R	CTTTCAGCAGTGGTGGAGAA	*ACT1*
Construction of the *FMP21*-overexpressing strain		
URA3-F	CAGGGTCCATAAAGCTTT	
URA3-R	TTTATAAAGGCCATGAAGCT	
TDH3 pro-F	GGAAAGAAAAAGCTTCATGGCCTTTATAAAAACACGCTTTTTCAGTTC	
TDH3 pro-R	TTTGTTTGTTTATGTGTGTT	
FMP21-A	AACGCACTCAAGGTTTTTGG	
FMP21-B	AAGATGAATTGAAAAGCTTTATGGACCCTGTAGTTCTTCTTTTTATTATT	
FMP21-C	GTTTCGAATAAACACACATAAACAAACAAAATGTTGTGCGCCATCAAAAG	
FMP21-D	GTATGAATGTGCCCAGTGTA	

the external standard curves for the quantification of each gene, the purified PCR products were diluted in a 10-fold series (1/10 to 1/10000) and analyzed with the new assay. The expression data and associated technical errors on duplicates were calculated by using LightCycler Software 4.0 (Roche). The expression levels for each sample were normalized for the expression of *ACT1.*

The gene expression rate is based on the expression levels of the target gene versus the reference gene (*ACT1*). We confirmed that the *ACT1* expression was stable under high temperature conditions by comparing it with the expression of *ALG9,* which is considered a suitable reference gene for real-time RT-PCR (Teste et al. 2009). The gene expression ratio is given as the ratio of the target gene expression rate after heat treatment to that in non-treated controls.

Assay for thermotolerance

Yeast strains were precultured in YPD medium until the log phase. Then, the preculture was inoculated into 5 ml of YPD medium and shaken at 70 rpm under aerobic conditions at 30°C or 39°C. The OD_{660} was automatically recorded with an Advantec TVS062CA biophotorecorder (Advantec Toyo Co. Ltd., Tokyo, Japan).

Construction of *FMP21*-overexpressing strains

We constructed an *FMP21*-overexpressing strain based on a previously reported method (Hasegawa et al. 2012) using the primer set listed in Table 1. Briefly, the promoter region of the glyceraldehyde-3-phosphate dehydrogenase gene (*TDH3*), which allows constitutive expression at a high level, was fused with the *URA3* marker gene and then inserted upstream of the start codon of the *FMP21* gene in BY4743. The resultant *FMP21*-overexpressing strains (*MATa/α ura3Δ0 URA3-*P_{TDH3}*-FMP21*) and control strains (*MATa/α ura3Δ0 URA3*) were analyzed for *FMP21* expression levels and thermotolerance. The expression levels of *FMP21* were determined by real-time RT-PCR as described above.

Results

Screening of thermotolerance-related genes by DNA microarray analysis

We used three yeast strains: a thermotolerant strain, NFRI 3236; a thermosensitive strain, NFRI 3155; and a standard strain, S288C (Figure 1). The doubling times (T_d) of NFRI 3236, NFRI 3155 and S288C at 39°C were shown to be 1.75, 2.90 and 2.11 h, respectively. On the other hand, the T_d values of NFRI 3236, NFRI 3155 and S288C at 30°C, a standard growth condition for the strains, showed no significant differences and were 1.15, 1.28 and 1.34 h, respectively.

We performed DNA microarray experiments in order to compare gene expression responses to the heat treatment at 39°C with the gene expressions in non-treated control cells. We first selected genes (441 genes) that showed a higher than 1.5-fold change in their expression in both NFRI 3236 and S288C, and then selected genes that showed a > 1.5 ratio of fold-change values of NFRI 3236 to S288C or of S288C to NFRI 3155. We then

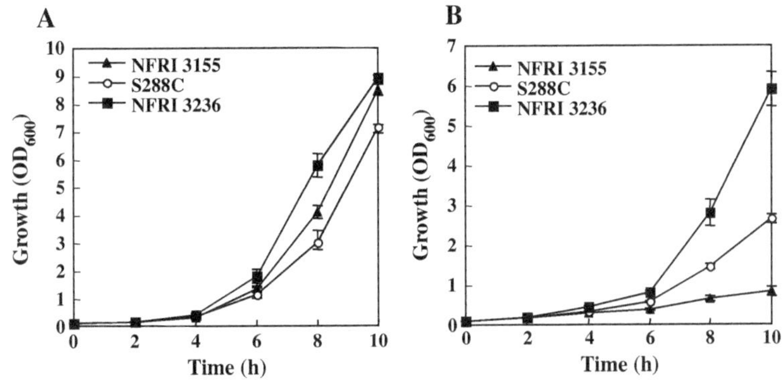

Figure 1 Effect of temperature on the growth of three yeast strains (NFRI 3236, NFRI 3155 and S288C). Each strain was grown in YPD with shaking at 30°C **(A)**, or 39°C **(B)**.

selected three genes whose expression ratios were correlated with the level of thermotolerance in the three strains (Table 2). The fold-change values of *YER034W*, *YIL117C* (*PRM5*) and *YBR269C* (*FMP21*) were more clearly correlated with the rank order of thermotolerance. We thus chose *YER034W*, *PRM5* and *FMP21* as candidate thermotolerance-related genes.

Expression analysis of the three candidates by real-time RT-PCR

We verified the expression of *YER034W*, *PRM5* and *FMP21* in NFRI 3236, NFRI 3155 and S288C under heat stress conditions. The expression ratio (i.e., the expression level at high temperature/the expression level at 30°C) of each candidate gene was measured by real-time RT-PCR (Figure 2). The *ACT1* gene was used as the expression standard. The expression ratio of *FMP21* was 3.47 in NFRI 3155, 4.50 in S288C and 8.50 in NFRI 3236, and it increased (statistically significant at $P < 0.05$) according to the thermotolerance of the yeast strains. However, the expression ratios of *YER034W* and *PRM5* showed no relationship with the thermotolerance of the yeast strains.

Relationship between the expression of *FMP21* and the thermotolerance of yeast strains

To assess the relationship between the expression of the *FMP21* gene and the thermotolerance of yeast strains, we measured the expression levels at 30°C, 37°C, and 39°C in five strains (NFRI 3122, NFRI 3145, NFRI 3205, NFRI 3225, and NFRI 3314) in addition to the three strains described above (eight strains total). Figure 3 is a graph plotting the expression ratios of *FMP21* against the T_d ratio. With the heat treatment of 37°C, the increase of the expression ratio of *FMP21* by heat treatment was correlated with the T_d ratio at a coefficient of determination of 0.787. In contrast, there was weak correlation at 39°C (coefficient of determination, 0.418).

Disruption and overexpression of the *FMP21* gene

To determine whether Fmp21 plays a role in the response to heat stress, we examined the growth ability of the *FMP21* deletion strain (*fmp21Δ*) in YPD medium at 39°C (Figure 4). The growth ability of *fmp21Δ* in YPD medium at 30°C was nearly the same as that of the wild-type strain; however, there was a reduction in the growth rate of *fmp21Δ* in the late growth phase. Under the high temperature condition, *fmp21Δ* showed a slower growth rate than the wild-type strain in the log-phase. This result revealed that the deletion of *FMP21* increased the sensitivity to high temperature.

To evaluate the phenotype of *FMP21* overexpression, we constructed *FMP21*-overexpressing strains and compared their thermotolerance with that of the control strains. *FMP21* was overexpressed using promoter sequences of the highly expressed *TDH3* gene. Then we

Table 2 Selected genes and their expression values in DNA microarray analysis

Gene	Fold change			Descriptions of gene products
	NFRI3236	S288C	NFRI3155	
FMP21	3.68	2.18	1.35	Putative protein of unknown function; the authentic, non-tagged protein is detected in highly purified mitochondria in high-throughput studies
YER034W	3.10	1.51	0.49	Putative protein of unknown function; non-essential gene; expression induced upon calcium shortage
PRM5	48.21	23.05	15.07	Pheromone-regulated protein, predicted to have 1 transmembrane segment; induced during cell integrity signaling

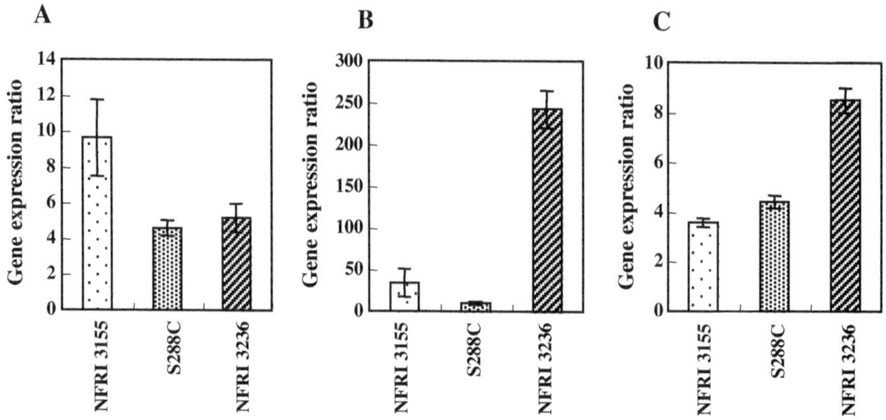

Figure 2 Comparison of the gene expression ratios of the selected genes. (A) *YER034W*, **(B)** *PRM5*, and **(C)** *FMP21*. Bars indicate the mean ratios for two independent biological replicates. Error bars = standard deviation.

confirmed the mRNA expression of *FMP21* by real-time RT-PCR (Figure 5). In the *FMP21*-overexpressing strains (BY4743/PTDH3-FMP21), the transcriptional expression level of *FMP21* was approximately 6-fold higher under non-stress conditions and was approximately 3-fold higher at 39°C (the differences are statistically significant at $P < 0.01$). Overexpression of *FMP21* had no effect on the growth rate at 30°C (Figure 6). The growth rate of *FMP21*-overexpressing strains was higher than that of the control strains when the cells were incubated at 39°C (Figure 6). These results suggested that yeast strains with enhanced *FMP21* expression exhibited the improved thermotolerance.

Discussion

We found a correlative increase of *FMP21* expression along with the increase in growth rate at high temperature in yeast strains. The results of our DNA microarray analysis using three strains showed increased expression ratios under heat conditions for *FMP21*, *YER034W* and *PRM5*. However, in the verification test, the expression rate of *YER034W* showed the lowest correlation among the three genes. In addition, the expression rate of *PRM5* was significantly increased by heat treatment only in the NFRI3236 strain. The expression levels in some of the genes were very different among the yeast strains, which had different genetic backgrounds. *PRM5* is also one of the genes whose expression strongly depends on the genetic backgrounds of the strains.

The expression ratio of *FMP21* induced by heat treatment was significantly correlated with the growth ratio at 37°C. However, there was a weak correlation at 39°C. Generally, *S. cerevisiae* strains show various levels of thermotolerance depending on their genetic backgrounds. The maximum temperature for growth is under about 40°C. The strains used in this study grew unstable at 39°C, and there is therefore a high possibility that the gene expressions at 39°C would also be unstable. Therefore, we

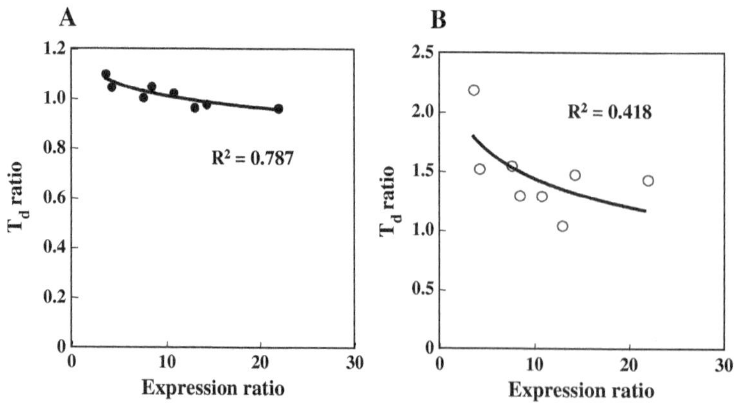

Figure 3 Relationship between the expression ratio of *FMP21* and the doubling time (T_d) ratio at 37°C (30°C/37°C) (A) and at 39°C (30°C/39°C) (B). The coefficients of determination (R^2) were 0.787 **(A)** and 0.418 **(B)**.

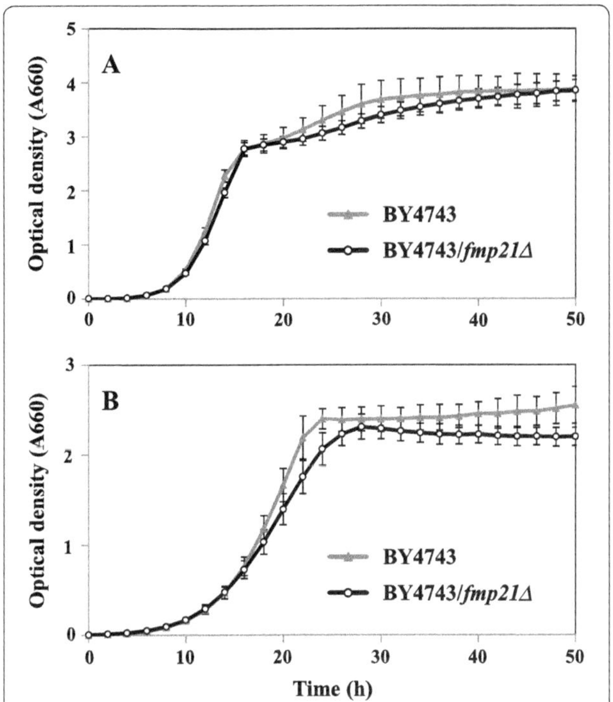

Figure 4 Growth of the wild-type (BY4743) and *fmp21*Δ strains under high temperature. Both strains were grown in YPD with shaking at 30°C **(A)** or 39°C **(B)**. Three experiments were repeated to ensure reproducibility and gave equivalent results. Each plot represents the average of triplicates.

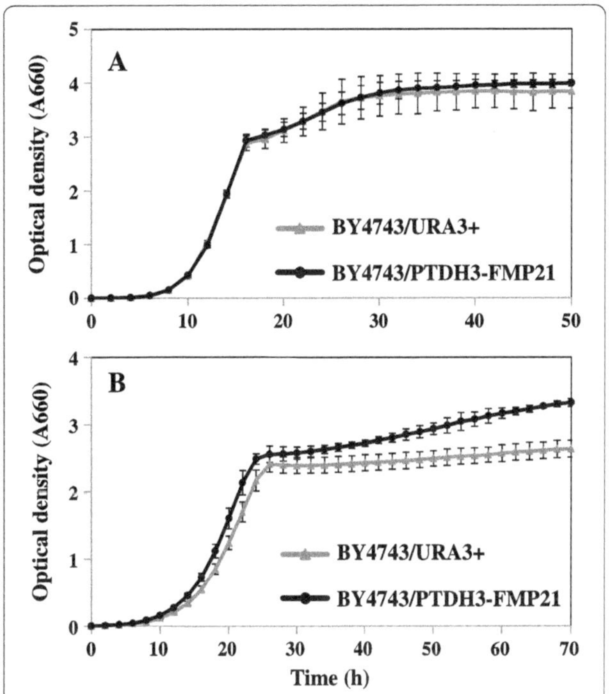

Figure 6 Growth of the *FMP21*-overexpressing strains under high temperature. Both the *FMP21*-overexpressing strains and the control strains were grown in YPD with shaking at 30°C **(A)** or 39°C **(B)**. The values are the means of results from three independent strains.

speculated that the unstable gene expressions at 39°C may have been responsible for the relatively low value of the coefficient of determination.

In yeasts, Hsf1 and Msn2/4 are major transcription factors to respond to heat stress. Hsf1 and Msn2/4 bind to heat shock elements (HSEs) and stress response elements (STREs), respectively, in the promoter region of

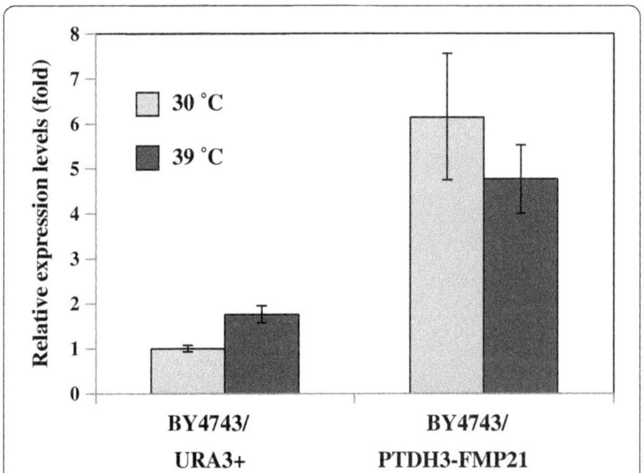

Figure 5 Transcriptional analysis of *FMP21* in the *FMP21*-over expressing strains. The expression data of the *FMP21*-overexpressing strains were compared with those of the control strains. The values are the means of results from three independent strains.

heat shock proteins and activate the expression of target gene transcription (Morano et al. 2012). Although the expression of *FMP21* was induced by heat stress, we did not find the typical consensus sequences of HSEs and STREs in the promoter region of *FMP21*. The *FMP21* expression may be controlled by unknown mechanisms. In contrast to the expression pattern of *FMP21*, Sakaki et al. (2003) reported that the expressions of mitochondrial genes were repressed in yeast cells grown at 37°C. We are interested in the functions of Fmp21—specifically whether this protein is involved in the repression of mitochondrial genes in 37°C cultivation.

FMP21 encodes a protein localized in mitochondria, but its function is still unknown (Reinders et al. 2006). Steinmetz et al. (2002) reported that the *FMP21* deletion variant showed a decreased rate of respiratory growth. We also confirmed that the *fmp21Δ* strains showed a decreased growth rate in medium containing glycerol as a sole carbon source (data not shown). The *fmp21Δ* strains showed a decreased growth rate after diauxic shift, suggesting that one of the functions of Fmp21 is to contribute to the respiratory metabolism in mitochondria of yeasts. After diauxic shift, yeasts are stressed by the lack of nutrients and by the accumulation of toxic metabolites such as ethanol. Outten et al. (2005) reported that the *FMP21* deletion variant was hyperoxia-sensitive. The absence of Fmp21 may render the mitochondrial function more stress-sensitive.

The *FMP21* deletion variant was a thermosensitive strain. As growth delay was observed in the late log phase, *fmp21Δ* strains might be sensitive to ethanol under high temperature conditions. However, the sensitivity to 5% (v/v) ethanol of *fmp21Δ* strains was similar to that of the wild-type strain at 30°C (data not shown). Thus it is unlikely that a function of Fmp21 is involved in ethanol tolerance. We therefore consider that the function of Fmp21 is important for the thermotolerance of yeasts.

The increased accumulation of thermoprotectants in yeasts occasionally leads to increased thermotolerance. The overexpression of heat shock proteins and the enzyme for trehalose synthesis have been shown to successfully enhance the thermotolerance of yeasts (Cheng et al. 1992; An et al. 2011). In this study, the *FMP21* gene was overexpressed using the promoter region of the *TDH3* gene. Although the expression of *TDH3* was decreased under high temperature (based on the DNA microarray analysis data), the expression of *FMP21* in the *FMP21*-overexpressing strains was higher than that in the control strains. The overexpression of the *FMP21* gene enhanced the thermotolerance of yeasts in aerobic cultures. As the overexpression of the *FMP21* gene also enhanced the growth in static cultures under high temperature conditions, there was no significant difference in ethanol productivity (data not shown). These results suggest that Fmp21 contributes to thermotolerance of yeasts, and that it may protect mitochondria against high temperature-related damages.

Thermotolerant yeast strains are still required in bioindustries. Various approaches have been adopted in an attempt to improve the thermotolerance of yeast strains, including the use of mating, fusion or genetic hybridization (Pasha et al. 2007; Marullo et al. 2009; Shi et al. 2009). However, these approaches were based on stochasticity and used various yeast strains with different genetic backgrounds. It is thus difficult to draw conclusions from these studies that would help in the acquisition of more thermotolerant strains. The results presented here are applicable to various strains that have different genetic backgrounds. It may also be possible to use these results to quantify and evaluate thermotolerance, and for the systematic construction of more thermotolerant strains.

Competing interests

The authors declare that they have no competing interests.

Acknowledgements

This work was partly supported by a grant from the Ministry of Agriculture, Forestry and Fisheries of Japan (Rural Biomass Research Project, BEC-BC050, BEC-BC051), and Advanced Low Carbon Technology Research and Development Program of Japan (ALCA) of Japan Science and Technology Agency (JST).

Author details

[1]National Food Research Institute, National Agriculture and Food Research Organization (NARO), 2-1-12 Kannondai, Tsukuba, Ibaraki 305-8642, Japan. [2]NARO Institute of Vegetable and Tea Science, 360 Kusawa, Ano, Tsu, Mie 514-2392, Japan. [3]Ryukoku University, Fukakusa Tsukamotohon-cho, Fushimi-ku, Kyoto, Kyoto 612-8577, Japan.

References

An MZ, Tang YQ, Mitsumasu K, Liu ZS, Shigeru M, Kenji K (2011) Enhanced thermotolerance for ethanol fermentation of *Saccharomyces cerevisiae* strain by overexpression of the gene coding for trehalose-6-phosphate synthase. Biotechnol Lett 33:1367–1374, doi:10.1007/s10529-011-0576-x

Auesukaree C, Damnernsawad A, Kruatrachue M, Pokethitiyook P, Boonchird C, Kaneko Y, Harashima S (2009) Genome-wide identification of genes involved in tolerance to various environmental stresses in *Saccharomyces cerevisiae*. J Appl Genet 50:301–310, doi:10.1007/BF03195688

Cheng L, Hirst K, Piper PW (1992) Authentic temperature-regulation of a heat shock gene inserted into yeast on a high copy number vector. Influences of overexpression of HSP90 protein on high temperature growth and thermotolerance. Biochim Biophys Acta 1132:26–34, doi:10.1016/0167-4781(92)90048-5

Eisen MB, Spellman PT, Brown PO, Botstein D (1998) Cluster analysis and display of genome-wide expression patterns. Proc Natl Acad Sci U S A 95:14863–14868, doi:10.1073/pnas.95.25.14863

Hasegawa S, Ogata T, Tanaka K, Ando A, Takagi H, Shima J (2012) Overexpression of vacuolar H^+-ATPase-related genes in bottom-fermenting yeast enhances ethanol tolerance and fermentation rates during high-gravity fermentation. J Inst Brew 118:179–185, doi:10.1002/jib.32

Marullo P, Mansour C, Dufour M, Albertin W, Sicard D, Bely M, Dubourdieu D (2009) Genetic improvement of thermo-tolerance in wine *Saccharomyces cerevisiae* strains by a backcross approach. FEMS Yeast Res 9:1148–1160, doi:10.1111/j.1567-1364.2009.00550.x

Morano KA, Grant CM, Moye-Rowley WS (2012) The response to heat shock and oxidative stress in *Saccharomyces cerevisiae*. Genetics 190:1157–1195, doi:10.1534/genetics.111.128033

Nakamura T, Mizukami-Murata S, Ando A, Murata Y, Takagi H, Shima J (2008) Changes in gene expression of commercial baker's yeast during an air-drying process that simulates dried yeast production. J Biosci Bioeng 106:405–408, doi:10.1263/jbb.106.405

Outten CE, Falk RL, Culotta VC (2005) Cellular factors required for protection from hyperoxia toxicity in *Saccharomyces cerevisiae*. Biochem J 388:93–101, doi:10.1042/BJ20041914

Pasha C, Kuhad RC, Rao LV (2007) Strain improvement of thermotolerant *Saccharomyces cerevisiae* VS strain for better utilization of lignocellulosic substrates. J Appl Microbiol 103:1480–1489, doi:10.1111/j.1365-2672.2007.03375.x

Reinders J, Zahedi RP, Pfanner N, Meisinger C, Sickmann A (2006) Toward the complete yeast mitochondrial proteome: multidimensional separation techniques for mitochondrial proteomics. J Proteome Res 5:1543–1554, doi:10.1021/pr050477f

Sakaki K, Tashiro K, Kuhara S, Mihara K (2003) Response of genes associated with mitochondrial function to mild heat stress in yeast *Saccharomyces cerevisiae*. J Biochem 134:373–384, doi:10.1093/jb/mvg155

Shi DJ, Wang CL, Wang KM (2009) Genome shuffling to improve thermotolerance, ethanol tolerance and ethanol productivity of *Saccharomyces cerevisiae*. J Ind Microbiol Biotechnol 36:139–147, doi:10.1007/s10295-008-0481-z

Singer MA, Lindquist S (1998) Thermotolerance in *Saccharomyces cerevisiae*: the Yin and Yang of trehalose. Trends Biotechnol 16:460–468, doi:10.1016/S0167-7799(98)01251-7

Sinha H, Nicholson BP, Steinmetz LM, McCusker JH (2006) Complex genetic interactions in a quantitative trait locus. PLoS Genet 2:e13, doi:10.1371/journal.pgen.0020013

Sinha H, David L, Pascon RC, Clauder-Münster S, Krishnakumar S, Nguyen M, Shi G, Dean J, Davis RW, Oefner PJ, McCusker JH, Steinmetz LM (2008) Sequential elimination of major-effect contributors identifies additional quantitative trait loci conditioning high-temperature growth in yeast. Genetics 180:1661–1670, doi:10.1534/genetics.108.092932

Steinmetz LM, Sinha H, Richards DR, Spiegelman JI, Oefner PJ, McCusker JH, Davis RW (2000) Dissecting the architecture of a quantitative trait locus in yeast. Nature 416:326–330, doi:10.1038/416326a

Steinmetz LM, Scharfe C, Deutschbauer AM, Mokranjac D, Herman ZS, Jones T, Chu AM, Giaever G, Prokisch H, Oefner PJ, Davis RW (2002) Systematic screen for human disease genes in yeast. Nat Genet 31:400–404, doi:10.1038/ng929

Teste MA, Duquenne M, François JM, Parrou JL (2009) Validation of reference genes for quantitative expression analysis by real-time RT-PCR in *Saccharomyces cerevisiae*. BMC Mol Biol 10:99, doi:10.1186/1471-2199-10-99

Watson K (1990) Microbial stress proteins. Adv Microb Physiol 31:183–223

24

Degradative actions of microbial xylanolytic activities on hemicelluloses from rhizome of *Arundo donax*

Licia Lama[1], Annabella Tramice[1], Ilaria Finore[1], Gianluca Anzelmo[1], Valeria Calandrelli[1], Eduardo Pagnotta[1], Giuseppina Tommonaro[1], Annarita Poli[1], Paola Di Donato[1,2], Barbara Nicolaus[1], Massimo Fagnano[3], Mauro Mori[3], Adriana Impagliazzo[3] and Antonio Trincone[1*]

Abstract

Polysaccharidases from extremophiles are remarkable for specific action, resistance to different reaction conditions and other biotechnologically interesting features. In this article the action of crude extracts of thermophilic microorganisms (*Thermotoga neapolitana*, *Geobacillus thermantarcticus* and *Thermoanaerobacterium thermostercoris*) is studied using as substrate hemicellulose from one of the most interesting biomass crops, the giant reed (*Arundo donax* L.). This biomass can be cultivated without competition and a huge amount of rhizomes remains in the soil at the end of cropping cycle (10–15 years) representing a further source of useful molecules. Optimization of the procedure for preparation of the hemicellulose fraction from rhizomes of *Arundo donax*, is studied. Polysaccharidases from crude extracts of thermophilic microorganisms revealed to be suitable for total degradative action and/or production of small useful oligosaccharides from hemicelluloses from *A. donax*. Xylobiose and interesting tetra- and pentasaccharide are obtained by enzymatic action in different conditions. Convenient amount of raw material was processed per mg of crude enzymes. Raw hemicelluloses and pretreated material show antioxidant activity unlike isolated tetra- and pentasaccharide. The body of results suggest that rhizomes represent a useful raw material for the production of valuable industrial products, thus allowing to increase the economic efficiency of *A. donax* cultivation.

Keywords: Biorefinery; Xylan; Thermophilic enzymes; Xylanases; Xylooligosaccharides; *Arundo donax*

Introduction

Among interesting thermostable enzymes isolated from microorganisms, polysaccharidases are remarkable for specific action, resistance to different reaction conditions and other biotechnologically interesting features. Microorganisms of the genus *Thermoanaerobacterium* and others of thermophilic or hyperthermophilic origin, are relevant for their abilities to bioconvert carbohydrates. In the framework of a research project focused on bioprospecting for thermophilic microorganisms, a new species, *T. thermostercus* (designated as Buff), was recently isolated from buffalo-dung and classified as a new member of *Thermoanaerobacterium*. The optimum growth temperature, pH and NaCl concentration were, respectively, 60°C, 6.5 and 0.5% (w/v). As carbon sources this strain utilized glucose, mannose, xylose, maltose, cellobiose, sucrose and xylan (Romano et al. 2010). *Geobacillus thermantarcticus*, a thermophilic bacterium isolated from Antarctic geothermal soil near the crater of Mount Melbourne, produced extracellular xylanase (1,4-β-D-xylan xylanohydrolase; E.C. 3.2.1.8) and β-xylosidase (1,4-β-D-xylan xylohydrolase; E.C. 3.2.1.37); the studied action of these two enzymes on commercial xylan gave only xylose (Lama et al. 2004). Glycosyl hydrolases in *Thermotoga neapolitana* resulted interesting in the enzymatic process for the cleavage of xylan to smaller oligosaccharides and also in their coupling to aglycones by transglycosylating activity achieved under certain reaction conditions (Tramice et al. 2007).

In biomass refinery, lignocellulosic material is fractionated to obtain the main components (cellulose, hemicelluloses, and lignin). Xylan is the major hemicellulosic component.

* Correspondence: antonio.trincone@icb.cnr.it
[1]Institute of Biomolecular Chemistry ICB-CNR, Consiglio Nazionale delle Ricerche, Via Campi Flegrei 34, 80072 Pozzuoli, Napoli, Italy
Full list of author information is available at the end of the article

Polysaccharidic chain is composed by a backbone of β-1,4-xylose units in pyranose form, decorated on different position with 4-O-methyl-D-glucuronic acid and L-arabinofuranose units. This skeleton can also be extensively esterified with acetic or ferulic acids; the latter in some species, provides a chemical link between hemicellulose and lignin. D-xylose and L-arabinose as end products are derived by biorefinery enzymatic processes from xylan of lignocellulosic biomasses. However, up until recently biomass-to-fuel investigation has neglected heteroxylan from lignocellulosic. This has been attributed to a number of reasons pointing out that (i) a sub-optimal settings for material extraction or (ii) the slow development of enzymatic processes and (iii) the scarce exploitation of derived products, characterize the field of pentose-specific conversion/utilization technologies. Specific xylanase actions on xylan could lead to important oligosaccharides (Dumon et al. 2012). Interesting bioactivities have been reported for both poly- and oligosaccharidic fragments. Immunostimulating effects and anti-inflammatory and antimicrobial activities have been reported for arabino(glucurono)xylans isolated from different sources (Christakopoulosa et al. 2003). A nutraceutical perspective has been also considered for these compounds (Saeed et al. 2011). A gastroprotective arabinoxylan from sugarcane bagasse was isolated (Mellinger-Silva et al. 2011). A structure-activity relationship study on the role of 4-O-methyl glucuronic acid (MeGlcUA) in regulating aggregation of β-polyxylosides of (9H-fluoren-9-yl)-methanol obtained via the transglycosylating action of *Thermotoga neapolitana* β-xylanase, was also studied. An anti-proliferative test of these compounds on human epithelial EFO 27 ovarian cancer cells indicated that the presence of MeGlcUA modulates biological activity, while its absence induced molecular aggregation (Tramice et al. 2009).

With the increasing demand of ligno-cellulosic materials in processes such as co-generation, production of biofuel and compost from municipal solid wastes, etc., a competition for land used for food production is observed with potential risks for food security in wide areas of the world (van der Horst and Vermeylen 2011). Among the different biomass crops for ligno-cellulosic materials, one of the most interesting is giant reed (*Arundo donax* L.), because it is tolerant to a wide range of environmental stresses so that it can be cultivated on marginal, degraded or contaminated lands thus reducing competition with food crops which generally require a better quality arable land (Di Nasso et al. 2013; Fiorentino et al. 2010; Diodato et al. 2009). *Arundo donax* is suggested in projects (i.e. Ecoremed 2011. LIFE11/ENV/IT/275), (Fagnano 2014) of phytoremediation of polluted soils (Bonanno et al. 2013). The most studied part of *Arundo donax* is represented by shoots (reeds) of interest in bioethanol, biodiesel (Pirozzi et al. 2010) or biopolymer production (Galletti et al. 2013). However at the end of cropping cycle of a giant reed stand (10–15 years), a huge amount of rhizomes remains in the soils and they have to be eliminated to afford further agricultural utilizations. These rhizomes could represent a further source of useful molecules for bioenergy or biochemicals (Di Nasso et al. 2013).

In the present article an overview of degradative actions of different enzymatic extract on hemicelluloses from rhizomes of *Arundo donax* is presented, comparing the processes acted by mesophilic and thermophilic biocatalysts. Convenient set up of a degradative processes with/without production of small useful oligosaccharides are reported.

Methods

General

Total carbohydrate content was determined by Bernfeld (1955) method using a xylose-based calibration curve. Xylose was determined by D-xylose assay kit (Megazyme), specifically designed to measure this sugar in samples containing up to 5 mg of glucose in the final assay reaction without any interference. Glucose was assayed enzymatically by GOPOD-FORMAT, Megazyme. Protein concentration was determined by the method of Bradford (1976) using bovine serum albumin as the standard. *Thermomyces lanuginosus* xylanase, (X2753 2500 U/g) was obtained from Sigma. *Thermotoga maritima* endo-1,4-β-xylanase, 22 U/mg of protein (wheat arabinoxylan at pH 5.0 and 40°C) was obtained from Megazyme.

^{1}H NMR spectra were recorded at 600.13 MHz on a Bruker DRX-600 spectrometer, equipped with a TCI Cryo Probe TM, fitted with a gradient along the Z-axis. Spectra in D_2O were referenced to internal sodium 3-(trimethyl-silyl)-(2,2,3,3-2H4) propionate (Aldrich, Milwaukee, WI). ^{13}C NMR, JMOD-^{1}H and COSY, TOCSY, HSQC, HSQC-EDITED, HMBC (^{3}J: 7 and 10 Hz), NOESY (mixing time at 100, 200, 300 msec) experiments were used for structural determination.

Mass spectra were acquired on a microQ-Tof mass spectrometer coupled with an Alliance HPLC (Waters, Milford, MA) equipped with an ESI source.

Microorganism and culture conditions

Geobacillus thermantarcticus (DSM 9572) isolated from Antarctic geothermal soil was grown in a flask (2 L) at 60°C on a medium containing xylan at pH 6.0 as previously reported (Nicolaus et al. 1996). *Thermoanaerobacterium thermostercoris* strain BUFF was grown at 60°C in a 40 L fermentor (B. Brown Biotech International Micro DCU400) as previously reported (Romano et al. 2010). *Thermotoga neapolitana* DSM 4359 obtained from DSMZ (Deutsche Sammlung von Mikroorganismen und

Zellkulturen) was grown at 80°C in a 40-L fermentor (B. Brown Biotech International Micro DCU400) as previously reported (Tramice et al. 2007).

Microbial growth was monitored turbidimetrically at 540 nm and cells were harvested in late exponential growth phase and collected by continuous centrifugation using an Alpha-Laval centrifuge and then pelleted by centrifugation at 10,000 rpm for 30 min and frozen.

Preparation of crude enzymatic extracts

Thermoanaeorbacterium thermostercoris (CFE-TT protein 11.4 mg/ml), and *Thermotoga neapolitana* cell free extracts (CFE-TN protein 34.4 mg/ml) were prepared in acetate buffer 50 mM pH 5.5 disrupting the cells by usual techniques (freeze-thawing, lysozyme 100 μg/ml). *Geobacillus thermantarcticus* extracellular suspension (ES-GTM1, protein 0.37 mg/ml) was prepared by adding ammonium sulfate to the cell-free broth (1 L) to 80% saturation. The precipitate was recovered by centrifugation (10,000 g, 1 h, 4°C), dissolved in 50 mM sodium acetate buffer (pH 5.6) and dialyzed overnight against the same buffer.

Biomass

Arundo donax was grown under Mediterranean climatic conditions: during the growth period (April-November 2012) rainfall was 355 mm and Reference Evapotranspiration was 923 mm and thus water deficit was very severe (568 mm). Rhizomes of *A. donax* were harvested at the experimental farm of the University "Federico II" Naples and were obtained from a 4 year experimental plantation located in the Sele river plain of Campania Region (Southern Italy, 40°61'N, 14°92'E, 30 m a.s.l.). Rhizome samples were harvested in a sample area of 2 m^2, and then chipped, dried at 60°C until constant weight and grinded at 4 mm. Successively, the material was grinded in order to obtain a homogenous fine powder (particle size of about 1 mm) by means of a kitchen mixer (Waring).

Preparation and optimization of hemicellulosic substrates from rhizome of *Arundo donax*

The hemicellulose fraction was solubilized in alkali medium by continuous magnetic stirring at room temperature; different 2% w/v solutions of rhizome of *Arundo donax* in potassium hydroxide (KOH, 0.5 N, 1 N and 2 N) were used at different time intervals, 24 h, 48 h, 72 h and 96 h. Experiments for the optimization of extraction procedure were conducted in triplicate; briefly, the grinded dry material was suspended in KOH solutions; afterwards the suspension was sieved (0.5 mm screen) and centrifuged at 10,000 rpm for 40 min at 4°C. The supernatant was treated with same volume of cold ethanol 96% (v/v) added drop wise under stirring. The alcoholic suspension was stored at −20°C overnight and then centrifuged at 10,000 rpm for 1 h at 4°C. The pellet was solubilized in hot distilled water, cooled at room temperature, the pH was adjusted to 7.0 with HCl and dialyzed against distilled water using Spectrapore dialysis tubes (12,000-14,000 MW cut-off), and finally freeze-dried. The lyophilized material (RK72h, RK48h, RK24h) was weighed using a Mettler Toledo analytical balance and the yield was expressed in weight (%, w/w) with respect to the initial dry biomass.

Enzymatic reactions

Hydrolysis reactions were conducted using various biocatalysts and hemicellulosic material prepared from *A. donax* (RK72h) as follows: *Thermomyces lanuginosus* xylanase, 22 units were dissolved in 3 ml phosphate buffer 50 mM, pH 7 containing 30 mg substrate. Reaction was performed at 34°C under stirring for 26 h and stopped by enzyme denaturation at 100°C (2 min). *Thermotoga maritima* endo-1,4-β-xylanase, 57 units were dissolved in 1 ml in acetate buffer 50 mM, pH 5.6 containing 10 mg of substrate. Reaction was performed under stirring at 70°C, 24 h and stopped by cooling. Cell free extract (CFE-TT) from *Thermoanaeorbacterium thermostercoris* (proteins: 11.4 mg/ml), was added to the needed amount of substrate (4 mg/ml of reaction) in acetate buffer 50 mM, pH 5.6. Reaction was performed at 50°C for 24 h under stirring. Cell free extract (CFE-TN) from *Thermotoga neapolitana* (proteins: 34.4 mg/ml), was added to the needed amount of substrate (4 mg/ml of reaction) in acetate buffer 50 mM, pH 5.6. Reaction was performed at 90°C for 24 h under stirring. Extracellular suspension of *Geobacillus thermantarcticus* 8 ml, ES-GTM1b, (proteins: 0.37 mg/ml) was reacted with 40 mg of substrate solubilized in 2 ml phosphate buffer 50 mM, pH 6, at 70°C for 24 h. In a different condition, ES-GTM1a, 1.5 ml were reacted with 20 mg of substrate solubilized in 4 ml phosphate buffer 50 mM, pH 6, or at 70°C for 24 h.

Transglycosylation reactions were performed using various biocatalysts as follows: 20 mg of *A. donax* (RK72h) in 1 ml of appropriate buffer/acetonitrile (87:13 v/v) containing 19.6 mg of 9-fluorene methanol; the reactions started by adding the different biocatalysts (57 U *Thermotoga maritima* endo-1,4-β-xylanase, at 70°C, 24 h; 500 μL *Thermotoga neapolitana* cell free extract (CFE-TN) at 70°C, 24 h; 500 μL cell free extract (CFE-TT) from *Thermoanaeorbacterium thermostercoris* at 50°C, 24 h; 4 ml of extracellular suspension *Geobacillus thermantarcticus* (ES-GTM1) at 70°C, 24 h) under stirring.

Monitoring of reactions, purification and analysis of products

Hydrolyses reactions were monitored by TLC in solvent A: n-BuOH/AcOH/H_2O 6:2:2 v/v, or B: EtOAc/AcOH/2-propanol/HCOOH/H_2O, 25:10:5:1:15 by vol, whereas transglycosylation reactions were monitored in solvent C:

EtOAc/MeOH/H_2O 70:20:10 by vol. Chemical determination of reducing sugars and enzymatic determinations of glucose and xylose were conducted in triplicate. Hydrolysis reaction mixture was concentrated under vacuum and loaded onto a Biogel P-100 column (1 × 35 cm) equilibrated with water and eluted at 0.1 ml/min with water. A second Biogel P-2 column (1 × 47 cm) was performed eluting at 0.3 ml/min with water. The fractions containing the carbohydrates were pooled and freeze-dried. Purified material was subjected to NMR spectroscopy. The reactions of transglycosylation were stopped by cooling and immediately subjected to reverse-phase column chromatography (Merck Lobar RP-18) eluting with water, and then with methanol thus efficiently separating total chromophoric xylosylated fraction and unreacted acceptor in methanol fraction from free saccharides in water. The methanolic fractions were collected and analysed by ESI-MS.

Monosaccharides composition of *Arundo donax* polysaccharides was determined by GC-MS of acetylated methyl glycosides (Silipo et al. 2012). Methanolysis was performed in 1.25 M HCl in methanol. In brief, sample (1.5 mg) was solubilized in anhydrous 1.25 M HCl/MeOH (1 ml) and then heated at 80°C overnight. Subsequently, air-dried samples were acetylated with acetic anhydride (50 µl) and pyridine (50 µl) at 100°C for 30 min. Then, samples were dried and washed with methanol in order to evaporate pyridine and dissolved in acetone (400 µl) and 1 µl was used for GC-MS analysis. All GC-MS runs were conducted on a Hewlett-Packard Series II Plus gas chromatograph (column: 30 m × 0.25 mm HP-5MS Hewlett-Packard Co.) coupled with a Hewlett-Packard 5989B mass spectrometer. Electron voltage was set at 70 eV and He was used as gas carrier at a costant flow rate of 1 ml/min. The temperature program used was 150°C for 3 min; from 150 to 280°C at 10°C/min; 280°C for 20 min.

Activity assays

Antimicrobial activity was carried out using liquid culture of *E. coli* (DSM 498), *B. subtilis* subsp. *spizizenii* (DSM 347), and *M. luteus* (DSM 348). The MIC was determined by a serial dilution, in duplicate, starting from 100 µg/mL to 0.01 µg/mL. Bacterial and yeast growths were observed after 48 h of incubation. Cytotoxic activity was evaluated by the brine shrimp (*Artemia salina*) test in triplicate with appropriate amounts of samples dissolved in DMSO (1% final volume) to reach final concentrations of 100, 10 and 1 µg/ml, using 10 freshly hatched larvae suspended in 5 ml of artificial seawater (Meyer 1982). Briefly, for each dose tested, surviving shrimps were left at RT and counted after 24 h, and the data statistically analyzed by the Finney program (Finney 1971), which affords LD_{50} values with 95% confidence intervals.

Free-radical scavenging activity was evaluated on pretreated grinded biomass in DMSO and on RK72h dissolved in H_2O at a concentration of 20 mg/mL, and assayed for DPPH test (Blois 1958). Different amounts (10, 50, 100 and 250 µL) of these solutions were added to 0.7 mL of DPPH in MeOH (6 mg/50 mL; 0.1 mM final concentration) and adjusted to 2 mL final volume with MeOH. The absorbance at 517 nm was determined after 30 min at room temperature and the percent of free radical inhibition was calculated. Fucoidan (Sigma) was used as standard. Free radical scavenging activity of samples was estimated as % inhibition of free radical DPPH. Antioxidant activity of both samples was also determined by DMPD method (Fogliano et al. 1999). The reaction mixture contained 1 mM DMPD, 0.1 mM ferric chloride in acetate buffer 0.1 M (pH 5.25) in a total volume of 1 mL. The assay temperature was 25°C. The reaction was monitored at 505 nm until absorbance became stable at a value of 0.900 ± 0.100. Samples were dissolved in H_2O at a concentration of 20 mg/mL. According to the method, the antioxidant activity of samples was carried out in triplicate on main solution (20 mg/mL) and on its diluted solutions 1:2, 1:5, 1:10. Then, 250 µL of solutions were added to the reaction mixture and the decrease in absorbance, which is proportional to the $DMPD^+$ quenched, was determined after 20 min at room temperature. Fucoidan was used as standard. The antioxidant activity was reported as % inhibition of radical cation DMPD.

Results

Hemicellulosic substrates from rhizomes of *Arundo donax*: preparation and optimization of the procedure

Aboveground biomass yield of *Arundo donax* was 39.5 ± 8.1 t/ha and rhizome yield was 40.0 t/ha. With this root: shoot ratio of 1 the results of other researches carried out in Mediterranean conditions on mature giant reed stands (Di Nasso et al. 2013), were confirmed.

Methods for the isolation of hemicelluloses from biomass include extraction in alkaline, alkaline peroxide and liquid hot water medium or steam explosion-based extraction, and other procedures based on slight modifications (Peng et al. 2012; Verbruggen et al. 1995). Under alkaline solubilization of hemicellulosic components, the native structure of xylan is extensively modified with saponification of ester-linkages and swelling of cellulose. Results of the optimization of the procedure for preparation of the hemicellulose fraction from rhizomes of *Arundo donax*, are reported in Figure 1.

The material extracted with 2 N KOH for 72 h at room temperature (RK72h) represented the 17.6% (w/w) of the dried biomass (Figure 1). In Table 1 is reported the monosaccharide composition of the extracted material. Arabinose to xylose ratio (Ara/Xyl) has been considered indicative

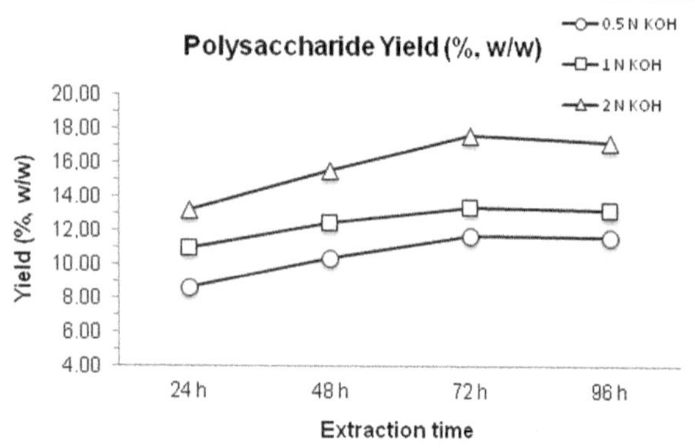

Figure 1 Optimization of polysaccharide extraction from *A. donax* rhizomes. Experiments were performed in triplicate with lyophilized material used in appropriate amounts; data are means of the values obtained that did not differ by more than 2%. Yield was expressed in weight (%, w/w) with respect to the initial dry biomass. See Methods for details.

of the degree of linearity or branching of hemicelluloses (Verbruggen et al. 1995). The arabinose/xylose ratio of polysaccharide extracted from rhizomes of *A. donax* was 0.12, thus implying that the hemicelluloses obtained occurred in a partially de-branched form. ^{13}C NMR spectrum (saturated solution, *d*-DMSO 40°C) of hemicellulosic substrate, RK72h, showed intense signals due to xylan main chain at 101.7, 75.3, 73.9, 72.5, 63.2 ppm, typical values of C1 (101.7) and C4 (75.3) linked β-xyloses. Small signals are present in accord to the monosaccharide composition (Table 1), in particular the percentage of arabinose paralleled in the NMR spectra to typical C1 (107.1) and others (87.3, 86.0, 80.2, 77.8, 61.8 ppm) signals of α-Ara*f* and C1 (103.5) signals of β-glucose.

The action of crude homogenates on rhizomes of *Arundo donax*

The action of crude extracts obtained from *Thermotoga neapolitana* CFE-TN, *Geobacillus thermantarcticus* ES-GTM1 and *Thermoanaerobacterium thermostercoris* CFE-TT has been compared to the action of known commercial xylanases, *Thermomyces lanuginosus* and *Thermotoga maritima*, to study a convenient set up for the degradative action with production of small useful oligosaccharides from hemicellulose fraction of *Arundo donax*. Reaction products were used as candidates in bioactivity assays. Analysis of enzymatic action was conducted comparing (i) total conversion (total yield of reducing sugars and xylose and glucose percentages of reaction mixture, Table 2), (ii) TLC patterns of reaction products and (iii) NMR results.

The reaction of endo-β-(1–4)-xylanase from *Thermomyces lanuginosus* with *A. donax* biomass (Table 2, entry 1) was stopped at 26 h when 23.5% total reducing sugars were formed; values for free xylose and glucose indicated small presence of these two free monosaccharides. Indeed, after purification of reaction mixture, 3.3 g/L of xylobiose, identified by NMR spectroscopy, and 2 g/L of an interesting pentasaccharide (1, Figure 2), were obtained. The pentasaccharidic nature of this compound was quickly evidenced by the peak at (m/z: 701 $[M + Na^+]$ in the mass spectrum. ^1H and ^{13}C NMR chemical shift values are reported in Table 3, they secured that the compound was constituted by four units of β-xylose 1,4 linked (Xyl_1 to Xyl_4 starting from non-reducing end) and one unit of α-L-arabinose. In the ^1H-NMR spectrum there were four anomeric signals of β-Xyl*p* at 4.474, 4.371, 4.330 and 4.406 ppm and one at 5.284 ppm indicating the presence of a *α*-L-Ara*f*. Positioning of *α*-L-Ara*f* was achieved by evaluating long-range correlations in HMBC experiments; in fact anomeric carbon of arabinose at 108.54 ppm correlated with proton signal at position 3 (3.65 ppm) of Xyl_2 unit. Long-range correlations were also recorded between carbon signal of Xyl_2 at position 4 and the anoneric proton of Xyl_1 (1, Figure 2). In NOESY experiments, interglycosidic NOE contacts were recorded between Xyl_2 anomeric proton (4.406 ppm) and the signal at 3.678 ppm of proton 4 of Xyl_3 and between anomeric proton signal at 4.37 ppm of Xyl_3 with the signal at 3.667 ppm, proton 4 of Xyl_4. All remaining chemical shift values of the pentasaccharide 1, (Figure 2), were in agreement with spectroscopic data of similar structure reported in literature (Pastell et al. 2008; Vliegenthart et al. 1992).

In a similar hydrolysis reaction (Table 2, entry 2) conducted using xylanase from *T. maritima* the production in high amount of xylobiose and small xylooligomers as main products from *Arundo* biomass, was confirmed. The absence of glucose was due to better purity of enzymatic activity (xylanase).

The analysis of reactions (Table 2, entries 3–5) using cell-free extracts from *T. thermostercoris* and *T. neapolitana* (Table 2, entry 6) indicated high conversions of the polysaccaridic constituent in terms of reducing sugars (entries 3,4 and 6). The presence of β-xylosidases and

Table 1 Monosaccharide composition of *Arundo donax* polysaccharides extracted with KOH 2 N at 72 h

Monosaccharide	*Arundo donax* %
Xyl	57.13
Glc	34.77
Ara	7.03
Man	0
Gal	0
Rha	0
GalA	0
GlcA	1.07

Abbreviations: *Xyl* Xylose, *Glc* Glucose, *Gal* Galactose, *Ara* Arabinose, *Rha* Rhamnose, *Man* Mannose, *GlcA* Glucuronic acid, *GalA* Galacturonic acid.

Table 2 Degradative actions of microbial and commercial xylanolytic activities on *A. donax* biomass and transglycosylation reactions

	Biocatalyst	S/B	RS%	Xyl%	Glc%	Useful results
1	TL-xylanase	3.3	23.5	2.26	8.6	xylobiose 3.3 g/L pentasaccharide 1 (Figure 2), 2 g/L
2	TM-xylanase	2.6	38	0	0	-
3	CFE-TT	3.5	90	19.2	16.9	-
4	CFE-TT	1.2	100	28.6	6.8	-
5	CFE-TT	3.5	20	8.7	7.8	tetrasaccharide 2 (Figure 2) 2.64 g/L
6	CFE-TN	1.2	75	40.5	15.2	see Figure 3
7	ES-GTM1b	13.5	73.5	26.5	0	see Figure 3
8	ES-GTM1a	36.3	63.1	9.85	0	see Figure 3
			Comment			
9	TM-xylanase	7.7	Formation of mono and dixylosylated acceptor			
10	CFE-TN	1.2	Formation of mono and dixylosylated acceptor			
11	ES-GTM1	13.5	Formation of mono, di and trixylosylated acceptor			
12	CFE-TT	3.5	Formation of mono and dixylosylated acceptor			

Data are expressed in percentages using means of the values obtained by chemical and enzymatic determinations conducted in triplicate that did not differ by more than 5%. S/B = amount of biocatalyst, mg substrate/mg total protein; RS = Reducing sugars yield % (DNS); TL = *Thermomyces lanuginosus* xylanase; TM = *Thermotoga maritima* xylanase; CFE-TT = cell free extract of *Thermoanaeorbacterium thermostercoris*; CFE-TN = cell free extract of *T. neapolitana*; ES-GTM1 = extracellular suspension *Geobacillus thermantarcticus*.

xylanase activities in these enzymatic preparations is evidenced by the formation of high amount of free xylose. Low amounts of different xylose tetra- and pentaoligomers were found at the proper Rf in TLC in the course of reaction. In addition, as evidenced by glucose enzymatic determination, substantial amounts of glucose are formed proving the presence of interesting β-glucosidases (cellulase) activities (entries 3–6, Table 2 and Figure 3).

The reaction using *T. thermostercoris* was conducted on a preparative scale and stopped at 20% conversion (Table 2, entry 5) for the recovery (2.64 g/L) of an interesting pure tetrasaccharide (2, Figure 2) isolated from the reaction mixture by Biogel P-2 purification. The compound showed a peak in the MS spectrum at m/z 569 $[M + Na]^+$ corresponding to a tetrasaccharide made of pentose sugars. ^1H and ^{13}C NMR signals are reported in Table 3. NMR signals of anomeric protons were found in the range 4.35-5.32 ppm and those of anomeric carbon atoms in the range 92.6-108.5 ppm. HMBC spectrum showed long-range correlations between ^{13}C anomeric signal of α-L-Ara*f* and signal of H3 proton of central xylose and the related proton anomeric signal of α-L-Ara*f* and C3 of central xylose (see Table 3 for values). Suitable resolution of spectra for signal assignment of both compounds in Figure 2 was achieved only in dilute conditions (3–4 mg/ml); more concentrated samples furnished unresolved spectra and time-dependent signal shifts. Despite the structural reporter group concept for these structures is widely known (Vliegenthart et al. 1992), structures in Figure 2 were established by analysis of ^1H-^{13}C NMR correlations on dilute samples.

1

2

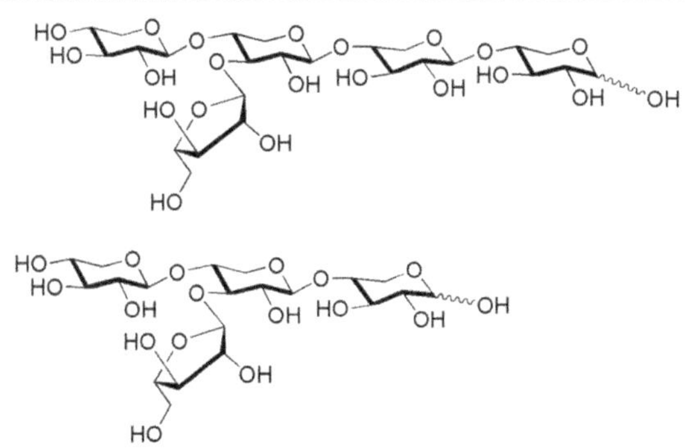

Figure 2 The pentasaccharide β-D-Xyl*p*-(1–4)-[α-L-Ara*f*-(1–3)]-β-D-Xyl*p*-(1–4)-β-D-Xyl*p*-(1–4)-β-D-Xyl*p* (1) from hydrolysis mixture using *T. lanuginosus* enzyme and β-D-Xyl*p*-(1–4)-[α-L-Ara*f*-(1–3)]-β-D-Xyl*p*-(1–4)-D-Xyl*p* (2) the tetrasaccharide, from the hydrolysis reaction mixture using crude extract of *T. thermostercoris*.

Amount of xylose was formed also in the case of *G. thermantarcticus* (Table 2, entries 7–8) in accord to the loading of the enzyme. Using high amount of substrate per mg of crude biocatalyst (Table 2, entry 8) interesting oligomers accumulated (from xylobiose to other high molecular oligosaccharides, Figure 3).

Transglycosylation reactions were performed using *A. donax* RK72h biomass as donor and different biocatalysts (entries 9–12, Table 1). At the end of each reaction, total chromophoric xylosylated fraction was obtained by reverse phase column and analyzed by TLC with authentic standards (Tramice et al. 2009), and MS spectroscopy.

Table 3 ^1H and ^{13}C chemical shifts of compounds in Figure 2

	1		2		3		4		5	
1 Figure 2	^1H	^{13}C	^1H	^{13}C	^1H	^{13}C	^1H	^{13}C	^1H	^{13}C
α-L-Ara*f*	5.289	108.54	4.055	81.63	3.795	78.3	4.164	85.90	3.684	62.36
									3.617	
1: β-D-Xyl*p*	4.330	**102.49**	3.142	74.35	3.304	76.61	3.480	70.19	3.806	66.14
									3.180	
2: β-D-Xyl*p*	4.406	102.67-102.69	3.331	74.68-74.59	**3.650**	**77.40**	**3.710**	**74.68-74.59**	4.016	63.84-64.00
									3.950	
3: β-D-Xyl*p*	4.371	102.67-102.69	3.195	73.70-73.96	3.440	75.09-74.93	**3.678**	**77.40**	4.016	63.84-64.00
									3.290	
4: α,β-D-Xyl*p*	β: 4.474	97.53	3.142	74.35	3.441	75.09-74.93	**3.667**	**77.40**	3.950	63.84-64.00
	α: 5.070	93.02							3.280	
	1		2		3		4		5	
2 Figure 2	^1H	^{13}C	^1H	^{13}C	^1H	^{13}C	^1H	^{13}C	^1H	^{13}C
α-L-Ara*f*	5.321	108.52	4.070	81.63	3.817	78.26	4.187	85.74	3.711	62.29
									3.635	
1: β-D-Xyl*p*	4.356	**102.40**	3.166	74.55	3.330	74.40	3.511	70.19	3.839	66.17-66.07
									3.190	
2: β-D-Xyl*p*	4.428	102.57	3.361	74.30	**3.658**	**77.39**	**3.697**	**76.78**	3.281	66.17-66.07
									3.975	
3: α,β-D-Xyl*p*	α: 5.115	92.97	3.175	74.25	3.460	75.00	**3.741**	**76.52**	4.049	63.95-61.93
	β: 4.496	97.58							3.315	

Indicative signals for interglycosidic linkage assignments are indicated in bold.

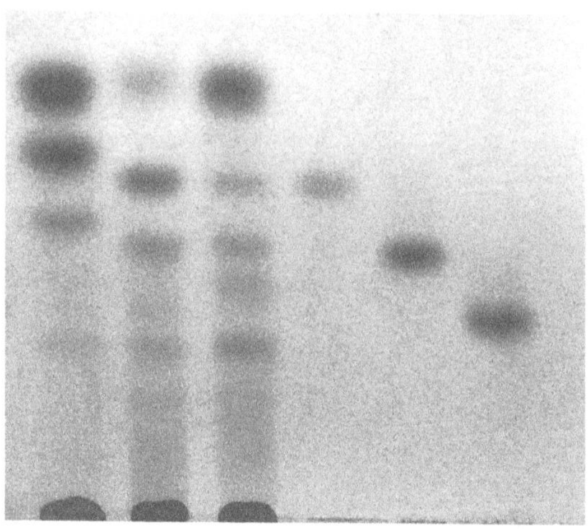

Figure 3 TLC pattern of enzymatic reactions using *A. donax* hemicelluloses extract as substrate. CFE-TN = cell-free extract *T. neapolitana* (Table 2, entry 6); ESGTM1a and ESGTM1b = extracellular suspension of *Geobacillus. thermantarcticus* (Table 2, entry 7 and 8); X2-X4 = β-1,4-xylooligomers.

Mono- (m/z 351 $[M + Na]^+$) and di-xylosylated (m/z 483 $[M + Na]^+$) 9-fluorene methanol were identified in the reaction using *T. maritima* xylanase. The same reaction was performed using *T. neapolitana*, *T. thermostercoris* and *G. thermantarcticus*, crude homogenates. In the latter case mass peak at m/z 615 $[M + Na]^+$ due to trixylosylated 9-fluorene methanol was also observed. Although not optimized for the yield, these reactions for the formation of the xylosylated products starting with *Arundo* biomass are of interest for enlarging the portfolio of pentose-specific conversion/utilization technologies (Dumon et al. 2012).

Biological activity

Samples of raw materials, extracts of rhizome and purified compounds from hydrolysis reactions were assayed for their antimicrobial, cytotoxic and antioxidant activities. Antimicrobial activity was evaluated on three bacterial strains, *E. coli* (DSM 498), *B. subtilis* subsp. *spizizenii* (DSM 347), and *M. luteus* (DSM 348). All samples did not show any significant antimicrobial activity. Brine shrimp assay was utilized as indicative of cytotoxicity (Meyer et al. 1982). The brine shrimp is a marine crustacean used at stage of larvae (48 h after hatching). LD_{50} is the minimum lethal dose for which we observe a 50% mortality of larvae. Results obtained by using this test showed that all samples were not toxic on brine shrimp larvae; in fact after 24 h of exposure to samples at concentration of 100, 10 and 1 μg/ml, all larvae were live and motile.

Antioxidant activities of the raw material and of the extract of rhizome RK72h compared to that observed using fucoidan were good both in DMPD and DPPH assays. Inhibition of DMPD radical cation (43 and 35%) and DPPH radical (68 and 45%) were observed for raw material and extract of rhizome RK72h, respectively at the maximum tested amount (4 mg). Dose dependent activity was observed at lower concentrations (2 to 0.1 mg). Unfortunately purified compounds (Figure 2) did not exhibit significant antioxidant activity.

Discussion

In this article the optimization of hemicellulosic substrate preparation from rhizome of *Arundo donax* has been reported. Hemicellulosic fraction represented ca. 20% of the dried biomass and occurred in a partially debranched form after alkaline extraction, as indicated by Ara/Xyl ratio (Verbruggen et al. 1995) and ^{13}C NMR spectroscopy.

The production of two different interesting oligosaccharides was possible with both commercial (xylobiose, pentasaccharide 1) and thermophilic microorganisms (tetrasaccharide 2) from this hemicellulose fraction. ^{1}H-^{13}C NMR correlations of products were studied for precise structural assignment. Significant amounts of low molecular mass compounds, xylobiose and trace amounts of xylose were liberated from birch wood xylan after 24 h (Lin et al. 1999) using *Thermomyces lanuginosus* enzyme; however cellulase activity could contaminate this endo-β-(1 → 4)-xylanase preparation (Subramaniyan and Prema 2002) explaining the results obtained. Xylanase from *Thermotoga maritima* is an extremely thermostable enzyme which hydrolyzed xylo-oligosaccharides and xylans to yield predominantly xylobiose as end product (Jiang et al. 2004). Xylobiose and small xylooligomers as main products in high amount from *Arundo* biomass confirmed this activity. The absence of glucose can be explained by better purity of enzymatic preparation (xylanase) with respect to *T. lanuginosus* reaction. As far as *T. neapolitana* CFE-TN and *T. thermostercoris* CFE-TT were concerned, presence of xylanases, xylosidases, glucuronidases and arabinofuranosidases were confirmed in the crude extracts by the action on suitable chromophoric and polysaccharidic substrates (data not shown). The results on *Arundo* indicated also that action of β-glucosidases is dependent on enzymatic load in the reaction mixture, and this phenomenon could be analyzed with respect to the known interrelation of cellulases and xylanases in the bioconversion of lignocellulosic material (Hu et al. 2011). All these data are of great interest in the view of research for biocatalyst suitable for complete saccharification of the cellulose and hemicellulose components in raw material to fermentable sugars.

The production of the transxylosylated products from *Arundo* was also possible with thermophilic enzymes and this is of further interest for enlarging pentose-specific conversion/utilization technologies.

In conclusion rhizomes represent a useful raw material for the production of valuable industrial products increasing the economic and energetic efficiency of *A. donax* cultivation. Antioxidant activities of the raw material and hemicellulose fraction from rhizomes compared to that observed using fucoidan are good both in DMPD and DPPH assays.

Competing interests
The authors declared that they have no competing interests.

Authors' contributions
LL conceived part of the study and VC carried out microbiological studies, ATra. and EP carried out the preparative enzymatic reactions, chromatographic purification and structural study of xylooligosaccharides, IF carried out the analytical enzymatic reactions, enzymatic assay and participated in the design of the study, GA and PdD carried out optimization of hemicellulose extraction from rhizome, MF is the responsible of the entire project and with MM and AI they were involved in the rhizome production and collection, GT carried out activity assay, BN and AP conceived part of the study and participated in the coordination, A. Tri. conceived part of the study, participated in the structural study of xylooligosaccharides and drafted the manuscript. All authors read and approved the final manuscript.

Acknowledgements
The work included in this article is part of the project PON01_01966 "Filiere agro-industriali integrate ad elevata efficienza energetica per la messa a punto di processi di Produzione Eco-compatibili di Energia e Bio-chemicals da fonte rinnovabile e per la valorizzazione del territorio (EnerbioChem)" 28/10/2011 prot. n. 881/Ric - Programma Operativo Nazionale (PON) "Ricerca e Competitività 2007 – 2013" Regioni Convergenza. We thank D. Melck, E. Castelluccio and A. Esposito (NMR service of Istituto di Chimica Biomolecolare, CNR, Pozzuoli, Italy) for their skilful assistance.

Author details
[1]Institute of Biomolecular Chemistry ICB-CNR, Consiglio Nazionale delle Ricerche, Via Campi Flegrei 34, 80072 Pozzuoli, Napoli, Italy. [2]Department of Environmental Sciences, University of Naples "Parthenope", Centro Direzionale, Isola C4, 80143 Naples, Italy. [3]Department of Agriculture, Naples University Federico II, Via Università 100, 80055 Portici, Naples, Italy.

References

Bernfeld P (1955) Amylases α and β. In: Kaplan N (ed) Colowick S. Methods in Enzymology Academic Press, New York

Blois MS (1958) Antioxidant determinations by the use of a stable free radical. Nature 26:1199–1200. doi:10.1038/1811199a0

Bonanno G, Cirelli GL, Toscano A, Lo Giudice R, Pavone P (2013) Heavy metal content in ash of energy crops growing in sewage-contaminated natural wetlands: potential applications in agriculture and forestry? Sci Total Environ 452–453:349–354. doi:10.1016/j.scitotenv.2013.02.048

Bradford MM (1976) A rapid and sensitive method for the quantitation of microgram quantities of protein utilizing the principle of protein-dye binding. Anal Biochem 72:248–254. doi:http://dx.doi.org/10.1016/0003-2697(76)90527-3

Christakopoulosa P, Katapodisa P, Kalogerisa E, Kekosa D, Macrisa BJ, Stamatisb H, Skaltsac H (2003) Antimicrobial activity of acidic xylo-oligosaccharides produced by family 10 and 11 endoxylanases. Int J Biol Macromol 31:171–175. doi:http://dx.doi.org/10.1016/S0141-8130(02)00079-X

Di Nasso NNO, Roncucci N, Bonari E (2013) Seasonal dynamics of aboveground and belowground biomass and nutrient accumulation and remobilization in giant reed (*Arundo donax* L.): a three-year study on marginal land. Bioenergy Res 6(2):725–736. doi:10.1007/s12155-012-9289-9

Diodato N, Fagnano M, Alberico I (2009) CliFEM – climate forcing and erosion response modelling at long-term sele river research basin (Southern Italy). Nat Hazards Earth Syst Sci 9:1693–1702. doi:10.5194/nhess-9-1693-2009

Dumon C, Song L, Bozonnet S, Fauré R, O'Donohue MJ (2012) Progress and future prospects for pentose-specific biocatalysts in biorefining. Process Biochem 47(3):346–357. doi:10.1016/j.procbio.2011.06.017

Fagnano M (2014) Implementation of eco-compatible protocols for agricultural soil remediation in litorale Domizio-Agro aversano NIPS. www.ecoremed.it, Accessed 9 June 2014

Finney DJ (1971) Statistical aspects of monitoring for dangers in drug therapy. Methods Inf Med 10:1–8

Fiorentino N, Impagliazzo A, Ventorino V, Pepe O, Piccolo A, Fagnano M (2010) Biomass accumulation and heavy metal uptake of giant reed on polluted soil in southern Italy. J Biotechnol 150:261

Fogliano V, Verde V, Randazzo G, Ritieni A (1999) Method for measuring antioxidant activity and its application to monitoring the antioxidant capacity of wines. J Agric Food Chem 47:1035–1040. doi:10.1021/jf980496s

Galletti AMR, Antonetti C, Ribechini E, Colombini MP, Di Nasso NNO, Bonari E (2013) From giant reed to levulinic acid and γ-valerolactone: a high yield catalytic route to valeric biofuels. Appl Energy 102:157–162. doi:10.1016/j.apenergy.2012.05.061

Hu J, Arantes V, Saddler JN (2011) The enhancement of enzymatic hydrolysis of lignocellulosic substrates by the addition of accessory enzymes such as xylanase: is it an additive or synergistic effect? Biotechnol Biofuels 4:36–49. doi:10.1186/1754-6834-4-36

Jiang ZQ, Deng W, Zhu YP, Li LT, Sheng YJ, Hayashi K (2004) The recombinant xylanase B of *Thermotoga maritima* is highly xylan specific and produces exclusively xylobiose from xylans, a unique character for industrial applications. J Mol Catal B Enz 27:207–213. doi:10.1016/j.molcatb.2003.11.012

Lama L, Calandrelli V, Gambacorta A, Nicolaus B (2004) Purification and characterization of thermostable xylanase and β-xylosidase by the thermophilic bacterium Bacillus thermantarcticus. Res Microbiol 155(4):283–289. doi:10.1016/j.resmic.2004.02.001

Lin J, Ndlovu LM, Singh S, Pillay B (1999) Purification and biochemical characteristics of β-D-xylanase from a thermophilic fungus. Thermomyces lanuginosus-SSBP Biotechnol Appl Biochem 30:73–79

Mellinger-Silva C, Simas-Tosin FF, Schiavini DN, Werner MF, Baggio CH, Pereira IT, Da Silva LM, Gorin PA, Iacomini M (2011) Isolation of a gastroprotective arabinoxylan from sugarcane bagasse. Bioresour Technol 102(22):10524–10528. doi:10.1016/j.biortech.2011.08.107

Meyer BN, Ferrigni NR, Putnam JE, Jacobsen LB, Nichols DE, McLaughlin JL (1982) Brine shrimps: a convenient general bioassay for active plant constituents. Planta Med 45(5):31–34. doi:10.1055/s-2007-971236

Nicolaus B, Lama L, Esposito E, Manca MC, di Prisco G, Gambacorta A (1996) *Bacillus thermantarcticus* sp. nov. from mount Melbourne, Antarctica: a novel thermophilic species. Polar Biol 16:101–104. doi:10.1007/s003000050034

Pastell H, Tuomainen P, Virkki L, Tenkanen M (2008) Step-wise enzymatic preparation and structural characterization of singly and doubly substituted arabinoxylo-oligosaccharides with non-reducing end terminal branches. Carbohydr Res 343(18):3049–3057. doi:http://dx.doi.org/10.1016/j.carres.2008.09.013

Peng F, Peng P, Xu F, Sun R-C (2012) Fractional purification and bioconversion of hemicelluloses. Biotechnol Adv 30:879–903. doi:10.1016/j.biotechadv.2012.01.018

Pirozzi D, Yousuf A, Fagnano M, Mori M, Fiorentino N, Impagliazzo A (2010) Lipids production by yeast grown on giant reed biomass. J Biotechnol 150S:167–168. doi:10.1016/j.jbiotec.2010.08.434

Romano I, Dipasquale L, Orlando P, Lama L, D'Ippolito G, Pascual J, Gambacorta A (2010) *Thermoanaerobacterium thermostercus* sp. nov., a new anaerobic thermophilic hydrogen-producing bacterium from buffalo-dung. Extremophiles 14(2):233–240. doi:10.1007/s00792-010-0303-x

Saeed F, Pasha I, Anjum FM, Sultan MT (2011) Arabinoxylans and arabinogalactans: a comprehensive treatise. Crit Rev Foods Sci Nutr 51 (5):467–476. doi:10.1080/10408391003681418

Silipo A, Sturiale L, De Castro C, Lanzetta R, Parrilli M, Garozzo D, Molinaro A (2012) Structure of the lipo-polysaccharide isolated from the novel species *Uruburuella suis*. Carbohydr Res 357:75–82. doi:10.1016/j.carres.2012.05.010

Subramaniyan S, Prema P (2002) Biotechnology of microbial xylanases: enzymology, molecular biology and application. Crit Rev Biotechnol 22:33–46

Tramice A, Melck D, Virno A, Randazzo A, Motta A, Trincone A (2009) Enzymatic synthesis and 3-D structure of anti-proliferative acidic (MeGlcUA) xylotetrasaccharide. J Mol Catal B Enzym 61(3–4):129–135. doi:10.1016/j.molcatb.2009.05.005

Tramice A, Pagnotta E, Romano I, Gambacorta A, Trincone A (2007) Transglycosylation reactions using glycosyl hydrolases from thermotoga neapolitana, a marine hydrogen-producing bacterium. J Mol Catal B Enz 47(1–2):21–27. doi:10.1016/j.molcatb.2007.03.005

van der Horst D, Vermeylen S (2011) Spatial scale and social impacts of biofuel production. Biomass Bioenerg 35(6):2435–2443. doi:10.1016/j.biombioe.2010.11.029

Verbruggen MA, Beldman G, Voragen AGJ (1995) The selective extraction of glucuronoarabinoxylans from Sorghum endosperm cell walls using barium and potassium hydroxide solutions. J Cereal Sci 21:271–282. doi:10.1006/jcrs.1995.0030

Vliegenthart JFG, Hoffmann RA, Kamerling JP (1992) A ^{1}H NMR spectroscopic study on oligosaccharides obtained from arabinoxylans. In: Vissert J (ed) Xylans and Xylanases Proceedings of an International Symposium Wageningen, December 1991. Elsevier Science Publisher, The Netherlands, p 17

Sterilization of hydrogen peroxide resistant bacterial spores with stabilized chlorine dioxide

Anthony Friedline[1], Malcolm Zachariah[1], Amy Middaugh[1], Matt Heiser[2], Neeraj Khanna[2], Parag Vaishampayan[3] and Charles V Rice[1*]

Abstract

Bacillus pumilus SAFR-032 spores isolated from a clean room environment are known to exhibit enhanced resistance to peroxide, desiccation, UV radiation and chemical disinfection than other spore-forming bacteria. The survival of *B. pumilus* SAFR-032 spores to standard clean room sterilization practices requires development of more stringent disinfection agents. Here, we report the effects of a stabilized chlorine dioxide-based biocidal agent against spores of *B. pumilus* SAFR-032 and *Bacillus subtilis* ATCC 6051. Viability was determined via CFU measurement after exposure. Chlorine dioxide demonstrated efficacy towards sterilization of spores of *B. pumilus* SAFR-032 equivalent or better than exposure to hydrogen peroxide. These results indicate efficacy of chlorine dioxide delivered through a stabilized chlorine dioxide product as a means of sterilization of peroxide- and UV-resistant spores.

Keywords: *Bacillus pumilus*; *Bacillus subtilis*; Chlorine dioxide; Spore killing; Spores

Introduction

During periods of environmental stress or nutrient depletion, certain Gram-positive bacteria use sporulation to protect its DNA within a protective shell. Germination into a vegetative cell occurs when conditions improve. These processes of sporulation and germination are well understood and reviewed in relevant literature (Errington 1993; Driks 2002; Setlow 2003; Piggot and Hilbert 2004; Paredes-Sabja et al. 2011; Higgins and Dworkin 2012). While *Bacillus subtilis* is studied as a model organism for sporulation, other species are also relevant due to healthcare or infectious disease reasons, such as *Bacillus anthracis* (Schmid and Kaufmann 2002; Spotts Whitney et al. 2003), *Clostridium botulinum* (Chaudhry 2011), and *Clostridium difficile* (Roberts et al. 2008; Best et al. 2010; Peniche et al. 2013). Each layer of the spore structure protects against DNA damage. The outer layer, the coat, is laminar, serving as a permeation barrier against lysozymes and other agents (Riesenman and Nicholson 2000; Takamatsu and Watabe 2002; Driks 2004). The cortex is a cross linked peptidoglycan structure surrounding the core, implicated in heat resistance of the spore (Mallidis and Scholefield 1987; Atrih and Foster 1999); it differs from that found in vegetative cells (Warth and Strominger 1971; Warth and Strominger 1972) by converting half of its *N*-acetyl-muramic acids into muramic lactams, which are targets for cortex lysis during germination (Atrih et al. 1996). The spore core houses the bacterial DNA, small acid-soluble spore proteins (SASPs) (Setlow 1988; Moeller et al. 2009), and a calcium-dipicolinic acid chelate (Setlow et al. 2006; Magge et al. 2008) which serve to protect the DNA from UV damage. These features, along with a lower water content in the spore core (Setlow 2006), allow spores to survive for long periods of time.

These features facilitate bacterial spores to be highly resistant to a wide variety of potentially lethal conditions (Setlow 2006; Nicholson et al. 2000). This has led to studies analyzing the possibility of survival of spores in extraterrestrial conditions, such as Martian regolith (Schuerger et al. 2003; Kerney and Schuerger 2011; Moeller et al. 2012). In 2004, Link *et al.* published a study discussing UV resistance of spores of *Bacillus pumilus* isolated from the Spacecraft Assembly Facility at the Jet Propulsion Laboratory (JPL) in Pasadena, CA (Link et al. 2004). The isolates had higher UV resistance than a standard strain of *B. subtilis* used for dosimetry

* Correspondence: rice@ou.edu
[1]Department of Chemistry and Biochemistry, Stephenson Life Sciences Research Center, University of Oklahoma, Norman, Oklahoma, USA
Full list of author information is available at the end of the article

and other similarly-isolated *B. pumilus* strains. *B. pumilus* SAFR-032 was ~340 times more resistant to 254 nm UV radiation than the *B. pumilus* type strain, ATCC 7061, and was remarked to be "...the most highly UV-resistant spores yet to be discovered" (Link et al. 2004). *B. pumilus* SAFR-032 was reported as being the most resistant spore strain to UV radiation under simulated Martian environmental conditions the following year (Newcombe et al. 2005). *B. pumilus* SAFR-032 was also observed to have peroxide resistance (Link et al. 2004; Gioia et al. 2007). In 2012, Vaishampayan *et al.* noted that spores of *B. pumilus* SAFR-032 exhibited survival rates of 10-40% on spacecraft-grade aluminum coupons exposed to space vacuum for 18 months during the EXPOSE-E mission onboard the International Space Station (ISS) (Vaishampayan et al. 2012). Additionally, limited numbers of spores were able to germinate after exposure to solar UV radiation onboard the ISS, both exposed to space vacuum and within a simulated Martian environment. Given the known ability for habitable space vehicles, including the ISS, to harbor diverse and potentially hazardous bacteria (Venkateswaran et al. 2014), the risk of contamination by bacterial spores—and thus the need for a method of inactivating them—cannot be ruled out.

Chlorine dioxide, ClO_2, is an oxidant known to exhibit antimicrobial properties (Beuchat et al. 2005; Gordon and Rosenblatt 2005; Novak et al. 2008; Rastogi et al. 2010). It has been demonstrated to have effects on many biological agents, such as *Bacillus anthracis* (Rastogi et al. 2010; Buhr et al. 2011), *Bacillus cereus* and *C. perfringens* (Foegeding et al. 1986), and *Legionella* sp. (Kim et al. 2002). It is unstable as a gas and is usually generated *in situ* for use, either electrochemically (Foegeding et al. 1986) or by reaction of sodium chlorite + chlorine gas, sodium hypochlorite + hydrogen chloride (Sharma and Sohn 2012), or sodium chlorate + peroxide and sulfuric acid (Gordon and Rosenblatt 2005). At sub-lethal concentrations, ClO_2 may induce different behaviors in bacteria, with a study indicating *B. subtilis* formed biofilms at sub-lethal exposures to chlorine dioxide (Shemesh et al. 2010).

A 2003 paper examined the efficacy and potential mechanism of action of a stabilized chlorine dioxide solution against spores of *B. subtilis* compared with hypochlorite solution (Young and Setlow 2003). It was reported that there was no difference in efficacy between hypochlorite and chlorine dioxide, and that the mechanism does not damage DNA but likely damage to the internal membrane of the spore. These results were similar for a wild-type spore strain and a strain lacking the core-associated SASPs. The authors did state in their work that "...very little is known about the mechanisms of spore killing by and resistance to chlorine dioxide" (Young and Setlow 2003).

Given the known resistance of *B. pumilus* SAFR-032 to traditional sterilization methods, we investigated the efficacy of stabilized chlorine dioxide against spores of *B. pumilus* SAFR-032 and *B. subtilis* ATCC 6051. We report that chlorine dioxide is effective at killing spores of *B. subtilis* ATCC 6051 in a dose-dependent fashion. Further, we report the ability of chlorine dioxide to inactivate spores of the peroxide-resistant *B. pumilus* SAFR-032.

Materials and methods

Spore preparation

Frozen stocks of *B. pumilus* SAFR-032 (provided by NASA JPL) and wild-type *B. subtilis* ATCC 6051 (purchased from American Type Culture Collection, Manassas VA, US) were inoculated overnight in 20 mL of lysogeny broth (LB) media, which was used to inoculate 200 mL of media at 37°C in a 1 L Erlenmeyer flask. The *B. pumilus* SAFR-032 was grown solely in LB media until sporulation occurred due to nutrient exhaustion, which typically occurred after three days. *B. subtilis* ATCC 6051 was initially grown in LB media for one day. Cells were isolated with centrifugation (10,000 X *g*, 20 min, 4°C) and the supernatant decanted before suspension in an equal volume (200 ml) of chemically-defined sporulation medium (CDSM) (Hageman et al. 1984); the culture sporulated in CDSM over a period of 3-10 days. The different media used were based on assessment in our lab that *B. subtilis* sporulated more efficiently in CDSM, while *B. pumilus* SAFR-032 sporulated best in LB through nutrient exhaustion. Sporulation was monitored via phase-contrast microscopy of 5-10 μL wet mounts (Olympus BX-50) at a magnification of 1000X.

Once a desired level (>80%) of free spores were observed via phase contrast microscopy, spores were purified using a modification of the protocol of Zhao *et al.* (Zhao et al. 2008). Cultures were centrifuged at 10,000 x *g* for 20 minutes at 4°C; the supernatant was decanted and the pellet then resuspended in 200 mL of sterile Milli-Q water. The pellet was washed a second time with centrifugation (10,000 X *g*, 20 min, 4°C); after decanting, the pellet was resuspended in 20 mL of sterile 50% (v/v) EtOH:H_2O in a centrifuge tube and allowed to shake (200 rpm) at 25°C for 2 hours to kill remaining vegetative cells. The suspension was spun down (10,000 X *g*, 20 min, 4°C), the supernatant decanted, and the pellet resuspended in 20 mL sterile Milli-Q water. The spores were washed by repeated (3x) centrifugation (10,000 X *g*, 20 min, 4°C) and resuspension in sterile Milli-Q water. After the final centrifugation and resuspension, the spore suspension was stored at 4°C overnight; the following morning the suspension was spun down, resuspended in sterile Milli-Q water, filtered through 3.1/1.2 μm glass fiber syringe filters connected in series, and stored at 4°C until used. The purified spore suspensions were then diluted with Milli-Q water to attain an optical density (OD_{600}) of 0.01 (~10^6 CFU mL^{-1}), measured using a spectrophotometer (Beckman-Coulter DU 640). Spore preparations in glass test tubes were heat-shocked at 80°C for 15 min, followed by rapid cooling on ice to 4°C.

Preparation of chlorine dioxide solutions

An aqueous solution of 2% stabilized chlorine dioxide (Oxine®; Bio-Cide International, Inc., Norman, OK, USA) was used in all exposure experiments. The product was activated by citric acid addition in a ratio of 0.2 g mL^{-1} of stabilized ClO_2 solution without stirring (30 min, room temperature), causing the formation of chlorine dioxide. Each experiment used a freshly activated ClO_2 solution. The total concentration of activated chlorine dioxide (after dilution) was measured using a proprietary iodometric titration method (Bio-Cide International). The concentration of free chlorine dioxide gas, which is a fraction of the total ClO_2 concentration, was determined using a Hach DR3900 spectrophotometer (Hach Co., Loveland, CO) at λ = 445 nm. The activated ClO_2 solution was then diluted in water to a total concentration of 500 µg/mL, followed by subsequent dilutions to approximately total concentrations of 200 µg/mL, 100 µg/mL, and 50 µg/mL for selected experiments. Experiments relying on a spray application of chlorine dioxide solution utilized 1 L of chosen dilutions in a hand-held pump-up sprayer (RL Flo-Master model 1401, Root-Lowell Manufacturing Co., Lowell, MI), which had been pressurized and flushed with at least 100 mL of the solution prior to use.

Exposure of dried spores to chlorine dioxide

To evaluate inactivation of dried spores on glass cover slips, spore cultures were mixed (200 rpm) in 1:1 EtOH: H_2O for 4 hours, and after the final centrifugation, the spore suspension was diluted to an OD_{600} of 1.0 (~10^8 CFU mL^{-1}). 100 µL of spore suspensions of both *B. subtilis* ATCC 6051 and *B. pumilus* SAFR-032 were each deposited onto three sterile microscope slide cover slips; these cover slips were allowed to dry at room temperature for 24 h in a laminar air flow chamber. After drying, the cover slips were exposed to a fine mist of stabilized chlorine dioxide solution at 47 or 187 µg/mL concentration for less than one second. The ClO_2-treated samples were undisturbed for a desired duration (10 min, 60 min, or 24 h) in a laminar air flow chamber. Cover slips were transferred to a 50 mL Falcon® tube containing 10 mL 0.1 mol L^{-1} sodium thiosulfate to neutralize remaining active chlorine dioxide on the cover slips. The tubes containing the thiosulfate solution and cover slips were placed in an ultrasonic water bath (Ultrasonic Cleaner B-220, Branson Cleaning Equipment, Shelton, CT) for 2 minutes to dislodge the spores, followed by vigorous vortex mixing for 10 seconds (previously used in removal of *B. cereus* spores from metal coupons) (Tauveron et al. 2006). The resulting spore suspensions were serially diluted with sterile phosphate-buffered saline (PBS; 8 g L^{-1} NaCl, 0.2 g L^{-1}KCl, 1.44 g L^{-1} Na_2HPO_4, 0.24 g L^{-1} KH_2PO_4, pH 7.4) and 100 µL portions were spread on TSA plates. Plates were incubated at 37°C to observe spore viability and the colonies counted after 48 h.

This procedure was also used with spores of *B. pumilus* SAFR-032 on spacecraft-grade aluminum coupons that had been autoclaved prior to use. The duration of exposure to chlorine dioxide was either 60 min or 24 h before neutralization. In addition to activated ClO_2, coupons with dried *B. pumilus* SAFR-032 spores were also exposed to inactivated ClO_2 (without citric acid) and 3.5% w/v hydrogen peroxide solutions for comparison of efficacy. A control set without exposure to any agent was also prepared. Two replicates (A and B) were prepared for both exposure durations. At the end of the exposure time, coupons were neutralized, sonicated, serially diluted and plated as described above; the resulting TSA plates were incubated at 37°C with colony counting performed 24 h after neutralization.

Results

Exposure of dried spores to chlorine dioxide

The exposure results of chlorine dioxide to dried spores on glass cover slips, both *B. pumilus* SAFR-032 and *B. subtilis* ATCC 6051, are shown in Table 1. As can be seen, chlorine dioxide is effective at inactivating spores of both species. This efficacy is dependent both on ClO_2 concentration and duration of exposure. Inactivation is not complete at the lower (47 µg/mL) concentration, but it does increase with longer exposure durations. At the

Table 1 Viability assay of stabilized chlorine dioxide solution against spores of *B. subtilis* ATCC 6051 and *B. pumilus* SAFR-032

		Colony forming units	
ClO_2 concentration (µg/mL)	Exposure time	*B. subtilis* ATCC 6051	*B. pumilus* SAFR-032
0	N/A	$4.59 \pm 1.71 \times 10^4$	$1.30 \pm 0.32 \times 10^5$
47	10 m	$4.70 \pm 3.11 \times 10^3$	$2.40 \pm 0.10 \times 10^4$
47	60 m	$3.50 \pm 2.12 \times 10^2$	$1.10 \pm 0.14 \times 10^3$
47	24 h	n.d.	n.d.
187	10 m	n.d.	n.d.
187	60 m	n.d.	n.d.
187	24 h	n.d.	n.d.

Notes: Viability assay of *B. subtilis* ATCC 6051 and *B. pumilus* SAFR-032 dried spores on glass cover slips after exposure to stabilized chlorine dioxide solution at 47 or 187 µg/mL concentration and 10 min, 60 min, or 24 h exposure. n.d. = no colonies detected and thus CFU could not be determined.

higher concentration (187 μg/mL), inactivation is much quicker, with no colonies having been detected on TSA plates even after a ten-minute exposure.

Spores of *B. pumilus* SAFR-032 were dried onto sterile aluminum coupons and exposed to both active and inactive chlorine dioxide. Additional data was collected for the control sample and after exposure to 3.5% w/v hydrogen peroxide, an agent for which *B. pumilus* SAFR-032 has resistance (Link et al. 2004; Gioia et al. 2007). Table 2 shows the results of exposure of these agents against *B. pumilus* SAFR-032 spores at one- and 24-hour time intervals, as determined by CFU mL^{-1} values. These values were calculated from the raw plate counts after 24 h growth following exposure at each time interval, as shown in Additional file: 1 Tables S1 and S2 in the supporting information.

At the shorter time interval, activated chlorine dioxide demonstrates a 4-log reduction in viability compared to the control. The presence of viable colonies after activated-ClO_2 treatment is plausible given the ClO_2 concentration measured (70.4 μg/mL) and detection of viable spores at a lower concentration as seen in Table 1. The activated stabilized-ClO_2 clearly has sporicidal properties against *B. pumilus* SAFR-032; however, inactivated ClO_2 has no measureable effect. Hydrogen peroxide treatment produces a one order-of-magnitude inactivation compared to the control, less effective than activated chlorine dioxide. At the longer duration of exposure, the peroxide and activated ClO_2 are comparable in near-total spore kill. Inactivated chlorine dioxide still has no efficacy, being no better than the unexposed control. The almost-total lack of viable spores at the most concentrated solution shows that the efficacy of chlorine dioxide increases with longer durations, as previously seen in Table 1. Additionally, while 3.5% peroxide solution does show near-total inactivation at 24 hour exposure, the much-reduced inactivation in the one-hour exposure test indicates peroxide is slower-acting against *B. pumilus* SAFR-032 than chlorine dioxide.

Table 2 Counting of *B. pumilus* SAFR-032 spores (in CFU mL^{-1})

Exposure agent	CFU mL^{-1}, 1 h exposure	CFU mL^{-1}, 24 h exposure
Nothing (control)	9.4×10^9	2.65×10^9
Peroxide, 3.5% v/v solution	1.47×10^8	$<5 \times 10^2$
Inactivated ClO_2 solution	2.79×10^8	6.3×10^9
Activated ClO_2 solution	1.6×10^5	$<1 \times 10^4$

Notes: Results of duplicate plate-counting experiments using 100 μL aliquots of exposed spores on aluminum coupons in serial dilutions of PBS following ultrasonication into sodium thiosulfate to inactivate chlorine dioxide. Inactive ClO_2 concentration = 108.5 μg/mL total ClO_2; active ClO_2 concentration = 70.4 μg/mL total ClO_2, 39 μg/mL free ClO_2. CFU mL^{-1} values based on the first plates in the serial dilution to not be too numerous to count (>300 total CFU or >100 CFU in one quarter of plate).

Discussion

Efficacy of chlorine dioxide on spore inactivation, comparison with hydrogen peroxide

Stabilized chlorine dioxide could be useful for inactivation of resistant spores. As Table 1 shows, there is no appreciable loss of efficacy between *B. subtilis* ATCC 6051 and *B. pumilus* SAFR-032 inactivation at 47 μg/mL. At higher concentrations, inactivation is rapid enough that no viable colonies were found in plate counting after a 10 minute exposure.

When chlorine dioxide was compared against hydrogen peroxide, as shown in Table 2, it demonstrated a greater inactivation at the one-hour exposure time than peroxide against *B. pumilus* SAFR-032. Considering the known peroxide-resistance traits in this species (Link et al. 2004; Gioia et al. 2007), this data is not unexpected. The three-log difference in the CFU mL^{-1} value indicates that chlorine dioxide is able to circumvent the resistance properties of the spore and inactivate it in some manner. The chlorine dioxide must be activated, however; it is clear from the results that inactivated ClO_2 has no measurable sporidical properties. Either inactive chlorine dioxide solution cannot permeate the spore, or chlorine dioxide is not activated within the spore environment. At the longer duration exposure of 24 hours, the chlorine dioxide and peroxide are roughly equal in sporicidal efficacy. The longer time necessary for peroxide to inactivate spores of *B. pumilus* SAFR-032 may result from a slow rate of penetration for peroxide, a slower rate of oxidative attack, or depletion of protective molecules in the dormant spore, such as peroxidases (Gioia et al. 2007; Checinska et al. 2012; Tirumalai et al. 2013). Another possibility arises from study of the oxidation potentials of the stabilized solution, which activated and in solution is primarily chlorine dioxide and chlorite (ClO_2^-), and hydrogen peroxide. While H_2O_2 has higher redox potential than ClO_2 or ClO_2^- (1.776 V vs. 0.954 V or 0.76 V), our results indicate less inactivation from the peroxide. It is possible that the different oxychloro species in the activated solution, mainly chlorine dioxide and chlorite, have a synergistic effect to enhance the overall sporicidal properties of the solution.

Despite the resistance mechanisms of *B. pumilus* SAFR-032 for peroxides, the chlorine dioxide treatment demonstrated efficacy against spores of this resistant strain as noted in Tables 1 and 2. The effects of ClO_2 appear equally valid for *B. subtilis* ATCC 6051 based on the results of Table 1. The efficacy of chlorine dioxide against *B. pumilus* spores in general has not, to our knowledge, been studied before, and in particular not against *B. pumilus* SAFR-032. This work could be of importance for inactivation and sterilization of surfaces contaminated with *B. pumilus* SAFR-032 in particular, and *Bacillus* spores in general.

Additionally, this work could form the basis for novel clinical cleaning procedures for pathogenic spore-forming bacteria in hospital environments. By showing chlorine dioxide-based sterilization of bacterial spores, including UV-resistant species such as *B. pumilus* SAFR-032, activated chlorine dioxide applied as a liquid could be used in a clinical setting for disinfecting and sterilization of surfaces contaminated by dormant bacterial spores, as gaseous chlorine dioxide has been shown capable of performing (Lowe et al. 2013). This treatment could also be effective against contamination with *C. difficile* spores; such spores are only effectively neutralized with chlorite-containing hospital cleaning agents, as alcohol-based agents have no efficacy and chlorite-containing solutions could be harmful to hospital equipment in the concentrations necessary to achieve spore inactivation (MacLeod-Glover and Sadowski 2010; Gould and McDonald 2008). Stabilized chlorine dioxide solutions applied to contaminated surfaces could serve as a useful tool in eliminating pathogenic spores from clinical settings with less negative impacts on equipment or individuals than bleach or other chlorite-based cleaning agents.

Additional file

Additional file 1: Table S1. Plate counting of *B. pumilus* SAFR-032 spores, one-hour exposure time. **Table S2.** Plate counting of *B. pumilus* SAFR-032 spores, 24-hour exposure time.

Competing interests
The authors declared that they have no competing interests.

Authors' contributions
AF, MZ, AM performed the experiments. NK, PV, CVR and AF participated in the design of the study. AF drafted the manuscript. CVR, NK, PV edited the manuscript. CVR and PV conceived the study. All authors read and approved the final manuscript.

Acknowledgments
This work is supported by the National Institutes of Health (1R01GM090064-01), a NASA EPSCoR Research Infrastructure Development (RID) grant NN07AL49A, and the University of Oklahoma. We also wish to express our gratitude to Dr. Kasthuri Venkateswaran, NASA JPL Biotechnology and Planetary Protection Group for sharing the *B. pumilus* SAFR-032 strain.

Supporting information
Tables of the raw plate counting CFUs for *B. pumilus* SAFR-032 exposed to stabilized chlorine dioxide solution and hydrogen peroxide are included as Additional file: 1 Tables S1 and S2, which were used to calculate the CFU mL^{-1} in Table 2.

Author details
[1]Department of Chemistry and Biochemistry, Stephenson Life Sciences Research Center, University of Oklahoma, Norman, Oklahoma, USA. [2]Bio-Cide International, Inc., Norman, Oklahoma, USA. [3]Jet Propulsion Laboratory, California Institute of Technology, Biotechnology and Planetary Protection Group, Pasadena, California, USA.

References
Atrih A, Foster SJ (1999) The role of peptidoglycan structure and structural dynamics during endospore dormancy and germination. Antonie Van Leeuwenhoek 75(4):299–307

Atrih A, Zollner P, Allmaier G, Foster SJ (1996) Structural analysis of *Bacillus subtilis* 168 endospore peptidoglycan and its role during differentiation. J Bacteriol 178(21):6173–83

Best EL, Fawley WN, Parnell P, Wilcox MH (2010) The potential for airborne dispersal of *Clostridium difficile* from symptomatic patients. Clin Infect Dis 50(11):1450–7, doi:10.1086/652648

Beuchat LR, Pettigrew CA, Tremblay ME, Roselle BJ, Scouten AJ (2005) Lethality of chlorine, chlorine dioxide, and a commercial fruit and vegetable sanitizer to vegetative cells and spores of *Bacillus cereus* and spores of *Bacillus thuringiensis*. J Ind Microbiol Biotechnol 32(7):301–8, doi:10.1007/s10295-005-0212-7

Buhr TL, Young AA, Minter ZA, Wells CM, Shegogue DA (2011) Decontamination of a hard surface contaminated with *Bacillus anthracis* Δ Sterne and *B. anthracis* Ames spores using electrochemically generated liquid-phase chlorine dioxide (eClO2). J Appl Microbiol 111(5):1057–64, doi:10.1111/j.1365-2672.2011.05122.x

Chaudhry R (2011) Botulism: a diagnostic challenge. Ind J Med Res 134:10–2

Checinska A, Burbank M, Paszczynski AJ (2012) Protection of *Bacillus pumilus* spores by catalases. Appl Environ Microbiol 78(18):6413–22, doi:10.1128/AEM. 01211-12

Driks A (2002) Overview: development in bacteria: spore formation in *Bacillus subtilis*. Cell Mol Life Sci 59(3):389–91

Driks A (2004) The *Bacillus* spore coat. Phytopathology 94(11):1249–51, doi:10.1094/PHYTO.2004.94.11.1249

Errington J (1993) *Bacillus subtilis* sporulation: regulation of gene expression and control of morphogenesis. Microbiol Rev 57(1):1–33

Foegeding PM, Hemstapat V, Giesbrecht FG (1986) Chlorine dioxide inactivation of *Bacillus* and *Clostridium* spores. J Food Sci 51(1):197–201, doi:10.1111/j.1365-2621.1986.tb10869.x

Gioia J, Yerrapragada S, Qin X, Jiang H, Igboeli OC, Muzny D, Dugan-Rocha S, Ding Y, Hawes A, Liu W, Perez L, Kovar C, Dinh H, Lee S, Nazareth L, Blyth P, Holder M, Buhay C, Tirumalai MR, Liu Y, Dasgupta I, Bokhetache L, Fujita M, Karouia F, Eswara Moorthy P, Siefert J, Uzman A, Buzumbo P, Verma A, Zwiya H, McWilliams BD, Olowu A, Clinkenbeard KD, Newcombe D, Golebiewski L, Petrosino JF, Nicholson WL, Fox GE, Venkateswaran K, Highlander SK, Weinstock GM (2007) Paradoxical DNA repair and peroxide resistance gene conservation in *Bacillus pumilus* SAFR-032. PLoS One 2(9):e928, doi:10.1371/journal.pone.0000928

Gordon G, Rosenblatt AA (2005) Chlorine dioxide: The current State of the art. Ozone-Sci Eng 27(3):203–7, doi:10.1080/01919510590945741

Gould CV, McDonald LC (2008) Bench-to-bedside review: *Clostridium difficile* colitis. Crit Care 12(1):203, doi:10.1186/cc6207

Hageman JH, Shankweiler GW, Wall PR, Franich K, McCowan GW, Cauble SM, Grajeda J, Quinones C (1984) Single, chemically defined Sporulation medium for *Bacillus subtilis*: growth, sporulation, and extracellular protease production. J Bacteriol 160(1):438–41

Higgins D, Dworkin J (2012) Recent progress in *Bacillus subtilis* sporulation. FEMS Microbiol Rev 36(1):131–48, doi:10.1111/j.1574-6976.2011.00310.x

Kerney KR, Schuerger AC (2011) Survival of *Bacillus subtilis* endospores on ultraviolet-irradiated rover wheels and Mars regolith under simulated Martian conditions. Astrobiology 11(5):477–85, doi:10.1089/ast.2011.0615

Kim BR, Anderson JE, Mueller SA, Gaines WA, Kendall AM (2002) Literature review–efficacy of various disinfectants against Legionella in water systems. Water Res 36(18):4433–44, doi:10.1016/S0043-1354(02)00188-4

Link L, Sawyer J, Venkateswaran K, Nicholson W (2004) Extreme spore UV resistance of *Bacillus pumilus* isolates obtained from an ultraclean spacecraft assembly facility. Microb Ecol 47(2):159–63, doi:10.1007/s00248-003-1029-4

Lowe JJ, Gibbs SG, Iwen PC, Smith PW, Hewlett AL (2013) Impact of chlorine dioxide gas sterilization on nosocomial organism viability in a hospital room. Int J Environ Res Public Health 10(6):2596–605, doi:10.3390/ijerph10062596

MacLeod-Glover N, Sadowski C (2010) Efficacy of cleaning products for *C difficile*: Environmental strategies to reduce the spread of *Clostridium difficile*-associated diarrhea in geriatric rehabilitation. Can Fam Physician 56(5):417–23

Magge A, Granger AC, Wahome PG, Setlow B, Vepachedu VR, Loshon CA, Peng L, Chen D, Yq L, Setlow P (2008) Role of dipicolinic acid in the Germination, stability, and viability of spores of *Bacillus subtilis*. J Bacteriol 190(14):4798–807, doi:10.1128/jb.00477-08

Mallidis CG, Scholefield J (1987) Relation of the heat resistance of bacterial spores to chemical composition and structure II. Relation to cortex and structure. J Appl Bacteriol 63(3):207–15

Moeller R, Setlow P, Reitz G, Nicholson WL (2009) Roles of small, acid-soluble spore proteins and core water content in survival of *Bacillus subtilis* spores exposed to environmental solar UV radiation. Appl Environ Microbiol 75(16):5202–8, doi:10.1128/aem. 00789-09

Moeller R, Schuerger AC, Reitz G, Nicholson WL (2012) Protective role of spore structural components in determining *Bacillus subtilis* spore resistance to simulated mars surface conditions. Appl Environ Microbiol 78(24):8849–53, doi:10.1128/AEM. 02527-12

Newcombe DA, Schuerger AC, Benardini JN, Dickinson D, Tanner R, Venkateswaran K (2005) Survival of spacecraft-associated microorganisms under simulated Martian UV irradiation. Appl Environ Microbiol 71(12):8147–56, doi:10.1128/aem. 71.12.8147-8156.2005

Nicholson WL, Munakata N, Horneck G, Melosh HJ, Setlow P (2000) Resistance of *Bacillus* endospores to extreme terrestrial and extraterrestrial environments. Microbiol Mol Biol Rev 64(3):548–72

Novak J, Demirci A, Han Y (2008) Novel chemical processes: ozone, supercritical CO_2, electrolyzed oxidizing water, and chlorine dioxide gas. Food Sci Technol Int 14(5):437–41, doi:10.1177/1082013208098815

Paredes-Sabja D, Setlow P, Sarker MR (2011) Germination of spores of *Bacillales* and *Clostridiales* species: mechanisms and proteins involved. Trends Microbiol 19(2):85–94, doi:10.1016/j.tim.2010.10.004

Peniche AG, Savidge TC, Dann SM (2013) Recent insights into *Clostridium difficile* pathogenesis. Curr Opin Infect Dis 26(5):447–53, doi:10.1097/01.qco.0000433318.82618.c6

Piggot PJ, Hilbert DW (2004) Sporulation of *Bacillus subtilis*. Curr Opin Microbiol 7(6):579–86, 10.1016/j.mib.2004.10.001

Rastogi VK, Ryan SP, Wallace L, Smith LS, Shah SS, Martin GB (2010) Systematic evaluation of the efficacy of chlorine dioxide in decontamination of building interior surfaces contaminated with anthrax spores. Appl Environ Microbiol 76(10):3343–51, doi:10.1128/AEM. 02668-09

Riesenman PJ, Nicholson WL (2000) Role of the spore coat layers in *Bacillus subtilis* spore resistance to hydrogen peroxide, artificial UV-C, UV-B, and solar UV radiation. Appl Environ Microbiol 66(2):620–6, doi:10.1128/aem. 66.2.620-626.2000

Roberts K, Smith CF, Snelling AM, Kerr KG, Banfield KR, Sleigh PA, Beggs CB (2008) Aerial dissemination of *Clostridium difficile* spores. BMC Infect Dis 8:7, doi:10.1186/1471-2334-8-7

Schmid G, Kaufmann A (2002) Anthrax in Europe: its epidemiology, clinical characteristics, and role in bioterrorism. Clin Microbiol Infect 8(8):479–88

Schuerger AC, Mancinelli RL, Kern RG, Rothschild LJ, McKay CP (2003) Survival of endospores of *Bacillus subtilis* on spacecraft surfaces under simulated martian environments: implications for the forward contamination of Mars. Icarus 165(2):253–76, doi:10.1016/s0019-1035(03)00200-8

Setlow P (1988) Small, acid-soluble spore proteins of *Bacillus* species: structure, synthesis, genetics, function, and degradation. Annu Rev Microbiol 42:319–38, doi:10.1146/annurev.mi.42.100188.001535

Setlow P (2003) Spore germination. Curr Opin Microbiol 6(6):550–6, doi:10.1016/j.mib.2003.10.001

Setlow P (2006) Spores of *Bacillus subtilis*: their resistance to and killing by radiation, heat and chemicals. J Appl Microbiol 101(3):514–25, doi:10.1111/j.1365-2672.2005.02736.x

Setlow B, Atluri S, Kitchel R, Koziol-Dube K, Setlow P (2006) Role of dipicolinic acid in resistance and stability of spores of *Bacillus subtilis* with or without DNA-protective α/β-Type small acid-soluble proteins. J Bacteriol 188(11):3740–7, doi:10.1128/jb.00212-06

Sharma VK, Sohn M (2012) Reactivity of chlorine dioxide with amino acids, peptides, and proteins. Environ Chem Lett 10(3):255–64, doi:10.1007/s10311-012-0355-5

Shemesh M, Kolter R, Losick R (2010) The biocide chlorine dioxide stimulates biofilm formation in *Bacillus subtilis* by activation of the histidine kinase KinC. J Bacteriol 192(24):6352–6, doi:10.1128/JB.01025-10

Spotts Whitney EA, Beatty ME, Taylor THJ, Weyant R, Sobel J, Arduino MJ, Ashford DA (2003) Inactivation of *Bacillus anthracis* spores. Emerg Infect Dis 9(6):623–7

Takamatsu H, Watabe K (2002) Assembly and genetics of spore protective structures. Cell Mol Life Sci 59(3):434–44

Tauveron G, Slomianny C, Henry C, Faille C (2006) Variability among *Bacillus cereus* strains in spore surface properties and influence on their ability to contaminate food surface equipment. Int J Food Microbiol 110(3):254–62, doi:10.1016/j.ijfoodmicro.2006.04.027

Tirumalai MR, Rastogi R, Zamani N, O'Bryant Williams E, Allen S, Diouf F, Kwende S, Weinstock GM, Venkateswaran KJ, Fox GE (2013) Candidate genes that may be responsible for the unusual resistances exhibited by *Bacillus pumilus* SAFR-032 spores. PLoS One 8(6):e66012, doi:10.1371/journal.pone.0066012

Vaishampayan PA, Rabbow E, Horneck G, Venkateswaran KJ (2012) Survival of *Bacillus pumilus* spores for a prolonged period of time in real space conditions. Astrobiology 12(5):487–97, doi:10.1089/ast.2011.0738

Venkateswaran K, Vaishampayan P, Cisneros J, Pierson D, Rogers S, Perry J (2014) International Space Station environmental microbiome — microbial inventories of ISS filter debris. Appl Microbiol Biotechnol 98(14):6453–66, doi:10.1007/s00253-014-5650-6

Warth AD, Strominger JL (1971) Structure of the peptidoglycan from vegetative cell walls of *Bacillus subtilis*. Biochemistry 10(24):4349–58

Warth AD, Strominger JL (1972) Structure of the Peptidoglycan from spores of *Bacillus subtilis*. Biochemistry 11(8):1389–96

Young SB, Setlow P (2003) Mechanisms of killing of *Bacillus subtilis* spores by hypochlorite and chlorine dioxide. J Appl Microbiol 95(1):54–67

Zhao J, Krishna V, Moudgil B, Koopman B (2008) Evaluation of endospore purification methods applied to *Bacillus cereus*. Sep Purif Technol 61(3):341–7, doi:10.1016/j.seppur.2007.11.002

26

Metabolism of alkenes and ketones by *Candida maltosa* and related yeasts

Andy Beier[1,2], Veronika Hahn[1], Uwe T Bornscheuer[2*] and Frieder Schauer[1]

Abstract

Knowledge is scarce about the degradation of ketones in yeasts. For bacteria a subterminal degradation of alkanes to ketones and their further metabolization has been described which always involved Baeyer-Villiger monooxygenases (BVMOs). In addition, the question has to be clarified whether alkenes are converted to ketones, in particular for the oil degrading yeast *Candida maltosa* little is known. In this study we show the degradation of the aliphatic ketone dodecane-2-one by *Candida maltosa* and the related yeasts *Candida tropicalis*, *Candida catenulata* and *Candida albicans* as well as *Trichosporon asahii* and *Yarrowia lipolytica*. One pathway is initiated by the formation of decyl acetate, resulting from a Baeyer-Villiger-oxidation of this ketone. Beyond this, an initial reduction to dodecane-2-ol by a keto reductase was clearly shown. In addition, two different ways to metabolize dodec-1-ene were proposed. One involved the formation of dodecane-2-one and the other one a conversion leading to carboxylic and dicarboxylic acids. Furthermore the induction of ketone degrading enzymes by dodecane-2-one and dodec-1-ene was shown. Interestingly, with dodecane no subterminal degradation products were detected and it did not induce any enzymes to convert dodecane-2-one.

Keywords: Hydrocarbon; alkene; ketone; *Candida*; yeast; biotransformation

Introduction

Aliphatic hydrocarbons (alkanes) are major components of natural gasoline and crude oil. However, alkanes can also be found in wax layers of plants and animals to reduce their loss of water by evaporation (Cheesbrough and Kolattukudy 1988). These mostly very hydrophobic substances can be used as sole carbon and energy sources for microbial growth. For this, oxygen serves as electron acceptor under aerobic conditions, whereas under anaerobic conditions sulfate and nitrite have this function for some bacteria (Ji et al. 2013). If oxygen is available, in most cases - especially in eukaryotic microorganisms - the degradation of alkanes is initiated by cytochrome P450 monooxygenases. The function of these enzymes is the oxidation of the *n*-alkanes to alcohols and the ω-hydroxylation of fatty acids (Scheller et al. 1998). In course of the oxidation of aliphatic hydrocarbons by yeasts a mono- or diterminal degradation is possible (Blasig et al. 1988; Käppeli 1986; Mauersberger et al. 1996; Schunck et al. 1987). Within the monoterminal pathway the alkanes are converted by terminal hydroxylation to the corresponding primary alcohols, which are subsequently oxidized by alcohol dehydrogenases (or alcohol oxidases) and aldehyde dehydrogenases to fatty acids, which are further degraded via β-oxidation (Watkinson and Morgan 1991). For diterminal degradation primary alcohols from the monoterminal pathway are hydroxylated at their ω-position resulting in the formation of diols. These are subsequently oxidized to ω-hydroxy fatty acids and then to dicarboxylic acids which are degraded by β-oxidation (Coon 2005; Kester and Foster 1963; Krauel et al. 1973; Watkinson and Morgan 1991). For some bacteria and filamentous fungi also a subterminal degradation is possible. Here, the initial hydroxylation happens intramolecularly and a secondary alcohol is formed (Forney and Markovetz 1970). The secondary alcohol is oxidized to the corresponding ketone, subsequently converted to an ester by a Baeyer-Villiger monooxygenase (BVMO) which can be hydrolysed by an esterase to an alcohol and a fatty acid (Forney and Markovetz 1970; Kirschner et al. 2007; Kotani et al. 2007; Lottmann et al. 1999). After the primary oxidation of the alkanes in all cases efficient fatty acid oxidation systems as well as

* Correspondence: uwe.bornscheuer@uni-greifswald.de
[2]Institute of Biochemistry, Department of Biotechnology & Enzyme Catalysis, Greifswald University, Felix-Hausdorff-Str. 4, 17487 Greifswald, Germany
Full list of author information is available at the end of the article

multiple anaplerotic metabolic sequences (e.g. acetyl-CoA metabolism, gluconeogenesis) are required (Schauer 2001). In general it is claimed that a mono- and partly di- and subterminal degradation of hydrocarbons is performed by filamentous fungi (Rehm and Reiff 1981). However, for yeast the subterminal degradation is unclear. Some yeasts possess excellent abilities to degrade *n*-alkanes and other aliphatic hydrocarbons. *Candida maltosa* is a yeast which is able to metabolize hydrocarbons fastly (Komagata et al. 1964). Thus, its growth with *n*-hexadecane is faster compared to that with glucose. That is the reason why *C. maltosa* is sometimes called "super oil yeast" and cause of the assumption that its alkane metabolizing enzymes have a high catalytic activity. The transfer of *C. maltosa* from a medium with glucose to one with *n*-alkanes results in the induction of several catabolic enzymes and further proteins as well as changes in the endoplasmatic reticulum and amount of peroxisomes in which these enzymes and proteins are located. Furthermore the degradation of *n*-alkanes, primary alcohols, secondary alcohols and ketones can be increased in *C. maltosa* in varying degrees according to their respective chain length with the addition of biotin (Schauer 1988). The availability of only short-chain hydrocarbons as carbon source results in a complete degradation to acetyl-CoA. This is amongst others used for the *de novo* fatty acid synthesis under requirement of biotin as cofactor for carboxylases (Tehlivets et al. 2007). However in case that long-chain hydrocarbons are available as well, these can be converted to fatty acids directly and in conclusion no carboxylation and therefore no biotin is required (Schauer 1988).

Unsaturated aliphatic hydrocarbons (alkenes) can be found in nature as gaseous and volatile compounds. Examples are the plant hormone ethene, isoprene from foliage and various monoterpenes as main components in plant oils (Abeles et al. 1992; Primrose 1979; Rasmussen 1970). Alkenes have received some attention as oxidable substrates for microorganisms. Investigations about the metabolism of hexadec-1-ene in *Y. lipolytica* indicated the formation of the corresponding ω-unsaturated primary and secondary alcohols and the respective fatty acids by oxidation of the saturated end of this alkene. Moreover the oxidation of the double bond resulted in the conversion to 1,2-epoxyhexadecane, hexadecane-1-ol, hexadecane-2-ol, 2-hydroxyhexadecanoic acid and hexadecane-1,2-diol. The latter was converted to the 2-hydroxy acid and after that oxidatively decarboxylated to the one carbon atom shortened fatty acid (Bruyn 1954; Klug 1969; Klug and Markovetz 1967, 1968; Stewart et al. 1960).

Apart from the formation from alkanes and alkenes by microorganisms, ketones are occurring in plant oils, insects and mammals (Forney and Markovetz 1971). Due to that ketones are common substances in the environment and therefore there are microorganisms that can convert those. Forney *et al.* showed the formation of undecane-1-ol and undecyl acetate from tridecane-2-one by *Pseudomonas multivorans* (Forney et al. 1967). In the course of further degradation undecyl acetate was hydrolyzed to undecane-1-ol and acetate. Moreover, it was reported that the strain *Gordonia* spec. TY-5 can convert propane to propan-2-ol and further to acetone (Kotani et al. 2007). This bacterium formed methyl acetate from the latter with a BVMO, which was subsequently hydrolyzed to methanol and acetate. This kind of ketone metabolism can not only be found in prokaryotes. One example from the fungus kingdom is the conversion of progesterone to testosterone acetate by *Cladosporium resinae* and the subsequent ester hydrolysis to the steroid alcohol and acetate (Fonken et al. 1960; Rahim and Sih 1966). Another one is the formation of ε-caprolactone from cyclohexanone by the cycloalkanone monooxygenase from *Cylindrocarpon radicicola* (Leipold et al. 2012).

However, there are no studies about the metabolism of ketones in yeasts. One can only suppose that it might be the same as in other fungi and bacteria but experimental evidence is missing. To understand the metabolism of aliphatic ketones in yeasts *Candida maltosa*, *Candida albicans*, *Candida tropicalis*, *Candida catenulata*, *Yarrowia lipolytica* and *Trichosporon asahii* were cultivated in this work with dodecane-2-one as sole carbon and energy source. Furthermore it is unclear, whether ketones could be formed from alkanes or alkenes by the hydrocarbon oxidizing yeast *Candida maltosa*. To answer this question, *C. maltosa* was also cultivated with dodecane and dodec-1-ene to investigate their degradation pathways. In addition, resting cells of *C. maltosa* were incubated with dodecane-2-one, dodec-1-ene and dodecane after pre-cultivation with one of these substrates to show their conversion rates and the formation of products resulting from ketone degradation.

Materials and methods

Chemicals

All chemicals were purchased from Fluka (Buchs, Switzerland) or Sigma-Aldrich (Steinheim, Germany).

Yeast strains and culture conditions

Experiments were carried out using *Candida maltosa* SBUG 700, *Candida tropicalis* SBUG 1019, *Candida catenulata* SBUG 512, *Yarrowia lipolytica* SBUG 1888, *Trichosporon asahii* SBUG 833 *and Candida albicans* SBUG 5121 which are deposited at the strain collection of the Department of Biology of the University of Greifswald (SBUG) from where they can be obtained upon request. Cultures were inoculated from an overnight malt agar culture.

Yeast cells were cultivated in a mineral salt medium (5 g $NH_4H_2PO_4$, 2.5 g KH_2PO_4, 1 g $MgSO_4 \times 7\ H_2O$, 0.02 g

$Ca(NO_3)_2 \times 7 H_2O$, 2.0 mg $FeCl_3 \times 6 H_2O$, 0.5 mg H_3BO_3, 0.4 mg $MnSO_4 \times 5 H_2O$, 0.4 mg $ZnSO_4 \times 7 H_2O$, 0.2 mg Na_2MoO_4, 0.1 mg $CuSO_4 \times 5 H_2O$, 0.1 mg $CoCl_2$, 0.1 mg KI in 1 L dest. H_2O, pH 5.4 (Hornei et al. 1972)) supplemented with 1% (v/v) vitamine solution (Van der Walt and van Kerken 1961) and 1% (v/v) of the respective substrate (dodecane-2-one, dodec-1-ene, dodecane). Cultures grew until an OD_{600nm} of 3 at 30°C and 250 rpm for 31.5 h (dodecane), 32 h (dodec-1-ene), 40 h (dodecane-2-one, *C. maltosa*), 290 h (*C. albicans*), 72 h (*C. catenulata*) 47 h (*C. tropicalis*), 189 h (*Y. lipolytica*) and 381 h (*T. asahii*). Cells were centrifuged at 4°C and 10000 × g for 15 min (Sorvall RC-5B Plus Refrigerated Superspeed Centrifuge). The supernatant was analyzed for secreted products (see section Liquid-liquid extraction). Controls contained either 1% (w/v) glucose, no substrate or no cells.

Preparation of resting *Candida maltosa* cells

Cell pellets from cultures of *Candida maltosa* grown with dodecane, dodec-1-ene or dodecane-2-one were washed three times with 100 mL sodium phosphate buffer (SPB, 67 mM, pH 5.4). Subsequently cell suspensions with an OD_{600nm} of 5 (approx. $5.8*10^7$ cfu/ml, total cell count of $2.4*10^9$ cells/ml, cell dry mass of 3.49 mg/ml) were prepared in SPB.

Biotransformations

For biotransformation experiments, resting cell preparations of *Candida maltosa* were used. Reactions were carried out in SPB at 0.25% (v/v) substrate concentration. Cells were centrifuged and suspended in SPB. The OD_{600nm} was adjusted to 5. No vitamines/cofactors were added. Cell suspensions were used in amounts of 2 mL. Controls contained no cells or no substrate. The total volume in every case was 10 mL. Incubation was performed in 500 mL flasks at 30°C and 250 rpm for 0, 1 and 4 hours. After these times the whole biotransformation approach of 10 mL was centrifuged for 5 min at 3857 g.

Liquid-liquid extraction

Before extraction the pH-value of the supernatants was first adjusted to pH 9 by addition of 25% NaOH (v/v). Extraction of samples was achieved by vortexing three times with 20 mL of diethyl ether (basic extracts). Next, the pH-value was set to pH 2 by addition of 32% HCl (v/v). Then, samples were extracted again as mentioned above (acidic extracts). Samples were dried over anhydrous sodium sulphate, concentrated in a rotary evaporator and desiccated with nitrogen. For GC/MS analysis the samples were dissolved in 500 µl hexane. The acidic extracts were derivatized by methylation using diazomethane (De Boer and Backer 1956). For investigation of metabolites formed during cultivation the respective supernatants were extracted three times with 50 mL diethyl ether first at pH 9 and then at pH 2. Further workflow was the same as mentioned above.

Structure elucidation of products by GC/MS

GC/MS analyses were performed on an Agilent gas chromatograph 7890A GC system (Waldbronn, Germany) equipped with a 30 m HP-5 ms column (0.25 mm by 0.25 µm film) and linked to a mass selective detector 5975C inert XL EI/CI MSD with a quadrupole mass spectrometer.

For separation of products a temperature program was used, starting with 5 min at 60°C followed by a ramp from 60–120°C at 20°C/min. The 120°C were maintained for 5 min and then followed by heating the column to 200°C at 3°C/min, to 280°C at 20°C/min and lastly 5 min at 280°C. For quantification of products formed commercially available standard substances were measured with concentrations from 0.5-10 mM for the creation of calibration curves.

Analytical data of the products can be found in the Additional file 1: Table S1.

Results

Biotransformation of dodecane-2-one, dodec-1-ene and dodecane by growing yeast cells

All the investigated yeasts *C. maltosa*, *C. albicans*, *C. catenulata*, *C. tropicalis*, *Y. lipolytica* and *T. asahii* and were able to grow with dodecane-2-one (**1**) as a sole source of carbon and energy. In the supernatants of the culture media a variety of metabolites was formed (Figure 1, see Additional file 1: Figure S1 for an exemplary GC chromatogram). The spectrum of products formed from **1** was very similar among the test organisms whereas the amount of formed products differed (Table 1). However, in every case dodecane-2-ol (**6**) and also (with the exception of *C. albicans*) decyl acetate (**4**) was detected.

In contrast to the incubation of *C. maltosa* with **1** in experiments with dodec-1-ene (**2**), only five were detected (Table 1). In cultures with dodecane (**3**) only two metabolites were formed, whereas it has to be highlighted that here in contrast to **2** no formation of ketones, esters and secondary alcohols could be observed. In control appoaches with glucose, no substrate or no cells none of these substances could be detected.

Biotransformation of dodecane, dodec-1-ene and dodecane-2-one by resting *C. maltosa* cells

Differences in the induction of enzymes involved in ketone metabolism, including BVMOs, in *C. maltosa* after cultivation with **1**, **2** and **3** as sole carbon and energy source were examined. Therefore the conversion of the substrates **1**, **2** and **3** (see section Conversion rate of **1–3** in dependence of the pre-culture substrate) and the formation of compounds

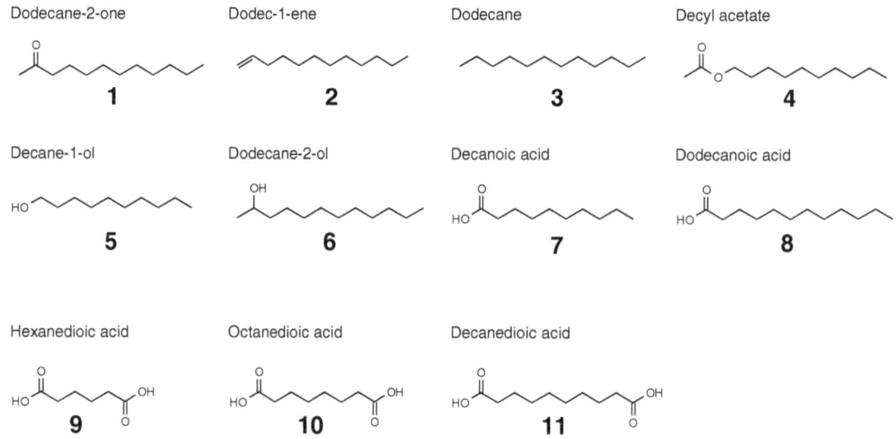

Figure 1 Substrates (1, 2, 3) and extracellular metabolites (1, 4–11) detected in cultures containing either dodecane-2-one, dodec-1-ene or dodecane. Carboxylic acids were detected as methyl esters.

(see section Product formation in course of biotransformation) were analysed by GC/MS.

Conversion rate of 1–3 in dependence of the pre-culture substrate

On the basis of the determined data the conversion of the substrates was calculated (Figure 2). The conversion started immediately with a high rate. Thus, approximately 20% were transformed in almost every sample within 30 min. Cells pre-cultured with **1** transformed 25.18 μmol (22.4%) of **1**. The conversion with cells cultivated with **2** was comparable to that. In this case 18.92 μmol (16.8%) of **1** were transformed. Cells cultured with **3** showed negligible conversion of **1** (1.03 μmol, 0.9%). Moreover, the conversion of **2** was 24.08 μmol (21.4%) and 24.58 μmol (21.8%) with cells grown on **2** and **3**, respectively (data not shown). Apart from that the conversion of **3** with cells pre-cultured with **3** was 18.07 μmol (16.4%) (data not shown).

Table 1 Metabolites identified in the yeast culture supernatants after growth until OD_{600nm} of 3 with 1% dodecane-2-one (1), dodec-1-ene (2) or dodecane (3) as sole carbon and energy source at 30°C and 250 rpm

Strain	Substrate	Detected metabolites
C. maltosa	1	4, 5, 6, 9, 10, 11
	2	1, 6, 7, 8, 9
	3	8, 9
C. albicans	1	6
C. tropicalis	1	4, 5, 6, 9, 10, 11
C. catenulata	1	4, 5, 6, 9, 10, 11
Y. lipolytica	1	4, 5, 6, 9, 10
T. asahii	1	4, 5, 6, 9, 11

Product formation in course of biotransformation

In case of *C. maltosa* cells incubated with 112.3 μmol **1**, which were pre-cultured with **1** or **2**, the metabolites **4**, **5** and **6** were detected (Figure 3). Cells grown on **3** did not secret these three compounds. Differences in product concentration were observed for **6** (Figure 3A). Thus, cells grown on **1** formed 0.18 μmol **6** while pre-cultured with **2** only 0.07 μmol were formed. The formation of **4** from **1** also differed in dependence of the substrate used for pre-cultivation (Figure 3B). Only cells grown on **1** were able to secrete an amount of 0.03 μmol **4**. Biotransformations with cells cultivated on **2** resulted in the formation of approx. 0.02 μmol **4**. The primary alcohol **5**, the decomposition product of **4**, was formed in similar rates as **4** (Figure 3C). Thus, amounts of **5** of approx. 0.02 μmol were detected with cells pre-cultured on **1** and **2**. In control experiments with *C. utilis*, a yeast which is not

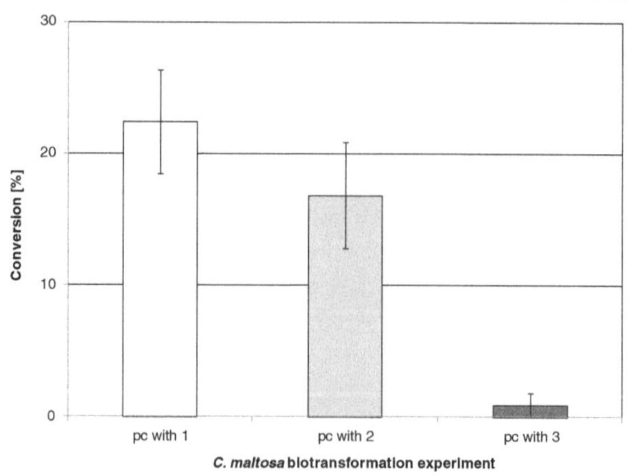

Figure 2 Conversion of dodecane-2-one (1) after 0.5 h by resting cells of *C. maltosa* either pre-cultured (pc) with 1, dodec-1-ene (2) or dodecane (3) for induction followed by biotransformation (bt). (For determination of conversion GC/MS was used).

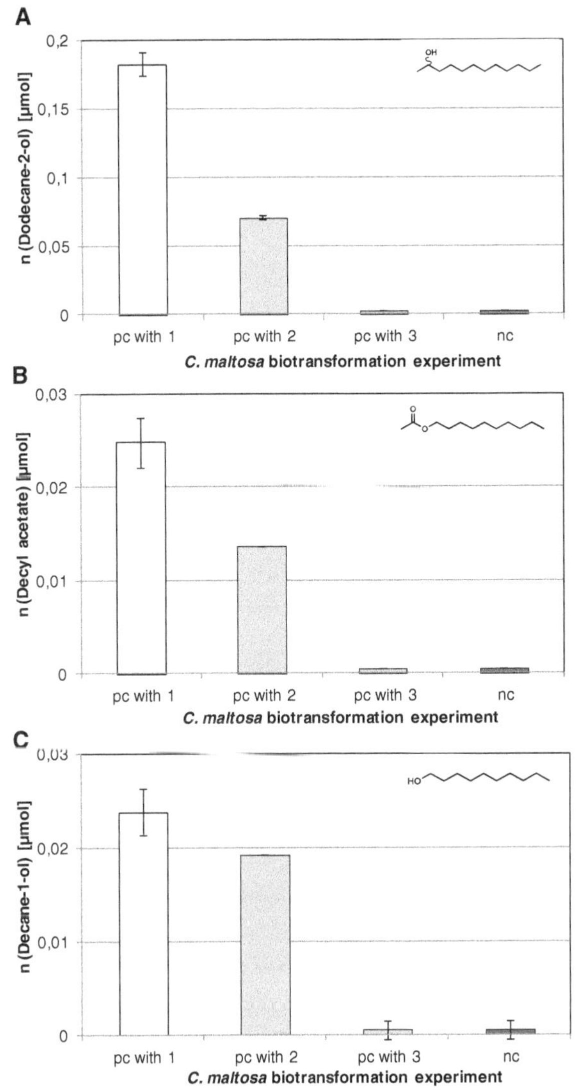

Figure 3 Formation of dodecane-2-ol (A), decyl acetate (B) and decane-1-ol (C) from dodecane-2-one (1) after 0.5 h by resting cells of *C. maltosa*, pre-cultured (pc) with 1, 2 or 3 as well as by resting cells of the hydrocarbon non-utilising yeast *Candida utilis*, pc with glucose (=negative control, nc).

able to metabolize aliphatic hydrocarbons, and **1** as substrate no product formation was observed. In addition, after incubation of *C. maltosa* cells pre-cultured with **2** or **3** and the biotransformation substrates **2** and **3** neither **4**, **5** nor **6** were detected (data not shown). In approaches without cells or substrate theses metabolites could not be detected.

Discussion

In this study we investigated the metabolism of alkenes and ketones in yeasts. We were able to detect the degradation products **1**, **6**, **7**, **8** and **9** from dodec-1-ene and **4**, **5**, **6**, **9**, **10** and **11** from dodecane-2-one in cultures of growing yeast cells and partially as well in biotransformation approaches with resting yeast cells. The production of some of these metabolites provided evidence of the induction of ketone degrading enzymes after growth on **1** or **2**, but not on **3**, in the investigated hydrocarbon oxidizing yeasts.

Biotransformation of dodecane-2-one by growing yeast cells

After cultivation of *C. maltosa*, *C. tropicalis*, *C. catenulata*, *C. albicans*, *Y. lipolytica* und *T. asahii* with **1** or **2** as sole carbon and energy source the culture media were used for the determination of formed metabolites to study the metabolism of aliphatic ketones and alkenes.

In culture media from *C. maltosa* with **1** as substrate in total six metabolites were detected. Although in quantity less compounds were secreted by some other yeasts, the products were the same. The conversion of **1** was either initiated by a keto reductase or a BVMO (Figure 4).

The first degradation pathway was identified for all investigated yeast strains with exception of *C. albicans* which is initiated by a BVMO with the formation of **4**. The enzymatic Baeyer-Villiger oxidation can lead to both regioisomeric products of the corresponding ketone (Alphand et al. 1989; Balke et al. 2012; Kelly et al. 1995; Leisch et al. 2011; Rehdorf et al. 2010; Song et al. 2013). In this case the "normal" product is **4** and undecanoic acid methyl ester would be the "abnormal" one. It seems that *C. maltosa* as a whole cell catalyst rather formed the "normal" product **4** than the "abnormal" undecanoic acid methyl ester as nothing of the latter could be detected. The prefered formation of the "normal" product from a ketone by a BVMO has been described already (Mihovilovic and Kapitán 2004). In addition, **5** was detected which was probably formed by conversion of **4** via a hydrolase. A similar reaction was reported by Forney *et al.* in which *Pseudomonas multivorans* converted tridecane-2-one to undecyl acetate and subsequently hydrolysed it to undecane-1-ol and acetate (Forney and Markovetz 1968; Forney et al. 1967). Due to the slow growth of *C. albicans* with **1** the accumulation of products may not have been sufficient to detect them in the cultures of this yeast. That could be the reason why just **6** was found as a metabolite.

In addition, the formed decanoic acid was probably hydroxylated at its ω-end resulting in the formation of 10-hydroxydecanoic acid which was further oxidized to the respective dicarboxylic acid decanedioic acid (**11**). Decandioic acid was degraded by β-oxidation which was confirmed by (De Boer and Backer 1956) detection of its C_2 shortened intermediates until hexandioic acid. Therefore a diterminal degradation took place what has been previously reported for the filamentious fungus *Mortierella isabellina*, the bacteria *Rhodococcus rhodochrous* and *Corynebacterium* spp. as well as for the yeasts *Candida guilliermondii* and *Candida* spp. (Coon 2005; Kester and

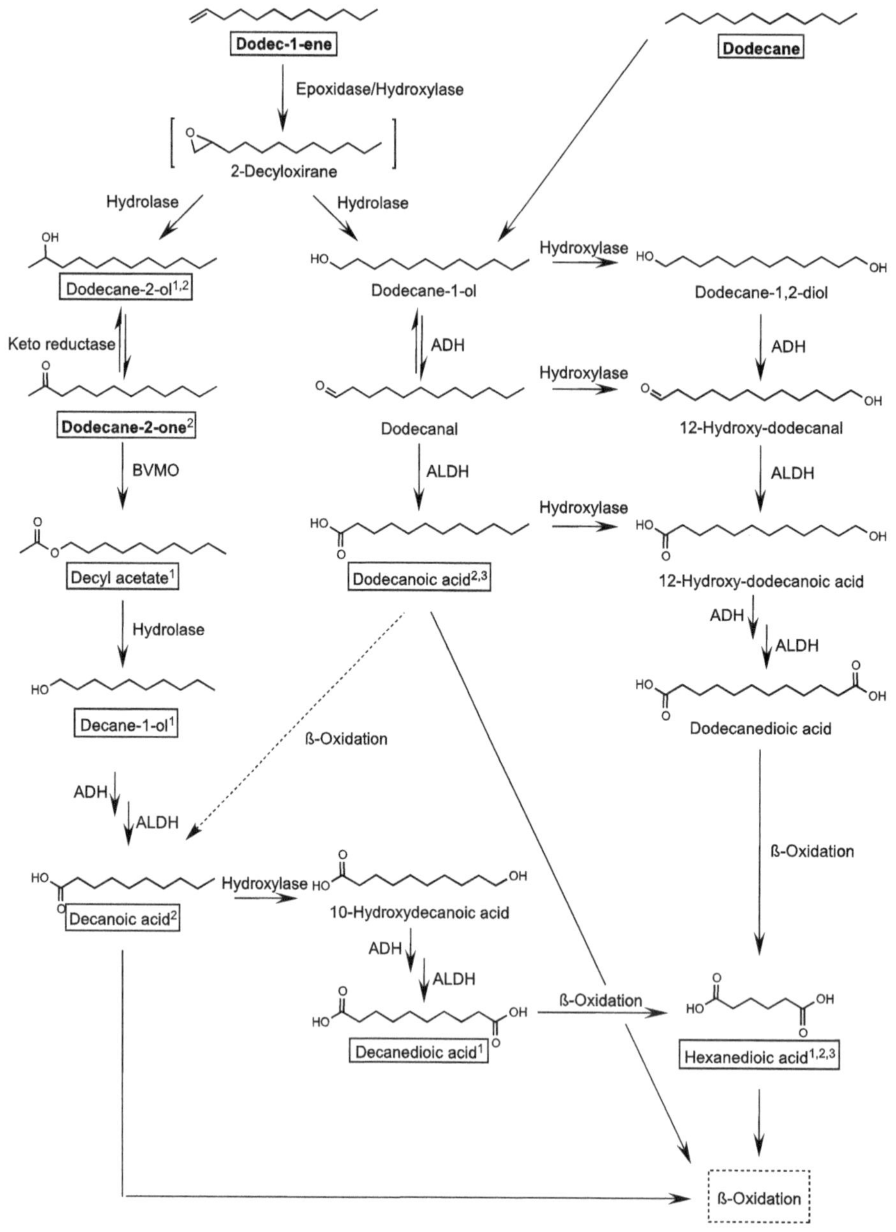

Figure 4 Suggested metabolism of dodecane-2-one, dodec-1-ene and dodecane in *Candida maltosa* SBUG 700. A black frame indicates intermediates detected from cultures of *C. maltosa* with **1**, **2** or **3** as sole carbon and energy source. Metabolites found in cultures 1: with **1**, 2: with **2**, 3: with **3**, ADH: Alcohol dehydrogenase, ALDH: Aldehyde dehydrogenase

Foster 1963; Krauel et al. 1973; Watkinson and Morgan 1991). At least some of the intermediates of the β-oxidation of this dicarboxylic acid were detected in the cultures of all investigated yeasts. The formation of dicarboxylic acids ranging from six to 12 carbons from dodecane for *C. tropicalis* and from 12 to 16 carbons from the respective alkanes for *Y. lipolytica* has already been reported (Picataggio et al. 1992; Smit et al. 2005).

Another degradation pathway is proposed by the detection of **6**. Therefore the yeasts have to contain at least one keto reductase that can reduce **1** to this secondary alcohol. The reduction of ketones like hexane-2-one, methyl acetoacetate and α-tetralone to the respective alcohols was described for other yeasts such as *Saccharomyces cerevisiae* and *Candida viswanathii* (Patil et al. 2013; Wolfson et al. 2013). It was assumed that this keto reductase has a relatively high activity due to the 7.4-fold and 7.7-fold higher amount of detected **6** in comparison to **4** and **5** respectively in approaches of **1** pre-cultured resting cells incubated with **1**.

Biotransformation of dodecane-1-ene and dodecane by growing yeast cells

In *C. maltosa* cultures with **2** as sole source of carbon and energy five compounds were detected. One of the products was **1**. For the formation of this compound **2** reacted possibly via 1,2-epoxidodecane to **6** and subsequently to **1** (Figure 4).

In addition, dodecane-1-ol possibly was formed also via 1,2-epoxidodecane from **2** and oxidized to dodecanal and subsequently to **8**. This acid was further metabolized via β-oxidation. Some of these reactions were already described for *Y. lipolytica*, *C. tropicalis* and *Candida* spec. (Bruyn 1954; Klug 1969; Klug and Markovetz 1967, 1968; Terasawa and Takahashi 1981). By ω-oxidation of **8** the dioic acid was formed and transformed by β-oxidation to **9**. The reason why there was no formation of dodecane-1-ol from **2** detectable could be that this primary alcohol was metabolized too fast to **7** and the subsequent products. Also, *C. tropicalis* and *Y. lipolytica* were able to form tetradecane-2-ol and tetradecane-2-one from tetradec-1-ene (Terasawa and Takahashi 1981).

In contrast, in *C. maltosa* culture media with **3** neither **4**, **5** nor **6** was formed. Instead, the degradation was accomplished by a monoterminal oxidation. Thus, dodecanoic acid was detected. In addition, a diterminal oxidation was observed supported by the detection of hexandioic acid. A similar study with *Candida rugosa* and decane indicated a mono- and diterminal oxidation as well (Mizuno et al. 1966).

Dependence of the enzyme induction on the substrate used for pre-cultivation

In order to reproduce the differences in product formation after cultivation of alkanes and alkenes the induction of ketone converting enzymes was studied and biotransformations with resting cells of *C. maltosa* SBUG 700 were performed. Cells grown on either **1**, **2** or **3** were examined regarding substrate conversion and product formation. Our results are matching the principle of the simultaneous adaptation by Stanier (1947). According to that the catabolism of all occurring intermediates is induced by incubation with the initial substrate if the participating enzymes of that pathway are inducible. In agreement with this principle **1** was immediately converted after pre-cultivation with **2** although the conversion rate was slightly reduced compared with cells grown with **1** (Figure 2). In contrast, cells pre-cultured with **3** transformed **1** just about 0.91% probably due to the lack of conversion of **3** to **1** during pre-cultivation. In addition, an enzyme with BVMO activity and a keto reductase were induced in cells pre-cultured with **1** or **2** as those formed **4** and its fission product **5** as well as **6** after incubation with **1** (Figure 3). The fact that with **1** pre-cultured cells formed higher amounts of all these 3 products than with **2** pre-cultured cells (e.g. 0.18 vs. 0.07 μmol dodecane-2-ol) indicates that the presence of the ketone leads to a stronger induction. Thus, the conversion of **2** to **1** during cultivation was probably not sufficient enough to form an adequate amount of **1** to induce the expression of ketone degrading enzymes as strong as with pure **1** alone. On **3** pre-cultured cells incubated with **1** formed none of the compounds **4**, **5** and **6**. Thus, the above mentioned lack of conversion of **3** to **1** during pre-cultivation was confirmed. This is leading to the conclusion that a subterminal oxidation of alkanes, in contrast to alkenes, does not take place in *C. maltosa*. Impurities of alkenes in alkane preparations can therefore lead to incorrect interpretations of a participation of subterminal reactions.

It was shown that *C. maltosa* can convert dodec-1-ene to dodecane-2-one and this aliphatic ketone to decyl acetate resulting from a BVMO activity which can also be found in all of the here investigated hydrocarbon oxidizing yeasts. The reduction of **1** to its corresponding secondary alcohol **6** was additionally shown.

In conclusion, the ketone metabolism in yeasts is comparable to that of bacteria. Ketones can either be oxygenated to esters or be reduced to the corresponding alcohol. These pathways seem to be quite ubiquitary among several hydrocarbon oxidizing yeasts. *C. albicans* was the only yeast not being able to form an ester out of the ketone in a detectable amount. Beyond this, alkenes can be converted to ketones and thus induce the production of ketone-degrading enzymes whereas this way seems to be invalid for alkanes.

Additional file

Additional file 1: Table S1. Overview of extracellular compounds detected in culture media with dodecane-2-one or dodec-1-ene as sole source of carbon and energy.

Competing interests

The authors declare that they have no competing interests

Authors' contributions

FS and UB initiated the project. AB performed the experiments with support by VH. All authors analyzed the data, wrote and approved the final manuscript.

Acknowledgements

We gratefully acknowledge financial support from the German Federal State Mecklenburg-Vorpommern through a state graduate scholarship (Landesgraduiertenstipendium) to Andy Beier.

Author details

[1]Institute of Microbiology, Department of Applied Microbiology, Greifswald University, Friedrich-Ludwig-Jahn-Str. 15, 17487 Greifswald, Germany. [2]Institute of Biochemistry, Department of Biotechnology & Enzyme Catalysis, Greifswald University, Felix-Hausdorff-Str. 4, 17487 Greifswald, Germany.

References

Abeles FB, Morgan PW, Saltveit ME Jr (1992) Ethylene in Plant Biology. Academic press, San Diego

Alphand V, Archelas A, Furstoss R (1989) Microbial transformations 16. One-step synthesis of a pivotal prostaglandin chiral synthon via a highly enantioselective microbiological Baeyer-Villiger type reaction. Tetrahedron Lett 30(28):3663–3664

Balke K, Kadow M, Mallin H, Saß S, Bornscheuer UT (2012) Discovery, application and protein engineering of Baeyer–Villiger monooxygenases for organic synthesis. Org Biomol Chem 10(31):6249–6265

Blasig R, Mauersberger S, Riege P, Schunck W-H, Jockisch W, Franke P, Müller H-G (1988) Degradation of long-chain *n*-alkanes by the yeast *Candida maltosa*. Appl Microbiol Biotechnol 28(6):589–597

Bruyn J (1954) An intermediate product in the oxidation of hexadecene-1 by *Candida lipolytica*. Koninkl Ned Acad Wetenschap Proc Ser C 57:41–44

Cheesbrough TM, Kolattukudy P (1988) Microsomal preparation from an animal tissue catalyzes release of carbon monoxide from a fatty aldehyde to generate an alkane. J Biol Chem 263(6):2738–2743

Coon MJ (2005) Omega oxygenases: nonheme-iron enzymes and P450 cytochromes. Biochem Biophys Res Commun 338(1):378–385

De Boer TD, Backer HJ (1956) Diazomethane. In: Leonard NJ (ed) Organic Synthesis, vol 36. Wiley, New York, pp 14–16

Fonken GS, Murray HC, Reineke LM (1960) Pathway of progesterone oxidation by *Cladosporium resinae*. J Am Chem Soc 82(20):5507–5508

Forney F, Markovetz A (1968) Oxidative degradation of methyl ketones II. Chemical pathway for degradation of 2-tridecanone by *Pseudomonas multivorans* and *Pseudomonas aeruginosa*. J Bacteriol 96(4):1055–1064

Forney FW, Markovetz AJ (1970) Subterminal oxidation of aliphatic hydrocarbons. J Bacteriol 102(1):281–282

Forney F, Markovetz A (1971) The biology of methyl ketones. J Lipid Res 12(4):383–395

Forney F, Markovetz A, Kallio R (1967) Bacterial oxidation of 2-tridecanone to 1-undecanol. J Bacteriol 93(2):649–655

Hornei S, Kohler M, Weide H (1972) Spectrum of fatty acids of a *Candida* strain following culture on *n*-alkanes. Z Allg Mikrobiol 12:19–27

Ji Y, Mao G, Wang Y, Bartlam M (2013) Structural insights into diversity and n-alkane biodegradation mechanisms of alkane hydroxylases. FMICB 4:58

Käppeli O (1986) Cytochromes P-450 of yeasts. Microbiol Rev 50(3):244

Kelly DR, Knowles CJ, Mahdi JG, Wright MA, Taylor IN, Hibbs DE, Hursthouse MB, Mish'al AK, Roberts SM, Wan PWH, Grogan G, Willetts AJ (1995) Model for the functional active site of Baeyer-Villigerases. X-ray crystal data for (1S,2R,5R,8S,1 [prime or minute]R)-8-endo-benzoyloxy-N-(1[prime or minute]-phenylethyl) bicyclo[3.3.0]octane-2-endo-carboxamide. J Chem Soc Perkin Trans 1(16):2057–2066, doi:10.1039/p19950002057

Kester A, Foster J (1963) Diterminal oxidation of long-chain alkanes by bacteria. J Bacteriol 85(4):859–869

Kirschner A, Altenbuchner J, Bornscheuer UT (2007) Cloning, expression, and characterization of a Baeyer–Villiger monooxygenase from *Pseudomonas fluorescens* DSM 50106 in *E. coli*. Appl Microbiol Biotechnol 73(5):1065–1072

Klug MJ (1969) The Metabolism of *n*-alkanes and *l*-alkenes by Members of the Genus *Candida*, Dissertation. University of Iowa

Klug MJ, Markovetz AJ (1967) Degradation of hydrocarbons by members of the genus *Candida*. II. Oxidation of n-alkanes and l-alkenes by Candida lipolytica. J Bacteriol 93(6):1847–1852

Klug MJ, Markovetz AJ (1968) Degradation of hydrocarbons by members of the genus *Candida* 3. Oxidative intermediates from 1-hexadecene and 1-heptadecene by *Candida lipolytica*. J Bacteriol 96(4):1115–1123

Komagata K, Nakase T, Katsuya N (1964) Assimilation of hydrocarbons by yeasts. J Gen Appl Microbiol 10(4):323–331

Kotani T, Yurimoto H, Kato N, Sakai Y (2007) Novel acetone metabolism in a propane-utilizing bacterium, *Gordonia* sp. strain TY-5. J Bacteriol 189(3):886–893

Krauel H, Kunze R, Weide H (1973) Bildung von Dicarbonsäuren durch *Candida guilliermondii*, Stamm H 17, aus *n*-Alkanen. Z Allg Mikrobiol 13(1):55–58

Leipold F, Wardenga R, Bornscheuer UT (2012) Cloning, expression and characterization of a eukaryotic cycloalkanone monooxygenase from *Cylindrocarpon radicicola* ATCC 11011. Appl Microbiol Biotechnol 94(3):705–717

Leisch H, Morley K, Lau PC (2011) Baeyer-Villiger monooxygenases: more than just green chemistry. Chem Rev 111(7):4165–4222

Lottmann J, Hammer E, Schauer F (1999) Methyl ketone formation during degradation of phenoxybutyric acid by *Penicillium canescens* SBUG-M 1139. Arch Microbiol 172(6):417–420

Mauersberger S, Ohkuma M, Schunck W-H, Takagi M (1996) *Candida maltosa* Non-conventional Yeasts in Biotechnology. Springer, Berlin Heidelberg, pp 411–580

Mihovilovic MD, Kapitán P (2004) Regiodivergent Baeyer–Villiger oxidation of fused ketone substrates by recombinant whole-cells expressing two monooxygenases from *Brevibacterium*. Tetrahedron Lett 45(13):2751–2754, doi:http://dx.doi.org/10.1016/j.tetlet.2004.02.036

Mizuno M, Shimojima Y, Iguchi T, Takeda I, Senoh S (1966) Fatty acid composition of hydrocarbon-assimilating yeast. Agric Biol Chem 30(5):506–510

Patil R, Banoth L, Singh A, Chisti Y, Banerjee UC (2013) Enantioselective bioreduction of cyclic alkanones by whole cells of *Candida* Species. Biocatal Biotransform 31(3):123–131

Picataggio S, Rohrer T, Deanda K, Lanning D, Reynolds R, Mielenz J, Eirich LD (1992) Metabolic engineering of *Candida tropicalis* for the production of long-chain dicarboxylic acids. Nat Biotechnol 10(8):894–898

Primrose S (1979) Ethylene and agriculture: the role of the microbe. J Appl Microbiol 46(1):1–25

Rahim M, Sih CJ (1966) Mechanisms of steroid oxidation by microorganisms XI. Enzymatic cleavage of the pregnane side chain. J Biol Chem 241(15):3615–3623

Rasmussen RA (1970) Isoprene: identified as a forest-type emission to the atmosphere. Environ Sci Technol 4(8):667–671

Rehdorf J, Mihovilovic MD, Bornscheuer UT (2010) Exploiting the regioselectivity of Baeyer–Villiger monooxygenases for the formation of β-amino acids and β-amino alcohols. Angew Chem Int Ed 49(26):4506–4508

Rehm H, Reiff I (1981) Mechanisms and Occurrence of Microbial Oxidation of Long-Chain Alkanes, vol 19. Springer Berlin, Heidelberg

Schauer F (1988) Zur Physiologie des Kohlenwasserstoffabbaus in *Candida maltosa*. Habilitation, Greifswald

Schauer F (2001) Abbau und Verwertung von Mineralölbestandteilen durch Mikroorganismen. Bodden 11:3–31

Scheller U, Zimmer T, Becher D, Schauer F, Schunck W-H (1998) Oxygenation cascade in conversion of *n*-alkanes to α, ω-dioic acids catalyzed by cytochrome P450 52A3. J Biolog Chem 273(49):32528–32534

Schunck W-H, Mauersberger S, Huth J, Riege P, Müller H-G (1987) Function and regulation of cytochrome P-450 in alkane-assimilating yeast. Arch Microbiol 147(3):240–244

Smit MS, Mokgoro MM, Setati E, Nicaud J-M (2005) α, ω-Dicarboxylic acid accumulation by acyl-CoA oxidase deficient mutants of *Yarrowia lipolytica*. Biotechnol Lett 27(12):859–864

Song JW, Jeon EY, Song DH, Jang HY, Bornscheuer UT, Oh DK, Park JB (2013) Multistep enzymatic synthesis of long-chain α, ω-dicarboxylic and ω-hydroxycarboxylic acids from renewable fatty acids and plant oils. Angew Chem Int Ed 52(9):2534–2537

Stanier R (1947) Simultaneous adaptation: a new technique for the study of metabolic pathways. J Bacteriol 54(3):339–348

Stewart JE, Finnerty WR, Kallio R, Stevenson D (1960) Esters from bacterial oxidation of olefins. Science 132(3435):1254–1254

Tehlivets O, Scheuringer K, Kohlwein SD (2007) Fatty acid synthesis and elongation in yeast. Biochim Biophys Acta - Mol Cell Biol L 1771(3):255–270

Terasawa M, Takahashi J (1981) Mode of metabolism of 1-tetradecene by *Candida* yeasts and citrates production. Agric Biol Chem 45(11):2433–2441

Van der Walt J, van Kerken AE (1961) The wine yeast of the cape. Antonie Van Leeuwenhoek 27(1):81–90

Watkinson RJ, Morgan P (1991) Physiology of Aliphatic Hydrocarbon-Degrading Microorganisms, Physiology of Biodegradative Microorganisms. Springer, Heidelberg, pp 79–92

Wolfson A, Dlugy C, Tavor D (2013) Baker's yeast catalyzed asymmetric reduction of prochiral ketones in different reaction mediums. Org Commun 6(1):1–11

Permissions

All chapters in this book were first published in AMBJ, by Springer; hereby published with permission under the Creative Commons Attribution License or equivalent. Every chapter published in this book has been scrutinized by our experts. Their significance has been extensively debated. The topics covered herein carry significant findings which will fuel the growth of the discipline. They may even be implemented as practical applications or may be referred to as a beginning point for another development.

The contributors of this book come from diverse backgrounds, making this book a truly international effort. This book will bring forth new frontiers with its revolutionizing research information and detailed analysis of the nascent developments around the world.

We would like to thank all the contributing authors for lending their expertise to make the book truly unique. They have played a crucial role in the development of this book. Without their invaluable contributions this book wouldn't have been possible. They have made vital efforts to compile up to date information on the varied aspects of this subject to make this book a valuable addition to the collection of many professionals and students.

This book was conceptualized with the vision of imparting up-to-date information and advanced data in this field. To ensure the same, a matchless editorial board was set up. Every individual on the board went through rigorous rounds of assessment to prove their worth. After which they invested a large part of their time researching and compiling the most relevant data for our readers.

The editorial board has been involved in producing this book since its inception. They have spent rigorous hours researching and exploring the diverse topics which have resulted in the successful publishing of this book. They have passed on their knowledge of decades through this book. To expedite this challenging task, the publisher supported the team at every step. A small team of assistant editors was also appointed to further simplify the editing procedure and attain best results for the readers.

Apart from the editorial board, the designing team has also invested a significant amount of their time in understanding the subject and creating the most relevant covers. They scrutinized every image to scout for the most suitable representation of the subject and create an appropriate cover for the book.

The publishing team has been an ardent support to the editorial, designing and production team. Their endless efforts to recruit the best for this project, has resulted in the accomplishment of this book. They are a veteran in the field of academics and their pool of knowledge is as vast as their experience in printing. Their expertise and guidance has proved useful at every step. Their uncompromising quality standards have made this book an exceptional effort. Their encouragement from time to time has been an inspiration for everyone.

The publisher and the editorial board hope that this book will prove to be a valuable piece of knowledge for researchers, students, practitioners and scholars across the globe.

List of Contributors

Edward J Prokop, John R Crigler, Claire M Wells, Alice A Young and Tony L Buhr
Naval Surface Warfare Center, Dahlgren Division, CBR Concepts and Experimentation Branch, 4045 Higley Road Suite 345, Dahlgren VA 22448-5162, USA

Parveen K Sharma, Jilagamazhi Fu and David B Levin
Department of Biosystems Engineering, University of Manitoba, Winnipeg, MB R3T 2N2, Canada

Xiangli Zhang and Brian Fristensky
Department of Plant Science, University of Manitoba, Winnipeg, MB R3T 2N2, Canada

Richard Sparling
Department of Microbiology, University of Manitoba, Winnipeg, MB R3T 2N2, Canada

Richard Hung, Samantha Lee and Joan W Bennett
Department of Plant Biology and Pathology, Rutgers, The State University of New Jersey, 59 Dudley Rd., New Brunswick, NJ 08901, USA

Cesar Rodriguez-Saona
Department of Entomology, Rutgers, The State University of New Jersey, 96 Lipman Drive, New Brunswick, NJ 08901, USA

Anni Määttänen, Petri Ihalainen and Jouko Peltonen
Laboratory of Physical Chemistry, Center of Excellence for Functional Materials, Abo Akademi University, Turku, Finland

Adyary Fallarero and Janni Kujala
Pharmaceutical Sciences, Department of Biosciences, Abo Akademi University, Turku, Finland

Pia Vuorela
Division of Pharmaceutical Biology, Faculty of Pharmacy, University of Helsinki, Helsinki, Finland

Lei Yang, Mette Lübeck and Peter S Lübeck
Section for Sustainable Biotechnology, Aalborg University Copenhagen, A. C. Meyers Vænge 15, DK-2450 Copenhagen, SV, Denmark

Pramod K Pandey, Philip H Kass and Sagor Biswas
Department of Population Health and Reproduction, University of California, Davis, California, USA

Michelle L Soupir
Department of Agricultural and Biosystems Engineering, Iowa State University, Ames, Iowa, USA

Vijay P Singh
Department of Biological and Agricultural Engineering & Zachry Department of Civil Engineering, Texas A & M University, College Station, Texas, USA

Naoki Izawa, Miyuki Kudo, Yukako Nakamura, Harumi Mizukoshi, Takahiro Kitada and Toshiro Sone
Yakult Central Institute, 5-11 Izumi, Kunitachi-shi, 186-8650 Tokyo, Japan

Catarina Barbosa, Patrícia Lage, Alice Vilela, Arlete Mendes-Faia and Ana Mendes-Ferreira
Institute for Biotechnology and Bioengeneering - Centre of Genomics and Biotechnology, Universidade de Trás-os-Montes e Alto Douro, Vila Real, Portugal

Domenico Schillaci, Maria Grazia Cusimano and Daniela Parrinello
Dip. STEBICEF, Università degli Studi di Palermo, Via Archirafi, 32-90123 Palermo, Italy

Angelo Spinello, Giampaolo Barone, Debora Russo and Vincenzo Arizza
Dip. STEBICEF, Università degli Studi di Palermo, Via Archirafi, 32-90123 Palermo, Italy
IEMEST, Istituto Euromediterraneo di Scienza e Tecnologia, Via Emerico Amari, 123-90139 Palermo, Italy

Maria Vitale
Istituto Zooprofilattico Sperimentale della Sicilia "A. Mirri", Via G. Marinuzzi, 3-90129 Palermo, Italy

Nayuta Furukawa, Misato Togawa, Masahiro Nakajima and Hayao Taguchi
Tokyo University of Science, 2641 Yamazaki, Noda, Chiba 278-8510, Japan

Akimasa Miyanaga
Tokyo Institute of Technology, 2-12-1 O-okayama, Meguro-ku, Tokyo 152-8551, Japan

Rieke Lohse, Desiree Jakobs-Schönwandt and Anant V Patel
Department of Engineering and Mathematics, University of Applied Sciences, Wilhelm-Bertelsmann-Str. 10, Bielefeld 33602, Germany

Aïcha Gharbi and Jean-Marc Berjeaud
Ecologie & Biologie des Interactions - UMR CNRS 7267, Microbiologie de l'eau, Université de Poitiers, 1 Rue Georges Bonnet, TSA 51106, 86073 Poitiers Cedex 9, France

Thibaut Legigan and Sébastien Papot
UMR-CNRS 7285, Groupe Systèmes Moléculaires Programmés, Université de Poitiers, Poitiers, France

Vincent Humblot
UPMC Université Paris 06, UMR CNRS 7197, Laboratoire de Réactivité de Surface, Sorbonne Universités, Paris, France

Zofia Stępniewska and Agnieszka Kuźniar
Department of Biochemistry and Environmental Chemistry, The John Paul II Catholic University of Lublin, Konstantynow 1I, 20-708 Lublin, Poland

Chang Liu, Yi-Ming Liu, Cheng-Ying Jiang and Shuang-Jiang Liu
State Key Laboratory of Microbial Resources, Chinese Academy of Sciences, Beijing 100101, China
Environmental Microbiology Research Center, Chinese Academy of Sciences, Beijing 100101, China

Qing-Lan Sun
Bioinformatic Center at Institute of Microbiology, Chinese Academy of Sciences, Beijing 100101, China

Febri Doni, Che Radziah Che Mohd Zain and Wan Mohtar Wan Yusoff
School of Biosciences and Biotechnology, Faculty of Science and Technology, Universiti Kebangsaan Malaysia, 43600 Bangi, Selangor, DE, Malaysia

Anizan Isahak
School of Environmental and Natural Resources Sciences, Faculty of Science and Technology, Universiti Kebangsaan Malaysia, 43600 Bangi, Selangor, DE, Malaysia

Vivian Jaskiw Szilagyi-Zecchin, Angela Cristina Ikeda, Douglas Adamoski, Vanessa Kava-Cordeiro, Chirlei Glienke and Lygia Vitória Galli-Terasawa
Departamento de Genética, Universidade Federal do Paraná (UFPR), Setor de Ciências Biológicas, Avenida Coronel Francisco Heráclito dos Santos, 210, Curitiba, Paraná 81531-970, Brazil

Mariangela Hungria
Embrapa Soja, Cx. Postal 231, Londrina, Paraná 86001-970, Brazil

Li Li
School of Environment and Energy, Peking University Shenzhen Graduate School, Shenzhen, 518055, China
Department of Biological Science, National University of Singapore, Singapore 117543, Singapore

Purnima Singh and Ying Liu
School of Environment and Energy, Peking University Shenzhen Graduate School, Shenzhen, 518055, China

Shenquan Pan
Department of Biological Science, National University of Singapore, Singapore 117543, Singapore

Guangyi Wang
School of Environment and Energy, Peking University Shenzhen Graduate School, Shenzhen, 518055, China
Tianjin University Center for Marine Environmental Ecology, School of Environmental Science and Engineering, Tianjin University, Tianjin 300072, China
Department of Microbiology, University of Hawaii at Manoa, Honolulu, HI 96822, USA

Steffi Noack-Schönmann, Ronald Banasiak, Nicole Knabe and William J Broughton
Department 4 (Materials & Environment), Federal Institute for Materials Research and Testing (Bundesanstalt für Material-forschung und -prüfung, BAM), Unter den Eichen 87, 12205 Berlin, Germany

Tanja Bus
Department of Biotechnology, University of Applied Sciences Jena, Carl-Zeiß-Promenade 2, 07745 Jena, Germany

H Den Dulk-Ras and Paul JJ Hooykaas
Institute of Biology Leiden, Universiteit Leiden, Sylviusweg 72, 2333 BE Leiden, The Netherlands

Anna A Gorbushina
Department 4 (Materials & Environment), Federal Institute for Materials Research and Testing (Bundesanstalt für Material-forschung und -prüfung, BAM), Unter den Eichen 87, 12205 Berlin, Germany
Department of Biology, Chemistry and Pharmacy & Department of Earth Sciences, Freie Universität Berlin, Malteserstrasse 74-100, 12249 Berlin, Germany

Deepika Sharma and Rajesh Kumari Manhas
Department of Microbiology, Guru Nanak Dev University, Amritsar, Punjab 143005, India

Santi M Mandal
Central Research Facility, Indian Institute of Technology, Kharagpur, West Bengal 721302, India

Massimiliano Marvasi and Ian A Durie
Soil and Water Science Department, University of Florida-IFAS, Gainesville, FL, USA

Charles Chen and Manuel Carrazana
Microbiology and Cell Science Department, University of Florida-IFAS, Gainesville, FL, USA

Max Teplitski
Soil and Water Science Department, University of Florida-IFAS, Gainesville, FL, USA
Cancer and Genetics Research Center, Genetics Institute, 2033 Mowry Road, Gainesville, FL 32611, USA

Roberto A Bobadilla-Fazzini and Pilar Parada
BioSigma S.A., Parque Industrial Los Libertadores, Lote 106, Colina, Santiago de Chile, Chile

Maria Paz Cortés
Laboratory of Bioinformatics and Mathematics of the Genome, Center for Mathematical Modeling (UMI2807-CNRS) and FONDAP Center for Genome Regulation, Santiago de Chile, Chile
Faculty of Mathematical and Physical Sciences, University of Chile, Santiago de Chile, Chile

Alejandro Maass
Faculty of Mathematical and Physical Sciences, University of Chile, Santiago de Chile, Chile
Department of Mathematical Engineering, Center for Mathematical Modeling (UMI2807-CNRS) and FONDAP Center for Genome Regulation, Santiago de Chile, Chile

Seiji Shibasaki
General Education Center, Hyogo University of Health Sciences, 1-3-6 Minatojima, Chuo-ku, Kobe 650-8530, Japan
Graduate School of Pharmacy, Hyogo University of Health Sciences, 1-3-6 Minatojima, Chuo-ku, Kobe 650-8530, Japan
Division of Rheumatology, Department of Internal Medicine, Hyogo College of Medicine, 1-1Mukogawa-cho, Nishinomiya 663-8501, Japan

Miki Karasaki and Hajime Sano
Division of Rheumatology, Department of Internal Medicine, Hyogo College of Medicine, 1-1Mukogawa-cho, Nishinomiya 663-8501, Japan

Torbjörn Gräslund and Per-Åke Nygren
Department of Molecular Biotechnology, School of Biotechnology, Royal Institute of Technology (KTH), AlbaNova University Center, SE-106 91 Stockholm, Sweden

Tsuyoshi Iwasaki
Graduate School of Pharmacy, Hyogo University of Health Sciences, 1-3-6 Minatojima, Chuo-ku, Kobe 650-8530, Japan
Division of Rheumatology, Department of Internal Medicine, Hyogo College of Medicine, 1-1Mukogawa-cho, Nishinomiya 663-8501, Japan
Division of Pharmacotherapy, Department of Pharmacy, School of Pharmacy, Hyogo University of Health Sciences, 1-3-6 Minatojima, Chuo-ku, Kobe 650-8530, Japan

Toshihide Nakamura, Mami Yamamoto and Katsuichi Saito
National Food Research Institute, National Agriculture and Food Research Organization (NARO), 2-1-12 Kannondai, Tsukuba, Ibaraki 305-8642, Japan

Akira Ando
NARO Institute of Vegetable and Tea Science, 360 Kusawa, Ano, Tsu, Mie 514-2392, Japan

Jun Shima
Ryukoku University, Fukakusa Tsukamotohon-cho, Fushimi-ku, Kyoto, Kyoto 612-8577, Japan

Licia Lama, Annabella Tramice, Ilaria Finore, Gianluca Anzelmo, Valeria Calandrelli, Eduardo Pagnotta, Giuseppina Tommonaro, Annarita Poli, Barbara Nicolaus and Antonio Trincone
Institute of Biomolecular Chemistry ICB-CNR, Consiglio Nazionale delle Ricerche, Via Campi Flegrei 34, 80072 Pozzuoli, Napoli, Italy

Paola Di Donato
Institute of Biomolecular Chemistry ICB-CNR, Consiglio Nazionale delle Ricerche, Via Campi Flegrei 34, 80072 Pozzuoli, Napoli, Italy
Department of Environmental Sciences, University of Naples "Parthenope", Centro Direzionale, Isola C4, 80143 Naples, Italy

Massimo Fagnano, Mauro Mori and Adriana Impagliazzo
Department of Agriculture, Naples University Federico II, Via Università 100, 80055 Portici, Naples, Italy

Anthony Friedline, Malcolm Zachariah, Amy Middaugh and Charles V Rice
Department of Chemistry and Biochemistry, Stephenson Life Sciences Research Center, University of Oklahoma, Norman, Oklahoma, USA

Matt Heiser and Neeraj Khanna
Bio-Cide International, Inc., Norman, Oklahoma, USA

Parag Vaishampayan
Jet Propulsion Laboratory, California Institute of Technology, Biotechnology and Planetary Protection Group, Pasadena, California, USA

Andy Beier
Institute of Microbiology, Department of Applied Microbiology, Greifswald University, Friedrich-Ludwig-Jahn-Str. 15, 17487 Greifswald, Germany
Institute of Biochemistry, Department of Biotechnology & Enzyme Catalysis, Greifswald University, Felix-Hausdorff-Str. 4, 17487 Greifswald, Germany

Veronika Hahn and Frieder Schauer
Institute of Microbiology, Department of Applied Microbiology, Greifswald University, Friedrich-Ludwig-Jahn-Str. 15, 17487 Greifswald, Germany

Uwe T Bornscheuer
Institute of Biochemistry, Department of Biotechnology & Enzyme Catalysis, Greifswald University, Felix-Hausdorff-Str. 4, 17487 Greifswald, Germany